P9-BIM-926

VERTEBRATE
DISSECTION

Fifth Edition

WARREN F. WALKER, JR., Ph.D.

Professor of Biology, Oberlin College

1975   W. B. SAUNDERS COMPANY   Philadelphia London Toronto

W. B. Saunders Company: West Washington Square
Philadelphia, Pa. 19105

12 Dyott Street
London, WC1A   1DB

833 Oxford Street
Toronto, Ontario M8Z 5T9, Canada

Library of Congress Cataloging in Publication Data

Walker, Warren Franklin.

Vertebrate dissection.

Bibliography: p.

Includes index.

1. Anatomy, Comparative—Laboratory manuals.
   2. Vertebrates—Anatomy—Laboratory manuals.
   3. Dissection.   I. Title.

QL812.W32 1975   596'.04   74–21017

ISBN 0–7216–9098–X

Vertebrate Dissection   ISBN   0-7216-9098-X

Last digit is the print number: 9   8   7   6   5   4   3   2   1

PREFACE TO FIFTH EDITION

Although the purpose and basic approach of this edition remain the same as for the previous ones, a number of changes have been made. Many of the directions for dissecting the mammal have been rewritten so that they can apply to the cat, mink, and rabbit. The mink is included because, as a by-product of the furrier, it is inexpensive, and it is becoming increasingly popular as a substitute for the cat. Both cat and mink are carnivores and therefore resemble each other closely. Unique features of the cat, mink, and rabbit are pointed out as necessary. Considerable economy of space has resulted from describing these mammals together. Since all mammals resemble each other in more ways than they differ, these directions are also applicable to many species not specifically included here. A common description of the mammals has been facilitated by applying the terminology recommended for quadrupeds by the International Committee on Veterinary Anatomical Nomenclature to all the species considered. Differences in terminology between homologous structures in a cat and rabbit, for example, have been eliminated. Another major change has been combining the descriptions of the arteries and veins of a given region, especially in mammals. Peripheral veins of mammals are often poorly injected; because of this, it is easier to find them when the arteries are being dissected than independently. Sections have been added on the lumbosacral plexus and on the major arteries and veins of the head. Apart from these, numerous smaller changes have been made throughout the book. Discussions of evolution and function have been updated and many new figures have been added.

The purpose of this book continues to be to present a manual on the systemic plan that covers the anatomy of a few animals with reasonable thoroughness. Studying comparative anatomy in the laboratory by tracing the changes that occur in one organ system through a series of animals, and then returning to consider another system has certain advantages over the study of one animal completely and then another. The laboratory work correlates better with the text and lectures, which usually follow the systemic approach. But more than this, the student has a better opportunity to visualize the transformations that occurred in a given organ system during the evolution of animals, and this is one of the main objectives of comparative anatomy. One valid criticism of the systemic approach, however, is that the student does not gain as thorough a concept of any vertebrate as a whole as he would through the systematic approach, but it seems to me that this is outweighed by the advantages gained. The systemic approach poses certain difficulties in the storage of specimens, but I have not found these critical for a one semester course. Mudpuppies, sheep brains, and other small specimens can be kept in preserving fluid in modest-sized plastic containers issued to each student; dogfish and mammals keep well in large plastic bags. After the mammal is skinned, it should first be wrapped in a cloth soaked in embalming solution to prevent drying. If for practical or other reasons, the instructor prefers to use the systematic approach, this manual can be adapted by selecting the appropriate sections from the table of contents.

The central theme of this manual is a study of the major anatomical transformations that have occurred in the vertebrates during their evolution from the fish to the mammalian stage, with the view of making the anatomy of the mammal meaningful. There is much about mammalian anatomy, including that of human beings, that is unintelligible unless one knows something of the anatomy of lower forms. One cannot, of course, examine the actual evolutionary sequence to mammals in living vertebrates, but one can simulate this sequential study to some extent by examining selected living lower vertebrates. For most organ systems, the more important stages can be seen by dissecting examples from three levels of organization—the fish, the primitive tetrapod, and the mammal. The dogfish (*Squalus*) is used as an example of a primitive, jawed fish; the mudpuppy (*Necturus*), as an example of a primitive terrestrial vertebrate; and the cat (*Felis*), mink (*Mustela*), and rabbit (*Lepus*), as examples of the mammal.

These directions not only give the instructor a choice of mammal to be studied, but also the option of introducing students to comparisons at one major evolutionary level. Different students can be issued different mammal species, including some not specifically included in these directions. Natural inquisitiveness, or formal comparisons of dissections after an exercise, will acquaint students with the species his or her neighbors may be dissecting. This will serve to emphasize the basic similarity of animals at one level of organization. The student sees that a carnivore and a rabbit, for example, have much in common, and can better realize that human beings too share many of these features. At the same time, the student has the opportunity to make comparisons between animals at the same level and to see the diverse features that are superimposed upon a common pattern. Although most of the differences between carnivores and the rabbit are the result of adaptation to different modes of life, some doubtless result from chance divergence. A few of the differences also illustrate primitive and advanced stages in the evolution of a structure within the mammalian level of organization. The duplex uterus of the rabbit and the bipartite uterus of carnivores are a case in point.

In addition to the jawed fish, primitive tetrapod, and mammal, directions are included for the study of representative lower chordates and an agnathous me, and I earnestly hope that they will continue to call my attention to any errors or deficiencies that they may find in this edition. Their past support and encouragement have greatly increased the merit of this manual. I am very grateful to Dr. Douglas B. Webster of the Louisiana State University Medical Center for permission to retain the excellent section on the dissection of the sheep cerebrum that he prepared for an earlier edition. New drawings have been made by Mr. J. P. Nail. His artistry and professionalism have greatly raised the quality of the figures. My wife has again been most helpful in lending encouragement and in proofreading the materials in their various stages of writing. Again I should like to pay tribute to my late teacher and mentor, Professor Alfred S. Romer, who has indirectly influenced this book in many ways. I also wish to thank the staff of the W. B. Saunders Company, who have been most helpful and encouraging in the preparation of all editions of this book. I am particularly grateful to Mr. Tyler Buchenau, retired Biology Editor of W. B. Saunders Company, for his encouragement and help in writing my first book, which was the first edition of this laboratory manual.

WARREN F. WALKER, JR.

fish (the lamprey), for those who have time to supplement the major sequence by examining these more primitive types.

Flexibility has been an aim throughout, for the length and emphasis of comparative anatomy courses are subject to much variation. Enough material is included for an intensive course. On the other hand, the directions are written so that much can be omitted. For example, the lower chordates and lamprey could be omitted, leaving the jawed fish, primitive tetrapod, and mammal. A still shorter course could also omit the primitive tetrapod. An alternate way of shortening the course would be by omitting parts of some organ systems—the muscles of the hind limb, certain of the sense organs, the lymphatic system, etc. Certain dissections could also be replaced by demonstrations. Regardless of what, if anything, is omitted, the sections left are rather thorough, for in my opinion it is a better educational experience for the student to do a limited amount of material well than to cover a lot of ground superficially.

Some textual material has been incorporated with the laboratory directions. These remarks are not intended to replace the text. Some point out the ways in which the species being studied resembles or departs from a generalized member of its group, but most are summaries of evolutionary trends or functions intended to make the laboratory work more meaningful. I have continued to add short sections explaining the structural and functional interrelationships of various parts. "Functional anatomy" gives new perspectives to our understanding of the structure and evolution of organisms. It is an aspect of anatomy receiving increasing attention nowadays, and it is one that can be introduced particularly well in the laboratory.

As the emphasis in biological teaching shifts more and more toward the experimental side of the field, less and less time seems to be available for certain of the essential classic courses, including comparative anatomy. Figures can save a considerable amount of the student's time, both in finding the structures and in serving as a record of his observations. But the student should be cautioned that figures should not serve as a substitute for careful dissection and observation. Many of the figures show not only the organs being considered but the relationship of these organs to surrounding parts. This should facilitate finding and remembering the structures concerned. A new edition has given me the opportunity to continue to upgrade the illustrations. Seven figures in the previous edition have been redrawn or replaced and 32 new ones have been added, bringing the new total to 186. Minor changes have been made in many other figures. Most of the drawings of the circulatory system show both the arteries and veins together, for this is the way they are seen by the student. The continued favorable reception of previous editions has made it possible to use two colors to distinguish arteries and veins.

I am indebted to many for help in preparing the fifth edition of this book. Students and colleagues have sent comments and useful suggestions to me or to the publishers. I wish to thank all of them, known and unknown to

CONTENTS

5

THE APPENDICULAR SKELETON 89

6

THE MUSCULAR SYSTEM 109

7

THE SENSE ORGANS

8

THE NERVOUS SYSTEM

9

THE COELOM AND THE DIGESTIVE AND RESPIRATORY SYSTEMS

Appendix 1

Appendix 2

A NOTE TO THE STUDENT:
ANATOMICAL TERMINOLOGY

Courses in Comparative Anatomy are sometimes compared to a foreign language in that a large vocabulary of often unfamiliar words is introduced. This cannot be avoided if one is to describe and discuss the parts of an animal and their locations. Most of the terms are based on Latin or Greek roots, and as you become familiar with the more common roots you will recognize the terminology for the shorthand that it is. The root "chondro," for example, always means cartilage, and it is used in many combinations: Chondrichthyes (cartilaginous fish), chondrocranium (cartilaginous braincase), perichondrium (connective tissue around cartilage), chondrocyte (cell in cartilage). There are also certain conventions, described below, that make the terminology more rational.

TERMS FOR ORGANS

As anatomists described the structure of animals, they used terms for organs that suited their fancy. Commonly an organ was described by its appearance in a human being, for our anatomy was one of the first to be studied. Over the centuries a plethora of names has been proposed, and many organs have a long list of synonyms. In order to bring some order to the developing chaos, human anatomists throughout the world have agreed upon codes of terminology, the most recent being the *Nomina Anatomica Parisiensia* (NAP, 1955). Terms are in Latin or Greek, but they are often translated into the vernacular of each language. Veterinary anatomists have agreed upon a *Nomina Anatomica Veterinaria* (most recently in 1973) in which they have brought most of the terminology for other mammals into agreement with human terminology, the major exception being certain terms for direction which are naturally different between a biped and quadruped (see below). No conventions have been agreed upon for nonmammalian vertebrates, but the tendency is to apply mammalian terms whenever applicable.

In this edition of *Vertebrate Dissection*, I have made a more consistent usage of the *Nomina Anatomica Veterinaria* for mammals, and I have applied these terms where possible to lower vertebrates. Favored terms are placed in boldface when first used for each animal. In the few cases where this is not a *Nomina Anatomica* term, the official term is given in italics, e.g., **kidney** (*ren*). The official term in this case is seldom used in English as a noun, but it forms the basis for the common adjective, renal. I have avoided introducing synonyms unless they are in common use, in which case the synonym is given in regular type, e.g., **cleidobrachialis muscle** (clavodeltoid). Terms based on the names of individuals are avoided in official terminology, but some in common use have been included, e.g., **auditory tube** (eustachian tube).

TERMS FOR DIRECTIONS, PLANES, AND SECTIONS

It is necessary when discussing the anatomy of animals to have terms that can be used to describe the location of parts. The following are the ones recommended in the *Nomina Anatomica Veterinaria* for use in a quadruped. Most of them are illustrated in the accompanying diagram.

Terms for Direction

Many terms for direction are the same in comparative and human anatomy, but there are certain differences occasioned by our upright posture. A structure toward the head end of a quadruped is described as **cranial** or **rostral** (e.g., cranial vena cava); one toward the tail, as **caudal.** Comparable

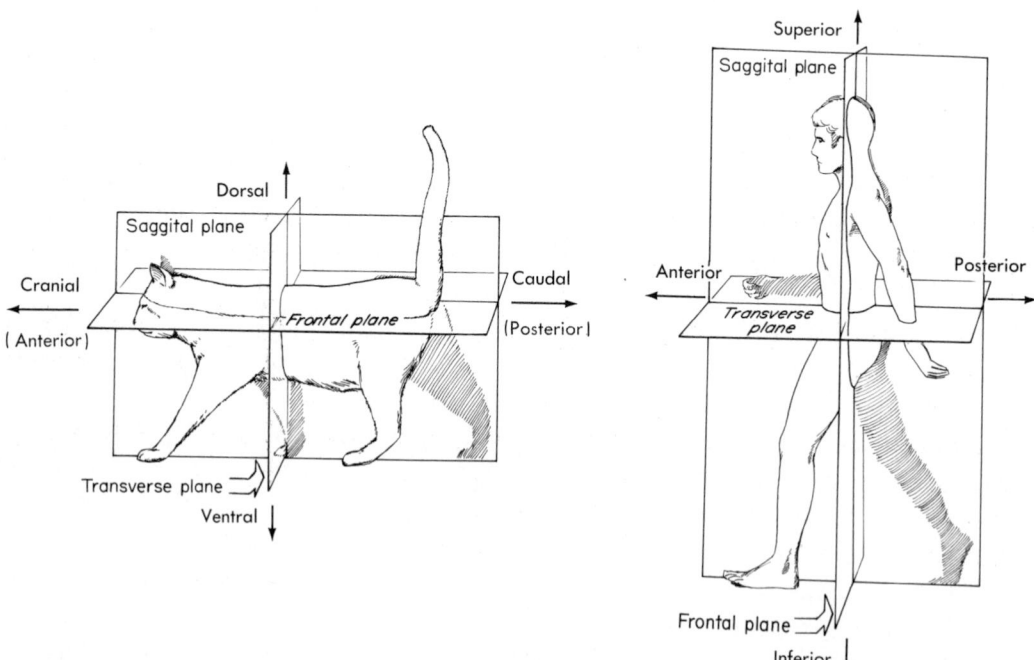

Figure N–1. Diagrams to show the planes of the body and the differences in the terms for direction in a quadruped and in a human being.

positions in human beings are described as **superior** (e.g., superior vena cava) and **inferior** (see Figure).

A structure toward the back of a quadruped is **dorsal**; one toward the belly is **ventral.** The terms dorsal and ventral are not used in human anatomy; comparable directions are referred to as **posterior** and **anterior.** It is recommended that the terms anterior and posterior not be used in a quadruped for the cranial and caudal ends of the body because of the possible confusion with the different usage in human anatomy. In general I have followed this recommendation except for well-established terms for certain organs (e.g., posterior cardinals in a fish).

Other terms for direction are used in the same way in all animals. **Lateral** refers to the side of the body; **medial** to a position toward the midline. **Median** is used for a structure in the midline. **Distal** refers to a part of some organ, such as an appendage or blood vessel, that is farthest removed from the point of reference, such as the center of the body or the origin of the vessel; **proximal**, to the opposite end of the organ, i.e., the part nearest the point of reference.

Left and **right** are self-evident, but it should be emphasized that in anatomical directions they always pertain to the *specimen's* left or right, regardless of the way the specimen is viewed by the observer.

Adverbs may be formed from the above adjectives by adding the suffix **-ly** or **-ad** to the root, in which case the term implies motion in a given direction. To say that a structure extends caudally, or caudad, means that it is moving toward the tail.

Planes and Sections of the Body

A body is frequently cut in various planes to obtain views of internal organs. A longitudinal, vertical section from dorsal to ventral that passes through the median longitudinal axis of the body is a **sagittal** section. Such a section lies in the sagittal plane. Sections or planes parallel with, but lateral to, the sagittal plane are said to be **parasagittal.**

A section cut across the body from dorsal to ventral, and at right angles to the longitudinal axis, is a **transverse** section, and it lies in the transverse plane.

A **frontal** (coronal) section or plane is one lying in the longitudinal axis, and passing horizontally from side to side.

HEMICHORDATES AND LOWER CHORDATES

Chordates, echinoderms (starfish and their relatives), and several phyla of worm-like creatures (Pogonophora, Chaetognatha, Hemichordata) are grouped together as **deuterostomes** because a second opening invaginates from the larval surface into the gut cavity and becomes the mouth. The original larval opening into the gut cavity—the blastopore—becomes the anus, or closes and a new anus invaginates nearby.

The phylum **Chordata** embraces about 35,000 living species, ranging in complexity from sessile sea squirts to human beings. Primitive chordates have a filter feeding mode of life that utilizes **pharyngeal gill slits** to permit excess water to escape from the pharynx and thereby concentrate food particles. The embryos of higher chordates have traces of such slits in the form of pharyngeal pouches that do not reach the body surface. Chordates are also distinguished from other deuterostomes by sharing two other characters at some stage of their life cycle: a **notochord** and a **single, dorsal, tubular nerve cord.** Both of these are related to the locomotor system. The nerve cord integrates the activities of longitudinal muscle fibers in the body wall, and the noto-chord, which is a semirigid yet bendable rod of cells, prevents the body from shortening when the muscle fibers contract. Contraction of longitudinal muscle fibers thus re-sults in a side-to-side bending of the body rather than a telescoping of the trunk as takes place in an earthworm. Variations in the degree of development of the gill slits, notochord, and nerve cord are among the criteria used to divide the phylum Chordata into its three subphyla: **Urochordata, Cephalochordata,** and **Vertebrata.**

Although chordates are related to other deuterostomes, the degree of relationship is uncertain. Any of the living deuterostome phyla are too specialized to be ancestral to any of the others, but presumably all have a common, remote ancestry. The hemi-chordates are probably more closely related to the chordates than are any of the other deuterostome groups; indeed, they have sometimes been regarded as a chordate sub-phylum. Hemichordates are also filter feeders with pharyngeal gill slits, but the other chordate characters are at best questionably developed.

Courses in comparative anatomy deal primarily with the vertebrates, but it is desirable for the student to have some idea of the nature of hemichordates and primi-tive chordates. In particular, the student should become acquainted with their general external features and understand how the three fundamental chordate characters are represented. The cephalochordates, which are closer in structure to the vertebrates, are often studied in more detail. If it is desirable to survey the lower chordates quickly, much of what follows could be studied on demonstration preparations.

PHYLUM HEMICHORDATA

Present-day hemichordates include a few rare, colonial, deep-sea forms, the pterobranchs, which resemble the graptolites of ancient Cambrian seas, and wormlike enteropneusts, or acorn worms. The latter are found burrowing in the sand and mud of tidal flats and shallow coastal waters, where they are sometimes abundant. *Balanoglossus* and *Saccoglossus (Dolichoglossus)* are common genera along North American coasts.

Examine the external features of one of the enteropneusts (Fig. 1–1). You may have to place the specimen in a pan of water and use a hand lens to see certain structures. The body is divided into three distinct regions: an anterior **proboscis,** a **collar,** and a long **trunk.** The proboscis attaches by a narrow stalk to the encircling collar located just posterior to it. The proboscis nesting in the collar often gives the appearance of an acorn in its cup—a fact that gives the common name to the group. The proboscis and collar assist the ciliated epidermis in burrowing. The collar coelom fills, thus inflating the collar and anchoring the worm in its burrow. Then the deflated proboscis is pushed forward by the action of muscles within it; its coelom is filled and it

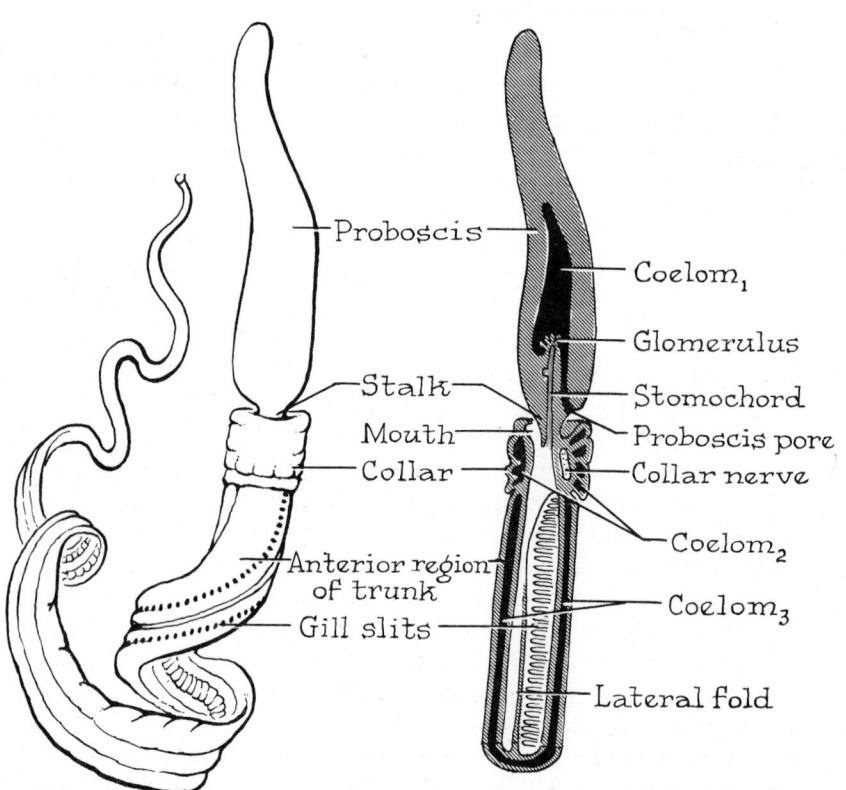

Figure 1–1. Left, external view of *Saccoglossus;* right, sagittal section of the anterior end. (From Villee, Walker, and Barnes, General Zoology; external view after Bateson.)

anchors the worm. Finally the collar coelom is emptied, and the worm is pulled forward to the proboscis.

When the animal is feeding, the proboscis is extended from the burrow, and cilia upon its surface carry minute food particles caudally to the **mouth,** which is situated inside the front of the collar ventral to the proboscis stalk. Food is carried into the ventral part of the pharynx (Fig. 1–1) and excess water enters the dorsal part to escape through numerous gill slits. The two parts of the pharynx are partially separated by a longitudinal fold. The **external gill slits** can be seen dorsolaterally on the anterior portion of the trunk. As many as 150 pairs have been counted on a 16-inch specimen. Each external gill slit leads to a gill pouch which connects with the pharynx by way of an unusual U-shaped internal gill slit. It is of interest that at one stage of development the gill slits of *Amphioxus* (a cephalochordate) have an identical appearance. Food continues from the ventral part of the pharynx into a simple, straight intestine. Material not digested and absorbed leaves through the terminal anus.

In mature individuals, prominent **genital ridges** will be found just ventral to the caudal external gill slits and extending a short distance caudad. The sexes are separate, but cannot be distinguished externally. In some, but not all species, conspicuous **hepatic ridges** will be seen posterior to the genital ridges. When present, they are the outward manifestation of digestive glands, the hepatic cecae, that bud off the intestine.

A longitudinal middorsal, and a similar midventral, ridge can also be seen on the trunk. Each contains a superficial and solid **nerve strand** that extends forward into the collar. Within the collar, the dorsal strand rolls up upon itself and thereby forms one or more cavities. Neurons from an echinodermlike subepidermal nerve plexus connect with the nerve strands.

As to the diagnostic chordate characteristics, pharyngeal gill slits are well represented. Nerve strands are present, but they are not in the form of a single, dorsal, tubular nerve cord. A diverticulum from the rostral end of the mouth cavity extends into and helps to stiffen the proboscis (Fig. 1–1). It consists partly of a chitinous plate and partly of vacuolated cells resembling those of a notochord. Some investigators consider it to represent a rudimentary notochord (the hemichordates get their name from this "half notochord"), but others question this interpretation and prefer to avoid any implications of homology by calling it a **stomochord.** Since all the chordate characteristics are not well represented at any stage of development, the Hemichordata are usually considered to be a distinct phylum related to, but not a part of, the phylum Chordata.

PHYLUM CHORDATA

Subphylum Urochordata

Members of the subphylum Urochordata are odd marine animals, most of which are encased in a leathery membrane called the tunic; hence, the animals are commonly referred to as tunicates. Although some are pelagic, the most familiar are the sessile sea squirts belonging to the class **Ascidiacea.** They are found attached to submerged objects in coastal waters, or occasionally partly buried in the sand or mud. Many sea squirts are colonial but some are solitary. Their anatomy can be studied conveniently on one of the latter types, such as *Molgula,* which is one of the most abundant along the Atlantic coast.

EXTERNAL FEATURES

Examine a specimen in a pan of water and notice its saclike appearance. Two spoutlike openings, or siphons, will be seen near the top of the animal (Fig. 1–2). When a living specimen is touched, water is expelled with considerable force through both openings. The name sea squirt is derived from this habit. Normally, however, water enters the organism through the topmost aperture (**incurrent siphon**) and is discharged through the opening that is set off on one edge (**excurrent siphon**). The margin of the incurrent siphon bears six small tentacles. Also notice the external covering, or **tunic.** It con-

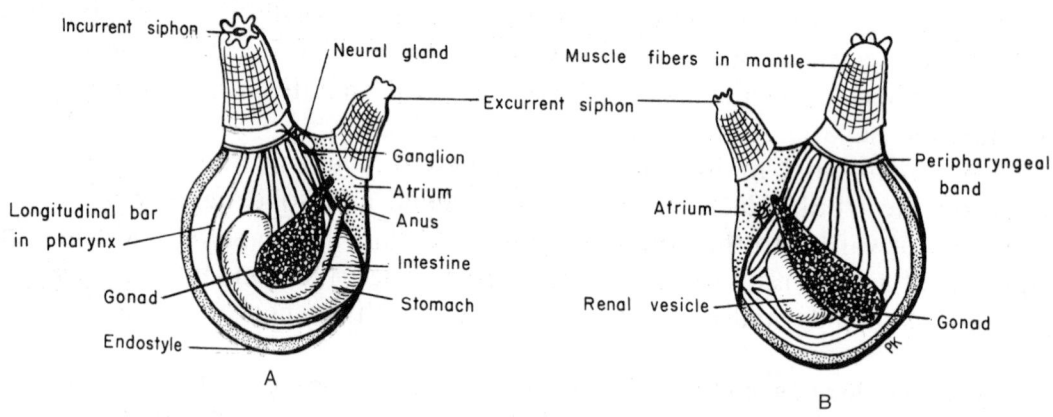

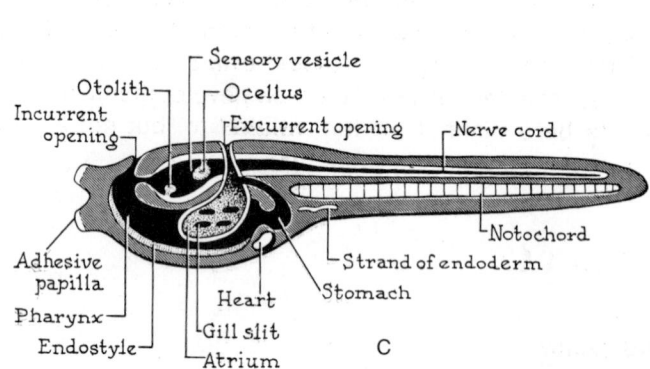

Figure 1–2. *A* and *B,* left and right side views, respectively, of a dissection of *Molgula.* All of the tunic and most of the mantle have been removed. *C,* diagrammatic lateral view of a larval ascidian. (*C* from Villee, Walker, and Barnes, General Zoology.)

tains a large amount of tunicin, a complex polysaccharide similar to cellulose found in plant cell walls. The tunic is secreted by cells derived from the
underlying body wall. Minute, hairlike processes extend out from the tunic
and help to anchor the animal to its substrate. The lower part of the tunic
may have sand grains adhering to the hairs, for *Molgula* often lies partly
buried in the sand.

Although the shape of the animal appears rather asymmetrical, the sea
squirt really has a modified bilateral symmetry. If one compares a larva with
an adult (Fig. 1–2), one will notice that the region between the two siphons
represents the dorsal surface, and the rest of the edge of the adult, the
ventral surface. The incurrent siphon lies anteriorly and the excurrent
siphon posteriorly.

DISSECTION

To expose the inside of the animal, cut through the tunic beneath the
excurrent siphon and extend the cut around the edge and base of the sac to a
point near the incurrent siphon. Reflect the tunic, observing that it is attached
to the rest of the body only at the siphons. Detach the tunic at these points.
A number of bundles of longitudinal and circular muscle fibers lie within the
thin body wall, or **mantle,** and aid in expelling water. The mantle is nearly
transparent, and many of the internal organs can be seen beneath it. To see
them more clearly, carefully peel off the mantle without unduly injuring
organs that may adhere to it. This part of the dissection should be done
beneath water.

Study the dissection and compare it with Figure 1–2. You may need a
hand lens to see certain structures. The incurrent siphon leads into a large,
thin-walled, vascular **pharynx,** which occupies most of the inside of the body.
In-many sea squirts the wall of the pharynx is divided into a number of
rectangular areas by longitudinal and transverse bars, and within each area
the wall is perforated by rows of gill slits. In *Molgula* the **longitudinal bars**
are grouped into conspicuous folds, transverse bars are not apparent, and the
gill slits are microscopic. To see them clearly it is necessary to remove a
piece of the pharynx wall and prepare a wet mount of it. The gill slits will
appear as arclike slits arranged in spirals. Although it will not be seen in
this type of preparation, the bars between the slits contain blood vessels in
which gas exchange with the environment occurs; the bars are covered with
cilia which create the current of water that passes through the animal. The
gill slits do not lead directly to the outside but into a delicate chamber,
the **atrium,** located on each side of the pharynx. The lateral portions of the
atrium may not be seen, but they converge posteriorly to form a more conspicuous median atrial chamber which opens to the surface through the
excurrent siphon.

The fold along the ventral surface of the pharynx is the **endostyle,** or
hypopharyngeal groove. It is generally more conspicuous, and more deeply
grooved on the inside, than other longitudinal folds. Certain cells of the

endostyle are ciliated; some are glandular and secrete a mucus which entraps minute food particles in the water; and some produce iodinated proteins as do the cells of the vertebrate thyroid gland. The food-containing mucous band is moved toward the incurrent siphon by the cilia. Near the anterior end of the pharynx it is carried to the dorsal side by lateral **peripharyngeal bands.** These may be hard to see. Then the mucous string moves posteriorly to the esophagus along a middorsal fold called the **dorsal lamina.** In some tunicates other lateral folds connect the endostyle with the dorsal lamina. Thus the pharynx of the sea squirt is primarily a food gathering device, although some gas exchange between the blood and environment does occur in its walls.

The rest of the alimentary canal of *Molgula* lies on the left side of the pharynx. A short **esophagus** leads from the posterior end of the pharynx to a slightly expanded portion of the gut, sometimes called the **"stomach,"** and this is followed by the **intestine** proper. The esophagus, stomach, and first part of the intestine form a C-shaped loop. Then the intestine doubles on itself, goes back beside the stomach and esophagus, and opens at the anus into the median portion of the atrial chamber. There is no stomach in the vertebrate sense; little food is stored in this "stomach" and it, together with minute glandular folds evaginated from it, secretes enzymes that act upon carbohydrates and fats as well as upon proteins. The intestine appears to be primarily absorptive.

In mature individuals, large **gonads** will be seen. In *Molgula* there is one on each side of the pharynx. Inconspicuous genital ducts lead from them to the median portion of the atrium. Tunicates are hermaphroditic, but generally not self-fertilizing. However, *Molgula* can fertilize itself.

An oval **renal vesicle** lies ventral to the right gonad. Some waste products, including concretions of uric acid, accumulate in the vesicle and stay there until the death of the animal, but most nitrogen is lost by diffusion through the pharynx wall.

A small oval-shaped structure will be seen on the dorsal edge of the pharynx between the two siphons. This is the **neural gland,** which has been homologized with the vertebrate hypophysis. It is connected with the pharynx by a minute duct visible only in special preparations. Extracts of the gland injected into vertebrates have many of the effects of pituitary extracts, and there is some evidence that it has gonadotropic effects in tunicates. It may also act as a chemoreceptor, testing the water current entering the pharynx. Carefully pull off the gland, and you will see beneath it an elongated nerve **ganglion.** Other internal organs are not usually seen in this type of dissection.

Aside from the abundant pharyngeal gill slits, there is little about an adult sea squirt that would suggest a chordate. However, the other diagnostic features of the phylum are represented in the free-swimming, tadpole-shaped, larval stage that most sea squirts pass through (Fig. 1–2, *C*). The tail of the larva, which contains longitudinal muscle fibers and is a locomotor organ, is supported by a notochord and above this is a single, dorsal, tubular nerve cord. The name of the subphylum is derived from the position of the larval notochord. Anteriorly the nerve cord expands into a sensory vesicle containing organs of equilibrium and light sense. At metamorphosis the larva attaches to the substratum by its anterior end, and its tail atrophies. The notochord and nerve cord are lost, and the sensory vesicle is reduced to the nerve ganglion.

Subphylum Cephalochordata

The subphylum Cephalochordata includes the lancelet or *Amphioxus (Branchiostoma[1])* and the related genus *Asymmetron.* All are superficially fishlike animals that have an extremely long notochord extending beyond the nerve cord to the very front of the animal. This extreme extension of the notochord is probably correlated with the burrowing habits of the animal. It also gives the name Cephalochordata to the subphylum. These animals are found in coastal waters, usually lying partly buried in the sand with only their front end protruding because, like the sea squirts, *Amphioxus* is a filter-feeder. At times they actively swim to new feeding sites, but their locomotion is not very efficient because stabilizing fins are poorly developed. In the United States, they are found south from Chesapeake and Monterey bays. Since many of the features of *Amphioxus* are believed to be very primitive, and hence may throw light on the origin of vertebrate structure, it will be considered in more detail than other lower chordates.

EXTERNAL FEATURES

Examine a preserved specimen of *Amphioxus* in a pan of water. You will need a hand lens to see certain structures. Its shape is streamlined, or fusiform, being elongate, flattened from side to side (compressed) and pointed at each end (Fig. 1–3). Segmental, V-shaped muscle blocks, the **myomeres,**[2] can be seen through the transparent epidermis. The apex of each V points rostrally. Note that the **myomeres** extend nearly the length of the body. Their number is a specific character ranging, in American species, from about 55 to 75. The lines of separation between the myomeres are connective tissue partitions called **myosepta** or myocommata.

A **dorsal fin** extends along the top of the body, and a **ventral fin** will be seen beneath the caudal quarter of the animal. The dorsal and ventral fins are continuous around the tail and expand slightly in this region to form a **caudal fin.** A pair of ventrolateral fins, or **metapleural folds,** continue forward from the rostral end of the ventral fins. (See Fig. 1–5.)

The metapleural folds end a short distance from the front of the body. In front of them, and ventral to the anterior few myomeres, you will see a transparent chamber called the **oral hood.** The mouth is located deep within this chamber and will not be seen at this time, but the opening of the oral hood on the ventral surface can be seen. It is fringed with small tentacles

---

[1]Although *Branchiostoma* (Costa, 1834) has priority over *Amphioxus* (Yarrel, 1836) as the technical name, the term *Amphioxus* is so familiar that it is customary to retain it, at least as a common name.

[2]**Myotome** and **myomere** are terms for the primitive muscle segments. They are often used synonymously, but I follow the usage of **myotome** for an embryonic muscle segment and **myomere** for an adult segment.

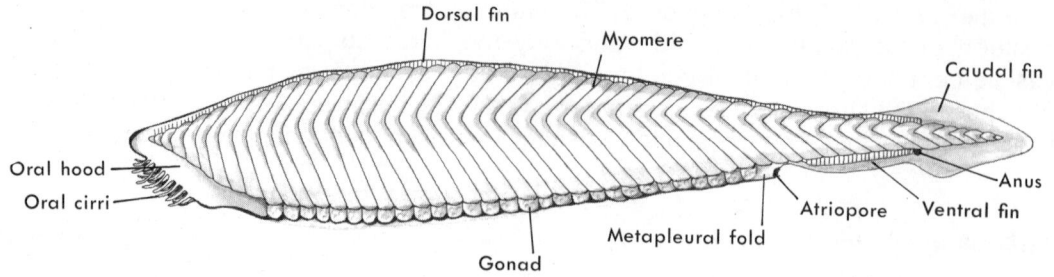

Figure 1–3. Lateral view of the external features of *Amphioxus.*

called **cirri,** which are often folded across its opening. The cirri contain chemoreceptive cells and also aid in excluding large material, permitting only water and small food particles to enter.

Water that enters the pharynx passes through gill slits into an atrial chamber whose opening (**atriopore**) you will see between the caudal ends of the two metapleural folds. The intestine opens by an **anus** located on the left side of the caudal fin, so there is a **postanal tail.** A postanal tail characterizes cephalochordates and vertebrates; other groups of animals have a terminal anus.

If the specimen is mature, you will see on each side a row of whitish, square **gonads.** They lie just ventral to the myomeres in the anterior half of the body. Their number ranges from about 20 to 35, differing slightly with the species. The gametes are discharged directly into the atrium. The sexes are separate.

WHOLE MOUNT SLIDE

Study a stained microscope slide of a small specimen of *Amphioxus* under the low power of the microscope (Fig. 1–4). Note its fusiform shape, and find the structures described earlier: myomeres; dorsal, ventral and

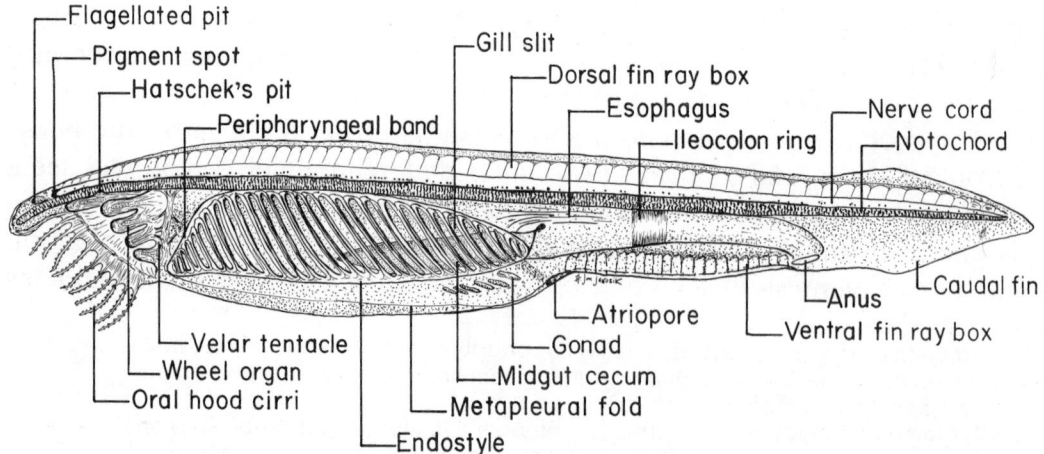

Figure 1–4. A lateral view of a whole mount slide of a young specimen of *Amphioxus.*

caudal fins; metapleural folds; oral hood and cirri; atriopore; anus; gonads. The **myomeres** have been cleared to render them somewhat transparent, so they will not be seen as plainly as in the preserved specimen. But you should see indications of them, at least, just ventral to the dorsal fin.

Observe that the **dorsal** and **ventral fins** are supported by small, transparent blocks, called **fin ray boxes.** In order to see the **metapleural folds** you will have to focus sharply on the surface. Each will appear as a horizontal line, parallel and slightly dorsal to the ventral edge of the body.

Since small specimens from which slides are made are generally not sexually mature, the **gonads** will not be fully developed, or may even be absent. If present, they will appear as a row of lightly staining, oval structures close to the ventral edge of the body. The largest ones are just rostral to the atriopore.

Notice the **notochord** located in the back, dorsal to the dark-staining alimentary canal. It extends nearly the entire length of the animal in the general position of the vertebral column of higher chordates and has a comparable function—namely to provide support and prevent the body from shortening when the myomeres contract.

The single, dorsal, tubular **nerve cord** will appear as a dark-staining band lying dorsal to the notochord. Its position may be recognized by the dark pigment granules along its ventral edge. Each granule represents parts of a simple photoreceptor. Notice that they are particularly numerous near the front of the animal. The nerve cord ends in a blunt point rostrally; there is no expanded brain. A prominent **pigment spot** of unknown function will be seen in front of the nerve cord. Sharp focusing will also reveal a clear, sac-like structure just dorsal to the front of the nerve cord. It is called the **flagellated pit,** and is believed to be a chemoreceptor. It occurs only on the left side of the snout. Embryologically, it connects with the nerve cord.

Examine the region of the **oral hood** in detail. The **cirri** have little processes along their edges, and each cirrus is supported by a skeletal rod. All the rods connect with a common basal rod. Ciliated grooves, or bands, are located on the inside of the lateral walls of the oral hood. In a lateral view they appear as large, dark-staining, fingerlike lobes extending forward from a common basal band. This complex is called the **wheel organ,** and it functions to draw a current of water into the organism. The dorsalmost lobe, which is called **Hatschek's groove,** is longer than the others. Slightly anterior to the middle of Hatschek's groove you will see a region where the groove is deeper and forms a pit that extends dorsally to overlap the right side of the notochord. This is **Hatschek's pit.** Besides aiding in the ciliary current, Hatschek's groove and pit secrete mucus that helps to entrap minute food particles in the water.

Caudal to the wheel organ you will notice a dark-staining line that is approximately in the transverse plane. This is the **velum**—a transverse partition that forms the posterior wall of the oral hood. The mouth, which cannot be seen in this view, is located in its center and is fringed with

tentacles. The tentacles can be seen extending either anteriorly or posteriorly from the velum. They, too, act as strainers and probably contain chemoreceptive cells.

The mouth leads into a large **pharynx,** most of whose lateral walls are perforated by numerous elongate **gill slits** with ciliated **gill bars** between them. In favorable specimens, supporting rods may be seen within the gill bars. In mature specimens there are over 200 gill bars. These provide a very large ciliated surface that plays the major role in moving water and food particles through the pharynx. Food is entrapped in mucus, and the water escapes through the gill slits. The pharynx is primarily a food concentrating mechanism; its role in respiration evolved later. Although blood vessels pass through the gill bars, no gills are present. Because of the great activity of the ciliated cells in this region, it is even possible that the blood leaving the bars contains less oxygen than that entering them. The major site of blood aeration appears to be the general body surface. As in the sea squirts, water does not pass directly to the outside, but through the gill slits into an **atrium.** The only part of the atrium to be seen in this view is the clear space ventral to the pharynx and continuing beneath the gut to the **atriopore.** The atrium is formed by the downgrowth of folds of the body wall around the gill slits, and it serves to protect the delicate gill bars.

The longitudinal band that extends along the entire floor of the pharynx is the **endostyle,** or hypopharyngeal groove, which was referred to in the section on urochordates. Its function is the same, for it secretes mucus in which minute food particles become entrapped, and its cilia carry the mucus anteriorly. The string of mucus passes to the dorsal side of the pharynx along the gill bars and peripharyngeal bands. A **peripharyngeal band** is located on each side of the front of the pharynx, and appears as a dark-staining line extending from the ventral edge of the velum diagonally dorsad and caudad just above the anterior gill slits. The mucus sheet is carried caudally along a middorsal **epipharyngeal groove.** Experiments have shown that certain endostylar cells also concentrate radioactive iodine, and Barrington has proposed that these cells are homologous to those of the vertebrate thyroid gland. In this case the iodinated proteins are discharged into the gut and absorbed further caudad.

The posterior end of the pharynx floor extends diagonally dorsad, and just behind the last gill slit the pharynx leads into a short, narrow **esophagus.** The outlines of the esophagus are often obscured by a large midgut cecum, but can be seen if you look carefully. The top of the esophagus lies just ventral to the notochord, and its floor will appear as a longitudinal line extending posteriorly a short distance from the bottom of the last gill slit.

The diameter of the alimentary tract increases two- or threefold just caudal to the esophagus, for the large **midgut cecum** has evaginated at this point. The midgut cecum extends toward the front of the animal, lying along the ventral side of the esophagus and right side of the pharynx. It is located within the atrium.

Smaller food particles are carried into the midgut cecum to be digested by enzymes secreted here and absorbed directly; larger food particles continue down the midgut, which becomes quite narrow caudal to the cecum. The deeply stained segment of the alimentary canal is called the **ileocolic ring.** At this point, cilia impart a rotary action to the cord of mucus and food in the entire midgut. This presumably aids in the discharge of enzymes by the midgut cecum and in mixing them with the food mass. A still narrower **hindgut** follows the ileocolic ring and opens on the body surface at the **anus.** Remains of microorganisms, including the shells of diatoms, are often seen in the hindgut.

Other internal structures are not seen in this type of preparation.

CROSS SECTIONS

(A) Common Features in the Sections

The anatomy of *Amphioxus* will be understood better if slides of representative cross sections can be examined. While studying such sections, compare them with the generalized diagram shown in Figure 1–5, and correlate the appearance of organs in this view with their appearance in the whole mount.

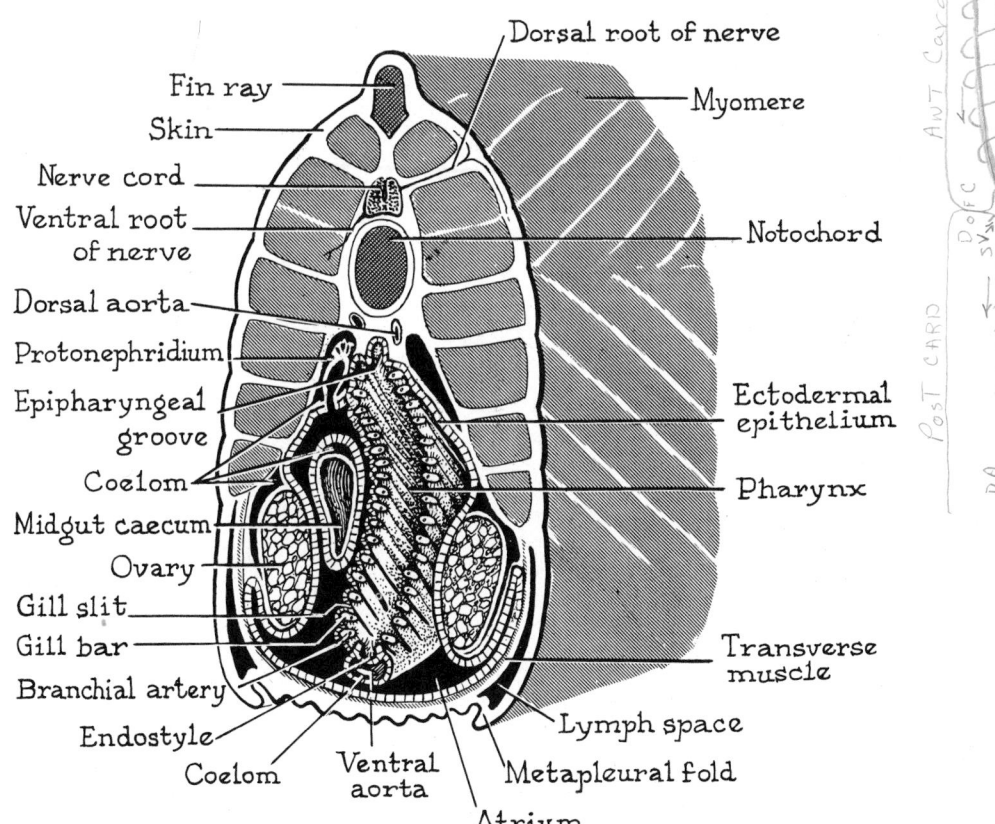

Figure 1–5. A diagrammatic cross section through the posterior part of the pharynx of *Amphioxus.* (From Villee, Walker, and Barnes, General Zoology.)

Many things will look much the same in any section. The **dorsal fin** will be recognized along with the hollow-appearing **fin ray box.** Actually, the fin ray box contains a gelatinous connective tissue. The skin consists of an **epidermis** of simple columnar epithelium supported by a thin layer of connective tissue (the **dermis**). **Myomeres** will appear as several oval chunks of tissue beneath the skin. They are separated from each other by the **myosepta.**

The **nerve cord** is the large tubular structure slightly ventral to the dorsal fin. Its cavity, the **neurocoele,** is very narrow. In favorable sections you will see lateral nerves arising from the cord. As in vertebrates, there are **dorsal** and **ventral roots** to the nerves; however, they do not unite in *Amphioxus,* but run directly to the tissues. The ventral roots carry motor fibers to the myomeres; the dorsal roots, sensory fibers to the integument and motor fibers to the ventral, nonmyotomal muscles.

You will see the **notochord** just beneath the nerve cord. It consists of vacuolated cells that are distended with fluid and held tightly together by a firm connective tissue sheath. You should be able to see the sheath, but the details of the cells will not be apparent.

(B) Section Through Oral Hood

In a section taken near the front of the animal you will see a space lying ventral to the notochord and myomeres and flanked laterally by the thin walls of the **oral hood.** The ciliated grooves of the **wheel organ** will appear as thicker patches of epithelium on the inside of the wall of the hood. **Hatschek's groove,** the most dorsal of these, is located a bit to the right of the median plane. The oral hood opens ventrally, but some sections may be taken just posterior to the opening. Pieces of **cirri** probably will be seen in the section.

(C) Section Through Pharynx

In a section through the pharynx the **metapleural folds** will be seen projecting from the ventrolateral portion of the body. There is a prominent **lymph space** in each. The wrinkled body wall between them contains a **transverse muscle** sheet that extends from the myomeres on one side to those on the other. This layer serves to compress the atrial cavity dorsal to it, and thus aids in expelling water. The **pharynx** occupies most of the center of the section and is surrounded laterally and ventrally by the **atrium.** Note the numerous **gill bars** that form its walls and the **gill slits** between them. The deeply grooved **endostyle** will be seen in the floor of the pharynx, and a similar **epipharyngeal groove** in its roof.

In certain sections, pieces of the **gonads** push into the atrium from the body wall, carrying the lining of the atrium before them. An ovary consists of many large nucleated cells; testis tissue appears as small dark dots or fine tubules.

If the section is taken near the posterior end of the pharynx, the hollow, oval-shaped **midgut cecum** will be observed lying on the right side of the pharynx. It first appears to be completely within the atrium, but actually is covered with a layer of atrial epithelium.

A coelom is present in *Amphioxus*, but in a highly modified form. Close examination of the section will reveal certain of its subdivisions. A pair of **dorsal coelomic canals** are located slightly lateral to several of the most dorsal gill bars. The atrium in this region is a narrow space between the gill bars and the dorsal coelomic canals. Another coelomic canal will be found ventral to the epithelium of the endostyle. The **subendostylar coelom** connects with the dorsal coelomic canals by small coelomic passages within every other gill bar. Portions of these may be seen. Another portion of the coelom will be seen lateral to the gonads. It, too, connects with the dorsal coelomic canals, but this connection often disappears in the adult. Finally, a very narrow coelomic space may be seen between the cells of the midgut cecum and the surrounding cells of the atrial epithelium.

Excretion is by way of clusters of flame cells or **protonephridia** of ectodermal origin that lead from the dorsal coelomic canals to the atrium. Portions of them may be found. Certain blood vessels may also be found, but the circulatory system will not be considered in detail. It is of interest, however, that the general course of blood flow is the same as in a vertebrate, i.e., from the tissues rostrally to the ventral side of the pharynx, dorsally through the gill bars, and caudally in a dorsal aorta to the body. There is no well developed heart, and many of the vessels are contractile. The tissues are supplied by open lacunae, for true capillaries are absent.

(D) Section Through Intestine

In a section through the midgut or hindgut the metapleural folds are absent and the **ventral fin** is present instead. Such a section will also show the **intestine** lying within a large **coelomic space,** for the separate coelomic passages of the pharyngeal region have coalesced. A posterior extension of the **atrium** lies on the right side of the intestine and coelom.

(E) Section Through Anus

A section through the anus passes through the **caudal fin.** Note that this fin is narrower and higher than either the dorsal or ventral fins which it replaces. The intestine opens at the **anus** on the left side of the fin.

All the three diagnostic characteristics of chordates are well represented in the cephalochordates. There can be no doubt of the affinities of the group. In addition, *Amphioxus* has many features that are found in the vertebrates, but not in the other lower chordates. Notable among these are the myomeres, a ventral, glandular diverticulum of the alimentary canal, and a postanal tail. *Amphioxus* is certainly closer to vertebrates than any of the other lower chordates, but it has certain peculiarities, including the atrial chamber, nephridialike excretory organs, and an extreme rostral extension of the notochord, that remove it from direct ancestry to the vertebrates. It may represent a somewhat parallel, active line of chordate evolution, or possibly a specialized and degenerate side branch of a stock that gave rise to vertebrates.

2

THE LAMPREY –
A PRIMITIVE VERTEBRATE

The vertebrates, or craniates, constitute the largest of the chordate subphyla. They are the most active and aggressive of all chordates, and many structural features reflect this mode of life. The evolution of navigational and feeding aids such as the **nose, eye** and **ear** has lead to the elaboration of the front end of the nervous system as a **brain,** which is enclosed in a skeletal brain case or **cranium.** The locomotor system is well developed. A **vertebral column** replaces the notochord in the adults of all but the most primitive vertebrates as the axial support for the body, and the myomeres and fins are better developed than in *Amphioxus.* Primitive vertebrates remain filter-feeders, but muscular action draws in water and food rather than ciliary currents; higher vertebrates gather their food with jawed mouths. Other distinctive features of vertebrates are their ventral **heart,** solid **liver, kidney** composed of tubules of mesodermal origin, and nonsegmented **gonads.**

Although the anatomy of vertebrates will be approached primarily by studying each organ system in a representative series, it is appropriate to begin by examining the overall structure of a primitive member of the group. This will acquaint the student with the basic organization of vertebrates and with the structure of a primitive living species. The most primitive vertebrates are jawless fishes of the class Agnatha. The most primitive agnathans, in turn, were the ostracoderms. These fishes became extinct about 300,000,000 years ago, but the modern cyclostomes (class Agnatha, subclass Cyclostomata) appear to be fairly direct descendants of this ancient vertebrate stock. Cyclostomes are represented today by the hagfishes (order Myxiniformes) and the lampreys (order Petromyzontiformes). A favorable species for study is the large sea lamprey, *Petromyzon marinus.* This species enters fresh water to breed, and has recently become established in the Great Lakes.

The lamprey shows a mixture of primitive and specialized characters. In studying the animal, the student should try to separate one from the other, for the aim is to learn the anatomy of the lamprey not just for its own sake, but also for the information it may provide concerning the structure of ancestral vertebrates. Among the major specializations of the lamprey are its eel-like shape, the absence of an armor of dermal bone, and its mode of feeding. Lampreys attach to other fishes by means of a suctorial buccal funnel, rasp away their prey's flesh, and suck its blood. A number of modifications of the digestive and respiratory systems are correlated with this habit.

EXTERNAL FEATURES

Examine the external features of a lamprey, noting its eel-like shape and the scaleless, slimy skin. The body can be divided into **head, trunk,** and **caudal** regions. The head extends through the gill or branchial area; the caudal region, or tail, from the cloacal aperture to the tip of the tail. The body is rounded in cross section cranially, but progressing caudally along the trunk and tail, the body becomes compressed, or flattened from side to side. This increases the surface area that can be pushed against the water as the body undulates during locomotion.

Observe that the only fins present are in the median plane—two **dorsal,** and a symmetrical **caudal fin** (Fig. 2–1). Lateral (paired) fins, usually found in other groups of fishes, are absent. The median fins are supported by slender, cartilaginous **fin rays,** which can be seen best if you cut across the fin in the frontal plane.

Notice the **buccal funnel** at the front of the head. It is fringed with **papillae,** and lined with **horny teeth.** The papillae are sensory, and also enable the lamprey to get a tight seal when it attaches to another fish. The "teeth" are composed of cornified cells, and hence differ from the true teeth of higher vertebrates. A protrusible **tongue** is situated near the center of the funnel. It is outlined by a ring of dark tissue, and is provided with small, rasplike, horny teeth. The **mouth opening** is just dorsal to the tongue. No jaws are present. This is a primitive characteristic, but the horny teeth and the tongue are specializations for the animal's blood sucking mode of feeding. A single, **median nostril** is located far back on the top of the head.

A median nostril is an unusual, but an ancient feature, for it is also found in certain of the ostracoderms. The manner in which the nostril is displaced during embryonic development from the more usual ventral position is shown in Figure 2–2. This condition derives from the tremendous enlargement of the upper lip of the embryo to form the buccal funnel.

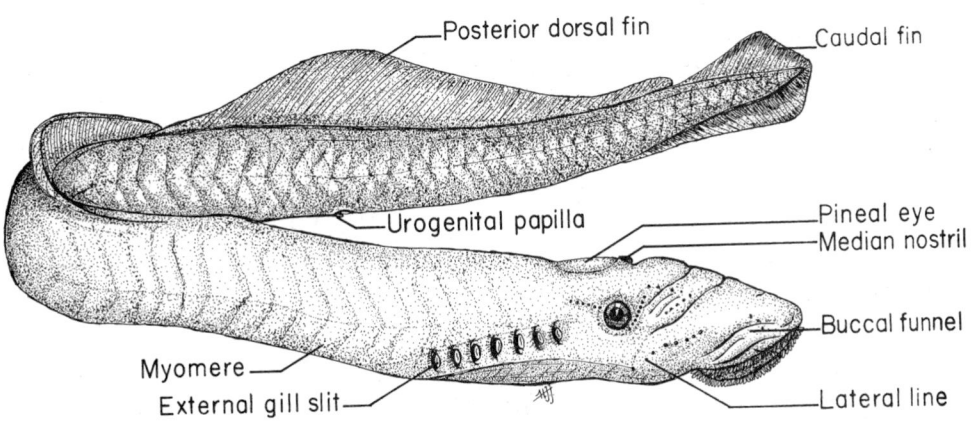

Figure 2–1. Lateral view of the sea lamprey, *Petromyzon marinus.*

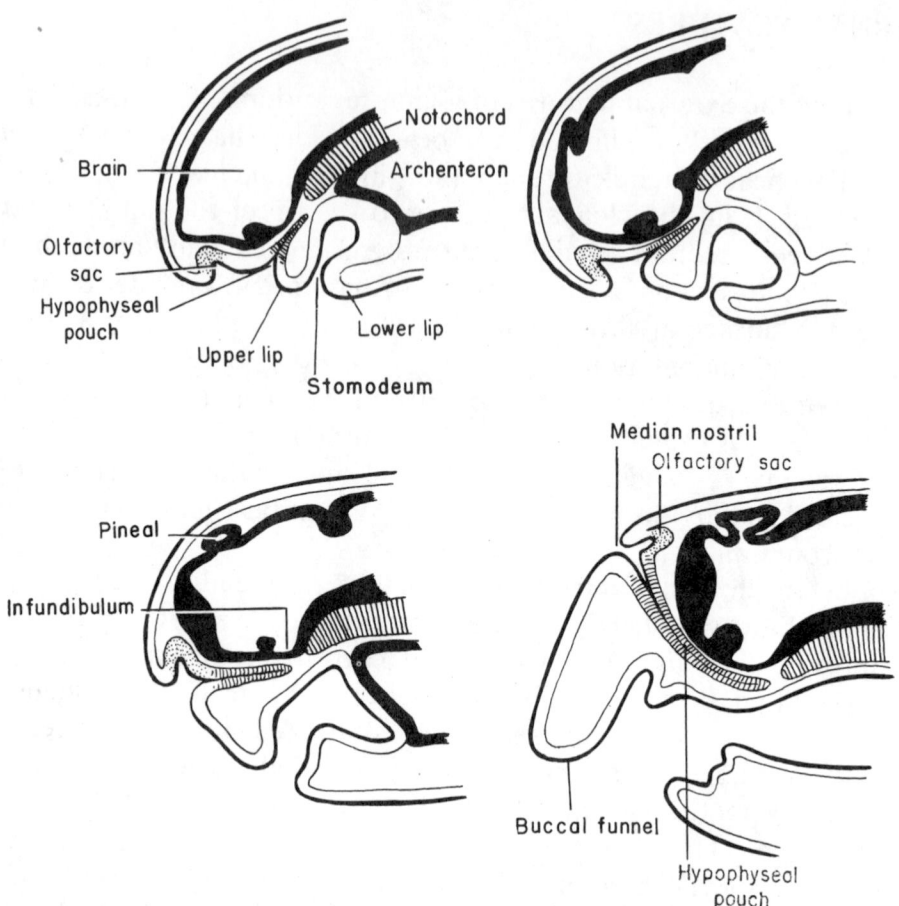

Figure 2-2. Diagrams in the sagittal plane of four stages in the development of the head region of *Petromyzon*. Note in particular how the originally independent hypophysis and olfactory sac become crowded together, and are pushed onto the top of the head by the enlargement of the upper lip. This region also differs from gnathostomes in that the hypophysis invaginates anterior to the stomodeum rather than from within the stomodeum. (From Parker and Haswell, A Text-Book of Zoology, St. Martin's Press. After Dohrn.)

Just caudal to the nostril, you will see an oval area that is often slightly depressed and generally a lighter color than the rest of the skin. The **pineal eye**, a primitive feature also found in the ancient ostracodums, lies beneath this depigmented skin. Experiments have shown that the pineal eye detects changes in light and initiates diurnal color changes in larval lampreys, probably by its effect upon the hypothalamus and hypophysis, or pituitary gland. It is possible that other physiologic activities are adjusted to the diurnal cycle in a similar way. A pair of conventional but lidless **eyes** are on the sides of the head. Seven oval **external gill slits** are located behind the eyes. A large number of gill slits characterizes primitive fishes.

If you let the head dry a bit and examine it with a hand lens, you may be able to detect groups of pores, or little bumps, arranged in short lines. One group is found caudal to the top of the lateral eye, another extends from the underside of the eye rostrally and dorsally. A third group is located on the ventral side of the head, caudal to the buccal funnel. These, together with other less conspicuous pores, are parts of the **lateral line system,** a group of sense organs associated with detecting vibrations and movements in the water (p. 169).

The body musculature consists chiefly of segmented **myomeres,**[3] whose outlines can be seen through the skin of the trunk and tail, especially in smaller specimens. Each myomere is roughly W-shaped, the top of the W being anterior. Furthermore, each is continuous from its dorsal to ventral end, there being no interruption near its middle as in higher vertebrates.

On the underside of the caudal end of the trunk you will see a shallow pit called the **cloaca.** It receives the excretory and genital products, which leave through the tip of a small **urogenital papilla,** and, just cranial to the papilla, the **anus,** or opening of the intestine. The opening of the cloaca to the surface is called the **cloacal aperture.**

SAGITTAL AND CROSS SECTIONS

Study the internal structure of the lamprey by examining a midsagittal section (Fig. 2–3) and a series of cross sections (Fig. 2–4). It is desirable to work in groups for this portion of the work. Certain students should prepare cross, and the rest sagittal sections. A good series of cross sections is as follows: (1) through the pineal and lateral eyes, (2) through a pair of external gill slits near the middle of the branchial region, (3) about one inch behind the branchial region, (4) near the middle of the trunk, (5) about three centimeters cranial to the cloaca, (6) through the tail. In preparing the sagittal sections use a large knife and be particularly careful to cut the head and branchial region as close to the sagittal plane as possible. It will probably be necessary to dissect the sagittal section a bit more as you proceed.

The Skeletal and Muscular Systems

Study the better, or larger, sagittal section, correlating the appearance of structures in this view with their appearance in the cross sections. Notice that the main skeletal axis is a long **notochord** extending from the caudal end of the body to a point beneath the middle of the **brain.** It has a gelatinous texture, but is enclosed in a strong fibrous sheath. It is firm, yet flexible, and probably serves primarily to prevent the body from telescoping when the myomeres contract. In favorable cross sections, cartilaginous blocks (**arcualia**) will be seen above the notochord on either side of the **spinal cord.**

---

[3]See footnote 2, p. 7.

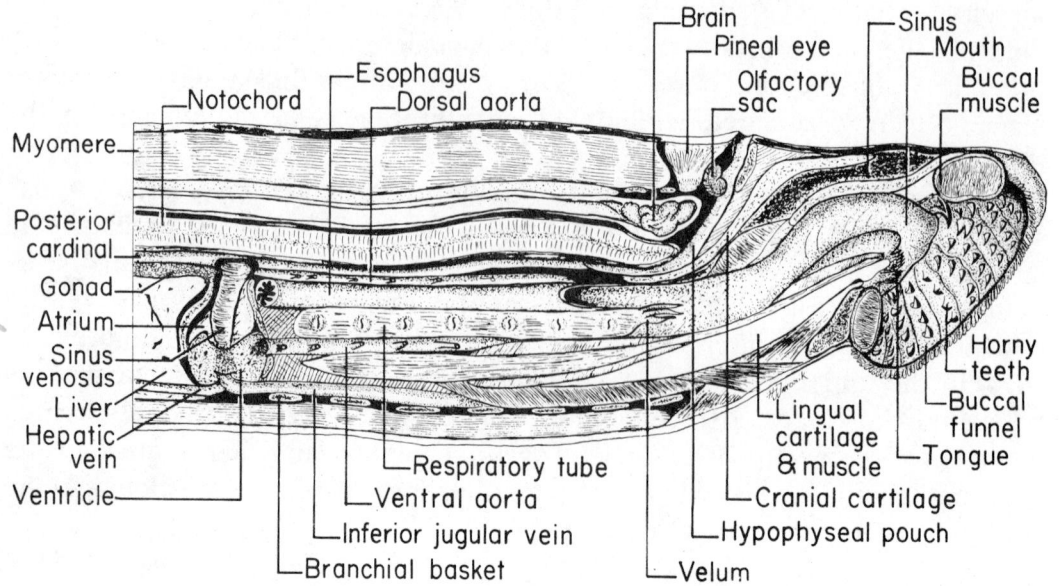

Figure 2–3. Sagittal section through the anterior portion of a lamprey.

They constitute the rudiments of vertebral arches. Other cartilages surround parts of the brain, and extend into the roof of the buccal funnel. These are parts of the primary brain case, or **chondrocranium.** A long median **lingual cartilage** extends into the tongue. Still other cartilages will be found beneath and lateral to the gill region and posterior to the heart. They form a **branchial basket** that supports the gill region. The branchial basket occupies a more superficial position than the visceral arches of other fishes, but its relationship to the various cranial nerves is the same, and it may be homologous to the visceral skeleton. Examine a special preparation of the skeletal system, if available, to better appreciate this system.

The muscular system consists primarily of the segmented myomeres already observed. Note their appearance in the sections. Each myomere consists of bundles of longitudinal muscle fibers that attach onto connective tissue **myosepta** between the myomeres. Waves of contraction passing alternately down the two sides of the body cause the lateral undulations of the trunk and tail. Jets of water expelled from the gill slits may also aid locomotion. The movements of the buccal funnel and tongue during feeding are caused by the intricate musculature you see associated with these structures.

The Nervous System and Sense Organs

The brain and the spinal cord have been seen lying above the notochord. The lamprey's brain has poorly developed acousticolateral centers and the cerebellum is small, but in other major respects it is similar to the dogfish's brain (p. 194).

Notice the connections of the nostril in the sagittal section. It first leads into a dark **olfactory sac** located anterior to the brain. The internal surface of the sac is greatly increased by numerous folds. Then a **hypophyseal pouch** continues from the entrance of the olfactory sac and passes ventral to the brain and the rostral end of the notochord. Respiratory movements of the pharynx squeeze the end of the hypophyseal pouch as one squeezes the bulb of a medicine dropper, thereby moving water in and out of the olfactory sac. The adenohypophysis, a part of the pituitary gland, is derived from an embryonic hypophysis, but in the higher vertebrates the hypophysis invaginates from the roof of the mouth (stomodeum, Fig. 2–2). The pituitary gland is difficult to see in this type of dissection.

The pineal eye is represented by the clear area caudal to the olfactory

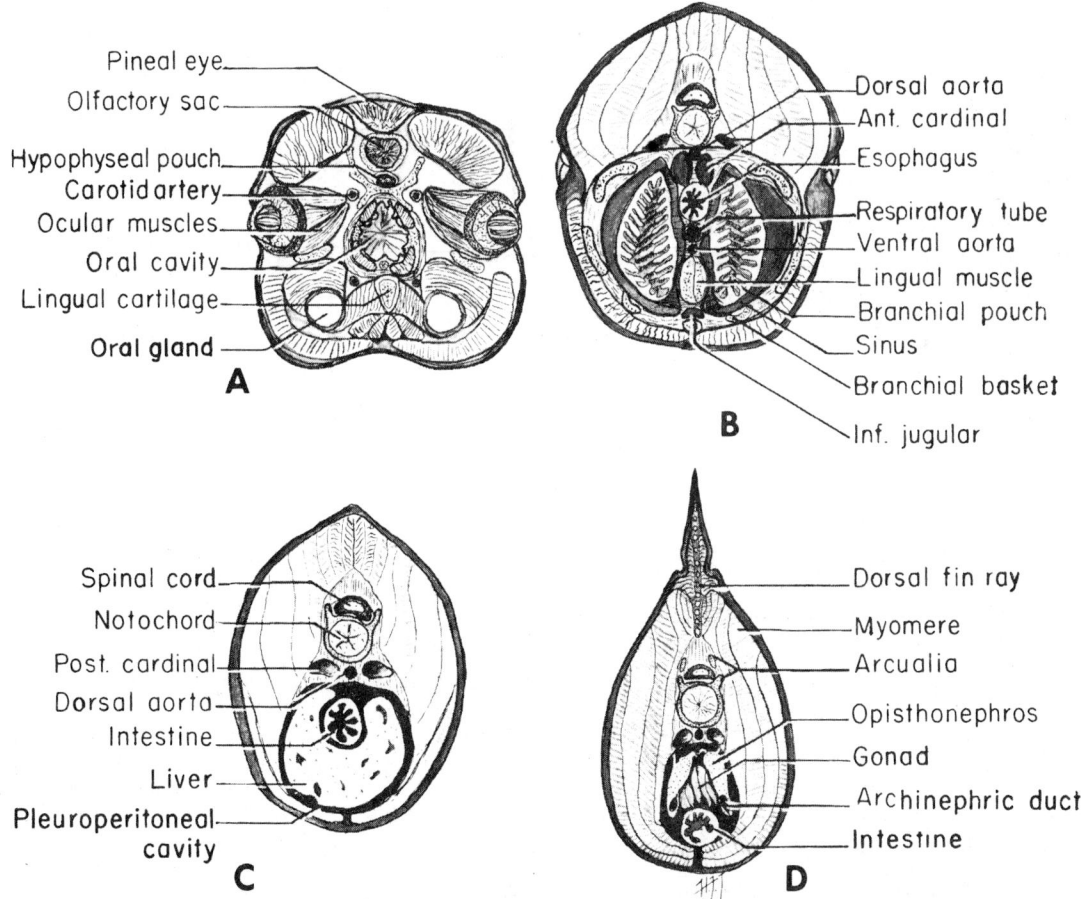

Figure 2–4.   Representative cross sections of a lamprey. *A,* through the pineal and lateral eyes; *B,* through a branchial pouch; *C,* through the liver; *D,* through the posterior part of the trunk.

sac. Actually, there are two median eyes here, for the pineal eye lies dorsal to a smaller parietal eye, but details cannot be seen. The well-developed **lateral eyes** show in one of the cross sections. If you make another cross section just caudal to the lateral eye, you will see part of the **inner ear.** It will appear as a bit of tissue imbedded in a cavity of the chondrocranium lateral to the brain.

The ear of fishes will be studied later, but only the inner part is present. As you will learn (Fig. 7–10), the inner ear normally has three semicircular canals. The lamprey, however, has only the two vertical canals. This condition was also found in certain of the ostracoderms. *May relate to 2 dimentional life style*

The Digestive and Respiratory Systems

In the sagittal section, it will be seen that the mouth opening leads into an **oral cavity,** which extends caudad to the level of the rostral end of the notochord. A pair of **oral glands,** which can be seen in the first cross section (Fig. 2–4, *A*), lie ventral and lateral to the oral cavity. Inconspicuous ducts lead from them to the underside of the tongue. Their secretion acts as an anticoagulant and is also hemolytic and cytolytic. The caudal end of the oral cavity leads into two tubes—an **esophagus** dorsally and a **respiratory tube** ventrally. The esophagus can be recognized by the numerous oblique folds in its lining. Slightly caudal to the gill region, the esophagus leads into a long, straight **intestine** that continues to the cloaca. The internal surface of the intestine is increased by longitudinal folds, one of which is particularly prominent, has a somewhat spiral course and is called the **spiral valve.** Otherwise there is little differentiation to the intestine. A true stomach is absent, and this is believed to be a primitive feature. *More likely related to blood feeding*

A large, and in the sagittal section, triangular-shaped **liver** is located beneath the cranial portion of the intestine. It grows out from the intestine embryologically and remains connected to it by an inconspicuous bile duct.

The **pancreas** of higher vertebrates is not present as a gross organ. But Barrington (1945) has found patches of cells comparable to the enzyme-secreting cells of the pancreas in the wall of that part of the intestine adjacent to the liver, and other groups of cells comparable to the endocrine portion of the pancreas (islets of Langerhans) imbedded in the intestinal wall and liver. Experimental destruction of the latter cells causes a rise in blood sugar.

The intestine and liver lie within a division of the **coelom** known as the **pleuroperitoneal cavity.** The intestine of vertebrates is usually supported by a long dorsal mesentery, but this is reduced in the lamprey to a few strands surrounding blood vessels that pass to the intestine.

Returning to the respiratory tube, it will be seen that its entrance is guarded by a series of tentacles that constitute the **velum;** its wall is perforated by seven **internal gill slits.** By looking at a cross section, and using a probe, you will note that each internal gill slit leads into an enlarged gill, or **branchial pouch** (Fig. 2–4, *B*), lined with gill lamellae. The pouches open to the surface through the **external gill slits.**

Both the respiratory tube and the esophagus of the adult develop from a longitudinal division of the larval pharynx. The resulting separation of digestive and respiratory tracts is correlated with the lamprey's mode of feeding. When attached to its prey, the lamprey respires by pumping water in and out of the external gill slits. Some water may seep into the respiratory tube, but this would not interfere with feeding. The most active phase of respiration is expiration, for then the branchial muscles constrict the pouches and water is forcibly expelled. Inspiration results primarily from the elastic recoil of the branchial basket. When the lamprey is not feeding, some water may enter the gill pouches through the mouth and internal gill slits—the usual situation in fishes.

The Circulatory System

Study the **heart** in the sagittal section. It consists of a thin-walled, tubular **sinus venosus** located between the atrium and ventricle (Fig. 2–5). The sinus venosus receives all the venous drainage of the body and leads into the large atrium through a **sinuatrial aperture.** The **atrium** is generally filled with hardened blood and, hence, is dark in color. It is located lateral to the sinus venosus and fills most of the left side of the pericardial cavity. It leads by an **atrioventricular aperture** into a muscular **ventricle** located in the right ventral portion of the pericardial cavity. Valves present in the apertures keep blood moving in the correct direction. A conus arteriosus, found in most fishes, is not present, so the **ventral aorta** leaves directly from the ventricle.

The heart lies in another division of the coelom, the **pericardial cavity,** which is separated from the pleuroperitoneal cavity by a **transverse septum.** In the lamprey this septum is stiffened by a large cartilage of the branchial

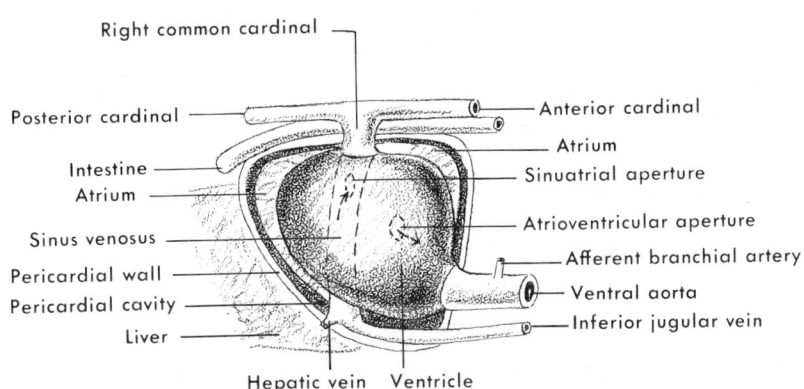

Figure 2–5. Lateral view of the right side of the heart of a lamprey. The position of the sinus venosus, which lies between the atrium and ventricle, is shown by dotted lines.

basket. Since the pericardial wall is stiff and will not collapse, contraction of the ventricle probably results in the increase in the "open" space within the cavity and, hence, a reduction of pressure around the atrium and sinus venosus. This reduced pressure would help "suck" blood in from the veins. Such a mechanism has been demonstrated in the dogfish (p. 279) where it plays an important role in the return of venous blood. Blood pressure in the veins is very low in primitive vertebrates.

The ventral aorta extends forward beneath the respiratory tube giving off eight paired **afferent branchial arteries** that lead to capillaries in the gills. The branchial arteries may not be seen; the first pair leads to a pair of vestigial branchial pouches located rostral to the first pair of well-developed pouches. The **dorsal aorta** can be found in the cross sections just ventral to the notochord. It receives aerated blood from the gills by way of **efferent branchial arteries** and continues caudally, supplying the body musculature and viscera. In the tail it is called the **caudal artery.** The head is supplied chiefly by a pair of **carotid arteries** (Fig. 2–4, A) that leave from the front of the dorsal aorta.

A **caudal vein** is located ventral to the artery. The characteristic renal portal system of fishes is absent, for the caudal vein does not go to the kidneys, but directly to the paired **posterior cardinals.** These veins can be seen in cross sections of the trunk on either side of the dorsal aorta (Fig. 2–4, C and D). At the level of the heart the right posterior cardinal unites with the right anterior cardinal coming from the head to form a **right common cardinal** (duct of Cuvier) that passes ventrally lateral to the gut to enter the sinus venosus (Fig. 2–5). Most fish have a comparable left common cardinal, but in the lamprey the left posterior and anterior cardinals curve medially dorsal to the gut to join the right common cardinal. An inconspicuous hepatic portal system runs from the alimentary canal to the liver. From here blood passes to the sinus venosus through a **hepatic vein.**

The head and the branchial region and their numerous venous sinuses are drained by a pair of **anterior cardinals** (Fig. 2–4, B), which lead caudally to the right common cardinal, and by a median, **inferior jugular vein** that enters the sinus venosus beside the entrance of the hepatic vein. The anterior cardinals are located lateral to the notochord; the inferior jugular vein, ventral to the prominent tongue musculature in the floor of the branchial region.

A spleen is absent from lampreys.

The Excretory and Genital Systems

The excretory organs consist of paired **opisthonephric kidneys** that appear as flaps suspended from the dorsal wall of the caudal half of the pleuroperitoneal cavity (Fig. 2–4, D). The gonad may have to be pushed aside to see them. Each is drained by an **archinephric duct,** which runs along its free border. Cut away the lateral body wall in the cross section segment that contains the cloaca and observe that the two archinephric ducts unite at the caudal end of the pleuroperitoneal cavity to form a **urogenital sinus** which opens at the tip of the urogenital papilla. You may be able to pass a bristle into an archinephric duct and out the urogenital papilla.

Although the gonads (**ovary** or **testis**) develop from paired primordia, that of the adult is a large median organ which fills most of the pleuroperitoneal cavity. It is supported by a **mesentery.** Genital ducts are absent in both sexes, the gametes being discharged directly into the coelom. They leave the coelom through paired **genital pores** located on either side of the urogenital sinus at the extreme posterior end of the pleuroperitoneal cavity. Search for the pores by carefully probing this area.

THE AMMOCOETES LARVA

After ascending rivers in the spring months to spawn, the marine lamprey dies. The eggs hatch into a larva that once was thought to be a distinct animal, named *Ammocoetes.* The larval stage lasts five to seven years, during which time the larva attains a length of 4 to 6 inches. It then undergoes a metamorphosis to the adult and descends the streams. The adult lives in the ocean or in larger inland lakes for a year or two.

Examine a whole mount slide of a small ammocoetes larva through the microscope (Fig. 2–6). Note its fusiform shape and the **dorsal** and **caudal fin** which are continuous with each other. The fine dark lines, or specks, on the body surface are pigment cells (**chromatophores**). If the pigment is dispersed, you can see that each chromatophore consists of a central area from which branching processes radiate. Although the **myotomes** have been rendered transparent, their outlines may show as faint lines on the surface.

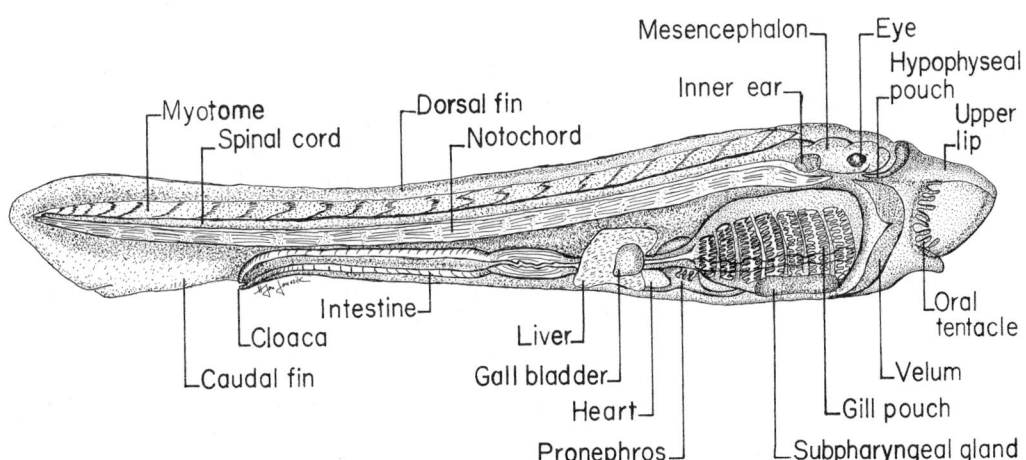

Figure 2–6. Lateral view of a whole mount slide of the *Ammocoetes* larva.

The **spinal cord** appears as a dorsal, dark-staining band that is enlarged rostrally to form the **brain.** In favorable specimens, you can see that the brain is composed of several lobes separated by constrictions, the largest and most caudal of which is the hindbrain (**rhombencephalon**); the next, the midbrain (**mesencephalon**); and the most rostral (sometimes subdivided), the forebrain (**prosencephalon**). The **notochord** appears as a light-staining, longitudinal band ventral to the spinal cord and the caudal two divisions of the brain.

The small surface protuberance rostral to the brain is the **median nostril,** which leads into the **hypophyseal pouch** lying ventral to the brain. If apparent, each **lateral eye** is represented by a round, dark spot lying between the mesencephalon and prosencephalon. An evagination from the caudal portion of the roof of the prosencephalon is the primordium of the **pineal eye;** the large, clear vesicle overlapping the front of the rhombencephalon is the primordium of the **inner ear.**

Many of the specializations of the adult digestive and respiratory systems are absent in the larva because it feeds by sifting minute food particles from the water. Observe that the **upper lip** has already enlarged to form the primordium of the buccal funnel. The **lower lip** appears as a transverse shelf. The **mouth opening** is surrounded by a series of **oral tentacles** which function as strainers and as sensory organs. Behind the mouth there is a clear chamber, the **oral cavity,** bounded caudally by a pair of large, muscular flaps, the **velum.** Movements of the velum bring a current of water and food into the **pharynx** behind it. It is of interest that the feeding current is caused by muscular action of the velum and pharynx rather than by ciliary action as it is in lower chordates. This is more efficient and is undoubtedly a factor that permits the ammocoetes larva to attain a considerably larger size than any of the lower chordates. Note the seven large **gill pouches** in the pharyngeal region. They are lined with **gill lamellae** and open through small, round, **external gill slits,** which may be seen by sharp focusing on the surface. Ventral to the pharynx you will find a large, dark-staining, elongate body called the **subpharyngeal gland.** It secretes mucus, which is discharged through a pore into the pharynx. Details of the feeding currents are not completely known, but the mucus probably rotates and entraps food particles as it passes caudally into the narrow **esophagus.** The subpharyngeal gland has many of the relationships of the lower chordate endostyle and may be its homologue. Certain of its cells also produce iodoproteins, and these cells transform into the thyroid gland of the adult. Posterior to the esophagus the alimentary canal widens to form the **intestine,** which continues to the **cloaca.**

The **liver** is located adjacent to the caudal portion of the larval esophagus and contains a large, clear vesicle, the **gall bladder.** Notice the **heart** lying ventral to the esophagus in front of the liver. Between the heart and esophagus you will see a few bell-shaped or fingerlike processes, or tubules, which are parts of the larval **pronephric kidney.**

The ammocoetes larva is of great phylogenetic interest, and may give some idea of the nature of the missing link between the vertebrates and lower chordates. It resembles *Amphioxus* in many fundamental characteristics, but lacks such specializations as the atrium; it also lacks many of the obvious specializations of the adult lamprey, yet has all of the essential vertebrate features. Its well-developed sense organs, brain, myomeres and caudal fin enable it to be more active than any lower chordate. It uses efficient muscular movements of the velum and pharynx in filter feeding. Being a larger and more active animal than *Amphioxus,* it no longer depends on cutaneous gas exchange but has evolved gills at a logical site where there is a good flow of water, i.e., in the gill pouches. These gill pouches, therefore, are now part of both the feeding and respiratory mechanisms.

3

THE EVOLUTION AND EXTERNAL ANATOMY OF VERTEBRATES

With the anatomy of a primitive vertebrate as a starting point, one can proceed to a consideration of the evolution of the higher vertebrates. But before examining the details of vertebrate anatomy, it is desirable to survey briefly the general course of vertebrate evolution. If possible, this should be illustrated by conducting the class through a museum or through a synoptic collection.

VERTEBRATE EVOLUTION

The subphylum **Vertebrata** is divided into eight classes, four of which are aquatic (superclass **Pisces**) and four essentially terrestrial (superclass **Tetrapoda**). Their relationships are depicted in Figures 3–1 and 3–2.

Class Agnatha.   As can be seen, the ancestral class was the **Agnatha**. This class is an ancient one, the earliest fossils being found in Ordovician deposits over 400,000,000 years old. The class consists of several extinct orders of heavily armored fishes collectively called the **ostracoderms,** and two living groups: the hagfish (order **Myxiniformes**) and lampreys (order **Petromyzontiformes**). Agnathans, as their name implies, lack jaws. They also lack typical paired appendages. Certain extinct groups had pectoral spines or lobes, but pelvic fins are never found. Although ostracoderms had well-developed bone in the skin, their internal skeleton was largly cartilaginous.

Class Placodermi.   Vertebrates above the agnathans have jaws and usually well-developed pectoral and pelvic appendages. Having jaws, they are often referred to as **gnathostomes.** Among the earliest gnathostomes were members of the class **Placodermi.** They had jaws formed from a rostral gill arch and associated superficial bone, and good pectoral and usually pelvic appendages. Large bony plates remained in the skin over the head, but bony scales were reduced or lost over the rest of the body. The deeper parts of the skeleton were mostly cartilaginous, although some bone was present. A swim bladder was probably absent in most groups. Placoderms became extinct about the end of the Devonian Period.

Class Chondrichthyes.   Cartilaginous fishes of the class **Chondrichthyes** appear to have evolved from early placoderms. Jaws and paired appendages are present, and the skeleton is entirely cartilaginous, for all bone has been lost except for minute scales in the skin. A hydrostatic swim bladder is absent, so these fishes tend to sink to the bottom when not actively swimming. There are two groups—the subclasses **Elasmobranchii** and **Holocephali.** Contemporary elasmobranchs are the sharks and dogfishes of the order **Selachii** and the skates and rays of the order **Batoidea.** Living holocepha-

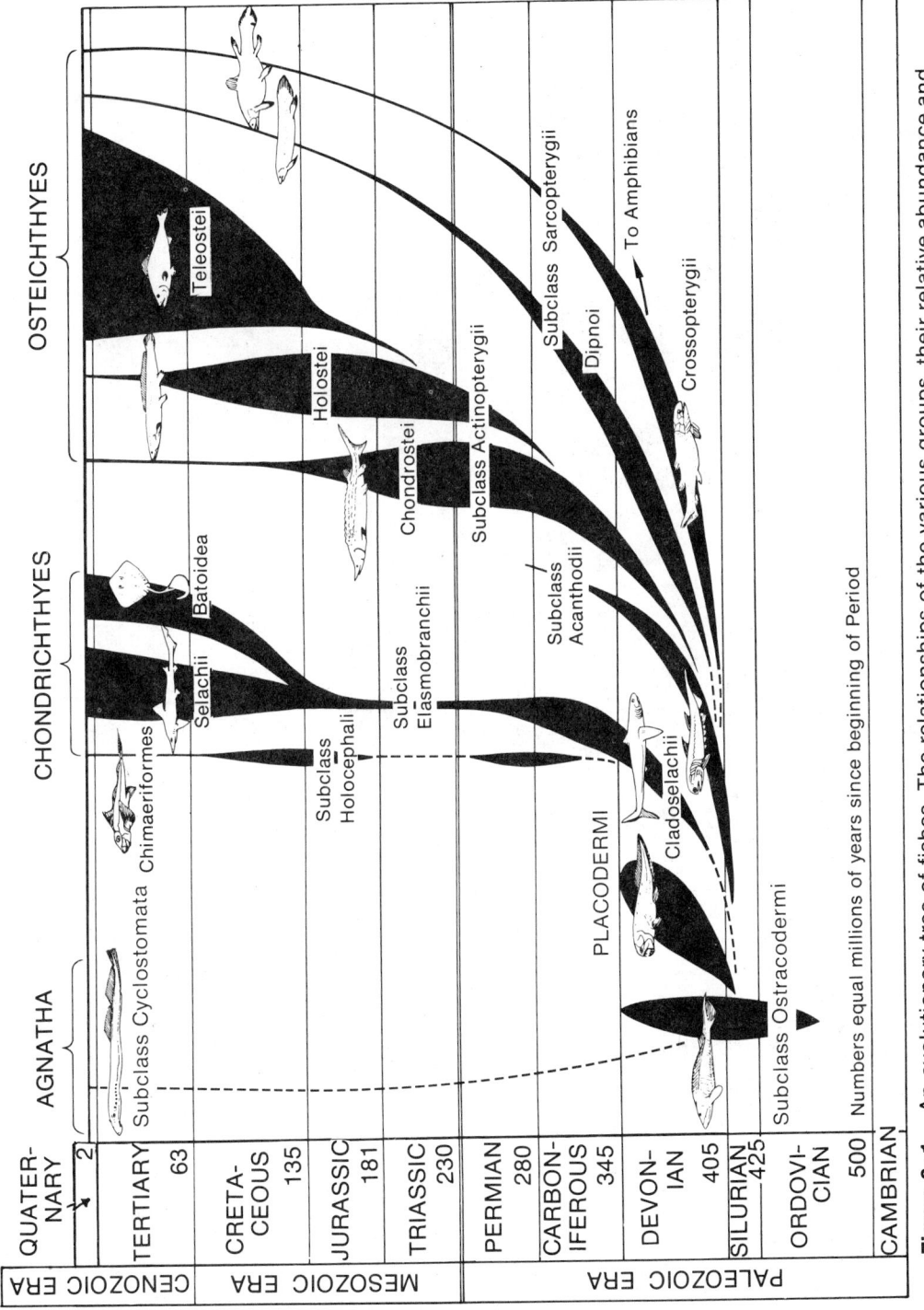

Figure 3–1. An evolutionary tree of fishes. The relationships of the various groups, their relative abundance and their distribution in time are shown. (From Villee, Walker, and Barnes, General Zoology.)

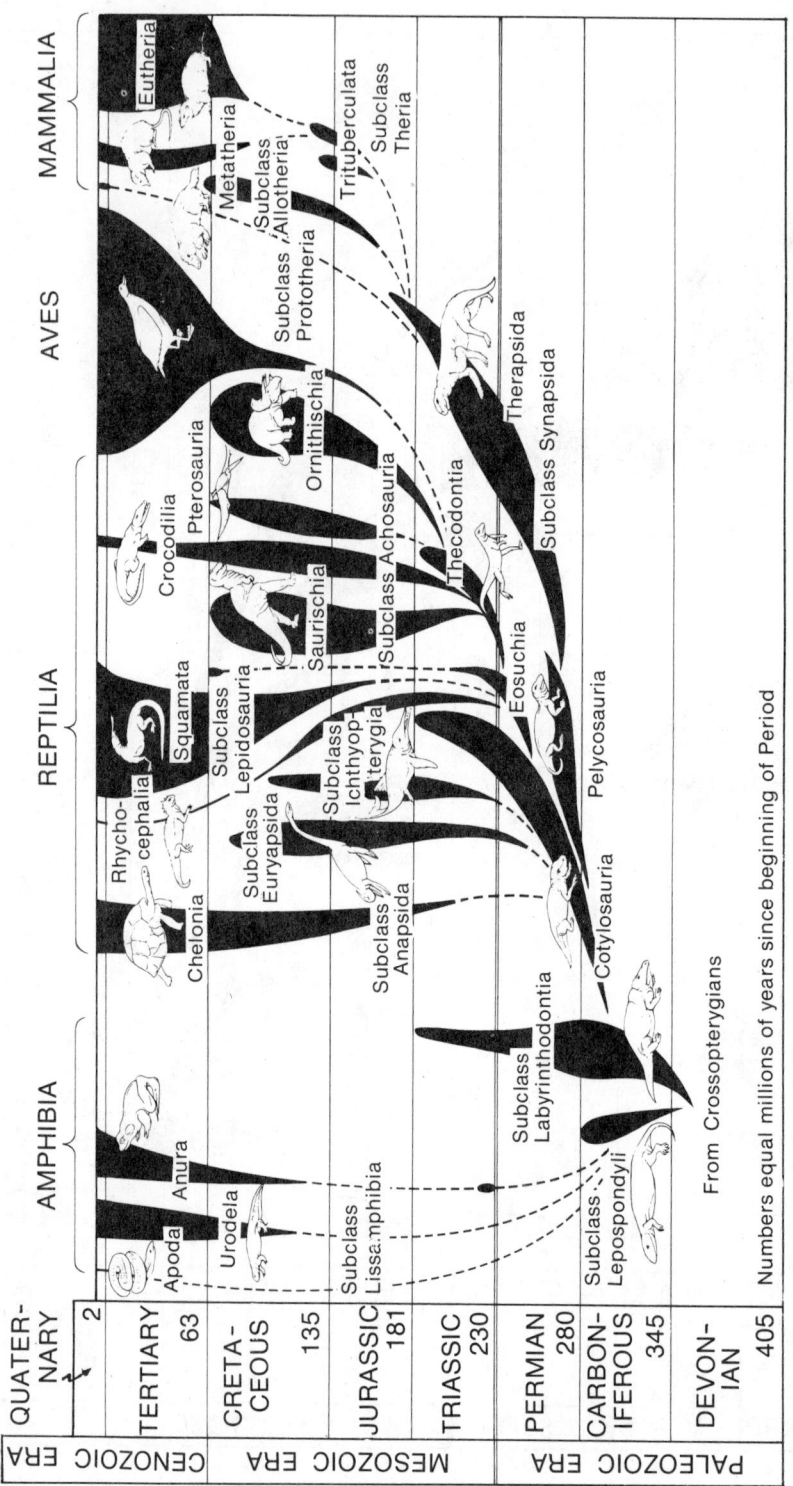

Figure 3–2. An evolutionary tree of terrestrial vertebrates. The relationship of the various groups, their relative abundance and their distribution in time are shown. (From Villee, Walker, and Barnes, General Zoology.)

lans are peculiar types known as chimaeras (order **Chimaeriformes**). With few exceptions, the cartilaginous fishes are marine.

Class Osteichthyes.   All remaining fishes belong to the class **Osteichthyes,** and are characterized by having jaws, paired appendages, usually well formed bony scales, at least a partially ossified skeleton, and lungs or lung derivatives (swim bladders). An internal sac of air, the volume of which can be regulated, enables these fishes to float at different depths in the water with a minimum of muscular effort. It is not clear whether bony fishes also diverged from placoderms or independently from ostracoderms. The class contains an extinct group characterized by numerous paired spines (subclass **Acanthodii**) and two subclasses that have survived and flourished: the **Actinopterygii** and the **Sarcopterygii** (Choanichthyes).The Actinopterygii is the larger. It includes such types as the African *Polypterus,* the paddlefish and sturgeon, belonging to the superorder **Chondrostei;** the bowfin *(Amia)* and gar pike, belonging to the superorder **Holostei;** and the great variety of trout, eels, minnows, perch, bass and other typical fishes that constitute the superorder **Teleostei.** Actinopterygians can be distinguished by their fanlike paired fins supported by numerous delicate rays. They also lack internal nostrils and generally have swim bladders. Sarcopterygians are the lungfishes of the order **Dipnoi** and the crossopterygians (order **Crossopterygii**). Crossopterygians were long believed to be extinct, but one surviving member was discovered in 1939 off the coast of South Africa and others have been found since then in the waters off the Comoro Islands near Madagascar. In contrast to actinopterygians, sarcopterygians have paired fins that are lobate in shape and supported by a central axis of flesh and bone. The extinct freshwater species also had internal nostrils and presumably lungs, but these structures were reduced in those, including the living species, that adapted to a marine environment.

Class Amphibia.   Freshwater crossopterygians gave rise to amphibians, the first tetrapods. Amphibians, which include the contemporary frogs and salamanders, are a transitional group between an aquatic and a terrestrial mode of life. They have acquired legs and other terrestrial adaptations but are found only close to water because their ability to conserve body water is rudimentary. Most reproduce in this medium since they have not evolved a cleidoic egg with its large store of yolk, extraembryonic membranes, fluid, and shell, which provide for all the requirements of the developing embryo and thereby obviate an aquatic larval stage. Amphibians resemble their piscine ancestors in being ectothermic, i.e., their body temperature and hence their rate of metabolism rises and falls with the ambient temperature. If aerial temperatures become too high or low, amphibians retreat to the relative thermal stability of the water or burrow into the ground. Three subclasses are recognized. Members of the extinct subclass **Lepospondyli** are not well known but may have been ancestral to modern amphibians, all of whom belong to the subclass **Lissamphibia:** frogs and toads (order **Anura**), salamanders (order **Urodela**), and tropical, wormlike caecilians (order **Apoda**). Lissamphibians have somewhat degenenerate skeletons with considerable cartilage remaining in the hand, foot, pelvic girdle (pubic region), and skull. They never have more than four fingers. The dominant amphibians during the late Paleozoic Era were members of the now extinct subclass **Labyrinthodontia.** Many labyrinthodonts were robust creatures with strong skeletons, and it is from certain of these that reptiles evolved.

Class Reptilia.   Reptiles have completed the transition from water to land by evolving a cleidoic egg and the ability to conserve body water. Reptiles and higher classes are often called **amniotes,** the amnion being a fluid-filled extraembryonic membrane that surrounds the embryo; amphibians and fishes are often called **anamniotes.** Although still ectothermic, many reptiles have considerable ability to maintain their daytime body temperature at a fairly high and constant level by regulating their degree of exposure to the sun. The stem group of reptiles were members of the order

Cotylosauria. They differed from labyrinthodonts in only a few structural details, but we believe that they had the cleidoic egg. Reptiles were the dominant terrestrial vertebrates for 200,000,000 years, and during this period adapted to many habitats. The dinosaurs became giants of the land and swamps. Other reptiles evolved true flight, and some returned to the sea. Of the numerous reptilian orders, only the **Chelonia** (turtles), **Rhynchocephalia** *(Sphenodon),* **Squamata** (lizards and snakes) and **Crocodilia** (crocodiles and alligators) have survived. Although these orders are reasonably successful, mammals and birds now occupy a position of dominancy on the land and in the air, respectively.

Class Aves. Both birds and mammals improved on the basic terrestrial adaptations of reptiles, becoming active and warm-blooded (endothermic). Body temperature is maintained at a relatively high and constant level by physiological control mechanisms rather than behaviourly. Birds, as a group, specialized for flight, evolving wings and feathers. Primitive birds had toothed jaws, clawed fingers, and long tails and, aside from their feathers, were very similar to, and probably evolved from, certain small bipedal reptiles closely related to early dinosaurs. More recent birds have lost these reptilian features but continue to lay reptilian-type eggs. Contemporary species are usually arranged in 27 orders, the perching and song birds (order **Passeriformes**) being among the highest.

Class Mammalia. The line of evolution to mammals early diverged from the cotylosaurs and passed through two extinct reptilian orders (**Pelycosauria** and **Therapsida**), collectively referred to as the mammal-like or **synapsid reptiles.** As the common name implies, these reptiles gradually came to resemble mammals, at least in their dentition and osteological features. The actual transition was made surprisingly early, even before the complete dominancy of the reptiles, but mammals remained inconspicuous in the fauna for millions of years. The most primitive of living mammals are the duckbilled platypus *(Ornithorhynchus)* and spiny anteater *(Tachyglossus),* belonging to the subclass **Prototheria,** order **Monotremata.** They resemble other mammals in having hair, mammary glands (although no nipples) and a characteristic type of jaw joint and auditory ossicle mechanism; however, they continue to lay a reptilian-type egg. Although practically unknown among fossils, monotremes must be an ancient group. It is now believed that they diverged from mammal-like reptiles independently of the other mammals.

All other mammals are placed in the subclass **Theria** and reproduce viviparously. That is, they retain the embryo in the uterus, where its needs are provided for by some sort of placenta, until it is ready for birth. The earliest known mammal fossils probably belong in this subclass. Little is known of these archaic mammals, but they seem to have been rat-sized, semiarboreal forms that must have spent much of their time eluding the reptiles. The more important ones are placed in the infraclass **Trituberculata.**

With the extinction of the dominant reptilian orders about 65,000,000 years ago, mammals came into their own. Two lines of evolution diverged from the trituberculates. One led to the infraclass **Metatheria,** which includes but one order, the **Marsupialia.** Marsupials are pouched mammals such as the kangaroo and opossum. They generally have a poorly formed placenta and so give birth to young at a relatively early stage. The young move into the pouch, or marsupium, where they attach to nipples of the mammary glands and complete their development. The period of intrauterine gestation is very short compared to that of pouch development. Opossums, for example, develop only 13 days in the uterus compared to 50 to 60 in the pouch.

The other line of evolution led to the infraclass **Eutheria,** or true placental mammals. Eutherians lack the marsupium and have a more efficient placenta, and the embryos are retained in the uterus until a much more advanced stage. They are now the dominant mammals. The order **Insectivora,** which includes the shrews and moles, is the most primitive of the numerous orders, and the others diverged directly, or indirectly, from it. Human beings belong to the order **Primates;** our laboratory examples to the orders **Carnivora** (cat and mink) and **Lagomorpha** (rabbit).

In studying the evolution to placental mammals, which is the main theme in this manual, one should ideally consider extinct groups through which the line of evolution passed—an ostracoderm, acanthodian, crossopterygian, labyrinthodont, cotylosaur, pelycosaur, therapsid, trituberculate and primitive insectivore. This can be done to some extent in lectures, but in the laboratory one must rely on available, living species. These inevitably have many peculiar features of their own, but if carefully selected will at the same time show a number of primitive features characteristic of the early members of their group. The lamprey, representing a primitive, jawless fish, has already been examined. In the sections that follow, the evolution of vertebrate anatomy will be considered by following the changes in each organ system through a representative series of animals. For the most part, this can be done in three stages—jawed fish, primitive tetrapod, and mammal.

EXTERNAL ANATOMY

Aside from the skin or **integument** *(cutis),* most of the external features of vertebrates are simply manifestations of other organ systems. Nevertheless, it is worthwhile studying them along with the skin at this time, for this will emphasize certain important general changes. The skin itself consists of two fundamental layers of tissue—an outer **epidermis** of stratified epithelial cells derived from the embryonic ectoderm, and an inner **dermis** *(corium)* of dense connective tissue derived from the mesoderm. The dermis is the thicker and also the more stable layer, for its basic structure remains much the same. But as vertebrates adapt to a terrestrial environment, the epidermis becomes thicker and its outer cells become cornified (keratinized); that is, the outer cells, as they die, become filled with a horny, water-soluble protein called **keratin** that renders them less pervious to water. The cornified cells, which often form a distinct epidermal layer called the **stratum corneum,** are continually being sloughed off, but they are replaced by mitosis in the basal cells of the epidermis, the **stratum basale.**

These changes cannot be seen grossly in the laboratory, but certain derivatives of the integument can be studied along with the general external features. Major derivatives found in various vertebrates are pigment, glands and a variety of bony and horny structures such as scales, feathers and hair. Although some pigment is contained within unspecialized epidermal cells in vertebrates, it is usually found within specific cells called **chromatophores** in fishes, amphibians and reptiles. The chromatophores are stellate-shaped cells in which the pigment may move about. When the animal is dark, the pigment is dispersed throughout the cell; when the animal is light, the pigment is withdrawn to the center. The chromatophores are located in the dermis despite their ectodermal origin. All the glands are derivatives of the epidermis, but they have usually invaginated into the dermis. Feathers and hair are also epidermal derivatives, but scales (of which there are a variety of types) may come from either the epidermis or the dermis. The horny and bony derivatives of the skin (various types of scales and plates, feathers and hair) are sometimes referred to as an "exoskeleton," but they should not be confused with the invertebrate exoskeleton which represents a noncellular secretion superficial to the epidermis.

Fishes

Even though the Chondrichthyes are on an evolutionary side line (Fig. 3–1), they represent the primitive jawed-fish stage better than any available bony fish, for they lack lungs, or lung derivatives, and certain other peculiarities found in most living Osteichthyes. A favorable species for study is the spiny dogfish, *Squalus acanthias* of the North Atlantic and northern Pacific (Fig. 3–3). The Pacific population is sometimes described as a separate species, *S. suckleyi*.

Dogfish are small sharks belonging to the order Selachii. Adult males range in length from 2 to 3 feet; females are slightly larger. They prefer water temperatures ranging from 6 to 15° C. and, hence, migrate north in the spring and south in the fall. A migratory school includes thousands of fishes. They are voracious and prey upon most species of fish smaller than themselves. They are considered edible in Europe, but North American fishermen consider them only a nuisance. They drive more favorable fish away and are rather destructive to fishing gear and hooked or netted fish.

(A) General External Features

Examine a specimen, noting the streamlined, or **fusiform,** shape that enables the animal to move easily through the water. The body can be divided into a **head,** which includes the gill region; a **trunk,** which continues to the cloaca; and a **tail** posterior to this. The **cloaca** is the chamber on the ventral side that receives the intestine and the urinary and genital ducts. The urinary ducts, and in the male the genital ducts as well, open at the tip of a **urinary papilla** which can be seen inside the cloaca. The **anus**—the opening of the intestine into the cloaca—lies cranial to the urinary papilla. The cloaca opens at the surface by way of the **cloacal aperture.** The body regions are not so well demarcated as in tetrapods, for all blend into one another. There is no neck, and the tail is a powerful organ of locomotion.

Note that the body is a dark color above, and light beneath. Such a distribution of pigment, referred to as **counter shading,** is common in vertebrates, especially aquatic forms. Optically it tends to neutralize the effect of natural lighting, which highlights the back and casts a shadow on the belly, and thus renders the organism less conspicuous.

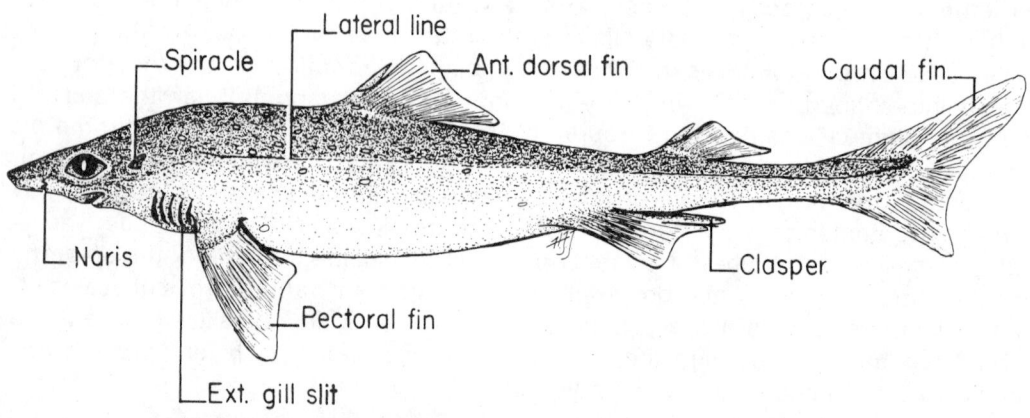

Figure 3–3. A lateral view of the dogfish, *Squalus acanthias.*

There are two **dorsal fins** (cranial and caudal), with a large **spine** in front of each. The spines are defensive and are associated with modified skin glands that secrete a slightly irritating substance. Ring like markings on the spines give an indication of the fish's age; the life span ranges from 25 to 30 years. Large, paired **pectoral fins** will be seen just behind the head, and paired **pelvic fins** at the caudal end of the trunk. Males have stout, grooved copulatory organs, called **claspers,** on the medial side of their pelvic fins. The tail ends in a large **caudal fin** of the **heterocercal** type; that is, the fin is asymmetrical, for the body axis turns up into its dorsal lobe and most of the fin rays are ventral to the axis. All the fins are supported by fibrous fin rays **(ceratotrichia)** which lie in the skin on each surface of the fin. Deep cartilages, which will be seen later, lie along the base of each fin and provide further support. In well preserved specimens a lateral keel will be seen on the trunk on each side of the base of the caudal fin.

Most primitive groups of fish have a heterocercal tail. This type of tail is associated with heavy bodied fishes that have large pectoral fins and are somewhat flattened along the cranioventral part of the body. When a fish of this type moves through the water, its front end tends to veer up. The movement of the caudal fin with its flexible sculling blade ventral to a stiff axis tends to give a compensatory lift to the caudal end, thereby keeping the fish on an even keel. More progressive fishes with lungs or swim bladder and more mobile paired fins have better control over their stability, and the caudal fin tends to become symmetrical.

The **mouth,** which is of course supported by jaws, is located on the underside of the head and is bounded laterally by deep **labial pockets.** There is a **labial fold,** containing a cartilage, between the mouth and pocket. A pair of large **eyes** will be seen set in deep sockets on each side of the head. The rim of each socket forms immovable **eyelids.** Paired external nostrils, technically called **nares,** are on the underside of the pointed snout. The opening of each is partially subdivided by a little flap of skin which separates the stream of water that flows through the nostril into, and out of, each **olfactory sac.** Pass a probe into a naris and notice that the olfactory sac does not communicate with the mouth cavity.

A row of five **external gill slits** is located in front of the pectoral fin. The term elasmobranch (the subclass to which sharks belong) means platelike gills and refers to the tissue between the gill slits. In most fishes the gills are covered by a common opercular flap. Caudal to the eye you will see a large opening called the **spiracle.** Little parallel ridges, representing a reduced gill called the **pseudobranch,** can be seen on a fold of tissue that is separated from the rostral wall of the spiracular passage by a deep recess. Probe to determine the extent of the recess. This fold of tissue is a **spiracular valve,** which can be closed to prevent water from leaving this way.

The spiracle is really the reduced, and modified, first gill slit of gnathostomes. Most fishes respire by taking water into the pharynx through the mouth and discharging it through the gill slits. When a spiracle is present, water also enters through it; in the bottom-dwelling skates and rays most of the water enters this way.

If you look at the top of the head between the spiracles with a hand lens, you will see a pair of tiny **endolymphatic pores,** one pore on each side of the midline. They communicate with the inner ear, which in most fishes is an organ of both equilibrium and hearing. Vibrations of low frequency and movements in the water are detected by the lateral line system. The position of one canal of this system (the **lateral line** in a restricted sense) is indicated by a fine, light-colored, horizontal stripe extending along the side of the body. It is nearer the dorsal than the ventral surface. You will also see patches of pores on the head through which a jellylike substance extrudes if the area is squeezed. They are the openings of the **ampullae of Lorenzini.** These ampullae are a modified part of the lateral line system now believed to be electroreceptors.

(B) Integumentary Derivatives

Major derivatives of the fish integument are chromatophores; glands, usually in the form of simple, scattered mucous cells; and hardened, dermal structures. Most of the last are bony scales. In the dogfish these are minute **placoid scales,** or dermal denticles, which cover the animal and can be felt by moving your hand anteriorly over the surface. To see them you will have to use a hand lens or, better still, observe a special microscopic preparation. (**Chromatophores** may be seen at the same time.) Each denticle consists of a basal plate imbedded in the dermis from which a spine perforates the epidermis and projects caudad (Fig. 3–4). At one time elasmobranch skin, sold as shagreen, was used as an abrasive for polishing wood.

Histologically a placoid scale resembles a tooth in many ways. It contains a **pulp cavity,** a thick layer of dentine and a surface covering of enamel-like material. **Dentine** is produced by dermal cells. It consists of inorganic crystals of hydroxyapatite and about 30 per cent organic material. **Enamel** which is harder, is produced by epidermal cells and consists nearly entirely of hydroxyapatite; only 3 per cent is organic material. The surface of a placoid scale is composed of a material resembling enamel, but many investigators believe that it is derived from the dermis, so it is often called **vitrodentine.**

Primitive fishes had an extensive armor of bony scales and plates, which consisted of basal layers of dermal bone overlaid first by a type of dentine **(cosmine)** and then by a type of enamel **(ganoine).** Depending on the relative amounts of cosmine and ganoine, and details of blood supply to the scale, they were called either **cosmoid** or **ganoid scales.** The scales of the Chondrichthyes represent essentially the surface parts of cosmoid scales. Living bony fishes have very thin bony scales called **cycloid** if they are smooth, or **ctenoid** if they have tiny processes on their surfaces. These represent part of the bony, basal layers of the ganoid or cosmoid scale. Primitive actinopterygians had a ganoid scale; primitive sarcopterygians, a cosmoid scale.

Other hard derivatives of the integument are the ceratotrichia and spines already observed. The ceratotrichia are composed of a fibrous material elaborated in the dermis. The spines are essentially enlarged scales. In the Osteichthyes the fins are supported by bony rays called **lepidotrichia,** which are also modified scales.

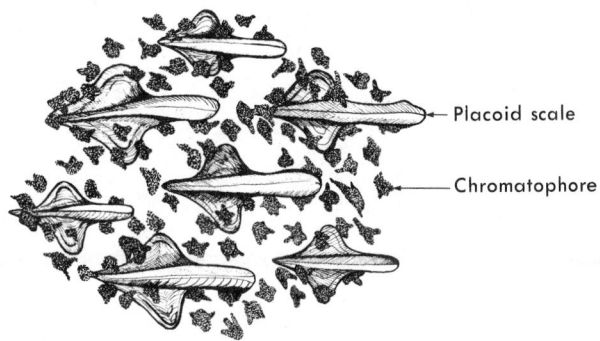

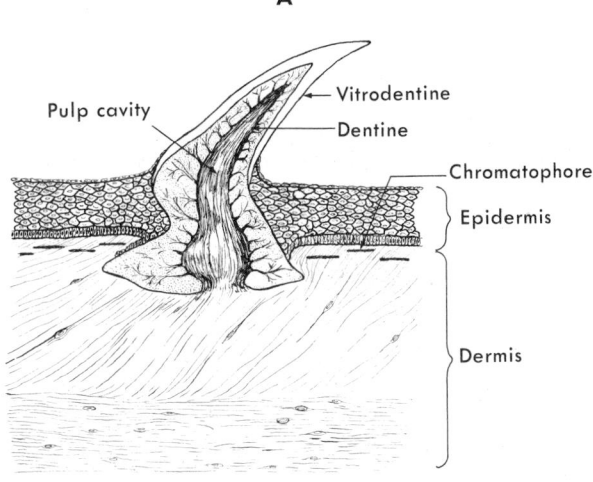

Figure 3–4. *A,* Magnified surface view of the placoid scales and chromatophores in dogfish skin; *B,* vertical section through the scale and the skin. (*B* redrawn from Dean.)

Primitive Tetrapods

Labyrinthodonts and cotylosaurs were, of course, the true primitive tetrapods. It is difficult to find a living amphibian or reptile that is a good representative of this stage of evolution, but any one of the species of mudpuppy *(Necturus),* which belongs to the class Amphibia, order Urodela, is reasonably satisfactory in most ways. However, it is unusual in one respect, for it is a permanent larva. That is, it reaches sexual maturity while in the larval stage and never completes its metamorphosis, a condition referred to as **neoteny.** This fact must be discounted in studying its anatomy. *Necturus* is distributed throughout most of the eastern half of the United States. The most widespread species is *N. maculosus.* It is most abundant in clear waters of lakes and larger streams, but it is also found in weed-choked, turbid and smaller bodies of water. It is most active at night, when it forages for small fish, crayfish, aquatic insect larvae and mollusks.

(A) GENERAL EXTERNAL FEATURES

Examine a specimen of *Necturus,* and compare it with *Squalus.* The body is elongate with a modest-sized, flattened **head;** an incipient **neck** region; a long **trunk;** and a powerful, laterally compressed **tail** (Fig. 3–5).

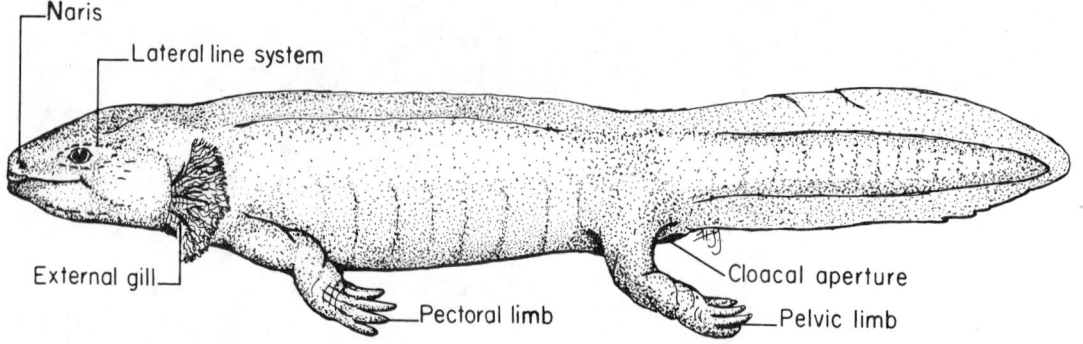

Figure 3-5. A lateral view of the mudpuppy, *Necturus maculosus.*

There are no median fins, except for traces on the tail, and these lack fin rays. The paired fins of fish have become transformed into pectoral and pelvic limbs. Each limb consists of three segments. In the pectoral appendage these are upper arm (**brachium**), forearm (**antebrachium**), and hand (**manus**). The elbow joint (**cubitus**) is between the brachium and antebrachium; the wrist (**carpus**) is in the proximal part of the manus. Corresponding parts of the pelvic appendage are the thigh (**femur**),[4] shank (**crus**), and foot (**pes**); corresponding joints are the knee (**genu**) and ankle (**tarsus**). Only four toes are present, the most medial probably being homologous to the second toe of amniotes. If the entire leg is pulled out to the side at right angles to the body with the palm of the hand (or sole of the foot) facing ventrally, the cranial border is said to be **preaxial;** the caudal, **postaxial.**

Since *Necturus* is aquatic, locomotion is still fishlike, with a lateral undulation of the trunk and tail playing an important role. The legs are used, more so in terrestrial salamanders than in *Necturus,* but are rather weak. When used, they are held in the primitive tetrapod position. The proximal segment of the leg projects horizontally at right angles to the body, the distal segment extends vertically downward, the manus points forward, the pes more or less laterally. Note that in such a position the preaxial border of the entire hind leg is still cranial, and there is a simple hinge joint at the knee. In the front leg the preaxial border of the brachium is cranial, but there has been a torsion, or rotation, at the elbow which brings the foot forward, so the preaxial border of most of the antebrachium is medial. The position of limbs in a primitive tetrapod is such that the body is not raised far off the ground, and the humerus and femur move back and forth in the horizontal plane. The limbs are rather inefficient mechanically for support and locomotion.

The **mouth** is terminal and is bounded by fleshy **lips.** Just above the upper lip is a pair of widely spaced external nostrils (**nares**). As will be seen later, they communicate with the front of the oral cavity by way of internal

---

[4]The term femur can be used for the thigh and for the bone within it. The *Nomina Anatomica* term for the thigh is femur and for the bone, os femoris.

nostrils (**choanae**), thus permitting air to be taken into the mouth. Small **eyes** are present, but are devoid of lids. The absence of eyelids is a larval feature not found in metamorphosed Amphibia. Most tetrapods have movable lids that help to protect, cleanse and moisten the eye. Unlike the situation in most amphibians and reptiles, salamanders also lack an external eardrum. There is an internal ear, however, and vibrations reach it primarily by way of skull bones. Another sense organ is the **lateral line system,** which will appear, when the specimen has dried a bit, as rows of depressed dashes above and below the eyes, on the cheek and on the ventral surface of the head. A less obvious row of dashes also extends caudad along the side of the trunk. The lateral line system, too, is a larval feature that is lost in metamorphosed amphibians.

There are three pairs of prominent **external gills** at the caudal end of the head. Although some gas exchange takes place through the highly vascular skin, and the animal occasionally comes to the surface to gulp air, these gills are the major respiratory organ. It should be emphasized that external gills are larval structures and are different from the internal gills, located inside the gill pouches, of adult fish. *Necturus* also has two **gill slits** which can be seen between the bases of the external gills. The fold of skin extending across the ventral surface of the head between the gills is called the **gular fold.**

Finally, observe the **cloacal aperture** at the caudal end of the trunk. It is bounded by lips bearing tiny papillae in the male, and small folds in the female.

(B) INTEGUMENTARY DERIVATIVES

Very early amphibians retained small, bony scales, but bony scales have been lost in most modern amphibians, leaving the skin naked. Only the peculiar caecilians have dermal scales hidden in folds of the skin. The skin is not well protected against drying. Numerous alveolar mucous glands help to keep the skin moist, but most amphibians cannot wander far from water without danger of desiccation. Chromatophores are abundant.

Since the reptilian stage of the evolution of the integumentary derivatives is essential to understand those of birds and mammals, it will be considered at this point. A lizard or a turtle is a good example. Chromatophores continue to be present, but most of the skin glands are lost, only a few scent glands being retained. The reduction of glands is correlated with an extensive cornification of the epidermis, so that the skin is covered with **horny scales.** Sometimes little nodules of bone lie beneath the horny scales. On the head and the outside of the lips, the horny scales are enlarged to form plates. The horny scales are connected to each other by less heavily keratinized areas of the stratum corneum. In turtles the scales are modified and form large horny plates that overlie plates of bone. Most of the bony plates represent a redevelopment of bone within the dermis, but some of the ventral plates (plastral plates) include remnants of the original bony armor of primitive

fish. The tips of the toes of reptiles bear **claws,** which are also keratin structures.

Mammals

Since all placental mammals resemble each other in more ways than they differ,
the basic features of a mammal can be seen by studying any convenient one. Directions
in this manual are written in such a way that they can apply to the domestic cat *(Felis
catus)* or the mink *(Mustela vison),* both of which belong to the order **Carnivora,** or to
the rabbit *(Lepus* sp.), which belongs to the order **Lagomorpha**. If different members
of the class are provided with one of the carnivores and a rabbit, comparisons can be
made between divergent species at the mammalian evolutionary level. This will illustrate the essential uniformity of mammalian anatomy and, at the same time, reveal
important differences that have been superimposed upon a common structural pattern
through a long independent evolution of these species and their adaptation to car-
nivorous and herbivorous modes of life.

(A) GENERAL EXTERNAL FEATURES

Examine either a carnivore or a rabbit and compare it with *Necturus.*
The diagnostic **hair** of mammals is at once evident (Fig. 3–6), and it will be
seen that the evolutionary trends which began in primitive tetrapods have
continued. The **head** *(corpus)* is large and separated from the **trunk** by a
distinct and movable **neck** *(collum).* The trunk itself can be divided into

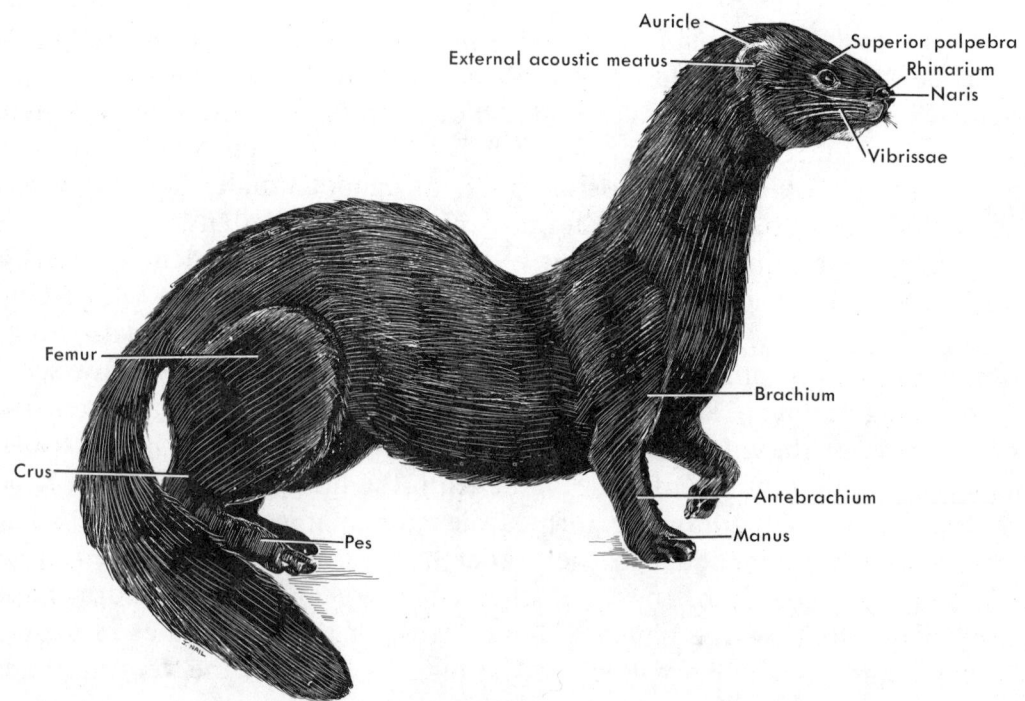

Figure 3–6.   A lateral view of the mink, *Mustela vison.*

the **back** *(dorsum)*, **thorax, abdomen,** and **pelvis.** A **tail** *(cauda)* is typically present in mammals, but, except in the whales and their allies, is greatly reduced in size in comparison with that of a primitive tetrapod. In some terrestrial mammals it is of use, often as a balancing organ in semiarboreal carnivores; in some other mammals it is essentially vestigial (rabbit); and in a few (man) it has been lost as an external structure.

The paired appendages consist of the usual parts—**brachium, antebrachium,** and **manus** in the pectoral; thigh or **femur,**[5] **crus,** and **pes** in the pelvic appendage. Mink have five toes, but in both the cat and the rabbit the most medial, or first, toe of the manus is vestigial, and the corresponding toe of the pes has been completely lost as an externally visible structure. Observe either on a mounted specimen or on a skeleton that a carnivore walks on its toes with the wrist and heel raised off the ground. This method of locomotion is referred to as **digitigrade,** in contrast to **plantigrade** (man), in which the entire sole of the foot is flat on the ground, or **unguligrade** (ungulates such as the horse), in which the animal walks on the tips of its toes. The rabbit is also digitigrade in respect to the front feet, but the hind legs are modified for hopping. Just before the leap, the pes is in the plantigrade position. The terminal segment of each toe has a **claw;** in the cat this segment is hinged in such a way that the claw can be retracted or extended.

Note that the limbs no longer occupy the primitive sprawled position but have rotated in such a way that they are directly beneath the body, and the legs move back and forth in the vertical plane. In the hind leg the change results from a 90 degree forward rotation. The knee and pes point cranially, and the original preaxial border is medial instead of cranial. In the front leg there has been a 90 degree backward rotation, so that the elbow points caudally. Because of a continued torsion at the elbow, seen beginning in primitive tetrapods, the manus still points cranially. The original preaxial border of the brachium is now lateral instead of anterior, but that of the antebrachium continues to be primarily medial. (In order to understand these changes, try to place your own appendages in the primitive tetrapod position and then slowly rotate them into the quadruped mammal position.) The new limb posture is more efficient for support. It also increases speed, for the legs are now capable of a powerful and rapid fore and aft drive. Walking on the toes, with the wrist and heel raised, increases the length of the stride. Although the trunk and tail are flexible, lateral undulations play no significant part in locomotion.

Turning to the head, notice that the **mouth** *(os)* is bounded by fleshy **lips** *(labia)*, the upper one being deeply cleft in the rabbit (harelip). The paired external nostrils (**nares**) are close together on the nose surrounded by moist, bare skin known as the **rhinarium.** The **eyes** *(oculi)* are large and are protected by movable upper and lower eyelids (**palpebrae**). Spread the palpebrae apart and observe a third lid, called the **nictitating membrane,** in the medial

---

[5]See footnote 4, p. 36.

corner of the eye. The nictitating membrane can be drawn across most of the eye, thus helping to moisten and cleanse this organ. Mammals have a prominent external ear consisting of a conspicuous external flap, called the **auricle,** and an external ear canal (**external acoustic meatus**) that extends into the head from the base of the auricle. The eardrum (**tympanum**) is located at the bottom of the meatus. It will not be seen at his time. That part of the head that includes the jaws, mouth, nose, and eyes is referred to as the **facial region;** the rest, containing the brain and ears, as the **cranial region.**

The cloaca of primitive vertebrates has become divided in therian mammals, so that the intestine and urogenital ducts open independently at the surface. The opening of the intestine, called the **anus,** will be found just ventral to the base of the tail. If the animal is a female, the combined opening of the urinary and reproductive ducts will appear as a second passage, the **vaginal vestibule,** bounded by small folds, ventral to the anus. If the animal is a male, the urogenital ducts open at the tip of a **penis.** Associated with this you will see the sac-shaped **scrotum** containing the testes. The testes may be retracted in the rabbit. The entire area of the anus and external genitals is called the **perineum** in both sexes. Further discussion of the details of the external genital organs will be deferred.

Carefully feel along the ventral surface of the thorax and abdomen on each side of the midline and you will find two rows of teats (**papillae mammae**) hidden in the fur. These bear the minute openings of the mammary glands. The teats are more prominent in females, but rudiments can sometimes be found in males. There are usually four or five pairs in the carnivores and six in the rabbit, but the number is subject to variation.

(B) Integumentary Derivatives

The integument of mammals is rich in derivatives, many of which can be seen or demonstrated in the laboratory. Chromatophores are largely absent, but pigment granules are present in the epithelial cells. There are many glands, which fall into three categories — **mammary glands,** tubular **sweat glands** (*sudoriferous glands*), and alveolar **sebaceous glands** usually associated with the hair follicles. The mammary glands resemble sweat glands in having contractile myoepithelial cells peripheral to the secretory cells and, for this reason some investigators consider them to be modified sweat glands. Wax and scent glands seem to be modified sebaceous glands. The openings of sweat glands can be seen with a hand lens on one's fingertips. The openings of the sebaceous glands and hair follicles are the more familiar "pores" of the skin.

The most conspicuous integumentary derivative is the insulating covering of **hair** (*capillus*). Hair replaces the horny scales of reptiles in most mammals, but scales may still be found on the tails of certain rodents, and they have redeveloped over the bony plates of the armadillo shell. Although hair is composed of keratinized cells, it is a new development of the epidermis.

It is not considered to be homologous with either horny scales or feathers, since details of its embryonic development are different. Moreover, the distribution of hair, as seen, for example, on the back of one's hand, leads to the conclusion that hairs evolved in small clusters between the scales, and then the scales were lost. The simultaneous presence of scales and hairs can be seen on the tails of some rodents. In most mammals the hair forms a dense fur over the body, being modified in certain places such as the **eyelashes** (*cilia*) and tactile whiskers (**vibrissae**) on the heads of carnivores and rabbits. But there are many departures from this pattern. Hair is reduced in man, and lost in the adults of such highly aquatic mammals as the whale. In some other mammals, the hair has become adapted for very specialized purposes. The quills of a procupine are a case in point, and the "horn" of a rhinoceros resembles a compact mass of hair.

Reptilian **claws** are retained in most mammals, but have been transformed into **nails** (*ungulae*) in certain primates, and into **hoofs** in the ungulates. Other common integumentary derivatives are the **foot pads** (*tori*) on the feet of most mammals. These are simply thickenings of the stratum corneum. The **friction ridges** (fingerprints) of primates are their homologue.

Aside from the widely distributed structures just mentioned, some mammals have still other derivatives. The redevelopment of **bone** in the dermis of the armadillo, the keratinized **whalebone plates** about the mouth of the toothless whales, and the horny covering of the **horns** of sheep, antelopes, and cattle belong in this category. The horns of these animals consist of a core of bone which arises from the skull and is covered with very heavily keratinized epidermis, i.e., horn. Horns of this type should not be confused with the **antlers** found in the deer group. Antlers are bony outgrowths of the skull that are covered with skin (the velvet) only during their growth. In contrast with horns, they generally are restricted to the male, they branch, and they are shed annually.

4

THE AXIAL AND VISCERAL SKELETON

The next organ systems to be studied will be a group concerned with the general functions of support and locomotion, namely, the skeleton, muscles, sense organs, and nervous system. The skeleton may appropriately be considered first, as it is a fundamental system about which the body is built.

Divisions of the Skeleton

The vertebrate skeleton consists of two basic parts—the **dermal skeleton** and the **endoskeleton.** Although these two become united in various degrees, they are distinct in their ontogenetic and phylogenetic origins. The dermal skeleton consists of bone that develops embryologically directly from the mesenchyme in, or just beneath, the dermis of the skin. This type of bone is called either **dermal** or **membrane bone.** It follows that the dermal skeleton is superficial. Bony scales and plates, and their derivatives, are dermal in nature. Among their derivatives are the dermal plates in the head region, teeth, and the dermal portions of the pectoral girdle. In addition, dermal bone has evolved independently of bony scales in the dermis of certain animals. Portions of the shell of the turtle and the armadillo are familiar examples.

The endoskeleton, on the other hand, arises in deeper body layers and consists of cartilage or bone that develops in association with cartilaginous rudiments. Although such bone has the same histological structure as dermal bone, it is convenient to differentiate it as **cartilage replacement bone.** The endoskeleton may be subdivided into visceral and somatic portions. The **visceral skeleton,** as the name implies, is associated with the "inner tube" (gut) of the body. It consists of skeletal arches (visceral arches) that form in the wall of the pharynx and, in fishes, support the gills and contribute to the jaws. The **somatic** portion of the endoskeleton is associated with the "outer tube" of the body (body wall and appendages). It may be further broken down into axial and appendicular subdivisions. The **axial skeleton** includes those parts of the somatic skeleton located in the longitudinal axis of the body—vertebrae, skeleton of the median fins, ribs, sternum, and those portions of the brain case composed of cartilage or cartilage replacement bone. The **appendicular skeleton** consists of the more laterally placed portions of the somatic skeleton—the skeleton of the paired appendages and those portions of their girdles composed of cartilage or cartilage replacement bone. The divisions of the skeleton are summarized in Table 1.

Table 1. Divisions of the Skeleton

DERMAL SKELETON (DERMAL BONE)	ENDOSKELETON (CARTILAGE REPLACEMENT BONE)
Bony Scales	Somatic Skeleton (in body wall)
Dermal Plates (become associated with parts of endoskeleton)	Axial Skeleton (chondrocranium, vertebrae, ribs, sternum)
Teeth	Appendicular Skeleton (girdles, bones of paired appendages)
	Visceral Skeleton (in gut wall)
	Visceral Arches

Evolutionary Tendencies in the Dermal Skeleton and Endoskeleton

At one time it was thought that the endoskeleton was more primitive than the dermal skeleton, but paleontological studies have shown that this is not the case. Such primitive fishes as the ostracoderms and placoderms had an extensive dermal skeleton consisting of thick, bony scales of the cosmoid type over the trunk and tail, and larger bony plates over the head. The endoskeleton, although present, was neither completely ossified nor so conspicuous. From these early ancestors the evolutionary tendency has been one of reduction of the dermal skeleton and increased development of the endoskeleton. As seen in the preceding chapter, the primitive, heavy, bony scales have become much thinner in living fishes, and are lost as such in tetrapods. However, the deeper parts of the original cephalic plates persist in Osteichthyes and tetrapods and become associated with the endoskeleton as integral parts of the head skeleton and pectoral girdle.

One must be continually aware of the difference between the endoskeleton and the dermal skeleton. But in studying the evolution of the entire skeleton, this dichotomy cannot always be followed. Bony scales were studied with the integument, and the rest of the dermal skeleton will be considered along with those portions of the endoskeleton with which it becomes associated. In studying the endoskeleton it is convenient to consider the axial and visceral portions together to some extent because they are combined in the formation of the head skeleton. After their evolution has been traced, that of the appendicular skeleton will be followed.

FISHES

In using the skeleton of *Squalus* as representative of the jawed-fish stage of evolution, one must remember that its skeleton is atypical in two major respects. For one thing, the endoskeleton is entirely cartilaginous. This represents the retention of an embryonic, not a primitive adult, condition. For another, the dermal skeleton is lost except for the small placoid scales. In view of the latter shortcoming, the bowfin, *Amia* (class Osteichthyes, superorder Holostei), will be used to illustrate the dermal cephalic bones that were present in primitive fishes ancestral to tetrapods.

Postcranial Axial Skeleton

(A) RELATIONSHIPS OF THE VERTEBRAL COLUMN

Make a fresh cross section of the tail of your specimen and observe that the vertebral column lies at the intersection of several connective tissue septa which separate the surrounding muscles. A **dorsal skeletogenous septum** extends from the top of the vertebrae to the middorsal line of the body; a **ventral skeletogenous septum,** from the bottom of the vertebrae to the mid-ventral line; and a **horizontal skeletogenous septum,** from each side of the vertebrae to the lateral surface of the body. The last is the hardest to see and it may be confused with portions of the **myosepta** which separate the muscle segments. The horizontal septum forms an arc that reaches the skin at the position of the lateral line, which appears as a small hole in the skin. Although it will not be seen at this time, the relationships in the trunk are much the same. The only difference is that the ventral septum has split, so to speak, and passes on each side of the body cavity (coelom) (Fig. 10–9, p. 292).

(B) VERTEBRAL REGIONS AND CAUDAL VERTEBRAE

The vertebral column of fishes consists of only **trunk** and **caudal verte-brae,** but these become somewhat modified at the attachment of the chondro-cranium and in the vicinity of the median fins. Study the structure of the caudal vertebrae of *Squalus* on a special preparation, or by dissecting the tail of your own specimen. If you dissect the tail, cut out a piece about two inches long, and carefully expose the vertebrae by cleaning away all the surrounding muscle and connective tissue. Do not select a piece adjacent to either the posterior dorsal or caudal fin.

Make a cross section through the joint between two vertebrae near one end of your piece and examine it along with the lateral aspect of other vertebrae. Each vertebra consists of a cylindrical, central portion, called the centrum, or **vertebral body,** which bears a dorsal and ventral arch of cartilage. The dorsal arch, which protects the spinal cord, is the **vertebral arch** (neural arch); its cavity, the **vertebral canal** (Fig. 4–1, *A* and *D*). The ventral arch, which surrounds and protects the caudal artery and vein, is the **hemal arch;** the passage through it, the **hemal canal.** You may see the vessels. The artery lies dorsal to the vein. The top of each arch may extend a short distance into the dorsal and ventral septa as a **spinous process** and a **hemal spine** respectively.

Make a sagittal section through several vertebrae and study it along with the lateral aspect of others (Fig. 4–1, *A* and *B*). Notice that each side of the vertebral arch is composed of two roughly V-shaped blocks of cartilage. That block located directly on top of the vertebral body, and having its apex pointing dorsally is called the **neural plate.** The other block located above the joint between vertebral bodies, and having its apex directed ventrally is the

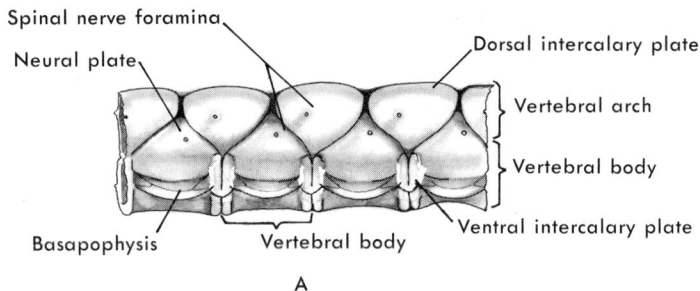

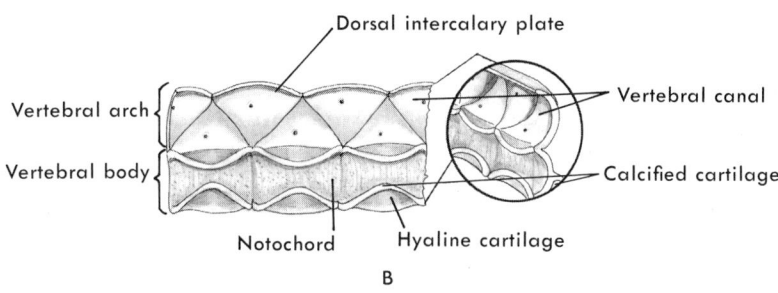

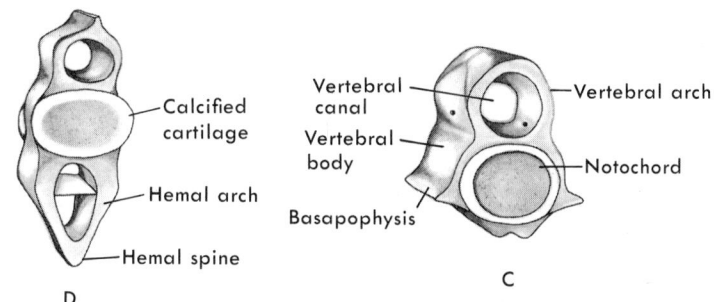

Figure 4-1. Vertebrae of *Squalus. A,* Lateral view of trunk vertebrae; *B,* sagittal section of trunk vertebrae; *C,* end view of a trunk vertebra; *D,* end view of a caudal vertebra.

dorsal intercalary plate. Each plate of the vertebral arch is perforated by a small hole (**foramen**) for either the dorsal or ventral root of a spinal nerve. A continuous vertebral arch is peculiar to cartilaginous fishes; such a condition is possible because cartilage is much more flexible than bone. In other vertebrates there is some space between successive vertebral arches, and this facilitates movement of the vertebral column. The vertebral arch of the other vertebrates may be considered homologous to the neural plate, for it develops primarily from a comparable group of cells.

The hemal arch is composed on each side of just one block (**hemal plate**). A small foramen for intersegmental branches of the caudal artery and vein can be found between the bases of the hemal arches.

Examine the vertebral body in the sagittal section. It is shaped like a biconcave spool, a shape termed **amphicelous.** The concavity at each end contains a gelatinous substance which is **notochordal tissue** that has persisted from the embryonic stage. A small strand of notochord also perforates each vertebral body. The cartilage immediately adjacent to the notochord has become calcified and thus appears white. Calcification of cartilage involves the deposition of calcium salts in the matrix. This strengthens the cartilage, but it is not the same as ossification. The area of calcified cartilage is hourglass shaped. The rest of the cartilage is glasslike, or hyaline.

(C) TRUNK VERTEBRAE

Compare the caudal vertebrae with a special preparation of the trunk vertebrae (Fig. 4–1, *A, B,* and *C*). The structure is basically the same, but the trunk vertebrae lack hemal arches. Instead, they have short, ventrolateral processes (**basapophyses**) that project from the sides of the vertebral body. Basapophyses are serially homologous to the proximal part of the hemal arches. Small **ventral intercalary plates** lie between successive basapophyses, but they are often lost in preparing the vertebrae. They are the ventral counterparts of the dorsal intercalary plates and, like them, are secondary developments.

Each vertebra develops embryonically from sclerotomal mesenchyme cells that migrate from the segmented, embryonic somites and accumulate around the nerve cord and notochord in an intersegmental position. A vertebra thus comes to lie between two segmental myomeres, which in turn have developed directly from the somites. Much of the vertebral mesenchyme forms cartilaginous vertebral arches, **arcualia.** Certain of these form the vertebral arch around the nerve cord; others, the hemal arch around the caudal vessels. The vertebral body develops between these arches and largely replaces the notochord. Its mode of development is extremely variable in vertebrates, but in fishes it appears to derive in part from arcualia bases, and in part from the direct deposition of cartilage, or in some cases of bone, in concentric rings around the notochord and even within the notochord sheath.

(D) RIBS

On a mounted skeleton it can be seen that short ribs articulate with the basapophyses and extend into the horizontal skeletogenous septum at the point where it intersects with the myosepta.

(E) MEDIAN FIN SUPPORTS

Look at the dorsal fins on a mounted skeleton and notice that each is supported proximally by several cartilages and distally by fibrous rods, the **ceratotrichia,** described earlier (p. 33). The cartilages, which collectively may be called **pterygiophores,** are in two series. Those that rest on the vertebral column are **basals;** the more distal ones, **radials.** There is some question whether the pterygiophores develop phylogenetically from spinous processes of the vertebrae, or independently within the dorsal septum. The caudal fin, which is of the heterocercal type (p. 33), is supported proximally by enlarged spinous processes and hemal spines; distally by ceratotrichia.

HEAD SKELETON

The head skeleton is a mixture of three groups of elements that are distinct in certain lower vertebrates, but become confusingly united and mixed in other vertebrates. These are (1) the **chondrocranium,** (2) the **visceral skeleton** (splanchnocranium), and (3) associated **dermal bones.** The chondrocranium[6] which is the anterior end of the axial skeleton, surrounds a variable amount of the brain and forms protective capsules about the olfactory sacs and inner ears. The visceral skeleton is composed of visceral arches, which primitively supported the gills, but in other vertebrates become involved in the jaw and ear mechanism, and even help to encase the brain, to mention but some of their transformations. The chondrocranium and visceral arches are, of course, part of the endoskeleton, but the associated dermal bones are part of the dermal skeleton. Primitively they covered the chondrocranium and visceral arches. Many are lost in higher vertebrates, but some persist to help to form the braincase, the jaws and the facial portion of the skull.[7]

(A) Composition and Structure of the Chondrocranium

The chondrocranium is a complex box of cartilage, or cartilage replacement bone, that can best be understood by describing briefly the major features of its embryonic development (Fig. 4–2). The basic elements in its formation in any vertebrate are two pairs of longitudinal cartilages that lie beneath the brain. The caudal pair, called **parachordals,** are located on each side of the cranial end of the notochord. (The notochord, itself, extends only as far forward as the hypophysis.) The cranial pair, called **trabeculae,** are situated in front of the notochord. The parachordals develop from mesenchyme derived from the sclerotome and hence are closely related to the vertebrae. But the trabeculae, like the visceral arches, develop from mesenchyme derived from the neural crest and are regarded as representing the remnant of a visceral arch that, in ancestral vertebrates, was located rostral to the first arch of gnathostomes. To these are added various other cartilages. An **otic capsule** develops around each inner ear; a **synotic tectum** connects the two otic capsules dorsally; a **nasal capsule** develops about each olfactory sac; a complex **orbital plate,** or latticework, forms medial to each eye; and a variable number of cranial vertebral elements **(occipital arches)** are added to the back of the otic capsules. The expansion and fusion of all these components forms the chondrocranium. Much of it ossifies in most vertebrates, but some of the rostral portion often remains cartilaginous. An optic capsule also develops embryologically around the eye, but it does not become a part of the adult chondrocranium. Rather, it contributes to the wall of the eye, often ossifying as sclerotic plates.

---

[6]The term chondrocranium is sometimes used to include all parts of the head skeleton derived from cartilage (chondrocranium proper and visceral arches), but it is here used in its narrower sense, which excludes the visceral arches.

[7]The term skull is one of those words in common usage that is difficult to define precisely for all vertebrates. It may be used in a broad sense for the entire skeletal complex of the head, but is more often restricted to those parts of the head skeleton that are frequently firmly united with one another, i.e., the braincase (cranium), facial region and upper jaw.

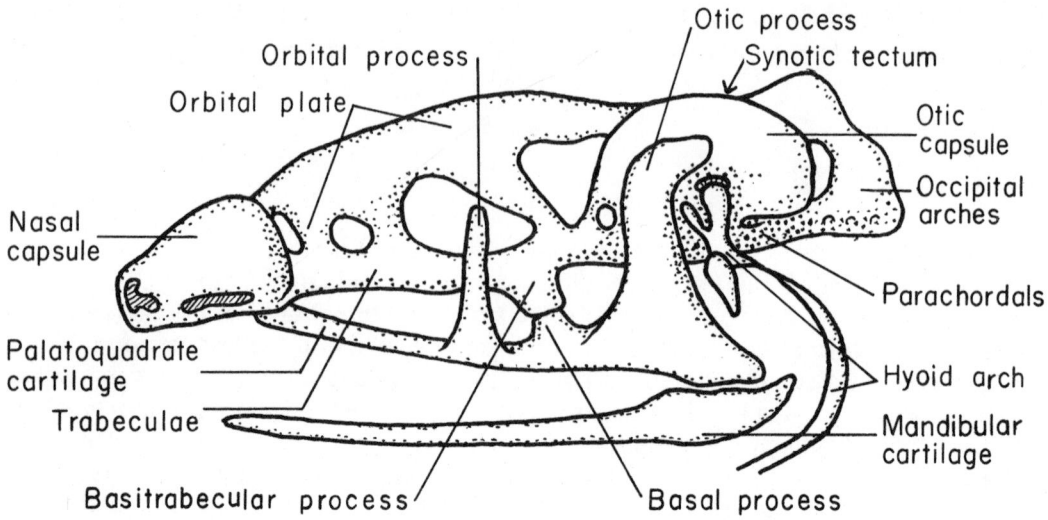

Figure 4–2. Lateral diagram of the embryonic chondrocranium and mandibular and hyoid arches of a tetrapod to show the components of the chondrocranium and the autostylic articulation of the palatoquadrate. (Modified after Goodrich, Studies on the Structure and Development of Vertebrates, The Macmillan Company.)

Study a preparation of the chondrocranium of *Squalus* (Fig. 4–3). The pointed, trough-shaped **rostrum** is at the front end; the caudal end is squarish. A pair of large sockets for the eyes, **orbits,** lies on the sides. The ventral surface of the chondrocranium is very narrow between the orbits; its dorsal surface is much wider in this region. The chondrocranium can be divided into several regions, which are not clearly demarcated in the adult, but do have distinct embryonic origins: occipital region (from occipital arches and synotic tectum), otic capsules (from otic capsules), basal plate (from parachordals), orbital region (from orbital latticework and caudal portions of the trabeculae), and nasal region (from rostral portions of trabeculae and nasal capsules).

The **occipital region** is the very caudal portion of the chondrocranium lying in the midline. It surrounds a large hole, the **foramen magnum,** through which the spinal cord enters the **cranial cavity** within the chondrocranium. A pair of bumps, the **occipital condyles,** will be seen ventral to the foramen magnum on each side of a centrumlike area. They develop from a pair of basapophyses of a vertebra that has been incorporated in the occipital region, and they help to articulate the chondrocranium with the vertebral column. There is little movement at the occipitovertebral joint in any fish, but a double occipital condyle of this type is unusual; most have a single, rounded condyle located directly ventral to the foramen magnum.

The paired **otic capsules** are the large, squarish, caudolateral corners of the chondrocranium that extend from the occipital region to the orbits. Between them, on the dorsal side, is a large depression called the **endolymphatic fossa.** Within the fossa you will see two pairs of openings which communicate with the inner ear. The smaller, cranial pair is the **endolymphatic foramina** for the endolymphatic ducts; the larger, caudal pair is the **perilymphatic foramina.** On each capsule you may find ridges, two dorsally and one laterally, beneath which lie the **semicircular canals** of the ear. Finally, there are two large foramina on the caudal edge of the chondrocranium lateral to the occipital condyles. The more medial is the **vagus foramen** for the vagus nerve; the more lateral, the **glossopharyngeal foramen** for a nerve of the same name.

Ventrally the otic capsules are connected by the flat, broad **basal plate.** Some of the notochord persists in the chondrocranium of the adult dogfish, and its position is indicated by a white strand of calcified cartilage which can be seen through the hyaline cartilage along the midventral line of the basal plate. The small hole in the midline, anterior to the strand of calcified cartilage, is the **carotid foramen** for the passage of internal carotid arteries.

The **optic region** is that area which includes and lies between the orbits. The cranial, dorsal, and caudal portions of each orbit are formed by walls of cartilage called the **antorbital process, supraorbital crest,** and **postorbital process,** respectively. Ventrally the orbit is open. Most of the floor of the chondrocranium is narrow between the two orbits, but near the caudal part of the orbit the floor is wider and bears a pair of prominent lateral bumps called the **basitrabecular processes.** As will be seen later, orbital processes of the upper jaws have a movable articulation with the floor of the chondrocranium just anterior to these processes. Note that the roof of the chondrocranium between the orbits is complete in *Squalus.* Primitively it was incomplete. The small, median hole near the rostral end of the roof is the **epiphyseal foramen.** In life, it contained a stalk of the same name which represents a vestige of the pineal eye. The series of foramina (one large, many small) that perforate the supraorbital crest are the **superficial ophthalmic foramina** for the passage of a similarly named nerve and its branches. A number of other foramina can be seen in the medial wall of the orbit. The

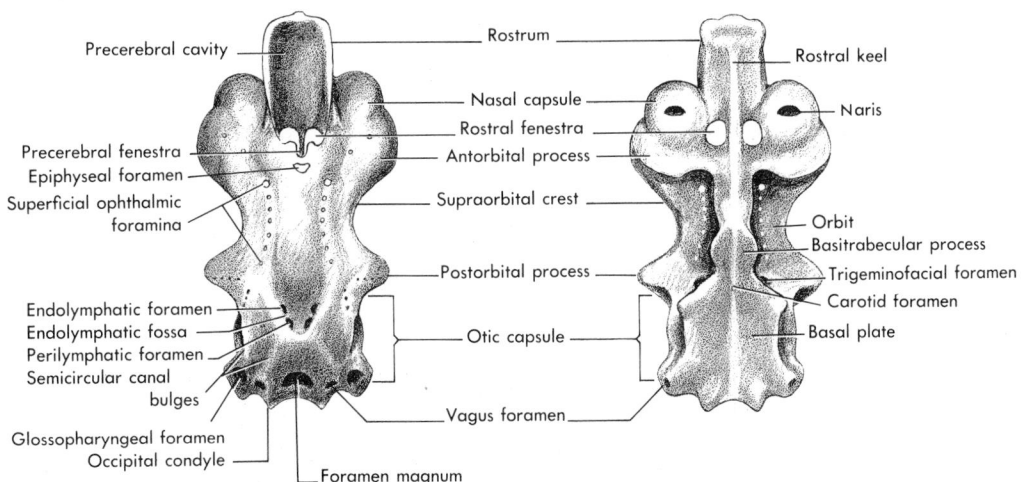

Figure 4–3. Chondrocranium of *Squalus* in (A) dorsal and (B) ventral views.

large rostral one is the **optic foramen** for the optic nerve; the large caudal one, the **trigeminofacial foramen** for the trigeminal and facial nerves. The remaining are for smaller cranial nerves and vessels. A small cartilaginous stalk resembling a golf tee is left in the orbit in some preparations. This is the **optic pedicle;** it abuts against the eyeball and helps to support it.

All the chondrocranium rostral to the antorbital processes may be called the **nasal region.** It consists of a pair of round **nasal capsules** attached to the front of each antorbital process, and a long, median **rostrum** which helps to support the snout. The wall of each nasal capsule is very thin and is generally broken. If complete, you will be able to see its external opening **(naris).** The opening within the capsule is for the passage of the olfactory tract. The rostrum is trough-shaped dorsally and keeled ventrally. Its dorsal concavity, called the **precerebral cavity,** communicates with the **cranial cavity** by way of a large opening, the **precerebral fenestra.** In life, the precerebral cavity is filled with a gelatinous material. Two other large openings **(rostral fenestrae)** into the cranial cavity will be seen ventrally lying between the nasal capsules on each side of the rostral keel.

(B) Sagittal Section of the Chondrocranium

Examine the inside of the cranial cavity in a sagittal section of the chondrocranium, if such a preparation is available. Certain structures seen earlier can be noted again in this view, but there are additional features of interest. Notice the depression in the floor dorsal to the basitrabecular processes. This is the **sella turcica,** a recess for the hypophysis. The large opening in the medial wall of the otic capsule, just caudal to the trigeminofacial foramen, is the **internal acoustic meatus.** Part of the glossopharyngeal nerve enters here to pass beneath the ear and emerge through the glossopharyngeal foramen, but most of the passage is occupied by the statoacoustic nerve from the inner ear.

(C) Visceral Skeleton

In primitive agnathous vertebrates the visceral skeleton is composed of **visceral arches** that pass between the gill slits and help to support the gills. These are generally fused to each other and to the chondrocranium, forming a compact unit. Recall the branchial basket of the lamprey. The arches are also numerous. *Petromyzon* has traces of nine, two being rostral to the first of the seven gill slits, and ostracoderms had even more. In gnathostomes there has been a reduction of the number of gill slits and arches and a mobilization of those arches that remain. No more than the last seven of the primitive series persist in higher fishes, and the first of these (the mandibular arch) forms, or helps to form, the jaws. Teeth, which are derivatives of the dermal skeleton, become attached to the jaws, and also frequently to the roof of the mouth and to other visceral arches.

In some placoderms there may have been a complete gill slit caudal to the jaws, but in most fishes the second visceral arch (the hyoid arch) has moved close to the jaws, sometimes aiding in their support. Thus the original first gill slit of gnathostomes is lost, or reduced to a spiracle. *Squalus* is a good example of this stage in the evolution of the visceral skeleton.

Examine a preparation of the visceral skeleton of *Squalus* (Figs. 4–4 and 4–5). The seven **visceral arches** of which it is composed show clearly. The first is modified to form the upper and lower jaws, and it therefore is called the **mandibular arch.** The second, called the **hyoid arch,** extends from the otic capsule to the angle of the jaws and ventrally into the floor of the mouth. The spiracle would be located between the dorsal portion of the hyoid arch and the upper jaw. The last five visceral arches, called **branchial arches,** support the interbranchial septa[8] and, hence, pass between the remaining gill slits. Note that the third visceral arch and first branchial arch are synonymous. In addition to the arches, small **labial cartilages** may be seen on the lateral surface of the jaws. They are located in the labial folds previously observed (p. 33), and are too superficial to be a part of the visceral skeleton.

Study the mandibular and hyoid arches in more detail. Each side of the upper jaw is formed by a **palatoquadrate cartilage;** each side of the lower jaw by the **mandibular cartilage** (Meckel's cartilage). Both cartilages bear several rows of sharp teeth which are loosely attached to the surface of the jaws and are similar to each other (Fig. 9–8, p. 250). A dentition in which the teeth are essentially the same is referred to as **homodont.** The teeth of selachians differ from those of other fishes primarily in their triangular shape. The teeth are conical in primitive fishes. Two prominent processes extend dorsally from the palatoquadrate. The one above the angle of the jaw is the **adductor mandibulae process** for the attachment of mandibular muscles. The one that extends up into the orbit (**orbital process**) passes lateral to the braincase and just anterior to a basitrabecular process. Orbital processes permit

---

[8]The tissue lying between successive gill slits supports the gill filaments and contains a number of structures associated with the gills—the skeletal arches, muscles, nerves and blood vessels. There is some confusion as to what the entire complex should be called. It has often been called gill, branchial or visceral arch, using the term arch in a broad sense. To avoid confusion I prefer to limit the term arch to the skeletal elements and to use the term interbranchial septum for the entire complex.

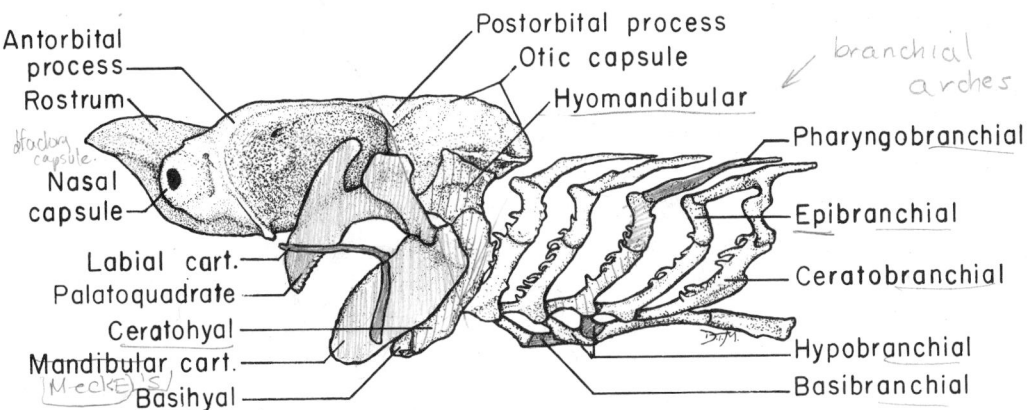

Figure 4–4. Lateral view of the chondrocranium and visceral skeleton of *Squalus.*

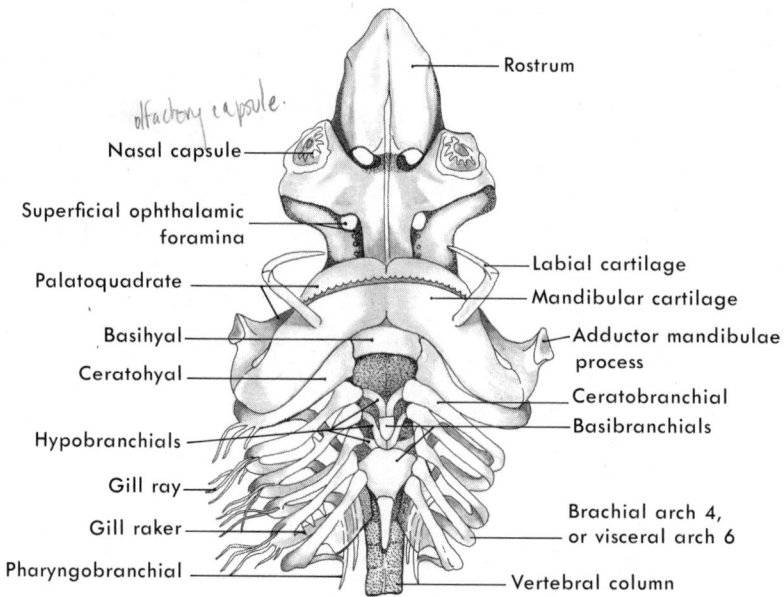

Figure 4–5. Ventral view of the chondrocranium and visceral skeleton of *Squalus*.

the upper jaw to move up and down slightly but they prevent lateral motion. The single midventral piece of the hyoid arch is called the **basihyal**. A **ceratohyal** extends from this element to the angle of the jaw, and a **hyomandibular** continues to the otic capsule. Ligaments unite the posterior end of the palatoquadrate cartilage and hyoid arch, so this arch provides the primary support for the upper jaw.

The type of jaw suspension seen in the dogfish, in which the palatoquadrate is loosely articulated to the chondrocranium by an orbital process and is buttressed posteriorly by the hyoid arch, is termed **hyostylic**. In primitive fishes ancestral to tetrapods, the palatoquadrate was suspended in part by the hyomandibular and in part by its own processes, which articulate rather securely to the chondrocranium, a condition referred to as **amphistylic**. In tetrapods (Fig. 4–2) the palatoquadrate articulates with the rest of the skull only by its own processes, a condition called **autostylic**.

Each of the branchial arches ideally consists of a midventral **basibranchial** and a chain of four additional elements extending dorsally on each side — a short, ventral **hypobranchial;** a longer **ceratobranchial** extending to the height of the angle of the jaws; an **epibranchial** continuing beyond this; and, finally, a **pharyngobranchial** which overlies the pharynx and points caudally but does not unite with the vertebrae. All these elements can be seen in the first branchial arch, but there has been some fusion and loss of certain of the elements in the caudal arches. Also notice that each branchial arch, and the hyoid arch as well, bears a number of laterally projecting **gill rays** which stiffen the interbranchial septa.

Certain of the relationships of the visceral skeleton to other organs of the head can be seen clearly in a transverse and sagittal section of the head (Fig. 9–9, p. 251, and Fig. 10–10, p. 294).

(D) Dermal Bones

The third component of the head skeleton is the dermal bones associated with the cranial portions of the endoskeleton as far caudally as the pectoral girdle. They have been lost in the evolution of the Chondrichthyes but were present in primitive fishes and in those ancestral to tetrapods. Some appreciation of the extent of the dermal bones in primitive forms can be gained by studying them in the bowfin, *Amia*.

Examine the skeleton of the head of *Amia,* noting the groups, or series, of dermal bones that sheathe the chondrocranium and visceral arches (Fig. 4-6). In some types of preparations the dermal bones have been removed on one side, exposing the endoskeletal structures. A **dermal roof** covers the top of the head, and most of the cheek region. It is pierced by tiny nares (two on each side) and by the large orbits. There is also a large gap, which is not found in more primitive fishes, between the cheek and upper jaw. The dermal bones on the margin of the jaw bear a single row of small, conical **teeth,** which are essentially similar to one another **(homodont).**

The caudal part of the palatoquadrate cartilage ossifies as the **quadrate,** but the part of it rostral to this is lost in the adult. A **palatal series** of dermal bones, on some of which there are several rows of teeth, forms much of the roof of the mouth in the general region of the missing portion of the palatoquadrate. A third "series" (actually composed of but a single element, the **parasphenoid**) lies in the midline of the roof of the mouth directly beneath the chondrocranium. It bears a number of very small teeth.

A **lower jaw series** covers the lateroventral and medial surfaces of the partly ossified mandibular cartilage. As with the upper jaw and palate, there is a single row of marginal teeth, and several rows of more medial teeth. The medial teeth in both the palate and lower jaw occupy the same general

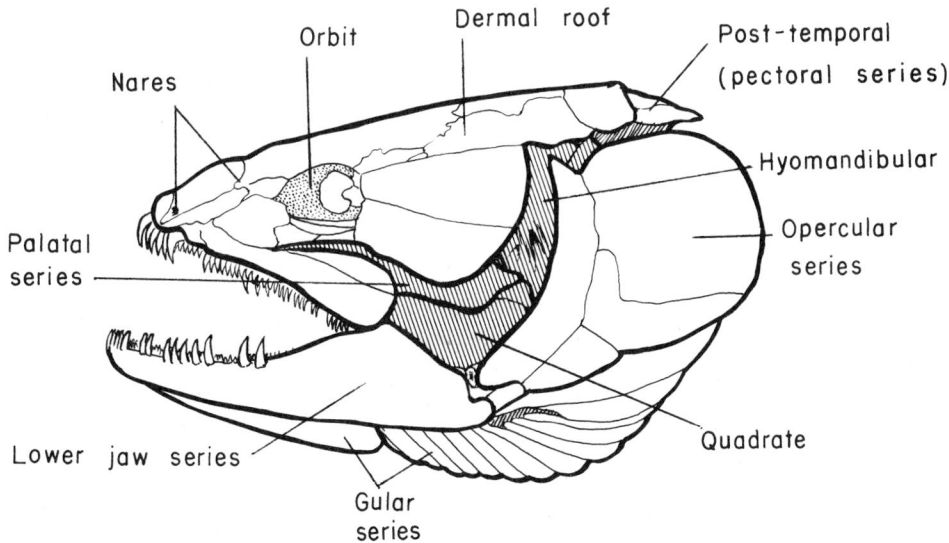

Figure 4-6. Lateral view of the skull of *Amia calva*. Series of dermal bones is shown in heavy outline. The quadrate and hyomandibular are parts of the visceral skeleton. (Modified after Goodrich, Vert. Craniata.)

position as the teeth on the mandibular arch of *Squalus* and are considered to be their homologues.

The remaining visceral arches are covered laterally by an **opercular series,** and ventrally by a **gular series** of dermal bones. Finally, the endoskeletal portions of the pectoral girdle are covered laterally by a **pectoral series** of dermal bones.

PRIMITIVE TETRAPODS

Since *Necturus* is neotenic, its skeleton is not so thoroughly ossified as is the skeleton in the ancestral tetrapods. This is particularly true for the head region and the appendicular skeleton. It is therefore desirable to supplement the study of *Necturus* by examining these portions of the skeleton in another reasonably primitive tetrapod. The snapping turtle, *Chelydra serpentina,* is a good example.

Postcranial Axial Skeleton

The vertebrae of tetrapods vary considerably, but the most primitive type seems to have been the **rhachitomous vertebrae** of primitive labyrinthodonts and cotylosaurs (Fig. 4–7). Such vertebrae consisted of a vertebral or neural arch beneath which was a notochord partly constricted by three blocks of bone that made up the vertebral body or centrum. A single **intercentrum** lay beneath the notochord at the front of each vertebra. A pair of small **pleurocentra** was located dorsal and caudal to the intercentrum. Hemal arches (chevron bones) were also present in the caudal region, where they were fused with the intercentra.

Numerous attempts have been made to compare the primitive tetrapod vertebra

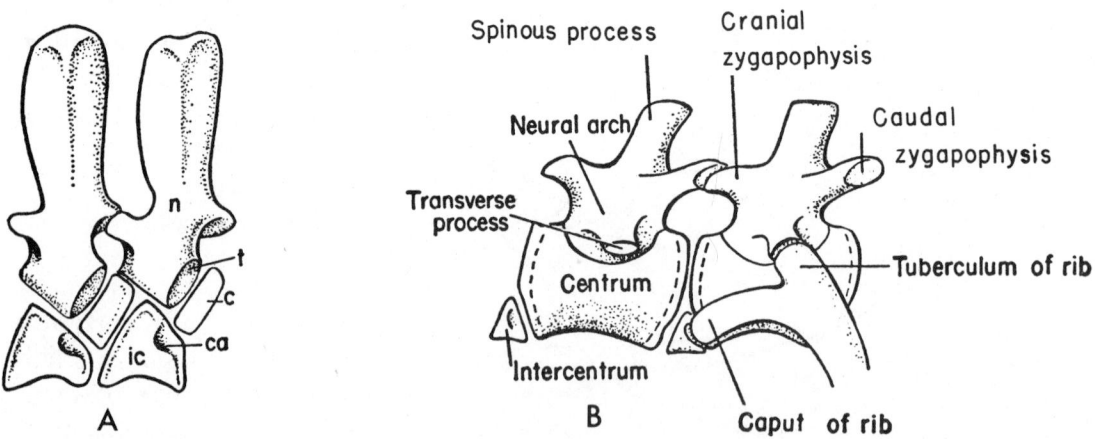

Figure 4–7. Diagrams to show: *A,* the adult rhachitomous vertebrae of a primitive labyrinthodont; *B,* the vertebrae and rib attachment of a primitive reptile. Both are lateral views with anterior toward left. Abbreviations: *c,* pleurocentrum; *ca,* attachment of caput of rib; *ic,* intercentrum; *n,* neural arch; *t,* transverse process (diapophysis). (From Romer, The Vertebrate Body.)

with that of fishes, and the subject has been reviewed by Williams (1959). The consensus is that the vertebral arch is homologous to the neural plate of the dogfish, and the tetrapod hemal arch represents the hemal arch of fishes. The central elements of the primitive tetrapod vertebra are believed to have developed independently of the arches from perichordal sclerotomal mesenchyme. Hot dam !

In the evolution from primitive amphibians through reptiles (Fig. 4–7) to birds and mammals the intercentrum decreases in size and the pleurocentrum enlarges to become the definitive centrum. The reduced intercentrum may contribute to the **intervertebral disc** when one is present. The notochord disappears in the adult, except for traces that lie within the disc.

Tetrapod ribs bear two processes that articulate with the vertebrae: a dorsal **tuberculum,** which articulates with the transverse process **(diapophysis)** on the neural arch, and a ventral **caput,** which primitively articulates with the intercentrum (Fig. 4–7, *B*). Sometimes there is a small lateral process in the latter region called a **parapophysis**. With the reduction and loss of the intercentrum the caput has an intervertebral articulation.

In ancestral amphibians and reptiles, ribs were found on all the vertebrae from the atlas to the base of the tail, those in the cranial trunk region being the longest. But in later tetrapods, many of these freely articulated ribs disappear as such by fusing onto the sides of the vertebrae to form a more complex transverse process known as a **pleurapophysis.** The ribs in the cranial half of the trunk, however, always remain free. Typically, most of the free ribs in amniotes connect ventrally with a median **sternum** formed of cartilage or cartilage replacement bone. Cranially the sternum also becomes associated with the coracoid region of the pectoral girdle. An unossified sternum was probably present in labyrinthodonts, but whether it connected with the ribs is unknown. It does not connect with the short ribs of living amphibians.

(A) VERTEBRAL REGIONS

Examine the vertebral column and ribs on a skeleton of *Necturus* and note that there is only slightly more differentiation into regions than in the skeleton of *Squalus*. The most cranial vertebra, called the **atlas,** lacks free ribs and is otherwise specialized for articulation with the skull. It is, of course, located in the incipient neck region and is the only one that can be called a **cervical vertebra. Trunk vertebrae** extend from here to the level of the pelvic girdle. All have free ribs. A single **sacral vertebra** (usually the nineteenth), and its pair of free ribs, is modified for the support of the pelvic girdle. The remaining are **caudal vertebrae;** they lack free ribs and most bear **hemal arches.**

(B) TRUNK VERTEBRAE

Study a trunk vertebra on the mounted skeleton and also examine an isolated specimen, if possible (Fig. 4–8). Dorsally it consists of a **vertebral arch** that overlies the spinal cord located in the **vertebral canal.** A small **spinous process** projects caudad from the top of the arch. Two pairs of small processes for the articulation of successive vertebrae project laterally from the front and back of the vertebral arch. The cranial pair of processes, whose smooth articular facets face dorsally, are called the **cranial zygapophyses;** the caudal pair, whose facets face ventrally, the **caudal**

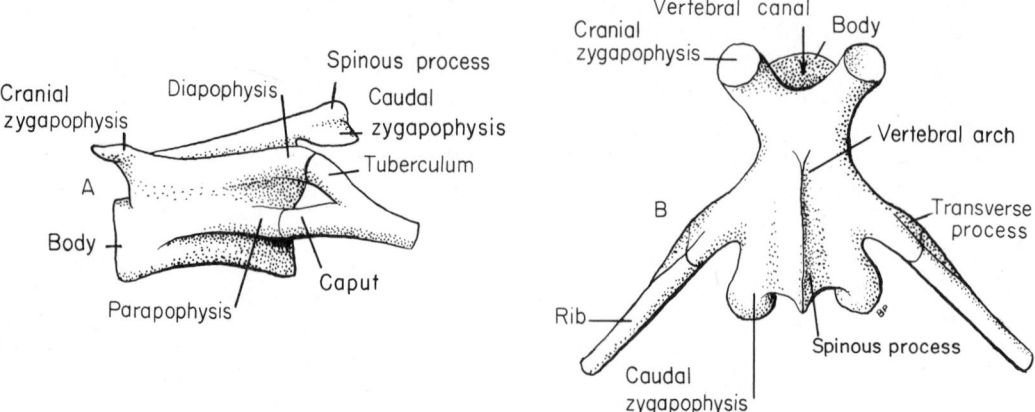

Figure 4-8. *A,* lateral, and *B,* dorsal view of a trunk vertebra and rib of *Necturus.*

zygapophyses. Note how the zygapophyses of adjacent vertebrae overlap. Zygapophyses strengthen the vertebral column and are a characteristic feature of tetrapod vertebrae. In fishes the body is partly supported by water and there are no zygapophyses.

Ventrally the vertebra consists of a biconcave (amphicelous) **vertebral body.** Notochordal tissue persists in the concavities. A prominent **transverse process** projects laterally and posteriorly from each side of the vertebral arch and body. Observe that it is rather high from its dorsal to ventral edge. The dorsal part represents a **diapophysis;** the ventral part a **parapophysis.**

(C) Ribs

The ribs of *Necturus* are shorter than the ribs of labyrinthodonts or cotylosaurs, but have the two heads (bicipital condition) characteristic of terrestrial vertebrates. The ventral **caput** attaches to the parapophysis; the dorsal **tuberculum,** to the diapophysis. The distal portion of the rib is its shaft, or **body.**

(D) Sternum

Necturus does not have the small, cartilaginous sternum found in most urodeles just posterior to the pectoral girdle. However, traces of cartilage, which may represent an incompletely developed sternum, are located in the ventral portions of the myosepta dorsal and posterior to the pectoral girdle, and in the midventral connective tissue septum. They may be noticed during the dissection of the muscles (p. 127).

Head Skeleton of Primitive Tetrapods and Necturus

The three groups of elements present in the head region of fishes (chondrocranium, visceral skeleton and dermal bones) are represented in amphibians and reptiles, but the opercular and gular series of dermal bones are lost, and the visceral arches are greatly reduced. These changes are correlated with the loss of the gills, and the beginning of neck formation. The elements that remain may be grouped, for purposes of description, into four units: (1) the skull in its restricted sense, (2) the lower jaw, (3) the teeth, and (4) the hyoid or hyobranchial apparatus.

The skull (Fig. 4–9) is composed of the chondrocranium, the palatoquadrate of the visceral skeleton and the dermal bones that more or less encase these two. The

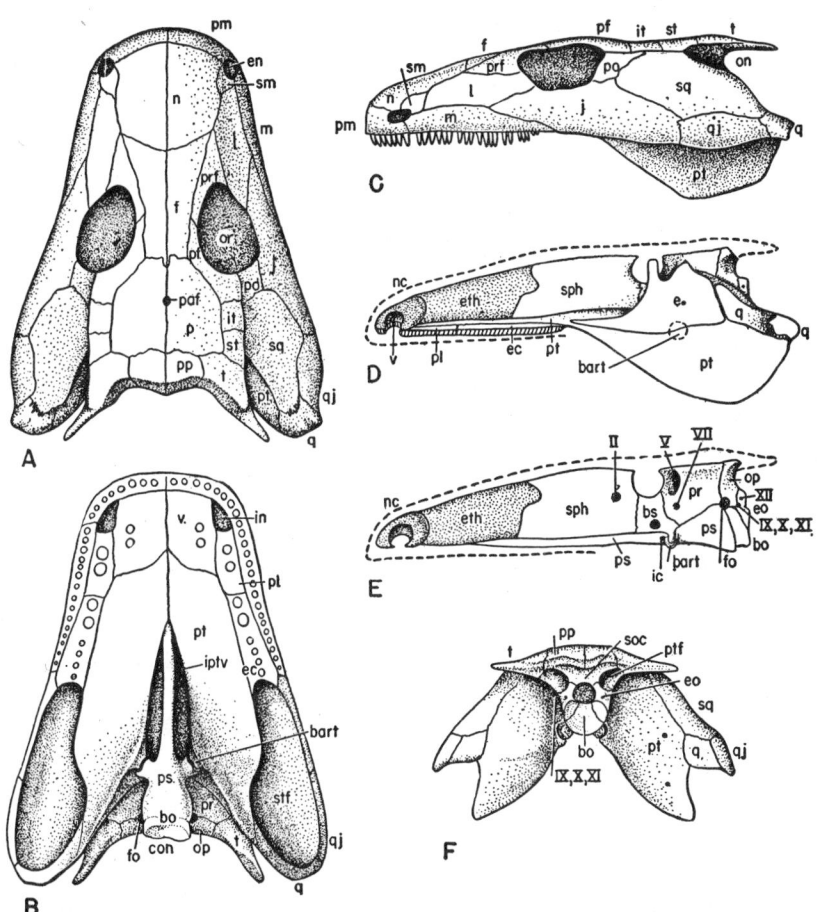

Figure 4–9. The skull of an ancestral land vertebrate, based primarily on the Carboniferous labyrinthodont, *Palaeogyrinus*. *A,* Dorsal view of the dermal skull roof; *B,* palate; *C,* lateral view; *D,* lateral view with dermal skull roof removed (outline in broken line); palatal bones—dermal and endoskeletal—of left side are shown; deep to them, the braincase. The hatched area is the sutural surface of palatal bones against the maxilla. *E,* Lateral view, and *F,* caudal view of braincase. Roman numerals indicate foramina for cranial nerves. Abbreviations: *bart,* basal articulation between palate and basitrabecular process of braincase; *bo,* basioccipital; *bs,* basisphenoid; *con,* condyle; *e,* epipterygoid; *ec,* ectopterygoid; *en,* naris; *eo,* exoccipital; *eth,* ethmoid region; *f,* frontal; *fo,* oval window; *ic,* foramen for internal carotid; *in,* choana; *iptv,* interpterygoid vacuity; *it,* intertemporal; *j,* zygomatic (jugal); *l,* lacrimal; *m,* maxilla; *n,* nasal; *nc,* nasal capsule; *on,* otic notch; *op,* opisthotic; *or,* orbit; *p,* parietal; *paf,* pineal or parietal foramen; *pf,* postfrontal; *pl,* palatine; *pm,* premaxilla; *po,* postorbital; *pp,* postparietal; *pr,* prootic; *prf,* prefrontal; *ps,* parasphenoid; *pt,* pterygoid; *ptf,* posttemporal fenestra; *q,* quadrate; *qj,* quadrato-jugal; *sm,* septomaxilla; *soc,* supraoccipital; *sph,* sphenethmoid; *sq,* squamosal; *st,* supra-temporal; *stf,* subtemporal fenestra; *t,* tabular; *v,* vomer. (From Romer, The Vertebrate Body.)

chondocranium covers the back, the underside and most of the lateral surfaces of the brain. It does not cover the top of the brain, and originally there was a gap in each of its lateral walls. It is usually ossified to a large extent, as shown in Figure 4–9 and Table 2, but the rostral ethmoid (nasal) region and nasal capsules were primitively unossified. As in most fishes, but not *Squalus,* the occipital condyle was originally single. The chondrocranium is perforated by foramina for nerves and vessels and by an **oval window,** located on the lateral surfaces of the otic capsule, for the stapes (see later).

The posterior portion of the palatoquadrate ossifies as the **quadrate** and continues to serve for the articulation with the lower jaw. The portion of it adjacent to the basitrabecular process of the chondrocranium generally ossifies as the **epipterygoid.** This bone articulates with the braincase and helps to fill in the gap in its side (Fig. 4–21, A). The rest of the palatoquadrate is lost in the adult.

The dermal bones constitute the largest component of the skull. A great many were present in labyrinthodonts and cotylosaurs, but the evolutionary tendency since then has been one of reduction. There were also a number of general features of importance in the dermal roof and palate of very primitive forms (Fig. 4–9). The external nostrils **(nares)** were widely separated, and the internal nostrils **(choanae)** entered the mouth at the very front of the palate. A foramen was present between the parietal bones for the pineal or parietal eye. An **otic notch,** which may have held a tympanic membrane, was located on the posterior part of the skull roof. Aside from this notch, the temporal portion of the roof was solid **(anapsid** condition). But posteriorly a pair of **posttemporal fenestrae** passed between the dermal roof and otic capsule. Mandibular muscles arose from the underside of the roof on each side and passed through a pair of lateral palatal openings **(subtemporal fenestrae)** to insert on the lower jaw. A **basal articulation** between the pterygoid and basitrabecular processes permitted some movement between the palate and braincase. **Interpterygoid vacuities** separated the palate and braincase cranial to the basal articulation, and a **cranioquadrate passage,** which transmitted important vessels, nerves and possibly an auditory tube, was situated just caudal to the basal articulation.

To these major components of the skull may be added the **stapes,** or **columella.** This is a small bone, found in most amphibians and reptiles, that transmits sound vibrations from the tympanic membrane, across the middle ear cavity, to the otic capsule. It is derived from the hyomandibular of fishes.

The lower jaw (Fig. 4–13) consists of a **mandibular cartilage** surrounded by a sheath of dermal bones. The caudal end of the mandibular cartilage (the region articulating with the quadrate) generally ossifies as the **articular** bone. The rest may either remain cartilaginous or disappear.

The **teeth** of primitive tetrapods are very similar to those of fishes, for they are small, conical, numerous and similar to one another (homodont). They are usually arranged in two series—a lateral series on the margins of the jaws and a medial series on the palate and medial side of the lower jaw.

As stated, the mandibular arch becomes incorporated into the skull and lower jaw and the dorsal part of the hyoid arch (hyomandibular) becomes the stapes. The rest of the hyoid arch unites with variable portions of the more cranial branchial arches to form the **hyoid apparatus.** This apparatus supports the newly evolved tongue and the floor of the pharynx. Portions of the more caudal branchial arches form the cartilages of the larynx.

(A) ENTIRE SKULL

Study a skull of *Necturus.* Unfortunately it has retained fewer of the primitive features and elements than many other amphibians or reptiles.

Table 2. Components of the Tetrapod Skull and Lower Jaw

The components of the skull and lower jaw of a labyrinthodont, together with the part of the skeleton to which they belong, are shown in the left hand column. The homologies between these elements and those of certain other tetrapods are shown in the right hand columns. An X indicates that an element is present; O, that it is absent; Unos., that the region is unossified. All the elements are paired unless indicated to the contrary by a number in parentheses: (1) indicates a median element; (2) or (3) that two or three of these are present on each side.

LABYRINTHODONT	NECTURUS	TURTLE	ALLIGATOR	MAMMAL
Skull				
Chondrocranium (Cartilage Replacement Bone)				
Basioccipital (1)	Unos.	X (1)	X (1)	X (1) ⎫ Occipital
Exoccipital	X	X	X Fused with opisthotic	X ⎭
Supraoccipital (1)	Unos.	X (1)	X (1)	X (1) ⎫
Opisthotic	X Operculum	X	X	X ⎬ Petrosal part of Temporal
Prootic	X	X	X	X ⎭
Basisphenoid (1)	Unos.	X (1)	X (1)	X (1) Body of Basisphenoid
Sphenethmoid (1)	Unos.	Unos.	X Laterosphenoid	X (1) Presphenoid
Unossified Ethmoid Region	Unos.	Unos.	Unos.	X Ethmoid
Unossified Nasal Capsule	Unos.	Unos.	Unos.	X Turbinates
Visceral Arches (Cartilage Replacement Bone)				
Palatoquadrate				
Quadrate	X	X	X	X Incus
Epipterygoid	O	X	O	X Wing of Basisphenoid
Hyomandibular				
Stapes	X	X	X	X
Dermal Bones				
Roof				
Tooth-bearing Marginal Bones				
Premaxilla (Incisive)	X	X	X	X
Maxilla	O	X	X	X
Median Series				
Nasal	O	O	X	X
Frontal	X	X	X	X
Parietal	X	X	X	X
Postparietal	O	O	O	X Interparietal, often a part of Occipital

Table 2. (Continued)

	LABYRINTHODONT	NECTURUS	TURTLE	ALLIGATOR	MAMMAL
Circumorbital Series					
Lacrimal	X	O	O	X	X
Prefrontal	X	O	X	X	O
Postfrontal	X	O	O	O	O
Postorbital	X	O	X	X	O
Zygomatic (Jugal)	X	O	X	X	X
Temporal Series					
Intertemporal	X	O	O	O	O
Supratemporal	X	O	O	O	O
Tabular	X	O	O	O	X ? Part of Occipital
Cheek Bones					
Squamosal	X	X	X	X	X Squamous part of Temporal
Quadratojugal	X	O	X	X	O
Palate and Underside of Chondrocranium					
Parasphenoid	X	X	X Fused with Basisphenoid	X Reduced	X ? Part of Basisphenoid
Vomer	X	X	X (1)	X	X (1)
Palatine	X	O	X	X	X
Ectopterygoid	X	O	O	X	X ? Part of Basisphenoid
Pterygoid	X	X	X	X	X Pterygoid process of Basisphenoid
Lower Jaw					
Visceral Arches (Cartilage Replacement Bone)					
Mandibular Cartilage		Unos.			
Articular	X		X	X	X Malleus
Dermal Bones					
Lateral Series					
Dentary	X	X	X	X	X
Splenials (2)	X	X	O	X	O
Surangular	X	O	X	X	O Tympanic part of Temporal (Endotympanic, a new cartilage replacement bone)
Angular	X	X	X	X	X
Medial Series					
Coronoids (3)	X	O	X	X	O
Prearticular	X	O	X	X Fused with articular	X Anterior process of Malleus

First examine the top of the skull (Fig. 4–10). Most of the bones that you see belong to the dermal roof, although other groups have been exposed through the loss of some of the original elements of the roof. The V-shaped, tooth-bearing bone on each side of the front of the upper jaw is the **premaxilla.** The **nasal cavity** is located in the notch posterior to its lateral wing. Continuing caudally along the middorsal line, the next elements are a large pair of **frontal** bones. Paired **parietals** extend from the frontals nearly to the **foramen magnum.** Each parietal also has a narrow process that extends cranially lateral to the frontals. The **orbits** are located ventral to these processes.

Lateral to the posterior half of the parietal, you will see two small bones. One forms the very caudal angle of the skull and continues forward to a tiny window of cartilage. The other is cranial to this window. These bones, the **opisthotic** and **prootic,** respectively, are a part of the chondro-cranium. The thin sliver of bone on the margin of the skull, lateral to the otic bones, is the **squamosal**—the last element of the original dermal roof. Cranial and ventral to it, at the point where the lower jaw articulates, you will see the partly ossified **quadrate**—the only part of the palatoquadrate present in *Necturus.*

The caudal end of the skull, lateral and ventral to the foramen magnum, is formed by the paired **exoccipitals.** Each bears a condyle. There are there-fore two **occipital condyles** in *Necturus,* unlike the single one of more primi-tive tetrapods.

Look at the underside of the skull (Fig. 4–11, *A*). The palate consists of two pairs of dermal bones. The cranial pair, **vomers,** are located caudal to the premaxillae and, like the premaxillae, bear a row of teeth. The caudal pair, **pterygoids,** continue back from the vomers to the otic region. The fronts of the pterygoids also have a row of teeth. Portions of the vomers and pterygoids can be seen from the dorsal side. The large median bone on the ventral sur-

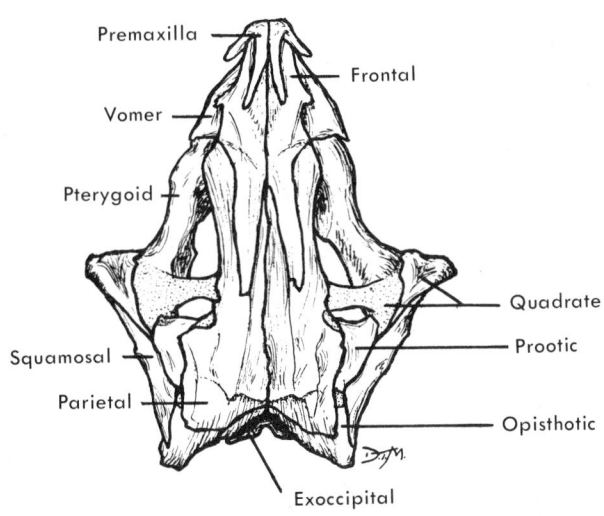

Figure 4–10. Skull of *Necturus,* dorsal view.

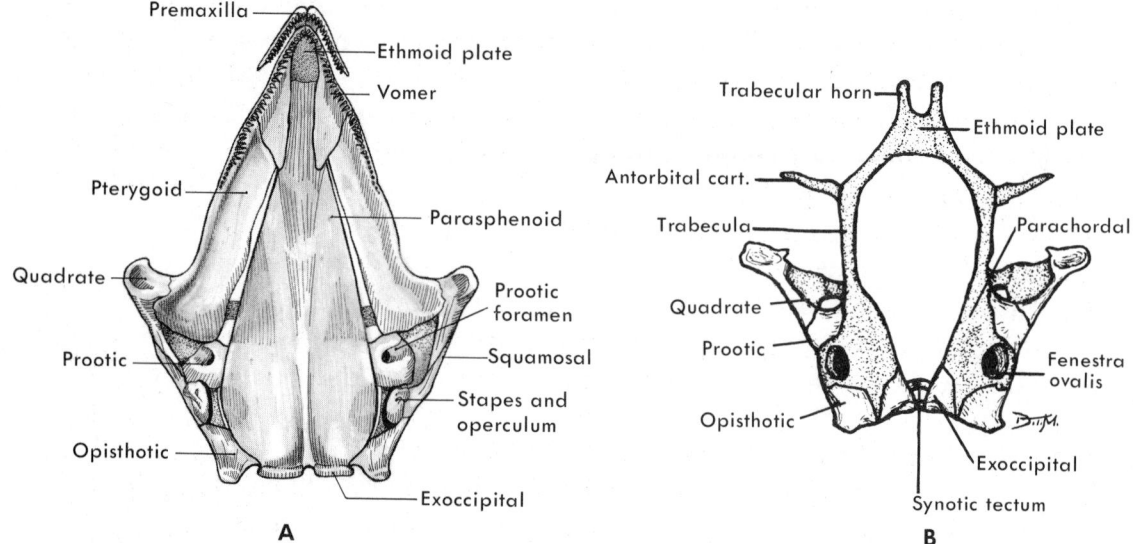

Figure 4–11. Skull of *Necturus. A,* Ventral view of entire skull; *B,* ventral view of chondrocranium.

face, lying between the vomers and the pterygoids and continuing to the exoccipitals, is the dermal **parasphenoid.** A bit of the unossified **ethmoid plate** of the chondrocranium is exposed anterior to the parasphenoid. The two otic bones can also be seen in the ventral view lying caudal to the pterygoids and lateral to the back of the parasphenoid. A cartilaginous area, containing the **oval window,** separates them. The oval window may be covered by a tiny disc-shaped bone bearing a little stem (stylus). This represents the **stapes** combined with the **operculum.** The opercular bone develops from the wall of the otic capsule, and is a unique part of the amphibian auditory apparatus (p. 187). It is connected by its stylus, and by ligaments, to the squamosal.

(B) CHONDROCRANIUM

Although parts of the chondrocranium can be seen in the complete skull, it can be seen more clearly by examing a preparation in which the dermal bones have been removed. The chondrocranium of *Necturus* is not too good a representative of the primitive adult stage since it is largely in the embryonic condition. (Compare what follows with the description of the development of the chondrocranium on page 47.) The pair of large, round caudolateral swellings are the **otic capsules.** You will see two ossifications in each — a cranial **prootic** and a caudal **opisthotic** (Fig. 4–11, *B*). The hole on the side of the capsule is the oval window, which may be covered by the stapes. The only other ossifications are a pair of ventral **exoccipitals** which form in the **occipital arch.** They are connected by a delicate cartilaginous bridge, the **basioccipital arch.** Dorsally the otic capsules are connected by another bridge, the **synotic tectum.** Often the quadrate (a part of the mandibular arch) is left on preparations of the chondrocranium and will be seen cranial and lateral to the prootic.

The shelves of cartilage united with the medioventral edge of each otic capsule are the **parachordals.** The pair of cartilaginous rods that continue forward from the parachordals are **trabeculae.** They are united cranially to form an **ethmoid plate,** from which a pair of **trabecular horns** continue between very delicate nasal capsules. The nasal capsules are generally destroyed. A small **antorbital cartilage,** representing a part of the orbital latticework of other vertebrates, extends laterally from each trabecular cartilage.

(C) LOWER JAW

Compare the lower jaw of *Necturus* with Figure 4–12. Only three dermal bones cover the mandibular cartilage in this animal. The largest of these is the **dentary,** which forms practically all the lateral surface of the jaw and a small portion of the medial surface near the front of the jaw. It bears the long anterior row of teeth. The short caudal row of teeth is on the **splenial,** most of which is on the dorsomedial surface of the jaw, but a bit of the bone shows laterally. Most of the medial surface of the jaw is formed by the **angular.** This bone is widest caudally and then tapers to a point which passes ventral to the splenial and continues cranially to the dentary. A small portion of the angular shows laterally at the caudoventral corner of the jaw. The surface of the lower jaw that articulates with the quadrate is formed by the **mandibular cartilage,** which is unossified in *Necturus.*

(D) TEETH

Notice that the teeth of *Necturus* are essentially fishlike. That is, they are small, conical and all the same type (**homodont**). They are also relatively numerous, and you can see parts of a **lateral** and **medial series.** The lateral series are on the premaxillae and dentaries; the medial, on the palatal bones and splenials. The teeth attach loosely to the jaws.

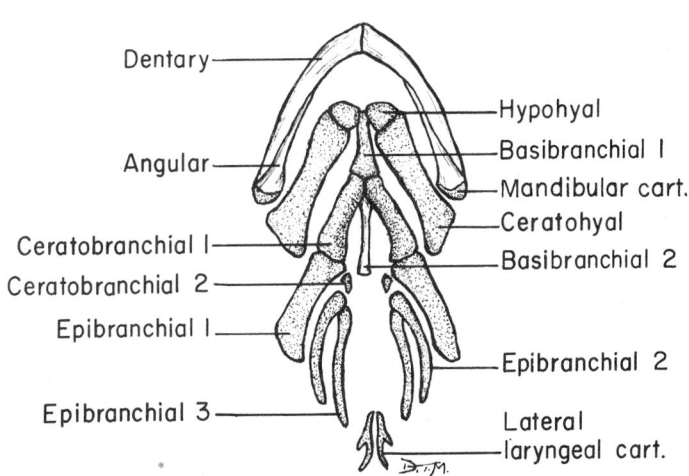

Figure 4–12. Ventral view of the lower jaw, hyoid apparatus, and laryngeal cartilages of *Necturus.* Cartilage is stippled.

(E) Hyoid Apparatus

Preserved skeletal material is better than dried for studying the hyoid apparatus. Notice that it is made up of parts of four visceral arches—the hyoid arch and the first three branchial arches (Fig. 4–12). The hyoid arch is the most cranial and the largest component. It consists on each side of a short **hypohyal,** which lies just lateral to the midventral line, and a longer **ceratohyal,** which extends toward the angle of the jaw. A median **basibranchial 1** extends caudally from the hypohyals to the first branchial. This arch, too, is composed of two cartilages on each side, the more medial being **ceratobranchial 1;** the more lateral, **epibranchial 1.** The next two arches (branchial arches numbers 2 and 3) are greatly reduced and at first sight appear to consist only of **epibranchials 2** and **3.** On closer examination, however, you will see a small cartilage (**ceratobranchial 2**) that connects them with the first branchial arch. The branchial arches support the three external gills. The two gill slits pass on either side of the second branchial arch. A small, median **basibranchial 2,** which is generally partly ossified, extends caudad from the base of the first branchial arch.

The last two branchial arches do not contribute to the hyoid apparatus and will not be seen. There is some doubt as to their fate, but traces of them may persist. A raphe in one of the branchial muscles, which contains a few cartilage cells, may represent a part of the fourth branchial (sixth visceral) arch. It is also probable that the rest of this arch plus the last arch have contributed to the lateral cartilages of the laryngotracheal chamber.

Head Skeleton of the Reptile

In many ways the skulls of certain living reptiles are better examples of a primitive tetrapod skull than is that of *Necturus.* The skull of the snapping turtle, *Chelydra,* is described and illustrated in Figures 4–14 to 4–17 and an illustration of an alligator's skull (Fig. 4–13) is included for comparison.

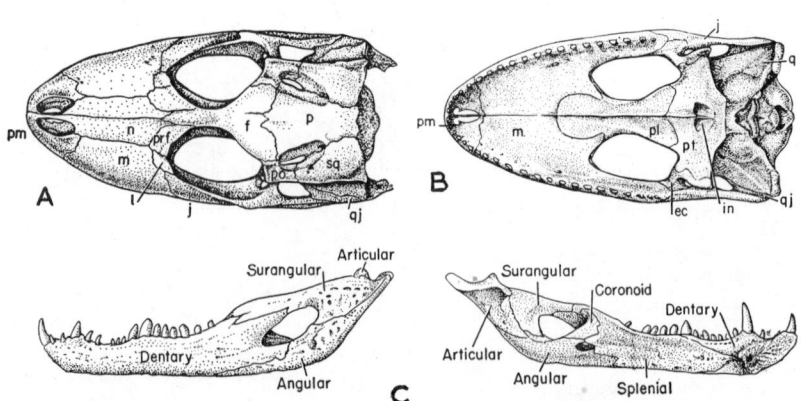

Figure 4–13. The alligator skull. *A,* Dorsal view; *B,* ventral view; *C,* lateral and medial views of the lower jaw. Abbreviations as in Figure 4–9. (From Romer, The Vertebrate Body.)

(A) GENERAL FEATURES OF THE SKULL

Examine a skull of *Chelydra*. Teeth are absent and are replaced functionally in life by a horny beak ensheathing the jaw margins. The **dermal roof** and **palate** can be seen surrounding the small, partly ossified **chondrocranium.** The originally paired **nares** have united to form one opening at the surface of the dermal roof, but the nasal cavities remain distinct; they open into the mouth by paired internal nostrils (**choanae**) located at the front of the palate. The **orbits** are situated far forward, for the snout is short. The dermal roof in the temporal area has been "eaten away," or emarginated, from behind in, most turtles but is complete in the sea turtles. Although this emargination is related to the bulging of the powerful jaw muscles, it is not comparable in its relationship to the temporal fenestrae of other reptiles. Turtles are usually considered to have the complete (**anapsid**) temporal region of primitive amphibians and reptiles. In any case, one can easily visualize the nature of such a roof by looking at the turtle. A pair of large **posttemporal fenestrae** can be seen in a caudal view of the sea turtle skull between the dermal roof and otic capsule. They are present in other turtles, too, but the loss of the overlying dermal roof in this region makes them less apparent.

The palatal bones have united solidly with the underside of the braincase, so interpterygoid vacuities are absent. The pair of large **subtemporal fenestrae** can be seen between the palatal bones and the lateral margins of the

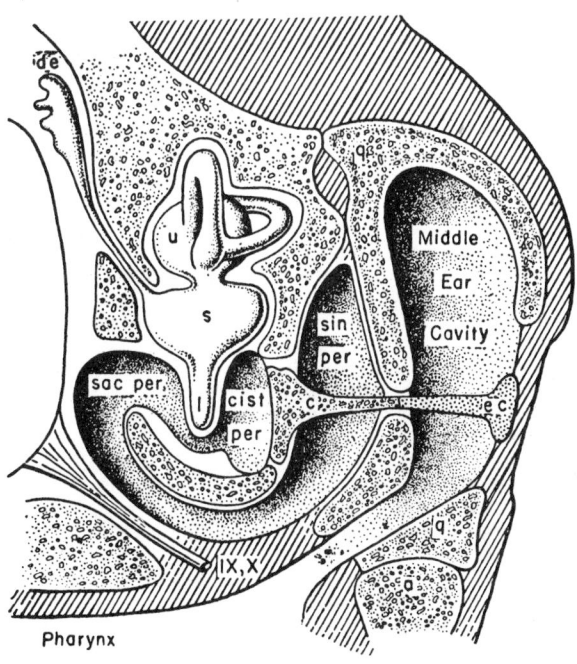

Figure 4-14. Diagrammatic vertical section through the ear and adjacent parts of a turtle skull. Abbreviations: *a*, articular; *c*, columella or stapes; *cist per*, perilymphatic cistern; *de*, endolymphatic duct and sac; *ec*, extracolumella; *l*, lagena; *q*, quadrate; *s*, sacculus; *sac per*, perilymphatic sac; *sin per*, pericapsular sinus; *u*, utriculus; *IX, X*, cranial nerves. (From Romer, Osteology of Reptiles, University of Chicago Press.)

dermal roof. Two pairs of **infraorbital fenestrae** can be found in the palate beneath the orbits.

Examine the back of the skull and note the position of the middle ear. It is quite close to the jaw joint. It has also been partly encased and constricted by bone, so that it appears as an hourglass-shaped cavity. The expanded lateral portion of the cavity lying dorsal to the jaw joint is the middle ear, or **tympanic cavity** proper; the expanded medial portion, known as the **tympanic recess**, contains an extension of the perilymphatic system of the inner ear (Fig. 4–14). The reduced cranioquadrate passage lies in the floor of the tympanic recess.

(B) COMPOSITION OF THE SKULL

The dermal elements along either lateroventral margin of the roof are a very small **premaxilla** ventral to the external naris, a large **maxilla** continuing beneath the orbit, a **zygomatic** (jugal) caudal to this, and a **quadratojugal** just cranial to the middle ear cavity. The zygomatic enters only the caudoventral corner of the orbit. The middorsal elements are a pair of large **prefrontals** caudal to the naris and dorsal to much of the orbit; a pair of small **frontals** that do not enter the orbit in *Chelydra;* and a pair of large **parietals** that extend to the prominent, middorsal, occipital crest. Each parietal also sends a wide flange ventrally that covers a part of the brain not covered by chondrocranial elements. Two other dermal elements complete that portion of the roof situated between the dorsal and marginal bones. A large **postorbital** lies between the orbit and temporal emargination, and a **squamosal** forms a cap on the extreme caudolateral corner of the skull.

Dermal bones also make up the palate. The very front of this region is formed by palatal processes of the premaxillae and maxillae. A median **vomer** (a fusion of originally paired elements) is located caudal to the pre-

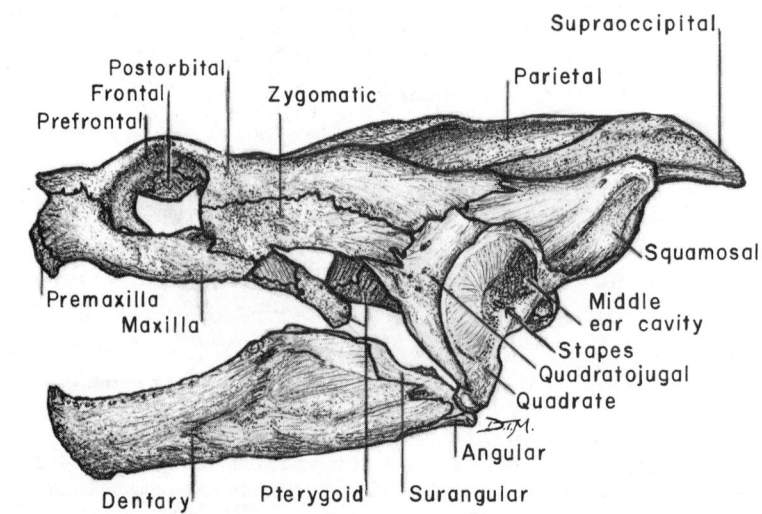

Figure 4–15.   Lateral view of the skull and lower jaw of the snapping turtle, *Chelydra.*

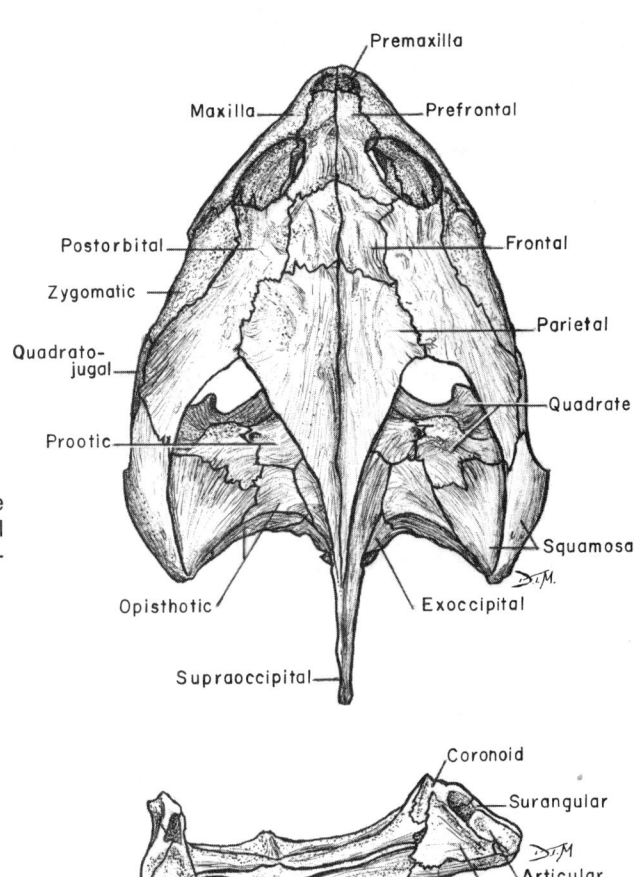

Figure 4-16. Snapping turtle skull and lower jaw. *Upper,* dorsal view of the skull; *lower,* postero-median view of the lower jaw.

maxillae, and **palatines** lateral to the vomer. The internal nostrils enter on each side of the front of the vomer in most turtles. But in the sea turtles, and a few others, the vomer and palatines have sent out flanges that unite with each other to form a small **secondary palate** ventral to the **primary palate.** This pushes the openings of the internal nares caudad. The rest of the palate is formed by a pair of large **pterygoids.**

The chondrocranium surrounds the base, some of the sides and the caudal portion of the brain, but it fails to cover the rest of the brain, which is covered instead by dermal bones of the roof and by the epipterygoid (see later). Four bones surround the **foramen magnum**—dorsally the **supraoccipital,** which forms the occipital crest; laterally the paired **exoccipitals;** and ventrally the **basioccipital.** The **occipital condyle** has begun to shift from its primitive position on the basioccipital to the exoccipitals. Since distinct portions of it are borne on all three of these bones, the condyle is tripartite. In a dorsal view an **opisthotic** can be seen extending laterally from the supra-occipital and exoccipital to the squamosal. A **prootic** is situated rostral to this.

The only other ossification of the chondrocranium is the **basisphenoid,** which can be seen ventrally lying between the pterygoids rostral to the basi-occipital. The dermal **parasphenoid** has united with it.

The caudal part of the palatoquadrate has ossified as the **quadrate.** This bone occupies the caudoventral corner of the skull, extending from the jaw joint dorsally to the squamosal. Most of it passes rostral to the middle ear, but a part of it goes caudal to the ear. A very slender, rodlike **stapes** may be seen extending from the tympanic cavity proper to the otic capsule. It passes through a small hole in the quadrate and crosses the tympanic recess. Its median end is enlarged to form a foot plate that fits into the **oval window.** The central portion of the palatoquadrate has ossified as the **epipterygoid.** This bone helps to fill in a gap in the side of the chondrocranium. It is difficult to see, but occupies a small triangular area between the pterygoid and ventral flange of the parietal rostral to a large **prootic foramen** for certain cranial nerves (the trigeminal and abducens).

(C) LOWER JAW

Compare the lower jaw of *Chelydra* with Figures 4–15 and 4–16. The caudal end of the mandibular cartilage has ossified as the **articular** and can be recognized by its smooth articular surface. The rest of the cartilage remains unossified, but it can often be found in the adult lying in a groove on the medial surface of the jaw. The largest of the dermal elements is the **dentary.** Most of the lateral surface of the jaw is formed by the dentary, but a small

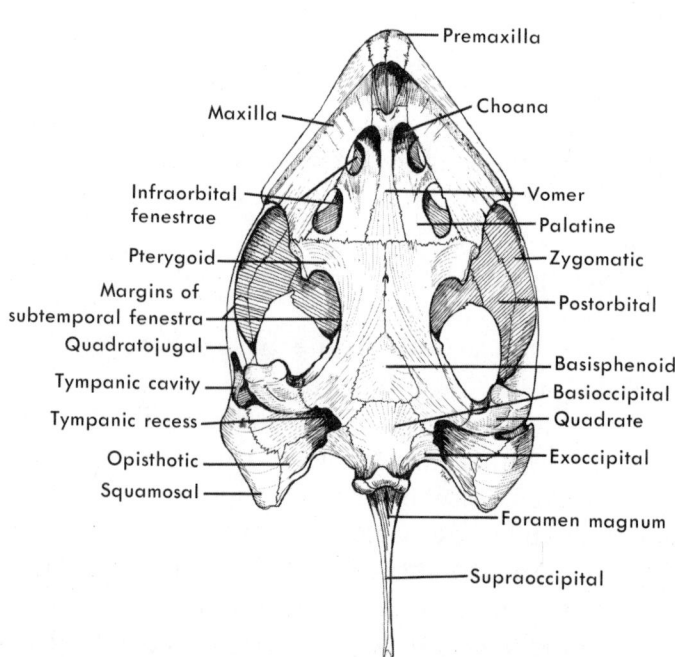

Figure 4–17.   Ventral view of the skull of the snapping turtle.

surangular lies on the lateral surface between the dentary and articular. The front half of the medial surface is also formed by the dentary, but three additional elements form the medial surface posteriorly—a **coronoid** dorsally; a **prearticular** ventral to it and bridging the groove for the mandibular cartilage; and an **angular** beneath the prearticular and forming the ventral border of the jaw. A bit of the coronoid and angular can also be seen from the lateral surface.

(D) Hyoid Apparatus

Study the hyoid apparatus of the turtle and observe that it consists of a median body (**corpus**) from which horns (**cornua**) project laterally. The corpus develops from the fusion of the basihyal and first two basibranchials. Most of it ossifies, but the front remains cartilaginous. A slender **lingual process** projects from its cranial end into the tongue. The two pairs of large, ossified horns represent the first and second ceratobranchials. In some cases a trace of the first epibranchial may be found. Ceratohyals are represented by a pair of short, cartilaginous horns arising from the corpus cranial to the ceratobranchials. More caudal visceral arches have contributed to the laryngeal cartilages.

MAMMALS

The following directions for study are based on the cat, but they could be applied to many other mammals. If material is available, it would be desirable to compare the skeleton of the cat with that of human beings and other species.

Postcranial Axial Skeleton

The structure of the individual vertebrae and ribs of mammals is very similar to that of some of the more advanced reptiles discussed earlier. Review the general remarks on page 55.

(A) Vertebral Groups

Examine a mounted skeleton of the cat, and a string of disarticulated vertebrae. There are five groups of vertebrae in mammals. The most cranial is, of course, the **cervical vertebrae.** There are more of them than in very primitive tetrapods since, with few exceptions, mammals have seven cervical vertebrae, all of which lack ribs freely articulated with them. **Thoracic vertebrae** bearing free ribs follow the cervical, and **lumbar vertebrae** follow the thoracic. The lumbar vertebrae lack free ribs but have very large transverse processes. The number of vertebrae in these two regions varies among mammals. The cat normally has 13 thoracic and seven lumbar vertebrae. **Sacral vertebrae** follow the lumbar and are firmly fused together to form a solid point of attachment, the **sacrum,** for the pelvic girdle. The number of vertebrae contributing to the sacrum varies between species. Most mammals, including the cat, have three; some, more than this. The remaining are **caudal vertebrae.** Their number varies with the length of the tail, but all mammals have some. Even in human beings there are three to five vestigial caudal vertebrae that are fused together to form a **coccyx.**

(B) THORACIC VERTEBRAE

The thoracic vertebrae should be studied first as they are more similar to the vertebrae of *Necturus* than are the vertebrae in other regions. Examine one from near the middle of the series. Identify the **vertebral arch** with its long **spinous process,** the **vertebral canal,** and the **vertebral body,** or centrum (Fig. 4–18). Each end of the vertebral body is flat, a shape termed **acelous.** In life, small, fibrocartilaginous **intervertebral discs** are located between successive bodies. Articulate two vertebrae and notice how the back of one vertebral arch overlaps the front of the one behind it. Disarticulate the vertebrae and look for smooth articular facets in the area where the vertebral arches came together. These are the zygapophyses (articular processes). The facets of the pair of **cranial zygapophyses** are located on the cranial surface of the vertebral arch and face dorsally; those of the pair of **caudal zygapophyses** are on the caudal surface of a vertebral arch and face ventrally. A lateral transverse process (**diapophysis**) projects from each side of the vertebral arch. Notice the smooth articular facet for the tuberculum of a rib on its end. In the anterior and middle portion of the thoracic series, the head of a rib articulates between vertebrae, so part of the facet for a head is located on the front of one centrum and part on the back of the next anterior centrum. Why does the head have an intervertebral articulation? The somewhat constricted portion of the vertebral arch between the diapophysis and vertebral body is called its **pedicle.** When several vertebrae are articulated, you will see openings for the spinal nerves, the **intervertebral foramina,** between successive pedicles.

Compare one of the thoracic vertebrae from the middle of the series with those near the cranial and caudal ends of the thoracic region and observe that the spinous processes, the zygapophyses, the transverse processes and the position of the facets for rib heads differ in details. For example, the most caudal thoracic vertebrae lack transverse processes, for their ribs lack a tuberculum and the facets for their rib heads are entirely on one vertebral body. Articulate the last two thoracic vertebrae and notice how the articulation of the zygapophyses is reinforced by a process of the pedicle that extends caudad lateral to the zygapophyses. This is an **accessory process.** A skilled observer can find differences among the vertebrae sufficient to identify each one precisely as the third, sixth or eighth thoracic, etc. However, it

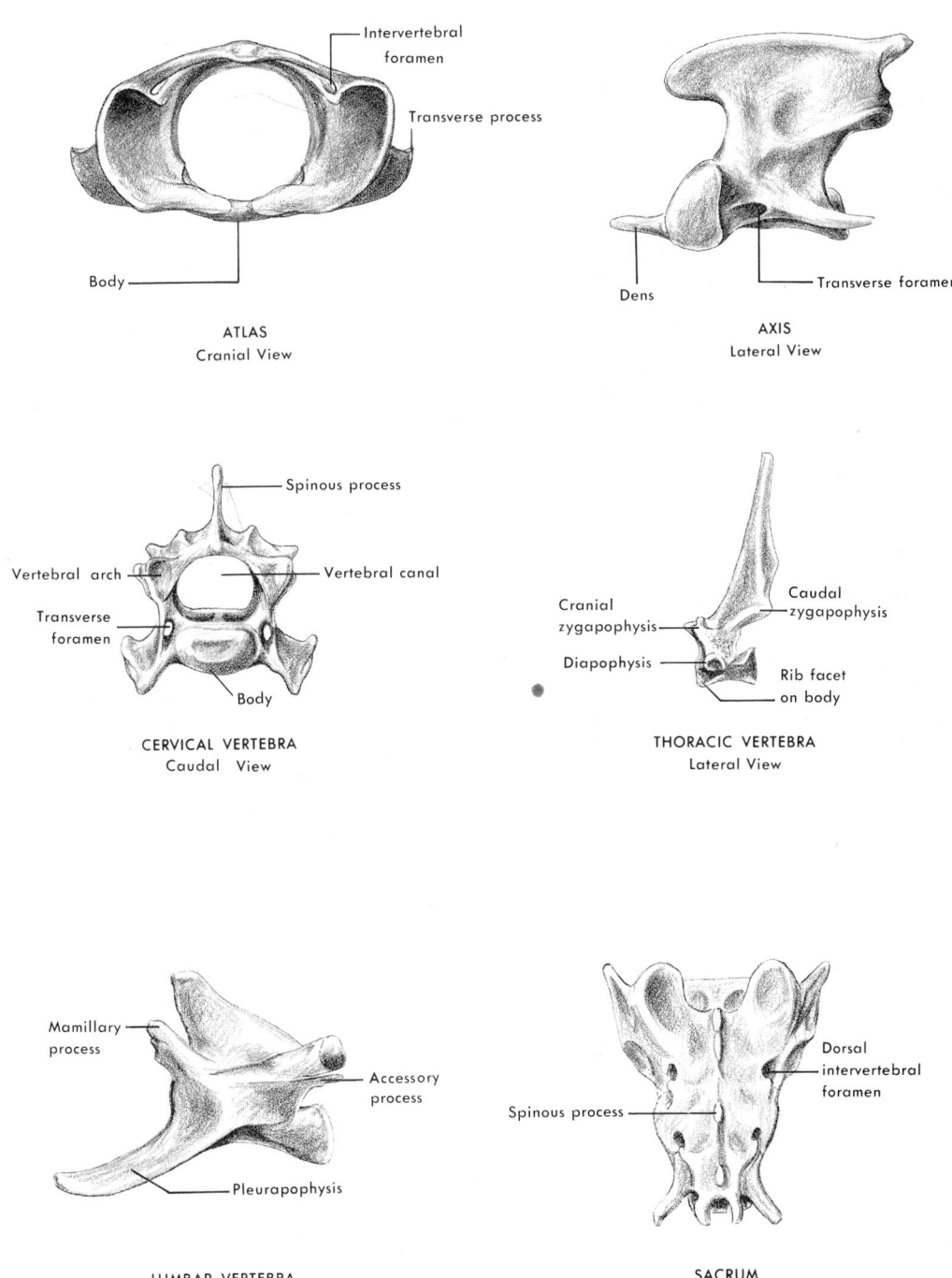

Figure 4–18. Drawings of selected vertebrae of the cat.

will be sufficient for students to be able to distinguish the major group to which a vertebra belongs. All thoracic vertebrae have at least one articular facet for a rib; no other vertebrae have such facets.

(C) LUMBAR VERTEBRAE

Lumbar vertebrae are characterized by their large size and by having prominent, bladelike transverse processes. The transverse process is a **pleurapophysis,** for a rib has united with the diapophysis. Lumbar vertebrae also have a small bump for the attachment of ligaments and tendons dorsal to the articular surface of each cranial zygapophysis. This is called a **mamillary process.** Traces of mamillary processes can also be seen on the more caudal thoracic vertebrae. Most of the lumbar vertebrae also have accessory processes.

(D) SACRAL VERTEBRAE

The sacral vertebrae can easily be distinguished by their fusion into a single piece (the sacrum), and by the broad surface they present for the articulation with the pelvic girdle. Examine the sacrum closely and you will be able to detect the spinous processes, zygapophyses, pleurapophyses, etc., of the individual vertebrae of which it is composed. Notice how the distal ends of the pleurapophyses have fanned out and united with each other lateral to the intervertebral foramina. This has produced separate dorsal and ventral foramina for the respective rami of spinal nerves.

(E) CAUDAL VERTEBRAE

Caudal vertebrae are characterized by their small size and progressive incompleteness. The more cranial ones have the typical vertebral parts, but soon there is little left but an elongated vertebral body. V-shaped **hemal arches,** which protect the caudal artery and vein, are found in some mammals. The cat has traces of such bones on the anterior caudal vertebrae, but they are usually lost in a mounted skeleton. However, the points of articulation of a hemal arch will appear as a pair of tubercles, **hemal processes,** at the front end of the ventral surface of a vertebral body.

(F) CERVICAL VERTEBRAE

Most of the cervical vertebrae can be recognized by their characteristic transverse processes. The transverse process is a pleurapophysis but, unlike those in the lumbar region, it is perforated in all the cervical vertebrae (except the most caudal one) by a **transverse foramen** through which the vertebral blood vessels pass. The last cervical vertebra normally lacks this foramen and, aside from the absence of rib facets, is much like the first thoracic vertebra. The transverse processes of many of the cervical vertebrae (the fifth or sixth is a good example) also have two parts—a dorsal, pointed transverse portion comparable to a diapophysis, and a ventral, platelike costal portion comparable to a rib. Most of the cervical vertebrae also have low spinous processes and wide vertebral arches.

The first two cervical vertebrae, the **atlas** and the **axis,** respectively, are very distinctive. The atlas is ring-shaped, with winglike transverse processes perforated by the transverse foramina. Its vertebral arch lacks a spinous process and is perforated on each side by an **alar foramen** through which the vertebral artery enters the skull and the first spinal nerve leaves the spinal cord. Cranially the vertebral arch has facets which articulate with the two occipital condyles of the skull; caudally it has facets that articulate with the vertebral body of the axis, which is reduced to a thin transverse rod.

The axis is characterized by an elongated spinous process that extends over the vertebral arch of the atlas; very small transverse processes; rounded articular surfaces at the anterior end of its body; and a median, tooth-shaped process **(dens)** that projects from the front of the body of the axis into the atlas.

Although an atlas is present in amphibians, an axis does not appear until reptiles. Together these two vertebrae form a universal joint which permits the free movement of the head characteristic of amniotes. The composition of the body of the atlas and axis is also somewhat different from that of other vertebrae. The atlas lacks the true centrum of amniotes (pleurocentrum), having instead a small intercentrum. The body

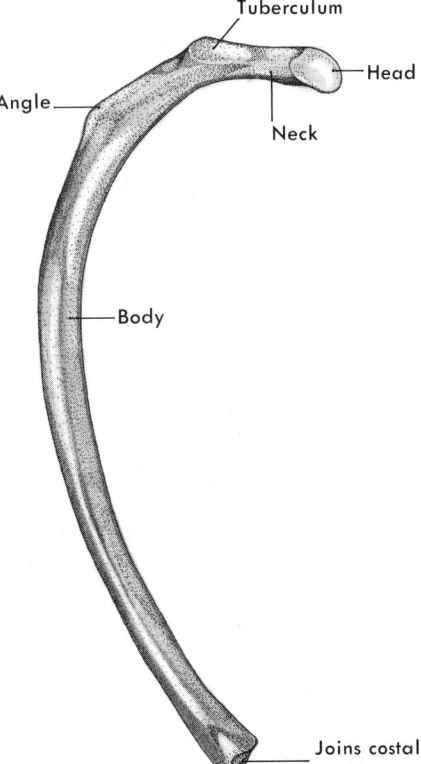

Figure 4–19. Left rib of a cat as viewed from the caudal surface.

of the axis is a typical pleurocentrum, but its dens is made up of the intercentrum of the axis and the missing pleurocentrum of the atlas.

(G) RIBS

Study the **ribs** (*costae*) of a cat from disarticulated specimens and on a mounted skeleton. Most of them have both heads characteristic of tetrapods —a proximal **head** articulating with the vertebral body and a more distal **tuberculum** articulating with the transverse process (Fig. 4–19). But the last three ribs have only the head. The portion of the rib between its two heads is its **neck;** the long distal part, its **body,** or shaft. A **costal cartilage** extends from the end of the shaft. Those ribs whose costal cartilages attach directly on the sternum are called **vertebrosternal ribs;** those whose costal cartilages unite with other costal cartilages before reaching the sternum are **vertebrocostal ribs;** those whose costal cartilages have no distal attachment are **vertebral ribs** (Fig. 4–20). Sometimes vertebrosternal ribs are referred to as **true ribs** and all the rest are referred to as **false ribs.** A floating rib is an alternate name for a vertebral rib. The cat normally has nine vertebrosternal, three vertebrocostal, and one vertebral rib.

(H) STERNUM

The **sternum** is composed of a number of ossified segments called the **sternebrae.** The first of these constitutes the **manubrium;** the last, the **xiphisternum;** and those between, the **body** of the sternum. A cartilaginous

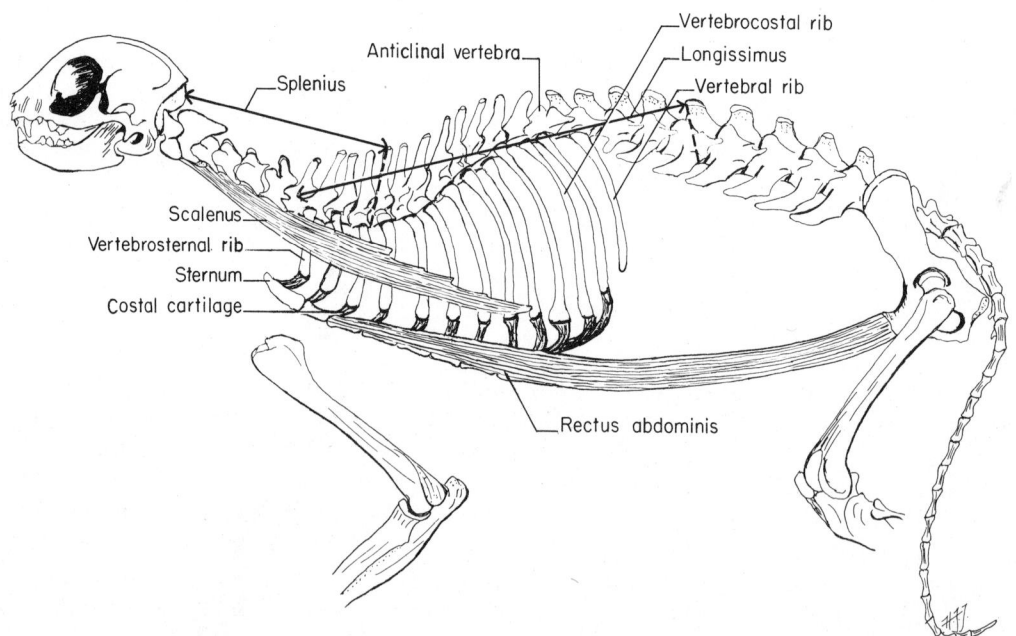

Figure 4–20. Diagram of a lateral view of a cat's skeleton to illustrate Slijper's views on the biomechanics of the trunk. The scapula has been omitted. Heavy arrows indicate the line of action of certain muscles upon the spinous processes. (Modified after Slijper.)

xiphoid process extends caudad from the xiphisternum. Where do the costal cartilages attach in relation to the sternebrae? In many mammals, but not in the cat, the clavicles of the pectoral girdle articulate with the cranial end of the manubrium.

The vertebral column of mammals is the primary girder of the skeleton. Not only does it carry the body weight and transfer this to the appendages but, through its extension and flexion, it participates in the locomotor movements of the body. In order to facilitate interpreting its structure, some authors have compared the vertebral column to a cantilever girder, but Slijper (1946) and others have taken a broader view and have compared the entire trunk skeleton and associated muscles to a bowstring bridge or to an archer's bow (Fig. 4–20). The vertebral bodies represent the main supporting arch, but this is a dynamic arch that has a tendency to straighten out because of the elasticity of certain dorsal ligaments and the tonus of the dorsal muscles. This is prevented by the "bow string"—the sternum and ventral abdominal muscles such as the rectus abdominis. The "bow" and "string" are connected posteriorly by the pelvic girdle and anteriorly by the stout cranial ribs, which are held in place by the action of such muscles as the scalenus. This hypothesis has the merit of viewing the vertebral column as a dynamic girder capable both of providing support and of participating in the movements of the body.

The spinous processes are viewed as muscle lever arms acting on the vertebral bodies. The direction of their inclination tends to be perpendicular to the major muscle forces acting upon them. Since several muscles may attach onto a single spinous process the situation becomes quite complex, but in carnivores and many other mammalian quadrupeds the tips of the spinous processes of the lumbar vertebrae point toward the head, partly in response to a very powerful longissimus muscle (Fig. 6–28, p. 162), whereas the spinous processes of the thoracic vertebrae slope in the opposite direction, partly in response to a powerful splenius. The vertebrae near the middle of the trunk, where the angle of inclination of the spinous processes reverses, is called the **anticlinal vertebra.**

If the spinous processes are lever arms, an increase in their height increases the length of the lever arms and the mechanical advantage of muscles acting upon them. The great length of the anterior thoracic spinous processes is considered to be related to the support of a relatively heavy head by the splenius and other muscles.

Head Skeleton

The touchstone of the evolution of mammals from primitive reptiles has been the development of an increased level of activity. Mammals are endothermic vertebrates that can maintain a relatively high and constant body temperature and level of metabolism. This would not be possible without concomitant structural changes in all organ systems. The skull has been affected primarily by the great enlargement of the brain needed in an active animal and by changes in feeding mechanisms that make possible the increased intake and processing of food. Major skull transformations related to these changes can be summarized at this time; details are set forth in Table 2 and will be seen during the examination of the cat skull.

Mammals need a large and strong skull to house the brain and resist the forces set up by the large and powerful jaw muscles. Mammals use their jaws not just to seize their prey but also for cutting up and, to some extent, chewing their food. This has contributed to the reduction in the number of individual bones throughout the skull. Some bones present in primitive tetrapods have been lost and many others have fused with neighboring ones to form complex elements.

The brain case is very large and the original chondrocranium only suffices to cover the back and underside of the brain. The anterior part of the chondrocranium, represented by cartilage in lower tetrapods, is ossified in mammals. Certain dermal roofing bones have developed extensions, or flanges, that have grown between the jaw muscles and expanding brain (Fig. 4–21). Although this bone is deeply situated it is important to remember that it is of dermal origin. The brain case is completed laterally by the epipterygoid, which becomes the wing of the basisphenoid (alisphenoid). The epipterygoid is derived from that part of the palatoquadrate cartilage that in more primitive vertebrates forms a movable articulation between the palatoquadrate cartilage and the chondrocranium.

As the brain case expanded and the jaw muscle enlarged during the evolution of mammals the space for the temporal jaw muscles, which lies between the temporal portion of the dermal roof and the brain case, became reduced (Fig. 4–21). This situation was met by the loss of some of the dermal roof on each side, thereby forming a window, or **temporal fenestra,** through which the muscles could bulge, especially during their contraction. Temporal fenestrae make their appearance in the mammal-like reptiles and are of the **synapsid** type; that is, there is only one on each side located ventral to the postorbital and squamosal bones (Fig. 4–22). At first the fenestra was small but it gradually enlarged. In primitive mammals most of the original lateral surface of the dermal roof has disappeared in the temporal region so the temporal muscles are lodged in a large **temporal fossa** that merges with the orbit. What appears to be the lateral dermal roof in a mammal skull is the extension of the dermal bones contributing to the brain case. All that is left of the original dermal roof in this region is a strip of bone in the middorsal line, another bordering the occipital region and the handlelike **zygomatic arch** lying ventral to the temporal fossa and orbit. Temporal muscles pass medial to the zygomatic arch to reach the lower jaw; cheek muscles arise on the arch and go to the lower jaw.

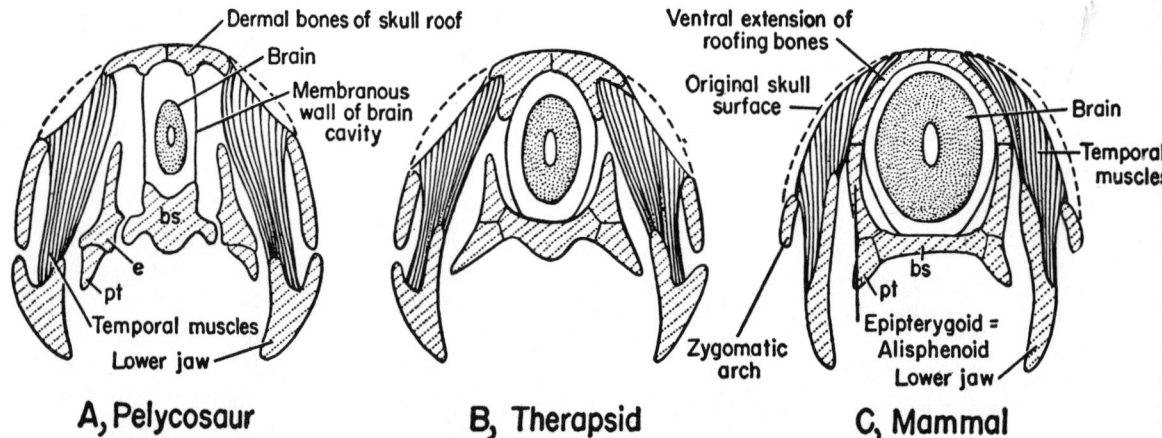

Figure 4–21. Diagrammatic cross section of the skull and jaws of *A*, a primitive mammal-like reptile; *B*, an advanced mammal-like reptile; and *C*, a mammal. These diagrams show the way in which the temporal fenestrae "erode" the dermal roof, and how part of the mammalian braincase is formed by the ventral extension of roofing bones, and by the epipterygoid. Abbreviations: *bs*, basisphenoid; *e*, epipterygoid; *pt*, pterygoid. (From Romer, The Vertebrate Body.)

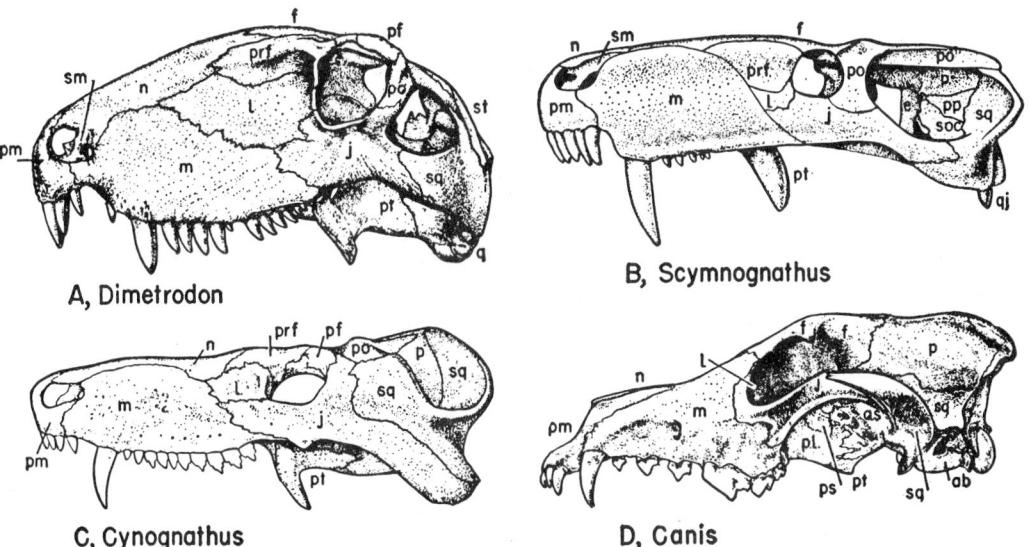

Figure 4-22. Lateral views of the skull of *A*, a primitive mammal-like reptile; *B* and *C*, advanced mammal-like reptiles; *D*, a dog. The enlargement of the temporal fenestra and the union of the temporal fenestra with the orbit show particularly well in this series. Abbreviations as in Figure 4-9, except in *D; ps*, presphenoid. (From Romer, The Vertebrate Body: *B* after Watson; *C* after Broili and Schroeder.)

A high level of metabolism requires mammals to eat a great deal of food and also to exchange a large volume of air in their lungs. Mammals must continue to breathe when they are feeding and chewing food. This is made possible by the evolution of a bony **hard palate** which in life is continued caudally by a fleshy **soft palate.** Processes of the premaxillary, maxillary and palatine bones extend toward the midline and unite with each other ventral to the original palate to form the hard palate. This displaces the openings of the internal nostrils posteriorly, thus permitting the animal simultaneously to breathe and manipulate food within its mouth. The rostral portion of the primitive palate, now situated dorsal to the hard palate, regresses to some extent and this area is occupied by enlarged nasal cavities.

During the evolution of mammals shifts occurred in the direction of pull of certain jaw muscles such that the resolution of the force of the muscle contration made for a stronger bite force and less force was wasted at the jaw joint. This was accompanied by important changes in the jaws. The dentary bone of the lower jaw, on which the major jaw muscles attached, enlarged and the postdentary bones became smaller. As forces at the jaw joint decreased, the joint bearing bones (quadrate and articular) could, and did, become much smaller (Fig. 4-23). Eventually they were little more than a nubbin of bone lying posterior to a part of the dentary that had reached the squamosal. A new joint evolved between the dentary and squamosal and the quadrate and articular disappeared from the jaw mechanism. The quadrate and articular were not lost, however, because they became associated with the auditory apparatus. There is evidence that a tympanic cavity and membrane were situated directly behind the jaw joint in mammal-like reptiles (see Fig. 7-9, p. 184), and some investigators have suggested that the quadrate and articular were part of a bone conducting system carrying sound waves from the lower jaw to the ear. The development of a new joint made it possible for certain of the primitive, posterior jaw elements to become specialized as parts of the auditory apparatus. The articular and quadrate have become additional auditory ossicles, the **malleus** and **incus,** respectively. The change in jaw joint and auditory

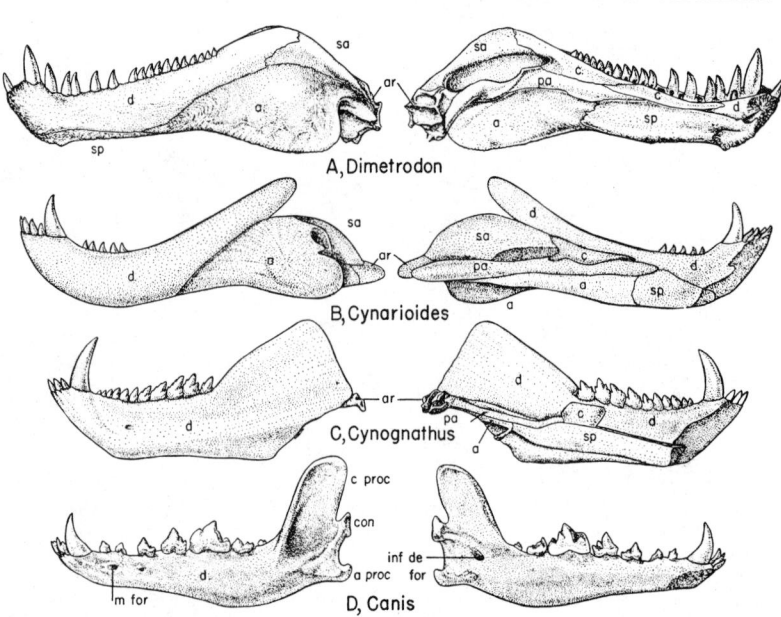

Figure 4–23. The left lower jaw of three mammal-like reptiles (*A, B,* and *C*) and a dog (*D*). The lateral surfaces are shown on the left side of the drawing and the medial on the right. Abbreviations: *a,* angular; *a proc,* angular process; *ar,* articular; *c,* coronoid; *con,* condyle; *c proc,* coronoid process; *d,* dentary; *inf de for,* mandibular foramen; *m for,* mental foramen; *pa,* prearticular; *sa,* surangular; *sp,* splenial. (From Romer, The Vertebrate Body.)

ossicles is the osteological feature diagnostic of mammals, but this change evolved independently several times during the evolution of different subclasses of mammals (Fig. 3–2, p. 28).

The delicate auditory ossicles of mammals (malleus, incus, stapes) are protected by the formation of a plate of bone beneath the tympanic, or middle ear, cavity. Two elements contribute to this bony encasement in most instances—a cartilage replacement **endotympanic** having no homologue in lower vertebrates, and a dermal **tympanic,** homologous to the angular of the lower jaw of amphibians and reptiles.

Changes in dentition accompanied the changes in the jaws. The teeth of mammals are no longer simple cones adapted for seizing and holding prey; advantage is taken of the stronger bite and the teeth are differentiated into types that can seize, cut and crush the food.

(A) GENERAL FEATURES OF THE SKULL

Examine the skull of a cat (Figs. 4–24 and 4–25). As with other vertebrates, it can be divided into a **facial region** containing the jaw, nose, and eyes, and a **cranial region** housing the brain and ear. Notice the **nares** at the rostral end of the skull and the large, circular **orbits.** Although the two nares may appear contiguous, they are separated in life by a fleshy and cartilaginous septum.

The **foramen magnum** can be seen in the occipital region at the caudal end of the skull. There is an **occipital condyle** on each side of it. The large, round swelling on the ventral side of the skull, cranial to each occipital condyle, is the **tympanic bulla,** which encases the underside of the tympanic cavity. In life the tympanum is lodged in the opening (**external acoustic**

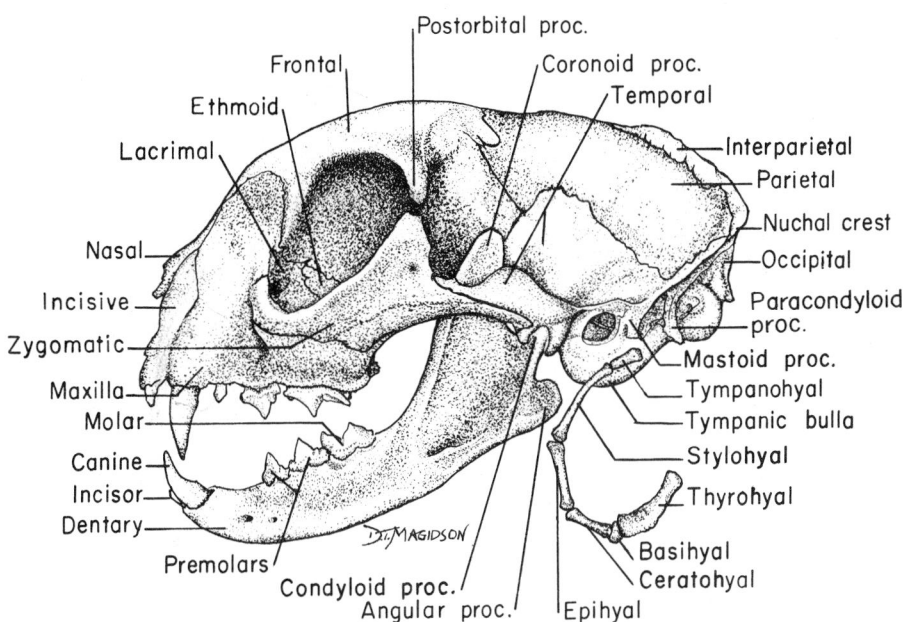

Figure 4–24. Lateral view of the skull, lower jaw, and hyoid apparatus of the cat.

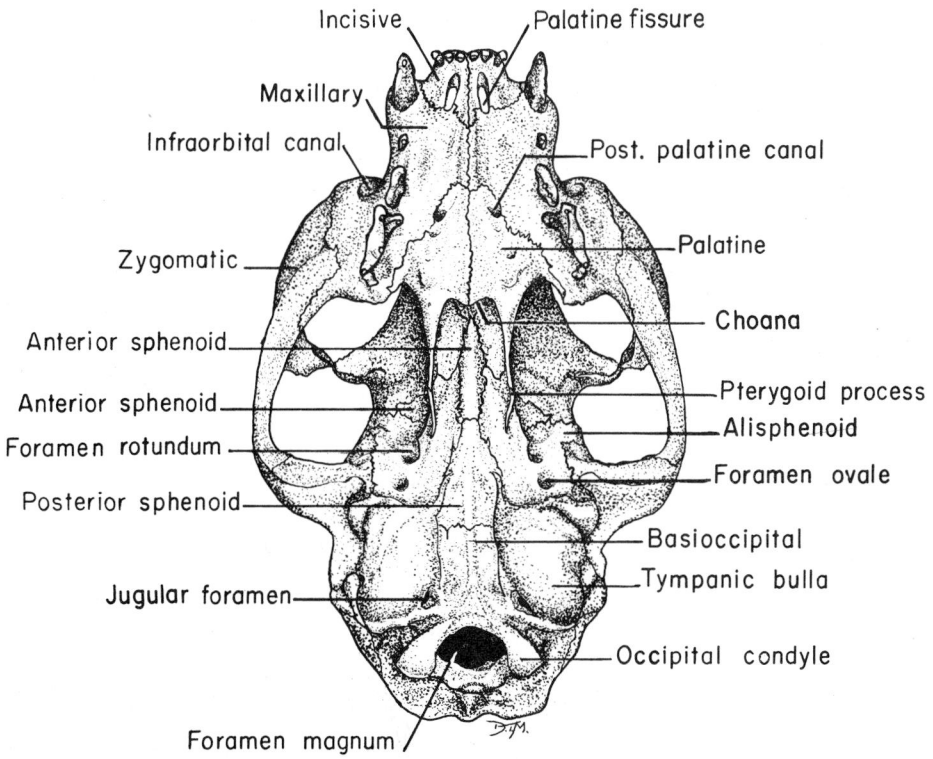

Figure 4–25. Ventral view of the skull of a cat.

meatus) on the lateral surface of the bulla. The tongue-shaped bump of bone on the lateral surface of the bulla just caudal to the external auditory meatus is the **mastoid process.** A similarly shaped **paracondyloid process** is located on the posterior surface of the bulla.

The handlelike bridge of bone on the side of the skull, extending from the front of the orbit to the external auditory meatus, is the **zygomatic arch.** All of it is included within the facial portion of the skull. A **mandibular fossa** for the articulation of the lower jaw appears as a smooth groove on the ventral surface of the caudal portion of the arch.

A large **temporal fossa,** which is exposed by the great enlargement of the temporal fenestra, is situated dorsal to the back half of the zygomatic arch and caudal to the orbit. Although temporal fossa and orbit merge to some extent, they tend to be separated by two processes, one of which projects down from the top of the skull, the other up from the zygomatic arch. The temporal fossa extends from the orbit and zygomatic arch dorsally and caudally over much of the side of the skull. Its caudal border is the **nuchal crest,** a ridge of bone that extends from the mastoid process on one side, over the back and top of the skull, to the mastoid process on the other side. Cervical muscles attach on the caudal surface of the nuchal crest, and temporal muscles on the cranial surface. The dorsal boundary of the fossa is the **sagittal crest,** a middorsal ridge of bone extending rostrally from the middle of the nuchal crest, and a faint ridge **(temporal line)** that curves from the cranial end of the sagittal crest toward the dorsal postorbital process.

The shelf of bone that extends across the palatal region from the teeth on one side to those on the other is the **hard palate.** In life it is continued caudally by a fleshy **soft palate.** The remains of the primitive primary palate lie dorsal to the hard palate. Internal nostrils, or **choanae,** will be seen dorsal to the caudal margin of the hard palate. A nearly vertical plate of bone, the **pterygoid process**, continues caudally from each side of the hard palate. The point of bone at its caudal end is called the **hamulus.** A small **pterygoid fossa,** for the attachment of certain mandibular muscles, appears as an elongated groove just lateral and posterior to the hamulus.

(B) COMPOSITION OF THE SKULL

To better visualize the elements of the skull, one should examine a set of disarticulated bones along with the entire skull. Most of the top and lateral sides of the skull are formed by dermal bones. A small, tooth-bearing **premaxilla** *(incisive)* is located ventral and lateral to each external nostril. It also contributes a small process to the hard palate. A **maxilla** completes the upper jaw. It, too, contributes a process to the rostral portion of the hard palate, sends one up rostral to the orbit, and forms the rostromost portion of the zygomatic arch. A small, delicate **lacrimal** is located in the medial wall of the front of the orbit just caudal to the dorsal extension of the maxilla. It is often broken. The portion of the zygomatic arch ventral to the orbit is formed by the **zygomatic** (jugal). A large **temporal** forms the caudal portion of the arch, the adjacent cranial wall, and encases the internal and middle ear.

The mammalian temporal bone has evolved through the fusion of a number of bones that are independent in lower vertebrates (Table 2). Its zygomatic process and that portion of it that helps to encase the brain (**squamous portion**) represent the squamosal. Of course, the part encasing the brain is an inward extension of the original squamosal of the dermal roof. The mastoid process and its inward extension into the cranial cavity, which will be seen in the sagittal section of the skull, represent the opisthotic and prootic of the chondrocranium. This part of the temporal is the **petrosal portion,** and the inner ear is contained within it. The thick part of the tympanic bulla adjacent to the external auditory meatus is the **tympanic** and is homologous with the dermal angular of the lower jaw of primitive tetrapods. Finally, a new cartilage replacement **endotympanic** forms the rest of the bulla and completes the encasement of the tympanic cavity. It has no homologue in lower forms.

You may be able to get glimpses of the auditory ossicles by looking in the external acoustic meatus, but you should examine a special preparation to see them clearly. The **malleus** is roughly mallet shaped. It has a long, narrow handle that attaches to the tympanic membrane and a rounded head that articulates with the incus. The **incus** is shaped like an anvil. It has a concave surface for the reception of the malleus and two processes that extend from the main surface of the bone. One of these articulates with the head of the stirrup-shaped **stapes.** A pair of narrow columns of bone extend from the head of the stapes to a flat, oval-shaped foot plate which fits into the oval window of the otic capsule. A stapedial artery passes between the columns of the stapes embryologically, and also in the adult of some species.

Dermal bones along the top of the skull are a pair of small **nasals** dorsal to the nares, a pair of large **frontals** dorsal and medial to the orbits, and a pair of large **parietals** posterior to the frontals. The parietals have a long suture with the squamous portion of the temporals. As was the case with squamosal, much of each parietal and frontal represents flanges of the original roofing bones that grew down median to the temporal muscles and helped to cover the brain.

A large, median **occipital** surrounds the foramen magnum and forms the posterior surface of the skull. The paracondyloid processes and most of the nuchal crest are on this element. Ventrally the occipital forms the floor of the braincase between the tympanic bullae. It is a compound bone containing the four separate occipital elements of the chondrocranium of primitive tetrapods and in most mammals certain dermal elements (Table 2). In the skulls of young cats a triangular-shaped <u>interparietal</u> will be seen in the mid-dorsal line in front of the occipital and between the caudal part of the two parietals. It represents a fusion of the paired postparietals present in the dermal roof of primitive tetrapods. In older individuals it often fuses with the occipital and parietals.

The occipital bone forms the posterior portion of the chondrocranium, and the otic elements are incorporated in the temporal. The rest of the primitive chondrocranium is represented by a part of the sphenoid, the ethmoid, and the turbinate bones. Although a tiny portion of the **ethmoid** can sometimes be seen in the medial wall of the orbit caudal to the lacrimal, most of this bone, and the turbinals too, can be seen best in a sagittal section described later.

The sphenoid can be observed in the entire skull. Roughly it forms the floor and part of the sides of the braincase rostral to the tympanic bullae and caudal to the hard palate. It is a single bone in human beings, but in the cat and many other mammals it is divided into a rostral presphenoid and a caudal basisphenoid. These should be seen in a disarticulated skull to appreciate their extent. The **basisphenoid** includes the plate of bone on the underside of the skull just rostral to the basioccipital portion of the occipital; the caudal portion of the pterygoid process; the hamulus; the three caudal foramina of a row of four at the back of the orbit; and a winglike process extending dorsally between the squamous portion of the temporal and the frontal. Most of the bone is homologous to the primitive basisphenoid. That portion of it which includes the three foramina and dorsal wing is homologous to the epipterygoid of more primitive vertebrates and is sometimes called the alisphenoid. There is some doubt as to the homologies of the pterygoid process and its hamulus, but the consensus is that this region includes the pterygoid of the primitive palate plus either the ectopterygoid or parasphenoid.

The **presphenoid** includes a narrow midventral strip of bone lying between the bases of the pterygoid processes, and lateral extensions that pass dorsal to the pterygoid processes to enter the medial wall of the orbits. The lateral portion, sometimes called the orbitosphenoid, contains the rostral foramen in the row of four referred to previously, and has a common suture with the ventral extension of the frontal bone. The presphenoid is homologous to the sphenethmoid of primitive tetrapods.

Certain of the original dermal palatal bones are incorporated in the sphenoid; two others can be seen in more rostral parts of the skull. The originally paired vomers have united to form a single element (the **vomer**), which appears as a midventral strip of bone rostral to the presphenoid and dorsal to the hard palate. You will have to look into the internal nostrils to see it, for it helps to separate the nasal cavities. Paired **palatine** bones form the caudal portion of the hard palate, the rostral portion of the pterygoid processes, and a bit of the medial wall of the orbits caudal and ventral to the lacrimals.

(C) SAGITTAL SECTION OF THE SKULL

Examine a pair of sagittal sections of the cat skull cut in such a way that the nasal septum shows on one half and the turbinals on the other. Note the large **cranial cavity** for the brain (Fig. 4–26). It can be divided into three parts—a **caudal cranial fossa** in the occipital-otic region for the cerebellum; a large **middle cranial fossa** for the cerebrum; and a small, narrow **rostral cranial fossa,** just caudal to the nasal region for the olfactory bulbs of the brain. In the cat a partial transverse septum of bone (the **tentorium**) separates the middle and caudal cranial fossae. The internal part of the petrosal portion of the temporal, containing the inner ear, can be seen in the lateral wall of the caudal cranial fossa dorsal to the tympanic bulla. It is perforated by a large foramen, the **internal acoustic meatus.** The fossa dorsal to the internal acoustic meatus lodges a lobule of the cerebellum. The saddle-shaped notch (**sella turcica**) in the floor of the cerebral fossa lodges the hypophysis (Fig. 9–16, p. 262). It is in the basisphenoid. A large **sphenoidal air sinus** can be seen in the presphenoid, and a **frontal air sinus** in the frontal bone dorsal to the rostral cranial fossa.

The rostral wall of the rostral cranial fossa is formed by a sievelike plate of bone whose foramina communicate with the nasal cavities. This plate of bone, the **cribriform plate** of the **ethmoid,** can also be seen by looking through the foramen magnum of a complete skull. A **perpendicular plate** of the ethmoid, which will be seen on the larger section, extends from the cribriform plate rostrally between the two nasal cavities. It connects with the nasal bones dorsally and with the vomer ventrally and forms much of the nasal septum. The very front of the septum is cartilaginous. Most of the ethmoid is shaped like the letter T, the top of the T being the cribriform plate, and its stem the perpendicular plate. In addition, the ethmoid has little lateral processes that often can be seen in the medial wall of the orbits caudal to the lacrimal.

Each nasal cavity is largely filled with thin, complex scrolls of bone called the **conchae** (turbinates). They can best be seen on the section that lacks the perpendicular plate of the ethmoid (Fig. 4–26). Although the

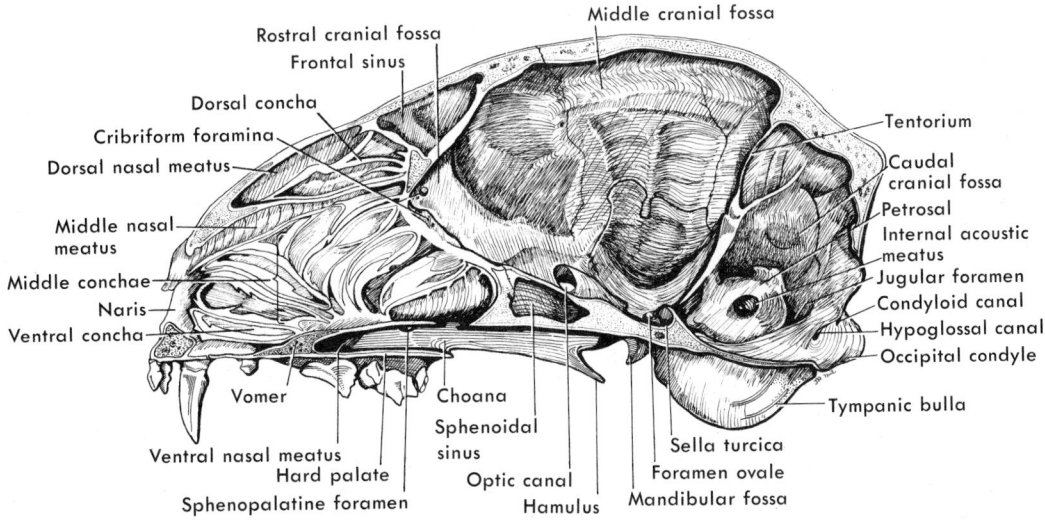

Figure 4–26. Sagittal section of the skull of a cat. The nasal septum has been removed.

turbinates ossify from the nasal capsules, they become attached to bones surrounding the nasal cavities. A small **dorsal concha** (nasoturbinate) lies rostral to the frontal sinus and lateral to a perpendicular septum of the nasal bone. A **ventral concha** (maxilloturbinate) lies in the rostroventral part of the nasal cavity. Its attachment to the maxillary bone can be seen by looking into the cavity through the naris. The large and very complexly folded **middle concha** (ethmoturbinate) lies between the others and fills most of the nasal cavity. Some of the air entering a nasal cavity passes directly caudally through an uninterrupted passage, the **ventral nasal meatus,** that lies between the turbinates and the hard palate, but much of the air filters back between and among the turbinates, which are covered by a nasal epithelium, and which serve to increase the surface area available for olfaction and for heating, cleansing, and moistening the inspired air. A **dorsal nasal meatus** leads from the dorsal part of the naris caudally to the caudal parts of the middle concha. A narrow **common meatus** lies between all of the conchae and the nasal septum.

(D) Foramina of the Skull

Numerous foramina for nerves and blood vessels perforate the skull. These may be considered at this time, or their consideration may be postponed until the nerves have been studied. In any case, the foramina should be studied on a sagittal section of the skull so that both their external and internal aspects can be seen. Certain of the passages should be probed to determine their course.

First, study the foramina for the 12 cranial nerves. The first cranial nerve, the olfactory, consists of many fine subdivisions that enter the cranial cavity through the **cribriform foramina** in the cribriform plate of the ethmoid. The **optic canal,** for the second cranial nerve, is the most rostral of the row of four foramina in the caudomedial wall of the orbit. Next caudad, and largest in this row, is the **orbital fissure** for the third, fourth, and sixth nerves going to the muscles of the eyeball, and for the ophthalmic division of the fifth nerve. The third and smallest foramen in this row (**foramen rotundum,** Fig. 4–25) transmits the maxillary division of the fifth nerve, and the last in this row (**foramen ovale**) the mandibular division of this nerve. The **internal acoustic meatus,** seen in the petrosal portion of the temporal, transmits both the seventh and eighth nerves. The eighth (vestibulocochlear) nerve comes from the inner ear, but the seventh (facial nerve) continues through a facial canal within the petrosal and emerges on the undersurface of the skull through the **stylomastoid foramen** located beneath the tip of the mastoid process. The ninth, tenth, and eleventh nerves, in company with the internal jugular vein, pass through the **jugular foramen,** which can be seen internally in the floor of the caudal cranial fossa caudal to the internal acoustic meatus. The jugular foramen opens on the ventral surface of the skull beside the posteromedial edge of the tympanic bulla. The **hypoglossal canal** for the twelfth cranial nerve can be seen in the floor of the caudal cranial fossa caudal and medial to the jugular foramen. Probe to see that the hypoglossal canal extends cranially and emerges on the ventral surface in common with the jugular foramen.

As the ophthalmic and maxillary divisions of the fifth (trigeminal) nerve are distributed to various parts of the head, certain of their branches pass through other foramina, often in company with blood vessels. A branch of the ophthalmic division traverses a small **ethmoid foramen,** or series of foramina, in the medial wall of the orbit to enter the nasal cavity. The ethmoid foramen lies in the frontal bone very near its suture with the presphenoid. One of the branches of the maxillary division emerges through an **infraorbital canal** located in the rostral part of the zygomatic arch. But before reaching the infraorbital canal, this branch has subsidiary branches that pass through several small foramina near the front of the orbit to supply the teeth of the upper jaw. A second branch of the maxillary division leaves the rostral portion of the orbit through a **sphenopalatine foramen** to enter the ventral part of the nasal cavity. The sphenopalatine foramen is the larger and more medial of two foramina that lie close together in the orbital process of the palatine bone. After entering the nasal cavity, part of this nerve continues forward through the nasal passages and finally drops down to the roof of the mouth through a **palatine fissure** which is located just lateral to the midline at the rostral end of the hard palate. In certain mammals the palatine fissure also carries a nasopalatine duct that leads from the mouth to the vomeronasal organ (Jacobson's organ, p. 181) located in the nasal cavity. The small foramen lateral and rostral to the sphenopalatine is the caudal end of the **palatine canal.** A third branch of the maxillary division runs through this canal to the roof of the mouth. The rostral end of the canal can be seen on the ventral surface of the hard palate in, or near, the suture between the palatine and maxillary bone. A fourth branch of the maxillary backtracks to enter the orbital fissure. It then passes into a small **pterygoid canal** whose rostral end may be seen in the floor of the orbital fissure. The caudal end of the canal appears as a tiny hole on the ventral side of the skull near the base of the pterygoid process of the basisphenoid. After emerging from the pterygoid canal, this branch enters the tympanic cavity through the osseous portion of the **auditory tube**—a large opening at the rostral edge of the tympanic bulla. A summary of the cranial nerves and the foramina through which they pass is given in Table 3.

The major foramina that remain do not carry nerves. A **nasolacrimal canal** for the lacrimal, or tear, duct will be seen in the lacrimal bone extending from the orbit into the nasal cavity. You may also be able to find parts of the small **carotid canal** for the vestigial (in the cat) internal carotid artery. The caudal end of the canal appears as a tiny hole in the rostromedial wall of the jugular foramen. From here the canal extends forward, dorsal to the tympanic bulla, and enters the cranial cavity. Its point of entrance can be seen in the caudal cranial fossa rostral to the petrosal and ventral to the tentorium. A **condyloid canal** for a small vein can be found in the caudal cranial fossa dorsal to the hypoglossal canal.

If you have access to a specimen in which the tympanic bulla has been

Table 3.   Mammalian Cranial Nerves and Their Foramina

The twelve cranial nerves of mammals are listed together with the foramina through which they pass. In those cases in which a nerve goes through two or more foramina before reaching the organ it supplies, the foramina are listed in sequence from the brain.

NERVE	FORAMEN
I. Olfactory	Cribriform foramina
II. Optic	Optic canal
III. Oculomotor	Orbital fissure
IV. Trochlear	Orbital fissure
V. Trigeminal	
Ophthalmic division	Orbital fissure
one branch	Ethmoid foramina
Maxillary division	Foramen rotundum
one branch	Infraorbital canal
one branch	Sphenopalatine foramen, palatine fissure
one branch	Palatine canal
one branch	Pterygoid canal
Mandibular division	Foramen ovale
one branch	Mandibular foramen, Mental foramina
VI. Abducens	Orbital fissure
VII. Facial	Internal acoustic meatus, Facial canal, Stylomastoid foramen
VIII. Vestibulocochlear	Internal acoustic meatus
IX. Glossopharyngeal	Jugular foramen
X. Vagus	Jugular foramen
XI. Accessory	Jugular foramen
XII. Hypoglossal	Hypoglossal canal, jugular foramen

removed, you will be able to see two openings on the underside of the periotic. The more dorsal is the oval window (**fenestra vestibuli**) for the stapes; the more ventral, the round window (**fenestra cochleae**) for the release of vibrations from the inner ear.

(E) LOWER JAW

With the transfer of certain of the original lower jaw bones to the ear region, and the loss of others, the pair of enlarged dentaries are left as the sole elements in the mammalian lower jaw.

Examine the lower jaw, or **mandible,** of the cat (Fig. 4–24) and notice that the **dentary bones** are firmly united anteriorly by a **mandibular symphysis.** The horizontal part of the dentary that bears the teeth is its **body;** the part caudal to this, its **ramus.** A large, triangular-shaped depression, the **masseteric fossa,** occupies most of the lateral surface of the ramus. Part of the masseter, one of the mandibular muscles, inserts here. Posteriorly the ramus has three processes—a dorsal **coronoid process,** to which the temporal muscle attaches; a middle, rounded **condyloid process** for the articulation with the skull proper; and a ventral **angular process** to which other mandibular muscles attach.

A large **mandibular foramen** will be seen on the medial side of the ramus, and two small **mental foramina** on the lateral surface of the body near its rostral end. A branch of the mandibular division of the trigeminal nerve, supplying the teeth and the skin covering the lower jaw, enters the mandibular foramen and emerges through the mental foramina. Blood vessels accompany the nerve

(F) Teeth

The teeth of mammals are quite different from those of lower vertebrates for they are limited to the jaw margins, are set in deep sockets **(thecodont),** and are differentiated into various types **(heterodont).** Most adult mammals have in each side of each jaw a series of nipping incisors at the front, a large canine behind these, a series of cutting premolars behind the canine, and finally a series of chewing, or grinding, molars. The number of each kind of teeth present in a particular group of mammals may be expressed as a dental formula. For primitive placental mammals this was $\frac{3.1.4.3}{3.1.4.3} \times 2$ = 44. Such an animal had three incisors, one canine, four premolars, and three molars in each side of each jaw. The number of teeth, and their structure, are adapted to the animal's mode of life. The molars especially are subject to much divergence among the groups of mammals.

Examine the teeth of the cat (Fig. 4–24 and 4–25). In each side of the upper jaw there are normally three **incisors** at the rostral end followed by one **canine,** three **premolars,** and one very small **molar.** In the lower jaw there are three incisors, one canine, two premolars, and one large molar. What would the dental formula be? In which bones are the teeth of the upper jaw located?

During its evolution, the cat has lost the first premolar in the upper jaw, the first two in the lower, and all molars posterior to the first. The gap left between each canine and the premolars is called a **diastema.** The first of the remaining premolars and the molar of the upper jaw are more or less vestigial. But the last premolar of the upper jaw (phylogenetically premolar number four) and the lower molar have become large and complex in structure. Articulate the jaws and note how these two teeth, which are known as the **carnassials,** intersect to form a specialized shearing mechanism. Carnassials are restricted to carnivores, and in contemporary species have the formula Pm 4/M 1.

(G) Hyoid Apparatus

A hyoid apparatus is present in mammals but is not so large as in more primitive tetrapods, since it is formed from only the ventral parts of the hyoid and first branchial arches. Insofar as more posterior arches are present, they are incorporated in the laryngeal and possibly the tracheal cartilages (p. 265).

Study the hyoid apparatus of the cat (Fig. 4–24). It may be in place on the skeleton, or removed and mounted separately. It consists of a transverse bar of bone, the **body of the hyoid** composed of a single bone **(basihyal),** from which two pairs of processes (horns) extend cranially and caudally. The caudal or **greater horns of the hyoid** are the larger, and each consists of but one bone, which, although it is called the **thyrohyal,** is derived from the first branchial arch. Each of the cranial or **lesser horns of the hyoid** consists of a small **ceratohyal.** A chain of ossicles connects the ceratohyal with the skull, attaching to the tympanic bulla medial to the stylomastoid foramen. From ventral to dorsal these are the **epihyal, stylohyal,** and **tympanohyal.** These ossicles and the ceratohyal are derivatives of the hyoid arch. The body of the hyoid develops partly from the hyoid arch and partly from the first branchial arch.

In some mammals (man, for example) the hyoid apparatus consists of a single bone, the hyoid, having a body, a lesser horn and a greater horn. A stylohyoid ligament, which extends from a styloid process at the base of the skull to the lesser horn, replaces the chain of ossicles seen in the cat as the support for the apparatus. The styloid process itself represents a part of the hyoid arch fused onto the temporal bone.

THE APPENDICULAR SKELETON

As stated in the preceding chapter, the appendicular skeleton consists of the bones of the paired appendages and the girdles to which the appendages attach. Most vertebrates have both **pectoral** (shoulder) and **pelvic** (hip) **girdles** and **appendages.** The appendicular skeleton is basically a part of the endoskeleton, so it consists primarily of cartilage or cartilage replacement bone. However, dermal bones have become intimately associated with the cartilaginous elements of the pectoral girdle in the majority of vertebrates. That portion of the girdle formed of cartilage, or cartilage replacement bone, is called the **endoskeletal girdle;** that portion formed by dermal bone, the **dermal girdle.**

FISHES

An appendicular skeleton is usually absent in primitive agnathous vertebrates— recall *Petromyzon*—but is typically present in gnathostomes. Although some ostracoderms had pectoral flaps, paired appendages were not commonly found until the placoderms and acanthodians. They were exceedingly variable in structure and number among the members of these groups, but in cartilaginous and other bony fishes they are in the form of pectoral and pelvic fins that help to keep the fish in the horizontal plane and aid in steering.

The Chondrichthyes and the actinopterygian group of Osteichthyes have fins that are somewhat fan shaped. Primitively, such fins had a wide base and were supported by numerous parallel bars (Fig. 5–1, *A*), but in living members of these groups, the base of each fin has become constricted, so that the basal elements are crowded together and reduced in number (Fig. 5–1, *B*). In the sarcopterygian group of bony fish, the fin is elongate and lobate in shape. It is supported by a single basal element and a central axis from which side branches arise (Fig. 5–1, *C*). This type of fin, called an **archipterygium,** is probably derived from the wider based fin and, in turn, is the type of fin from which the tetrapod limb evolved (Fig. 5–1, *D*).

Pectoral Girdle and Fin

(A) SQUALUS

The appendicular skeleton of *Squalus,* which will be used as the chief example of the fish condition, is reasonably representative of the type found

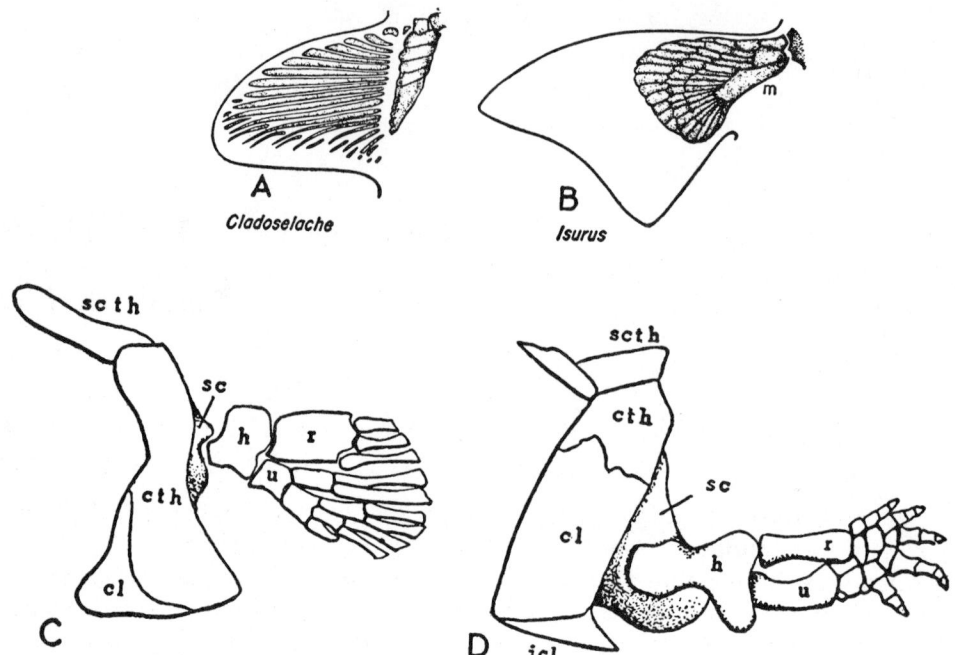

Figure 5-1. Pectoral fin of *A*, a primitive fossil shark, and *B*, a modern shark; pectoral fin and shoulder girdle of *C*, a crossopterygian, and *D*, a labyrinthodont. All figures are of the left side. Abbreviations: *cl*, clavicle; *cth*, cleithrum; *h*, humerus; *icl*, interclavicle; *m*, metapterygium: *r*, radius; *sc*, scapulocoracoid (endoskeletal girdle); *scth*, supracleithrum; *u*, ulua. The metapterygium of the tribasic selachian fin is believed to be homologous to the single basal of the archipterygium fin of crossopterygians. (From Romer, The Vertebrate Body; *A* after Dean; *B* after Mivart; *C* after Gregory.)

in many living fishes. But like other parts of the skeleton, it is atypical in being entirely cartilaginous. Also, the dermal bones that early became associated with the pectoral girdle are absent but can be seen in one of the primitive bony fishes such as *Amia*.

Examine the pectoral girdle and fin on a skeleton of the dogfish. The girdle is located just caudal to the gills (Fig. 5-2), for there is no neck region in fishes. It consists of a U-shaped bar of cartilage, which represents the endoskeletal pectoral girdle (Fig. 5-3). A dermal girdle, as stated, is absent. At the top of each limb of the U there is a separate **suprascapular cartilage** about one half an inch long. The rest of the girdle is formed of a single piece, the **scapulocoracoid.** In some sharks, but not in *Squalus,* it is obvious that the scapulocoracoids are paired elements that have fused in the midventral line. That part of the scapulocoracoid which is located ventral to the point where the fin attaches (**glenoid surface**) may be called the **coracoid bar** since it has the same position as the coracoid of tetrapods. The rest of it may be called the **scapular process.** A small **coracoid foramen** for vessels and nerves can be found just cranial to the glenoid surface.

The pectoral fin is narrow at the base but widens distally. It is supported proximally by a series of cartilages, collectively called the **pterygiophores,** and distally by fibrous fin rays (**ceratotrichia**) that develop in the dermis on each surface of the fin. The three large pterygiophores that articulate with the girdle are the **basals;** the rest are **radials.** The three basals are, from cranial to caudal, the **propterygium, mesopterygium,** and **metapterygium.** Note that the metapterygium is the longest.

In many cartilaginous fishes and actinopterygians, the metapterygium forms the axis of the fin. It is probable that the metapterygium is the element that persists as the single "basal" in the archipterygium fin of sarcopterygians.

(B) AMIA

omit

Examine a skeleton of the bowfin, *Amia.* Notice that the **endoskeletal girdle** consists, on each side, of a small area of unossified cartilage which lies between the pectoral fin and a conspicuous arch of bone posterior to the **operculum** (dermal gill covering). This arch of bone, which is of dermal origin, consists of four elements (Fig. 5-4). A large, ventral **cleithrum** begins ventral to the caudal visceral arches and continues dorsally a bit beyond the fin and endoskeletal girdle. A **supracleithrum** continues from the cleithrum nearly to the roof of the skull; a **posttemporal,** which is a part of the girdle even though it has surface sculpturings similar to the skull bones, joins the supra-cleithrum and skull. Finally, a small **postcleithrum** is located caudal to the junction of cleithrum and supracleithrum.

The **clavicle,** a ventral, dermal girdle element characteristic of many fishes and terrestrial vertebrates, has long been thought to be absent in *Amia.* However, Leim and Woods (1973) propose that the clavicle is represented by a slender, rather superficial sliver of bone that overlaps the front of the ventral part of the cleithrum. It is not usually retained in commercial preparations of the skull and pectoral girdle of *Amia.*

Pelvic Girdle and Fin

The pelvic girdle of *Squalus* consists of a simple transverse rod of cartilage, the **puboischiadic bar,** located in the ventral abdominal wall just

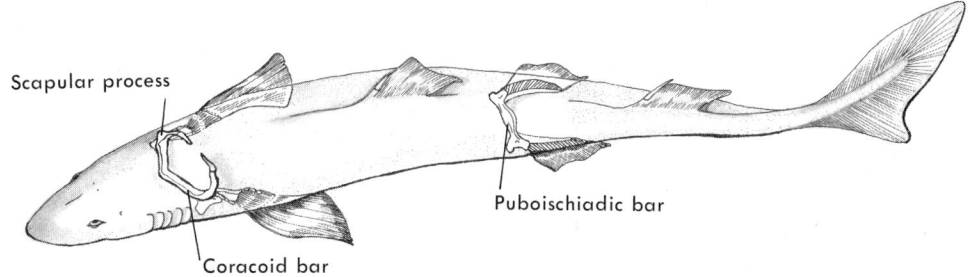

Figure 5-2. Transparent view of *Squalus* to show the location and orientation of the pectoral and pelvic girdles.

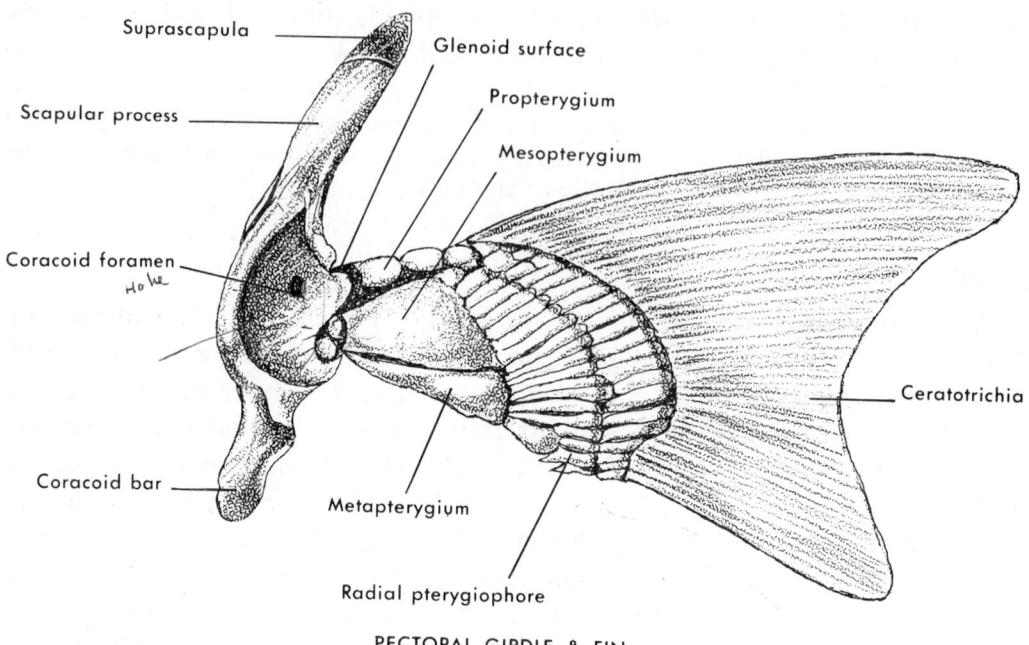

Suprascapula

Scapular process

Coracoid foramen

Coracoid bar

Glenoid surface

Propterygium

Mesopterygium

Ceratotrichia

Metapterygium

Radial pterygiophore

PECTORAL GIRDLE & FIN
Ventrolateral View

Puboischiadic bar

Acetabular surface

Propterygium

Iliac process

Radial pterygiophore

Metapterygium

Ceratotrichia

PELVIC GIRDLE & FIN
Dorsal View

Figure 5–3. Drawings of the girdles and fins of *Squalus*.

cranial to the cloaca (Fig. 5–2). Each end of it extends dorsally as a very short **iliac process,** but these processes do not reach the vertebral column (Fig. 5–3). There are often two small nerve foramina near the base of each iliac process.

Like the pectoral fin, the pelvic girdle consists of a series of proximal cartilaginous pterygiophores and distal ceratotrichia. There are only two **basal cartilages** in *Squalus*—a long **metapterygium** extending caudally from the girdle, and clearly forming the main support of the fin, and a short **propterygium** projecting laterally from the girdle. A number of **radials** extend into the fin from the two basals. In males a **clasper,** the skeleton of which is formed by enlarged and modified radials, extends caudad from the end of the metapterygium. The fin attaches to the **acetabular surface** of the girdle.

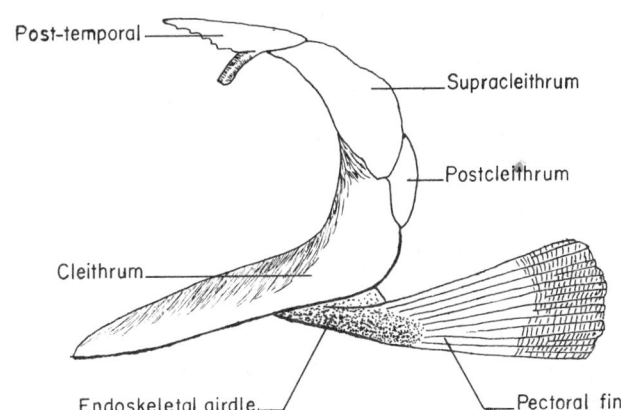

Figure 5–4. A lateral view of the pectoral girdle and fin of *Amia;* cartilage is stippled.

Post-temporal

Supracleithrum

Postcleithrum

Cleithrum

Endoskeletal girdle

Pectoral fin

PRIMITIVE TETRAPODS

In the evolution from crossopterygian fishes to tetrapods, the archipterygium is transformed into a pentadactyl limb (Fig. 5–1, *C* and *D*) consisting of three segments. In the pectoral appendage these segments are the **brachium** (containing the **humerus**), **antebrachium** (containing the **radius** and **ulna**), and **manus** (**carpals, metacarpals, and phalanges**). Corresponding segments and bones in the pelvic appendage are **femur** (**os femoris**),[9] **shank** (**tibia** and **fibula**), and **pes** (**tarsals, metatarsals, and phalanges**). The number of phalanges present in the toes of different groups of tetrapods varies. In early reptiles (cotylosaurs) there were two phalanges in the first toe, three in the second, four in the third, five in the fourth, and either three or four in the fifth toe. This can be expressed as a **phalangeal formula,** i.e., 2-3-4-5-3 (or 4).

In primitive amphibians and reptiles the limbs were in a sprawled position at right angles to the body and were used as a supplement to lateral undulations of the body in locomotion. In such a limb the humerus and femur move back and forth in the horizontal plane. But in the evolution through more advanced reptiles to birds and mammals the limbs became increasingly important in support and locomotion. In this connection the limbs of mammals, and the pelvic limbs of birds, have rotated beneath

---

[9]See footnote 4, p. 36.

the body so that the humerus and femur move back and forth in the vertical plane. Review the changes in limb posture described in the section on External Anatomy (p. 39).

Correlated with the increased importance of the appendages, the girdles became larger and stronger. In the endoskeletal pectoral girdle of labyrinthodonts (Fig. 5–1, D) there was only one ossification on each side (a **scapulocoracoid**), but in cotylosaurs (Fig. 5–5, A) a **scapula** ossified dorsal to the glenoid cavity, and an **anterior coracoid** (**procoracoid**) ventral to the fossa. Although these elements were large and platelike, the endoskeletal girdles of opposite sides neither united with each other ventrally nor connected with the vertebral column. Correlated with the evolution of a distinct neck, the dermal pectoral girdle lost its connection with the back of the skull. However, the cleithrum and clavicle persisted (Fig. 5–5, A), and a new dermal element, the **interclavicle,** became associated with the girdle. The interclavicle is a median ventral element which connects the clavicles of opposite sides.

The endoskeletal pectoral girdle of living amphibians, reptiles and birds is derived from the type just described and is usually close to it in essential pattern. But in the evolution toward mammals a third ossification, the **posterior coracoid** (or simply coracoid), appeared in the endoskeletal girdle caudal to the anterior coracoid (Fig. 5–5, B). In mammals, the scapula area is greatly expanded and the coracoid region is reduced. The anterior coracoid is completely lost in placental mammals, and the posterior coracoid is represented only by a small coracoid process of the scapula. In all living tetrapods the dermal portions of the girdle have become reduced, the cleithrum being completely lost except in certain primitive frogs.

The pelvic girdle similarly enlarged during the transition from water to land, and in tetrapods typically consists of three cartilage replacement bones—a **pubis** and **ischium** ventral to the acetabulum and an **ilium** that extends from the acetabulum to the sacral ribs and vertebrae (Fig. 5–5, C and D). The ventral elements of opposite sides unite with each other. Primitively the ventral elements formed a broad plate and the smaller ilium connected with only a single sacral vertebra. But in birds and mammals the ventral elements have become relatively smaller and the ilium expands, turns forward, and unites with more sacral vertebrae.

Appendicular Skeleton of Necturus

As explained in the preceding chapter, *Necturus* is neotenic and parts of its skeleton are unossified. This is true for many parts of the girdles and appendages. Since it is an aquatic animal, it does not use its limbs much and they do not need to be as strong as those of a terrestrial species.

(A) PECTORAL GIRDLE AND APPENDAGE

Examine the pectoral girdle on a skeleton of *Necturus,* comparing it with Figure 5–6, A. The two halves of the pectoral girdle overlap slightly ventrally but are not united. On each side an ossified **scapula** extends dorsally cranial to the **glenoid cavity** (a depression for the articulation of the girdle with the humerus). The scapula is capped by a **suprascapular cartilage.** The ventral part of the girdle remains unossified and may be called the **coracoid plate.** A **procoracoid process,** not to be confused with the anterior coracoid bone of other tetrapods, extends forward from the coracoid plate. A **coracoid**

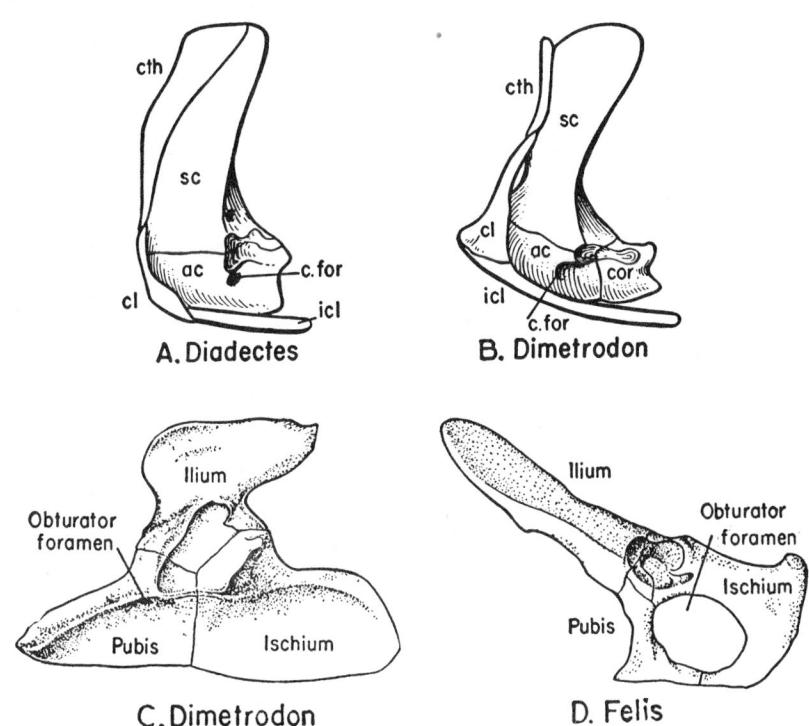

Figure 5–5. *A*, Pectoral girdle of a cotylosaur; *B*, pectoral girdle of a mammal-like reptile; *C*, pelvic girdle of a mammal-like reptile; *D*, pelvic girdle of a mammal. All are lateral views of the left side. Abbreviations: *ac*, anterior coracoid; *c. for*, coracoid foramen; *cl*, clavicle; *cor*, posterior coracoid (or coracoid of mammals); *cth*, cleithrum; *icl*, interclavicle; *sc*, scapula. The acetabular bone, although not labeled, can be seen in the acetabulum of *Felis* above the pubis. The clavicle, interclavicle, and cleithrum are dermal elements. All others are cartilage replacement bone. (From Romer, The Vertebrate Body.)

foramen, for vessels and nerves, may be seen in the coracoid plate ventral to the scapula.

Study the pectoral skeleton, noting first the position of the different segments of the limb and the preaxial and postaxial surfaces (see page 36). A single bone, the **humerus,** extends from the glenoid cavity to the elbow joint. Two bones of approximately equal size compose the forearm—a **radius** on the preaxial (medial) side, and an **ulna** on the postaxial (lateral) side. The manus consists of a group of six cartilaginous **carpals** in the wrist, four ossified **metacarpals** in the palm of the hand and the ossified **phalanges** of the digits or toes. Terms for the individual carpals are derived from their positions relative to other bones (see Fig. 5–10), but the individual carpals are usually not distinct in dried skeletons. Note that only four toes are present, a number characteristic of living amphibians. It is uncertain whether this is a retention of the condition found in many primitive amphibians, or whether it resulted from the loss of a toe present in certain five-toed, primitive amphibians. In any case, the concensus is that the toes present are homologous to the second through the fifth toes of amniotes. What is the phalangeal formula?

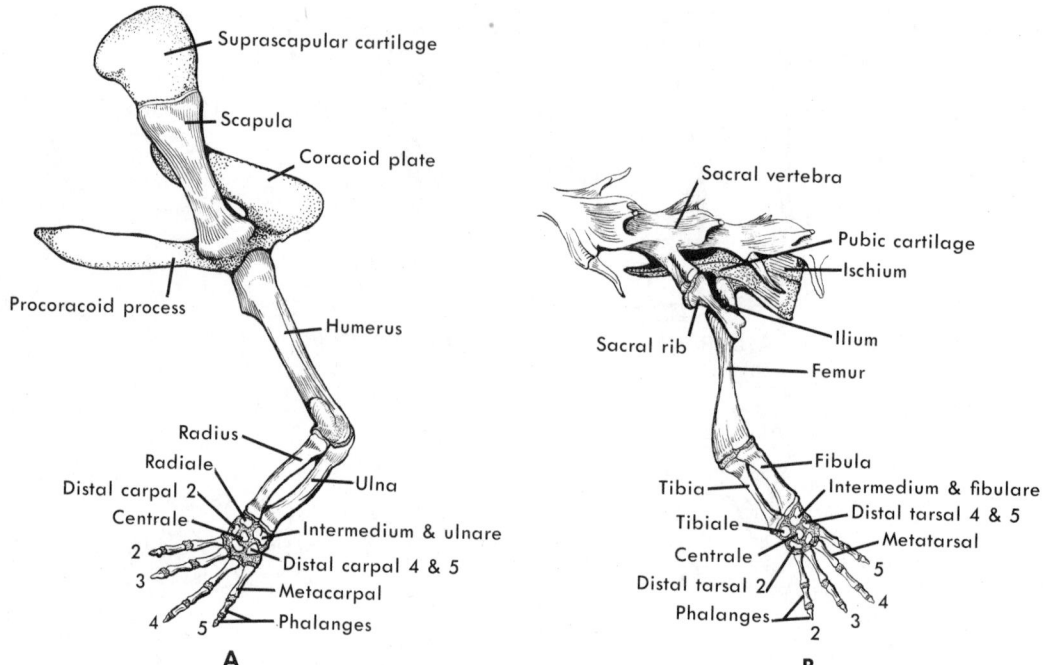

Figure 5–6. Dorsolateral views of the pectoral girdle and appendage (*A*) and pelvic girdle and appendage (*B*) of *Necturus*.

(B) PELVIC GIRDLE AND APPENDAGE

Each half of the pelvic girdle has a narrow ossified **ilium** that extends dorsally from the socket for the leg articulation (**acetabulum**) to attach on a single sacral rib and vertebra (Fig. 5–6, *B*). Ventrally there is a broad **pubo-ischiadic** plate, which contains a pair of ossified **ischia** caudally and a **pubic cartilage** cranially. An **obturator foramen,** for a nerve of the same name, may be seen in the pubic cartilage. Note that the pelvic girdle, together with the sacral rib and vertebra, forms a ring of bone around the caudal end of the trunk. The passage through this ring is called the **pelvic canal.** Structures that lead to the cloaca pass through this canal.

Study the pelvic limb, noting its position and its preaxial and postaxial surfaces. A **femur** forms the upper segment of the limb; a **tibia** and **fibula** are contained within the shank (the former lying along the preaxial surface); and the pes consists of a group of six cartilaginous **tarsals** and ossified **metatarsals** and **phalanges.** The individual tarsals cannot be distinguished in dried skeletons. Many salamanders have five toes, but *Necturus* has only four, the most medial probably being homologous to the second toe of amniotes. What is the phalangeal formula?

Appendicular Skeleton of the Turtle

Although the appendicular skeleton of the turtle is specialized in some respects, certain primitive features can be seen more clearly in it than in the appendicular skeleton of *Necturus.*

(A) PECTORAL GIRDLE AND APPENDAGE

Examine a mounted skeleton and an isolated pectoral girdle of the turtle (Fig. 5–7). The endoskeletal pectoral girdle, which has an unusual triradiate shape, is located between the bottom (**plastron**) and top shell (**carapace**); hence, it lies partly beneath the ribs. This is a feature peculiar to turtles. The dorsal prong of the endoskeletal girdle, which articulates with the carapace, represents the **scapula**; the cranioventral prong, a part of the scapula known as the **acromion**; and the expanded caudoventral prong, the **anterior coracoid.** In life the acromion is connected by a ligament to the **entoplastral plate** of the plastron and the scapula by a ligament to the cara-pace. An **acromiocoracoid ligament** may be seen extending between the tips of the acromion and coracoid. A **glenoid cavity** is present for the articulation of the arm. The entire girdle rotates forward and backward during locomo-tion, and also during breathing movements of the animal.

Parts of the dermal girdle are present but are incorporated in the cranial plates of the plastron. Examine a plastron in which the epidermal scutes (**lamina**) have been removed and the **dermal plates** exposed (Fig. 5–8). The front of the plastron is formed by a pair of **epiplastra,** caudal to which are a median **entoplastron** and three additional paired plates—**hyoplastra, hypo-plastra,** and **xiphiplastra.** All these plates represent in part an ossification in

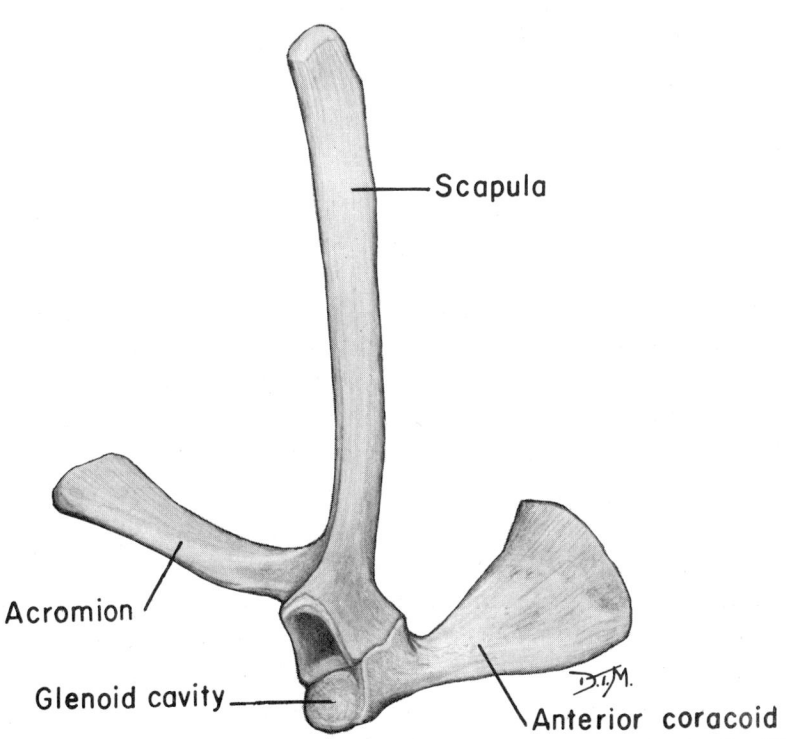

Figure 5–7. Lateral view of the left pectoral girdle of the snapping turtle, *Chelydra.* An-terior is toward the left.

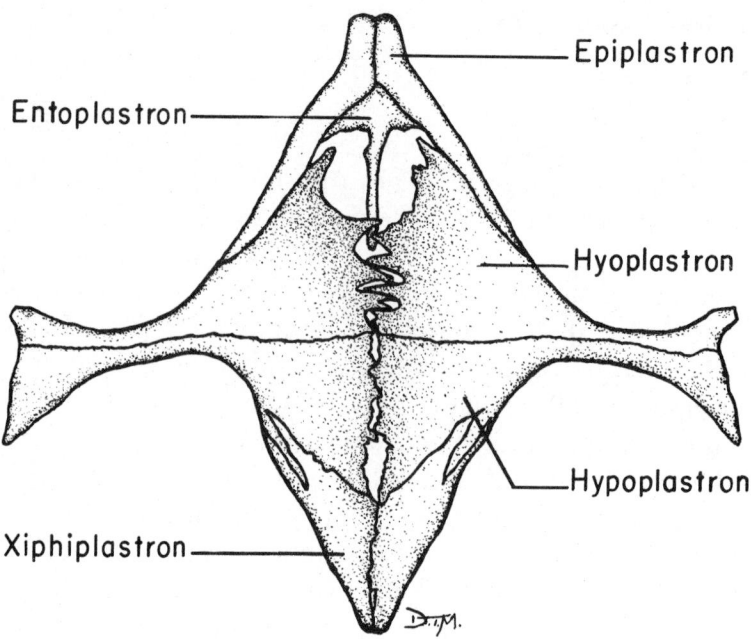

Figure 5–8. Ventral view of the plastron of the snapping turtle. The epidermal scutes have been removed.

the dermis of the skin of the underside of the body. But during embryonic development the originally separate primordia of the clavicles and inter-clavicle become incorporated in the first three plates, which are therefore compound plates. The epiplastra include the paired **clavicles;** the entoplas-tron includes the **interclavicle.** The remaining plastral plates may include the **gastralia** of early amphibians and reptiles. In these primitive tetrapods the gastralia were riblike rods of dermal bone found on the ventral abdominal wall. They were remnants of the piscine dermal scales and may have had a protective function. They are retained in a few living reptiles—*Sphenodon,* crocodiles and, possibly, turtles.

Study the pectoral appendage (Fig. 5–9). How does the position of the limbs of a turtle compare with that of one of the early amphibians or reptiles (labyrinthodont or cotylosaur)? Identify the preaxial and postaxial surfaces. The long bones of the appendage are a **humerus** in the upper arm and a **radius** and **ulna** in the forearm. The proximal end of the humerus has a round **head** that fits into the glenoid cavity and two prominent enlargements (processes) for the attachment of muscles. Of the forearm bones, the radius is the one on the preaxial (cranial or medial) surface. Both radius and ulna are about equal in size, but the ulna tends to extend over the distal end of the humerus, whereas the radius articulates on the underside of the end of the humerus. As in primitive tetrapods generally, both the radius and ulna ar-ticulate with the wrist bones and there is no distal radioulnar joint.

The manus consists of a group of **carpals** in the wrist, a row of five metacarpals in the palm, and the **phalanges** in the free part of the toes. There are five toes, the first being the most medial. What is the phalangeal formula?

The carpus of the turtle is very similar to that of more primitive tetrapods, and the individual components should be identified. The carpals can be grouped into a proximal and distal row. The proximal row consists of three bones — an **ulnare** adjacent to the ulna, an **intermedium** lying between the distal ends of the radius and ulna, and an elongated element distal to the radius and intermedium, which represents a fusion either of two centralia or of a **centrale** and **radiale** (Fig. 5–10). The distal row consists of five distal carpals which are numbered according to the digit to which they are related — distal carpal 1, distal carpal 2, etc. In addition there may be a small **sesamoid** bone on the lateral edge of the carpus. Sesamoid bones develop in the tendons of muscles and are rather variable. However, the one on the lateral edge of the wrist adjacent to the ulnare is consistent enough to be given a name — the **pisiform.**

(B) PELVIC GIRDLE AND APPENDAGE

Study the pelvic girdle of the turtle on a mounted skeleton and from an isolated specimen (Fig. 5–11). Each half of the girdle consists of a dorsal **ilium** that inclines posteriorly and articulates with two sacral ribs and vertebrae, a cranioventral **pubis,** and a caudoventral **ischium.** All three elements share in the formation of the **acetabulum,** the socket for the leg articulation. Pubis and ischium of opposite sides unite by a **symphysis.** An **epipubic cartilage,** which may be partly ossified, extends forward in the mid-ventral line from the pubic bones. Both the pubis and ischium have a lateral process that is directed ventrally and, in life, rests on the plastron. A large **puboischiadic fenestra,** which develops in association with the origin of a muscle, lies between the pubis and ischium on each side. A separate obturator

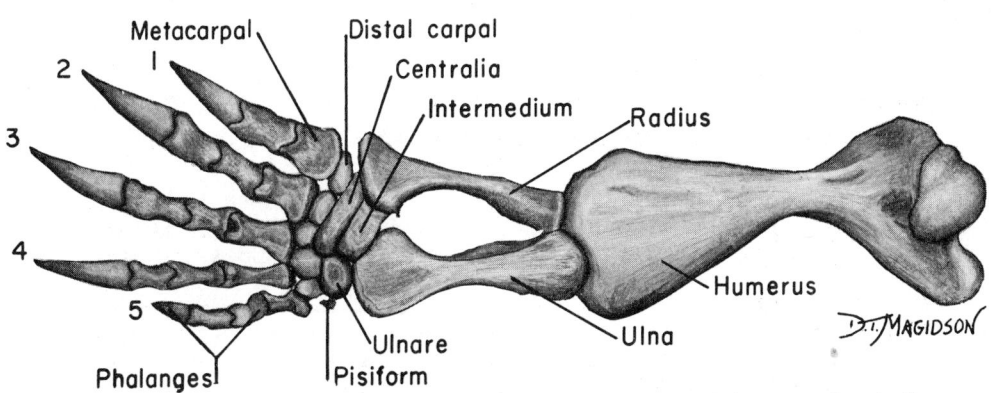

Figure 5–9. Dorsal view of the left pectoral appendage of the snapping turtle.

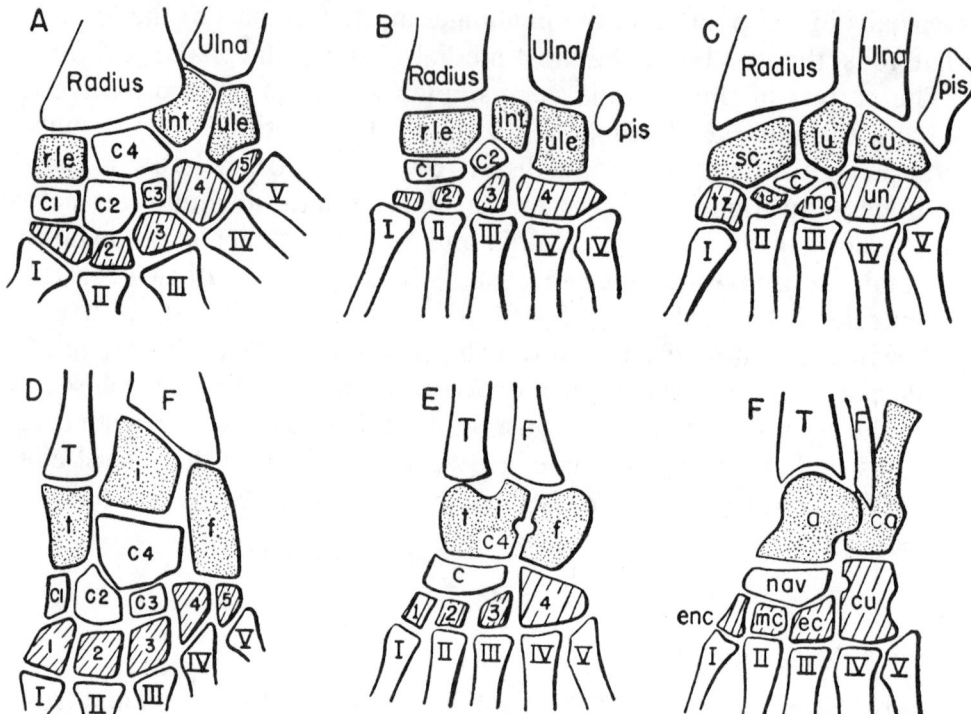

Figure 5–10. Diagrams of the left carpus (top row) and tarsus (bottom row) of a labyrintho-dont (*A, D*), primitive reptile (*B, E*), and mammal (*C, F*), to show the homologies between the elements. Roman numerals indicate the metapodials of the digits; Arabic numerals the distal carpals and tarsals. Other abbreviations: *a,* talus; *c,* centralia; *ca,* calcaneus; *cu,* triquetrum in carpus, cuboid in tarsus; *ec,* lateral cuneiform; *enc,* medial cuneiform; *f,* fibulare; *F,* fibula; *i, int,* intermedium; *lu,* lunar; *mc,* intermediate cuneiform; *mg,* capitate; *nav,* navicular; *pis,* pisi-form; *rle,* radiale; *sc,* scaphoid; *t,* tibiale; *T,* tibia; *td,* trapezoid; *tz,* trapezium; *ule,* ulnare; *un,* hamate. (From Romer, The Vertebrate Body.)

foramen seen in many lower vertebrates is not present in turtles since the obturator nerve also passes through the puboischiadic fenestra.[10]

Examine the pelvic limb, noting its position and its preaxial and post-axial surfaces (Fig. 5–12). The long bone of the thigh is the **femur.** Its proximal end bears a round **head** that fits into the acetabulum, and two prominent processes for the attachment of muscles. The long bones of the shank are the **tibia** and **fibula,** the former being the larger and more cranial or medial.

The pes consists of a group of **tarsals** in the ankle region, a row of five **metatarsals** in the sole of the foot, and **phalanges** in the free part of the five

---

[10]The terminology and homology of the various pelvic openings are unfortunately confused. In most amphibians and reptiles an obturator foramen (pubic foramen) for an obturator nerve perforates the pubis cranial to the acetabulum (Fig. 5–5, *C*). The rest of the puboischiadic plate is solid. In some reptiles (lizards) an additional opening, known as the puboischiadic fenestra (thyroid fenestra), develops between the pubis and ischium in association with the origin of certain pelvic muscles. In mammals the fenestrations of the puboischiadic plate includes both the primitive obturator foramen and the puboischiadic fenestra. Such an opening is also termed an obturator foramen. The turtle would seem to parallel this condition, but most authorities call the opening a puboischiadic fenestra.

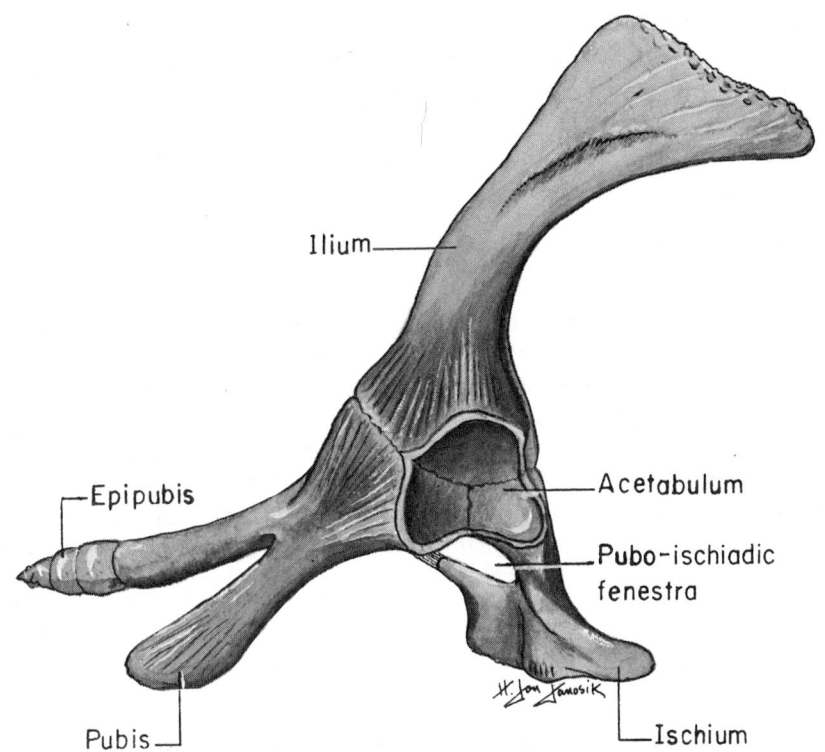

Figure 5–11. Lateral view of the left side of the pelvic girdle of the snapping turtle, *Chelydra.* Anterior is toward the left.

digits. What is the phalangeal formula? The metatarsal of the fifth toe is flat and broad rather than round and elongate like the others.

The individual tarsals should be studied and compared with Figure 5–12, *D*. There is a row of four **distal tarsals** next to the metatarsals. The fourth distal tarsal is larger than the others and is associated with the fourth and fifth toes. All the remaining elements of the tarsus tend to fuse into a single bone, but one can often see the lines of union between the three major elements: **talus** (tibiale, intermedium and proximal centrale), **distal centrale,**

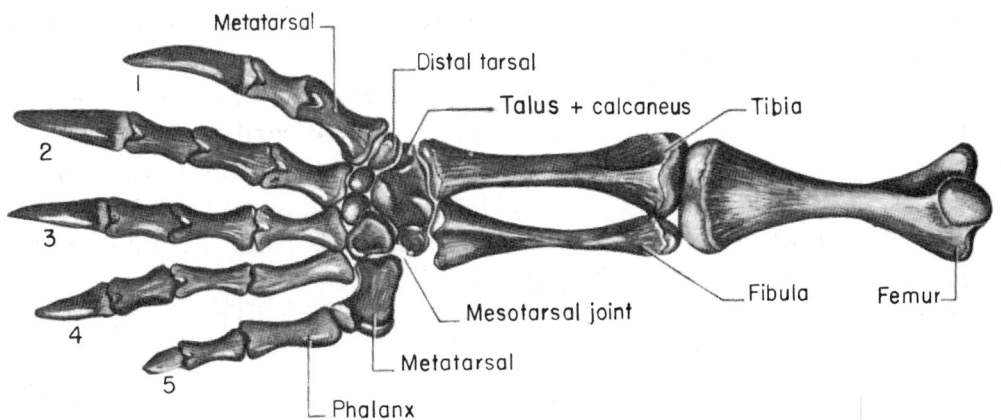

Figure 5–12. Dorsal view of the left pelvic appendage of the snapping turtle.

and **calcaneus** (fibulare). Sometimes the calcaneus remains distinct. The main ankle joint of the turtle, as in many reptiles, is a **mesotarsal joint,** for it lies between the large proximal element(s) and the distal tarsals.

MAMMALS

The appendicular skeleton of mammals is much better suited than that of primitive tetrapods for terrestrial locomotion. Note the position of the limbs beneath the body on a mounted skeleton of the cat. How has the changed position of the limbs come about? Where are the original preaxial and postaxial surfaces now located? Primitive mammals walked with the soles of their feet on the ground (**plantigrade**). But in such mammals as the cat, a longer stride is made possible by walking on the digits with the soles of the feet off the ground (**digitigrade**). Ungulates, such as the horse and cow, carry the tendency further and walk on their toe tips (**unguligrade**).

Pectoral Girdle and Appendage

Study the appendages and girdles on mounted specimens of the cat and from disarticulated bones. Learn to distinguish the individual girdles and the long bones of the appendage, including a recognition as to whether they are from the left or right side. The pectoral girdle consists primarily of an expanded triangular-shaped **scapula** (Fig. 5–13). In therian mammals the anterior coracoid has been lost and the posterior coracoid is reduced to a small, hooklike **coracoid process** which can be seen medial to the cranial edge of the **glenoid cavity** (socket for the articulation with the front leg). Correlated with the shift of the limbs under the body, the glenoid cavity is directed ventrally rather than laterally. Of the elements of the dermal girdle, only the **clavicle** is present. In many mammals, human beings included, the clavicle extends from the scapula to the sternum. But in some mammals, including the cat, the clavicle loses its connections with the rest of the skeleton and is reduced to a sliver of bone imbedded in the muscles cranial to the shoulder joint. This permits the scapula to participate in the fore and aft swing of the front leg during locomotion.

If one pictures the scapula as an inverted triangle, the glenoid cavity is at the apex, and the curved top of the scapula (**dorsal border**) is at the base of the triangle. The cranial edge of the scapula is its **cranial border,** and the straight caudal edge, which is adjacent to the armpit, its **caudal border.** A prominent ridge of bone, the **scapular spine,** extends from the dorsal border nearly to the glenoid cavity. The ventral tip of the spine continues lateral to the glenoid cavity as a process known as the **acromion.** The clavicle articulates with this process in those mammals having a prominent clavicle. A **metacromion** *(hamate process)* extends caudally from the spine dorsal to the

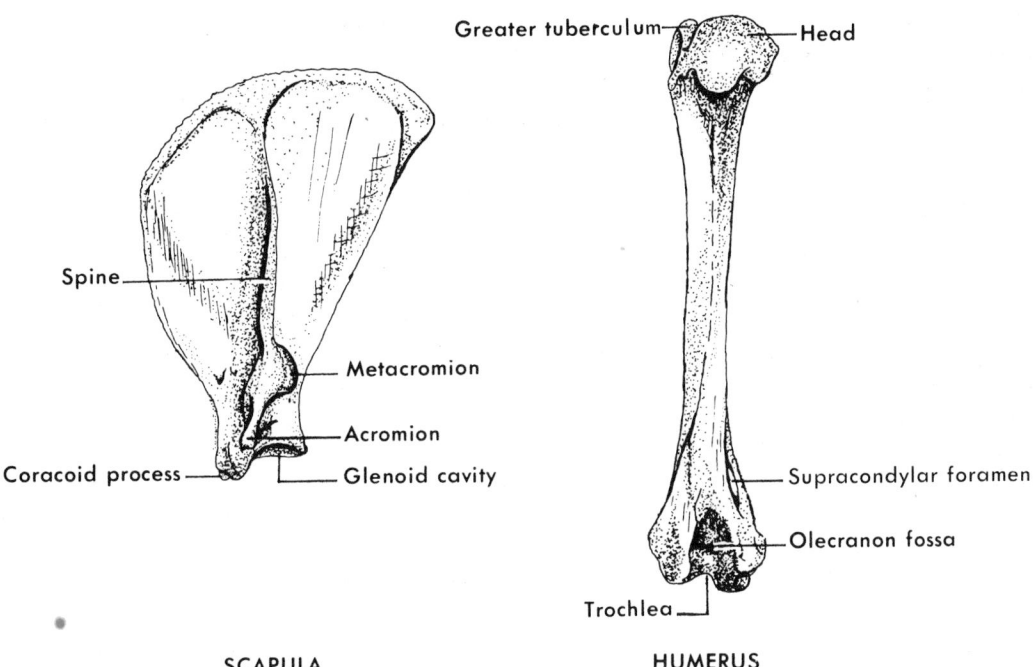

Greater tuberculum — Head

Spine

Metacromion

Acromion

Coracoid process — Glenoid cavity

Supracondylar foramen

Olecranon fossa

Trochlea

SCAPULA   HUMERUS

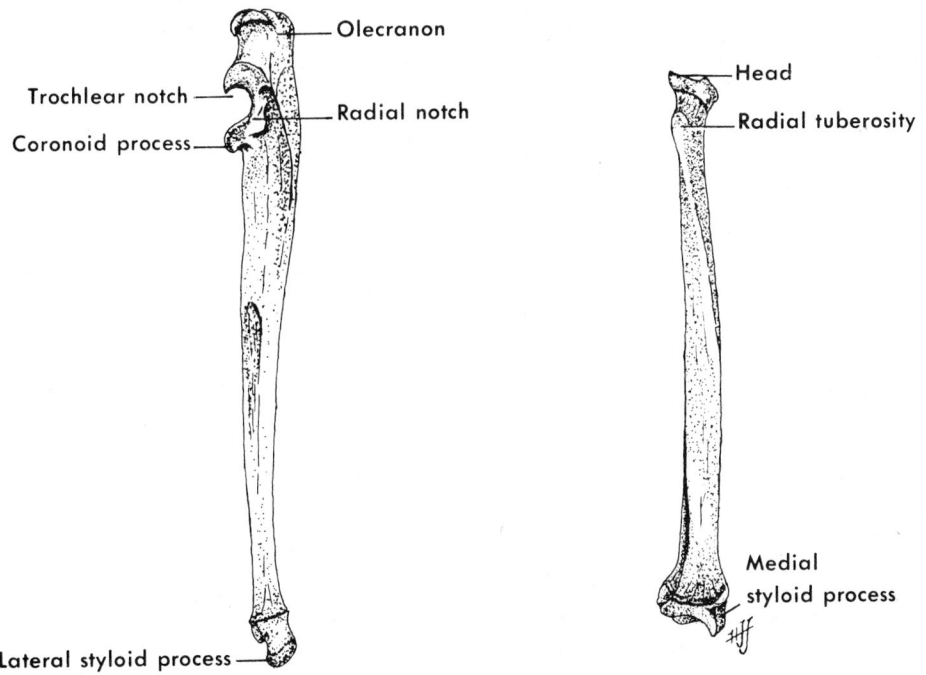

Olecranon

Trochlear notch

Radial notch

Coronoid process

Head

Radial tuberosity

Medial
styloid process

Lateral styloid process

ULNA   RADIUS

Figure 5–13. Pectoral bones from the left side of the cat.

acromion. That portion of the lateral surface of the scapula caudal to the spine is called the **infraspinous fossa**; that portion cranial to the spine, the **supraspinous fossa.** Since the spine represents the primitive anterior edge of the scapula, the portion of the scapula cranial to the spine is added during the course of evolution. The medial surface of the scapula is called the **subscapular fossa.** In human beings, from whom much of our anatomical terminology is derived, there is a prominent fossa here, but this surface is flat in the cat.

The bone of the brachium is the **humerus.** Its expanded proximal end has a smooth rounded **head** that articulates with the glenoid cavity, and processes for muscular attachment — a lateral **greater tuberculum** and a medial **lesser tuberculum.** An **intertubercular groove** for the long tendon of the biceps muscle lies between the two tubercles on the craniomedial surface of the humerus. The distal end of the bone is also expanded and bears a smooth articular surface known as a **condyle.** Although there appears to be but a single condyle, it can be divided into a medial pulley-shaped portion (the **trochlea**) for the ulna of the forearm (the bone which comes up behind the elbow), and a lateral rounded portion (the **capitulum**) for the radius. You may have to articulate the ulna and radius with the humerus to determine the extent of trochlea and capitulum. An **olecranon fossa,** for the olecranon of the ulna, is situated proximal to the trochlea. The enlargements to the sides of the articular surfaces are the **medial** and **lateral epicondyles,** respectively. A **supracondylar foramen** for a nerve and vessel is located above the medial epicondyle. This foramen is a primitive feature found in early reptiles but lost in most mammals, including human beings. That portion of the humerus, or of any long bone, lying between its extremities is its **body** or shaft. The faint ridges and rugosities upon it mark the attachments of certain muscles.

The **ulna** is the longer of the two forearm bones; a prominent, semilunar-shaped **trochlear notch,** for the articulation with the humerus, will be seen near its proximal end. The end of the ulna lying proximal to the notch is the **olecranon** or "funny bone." A **coronoid process** forms the distal border of the notch, and a **radial notch,** for the head of the radius, merges with the trochlear notch lateral to the coronoid process. The bone terminates distally in a **lateral styloid process** which articulates with the lateral surface of the wrist. Note that the ulna and radius articulate distally and that the ulna plays a relatively insignificant role in the formation of the wrist joint. In some mammals the distal half of the ulna is lost.

The other bone of the forearm is the **radius.** The articular surfaces on its **head** are of such a nature that the bone can rotate on the humerus and ulna. Slightly distal to the head is a prominent **radial tuberosity** for the attachment of the biceps muscle. The distal end of the radius is expanded, has articular surfaces for the ulna and carpus, and a short **medial styloid process.**

In primitive tetrapods the ulna extended straight down the lateral edge of the forearm and the radius straight down the medial edge. The articulation of the radius with the humerus, besides being medial, was slightly cranial to that of the ulna. The manus pointed forward. If the elbow rotated posteriorly only 90 degrees during the evolution of mammals, the manus would extend laterally. To bring the manus forward, the radius rotated at the elbow. The end result is that the radius continues to be the more medial bone at the wrist but is cranial and lateral to the ulna at the elbow. One can put the arm of a mounted human skeleton into the various positions to better visualize the changes.

Study the hand (**manus**) of the cat. Its first portion, the **carpus** or wrist (Fig. 5–14), consists of two rows of small **carpal bones.** The proximal row contains a large medial **scapholunar,** which represents the radiale, intermedium and a centrale fused; a smaller **triquetrum,** which represents the ulnare; and, on the lateral edge, a large, caudally projecting **pisiform.** The pisiform is one of many small sesamoid bones found in the appendages. They are not supporting elements, rather they are associated with the attachments of muscle tendons. Most are not named. The four elements of the distal row are, from medial to lateral, the **trapezium,** representing distal carpal 1; **trapezoid,** representing distal carpal 2; **capitate,** representing distal carpal 3; and **hamate,** representing distal carpal 4. Five **metacarpals** form the palm of the hand, and the free parts of the toes are composed of **phalanges.** The first toe is the most medial. Note that the number of phalanges has been reduced from that of primitive tetrapods. The terminal (**ungual**) phalanx of the cat-like carnivores is articulated in such a way that it, and the claw which it bears, can be either extended or pulled back over the penultimate phalanx.

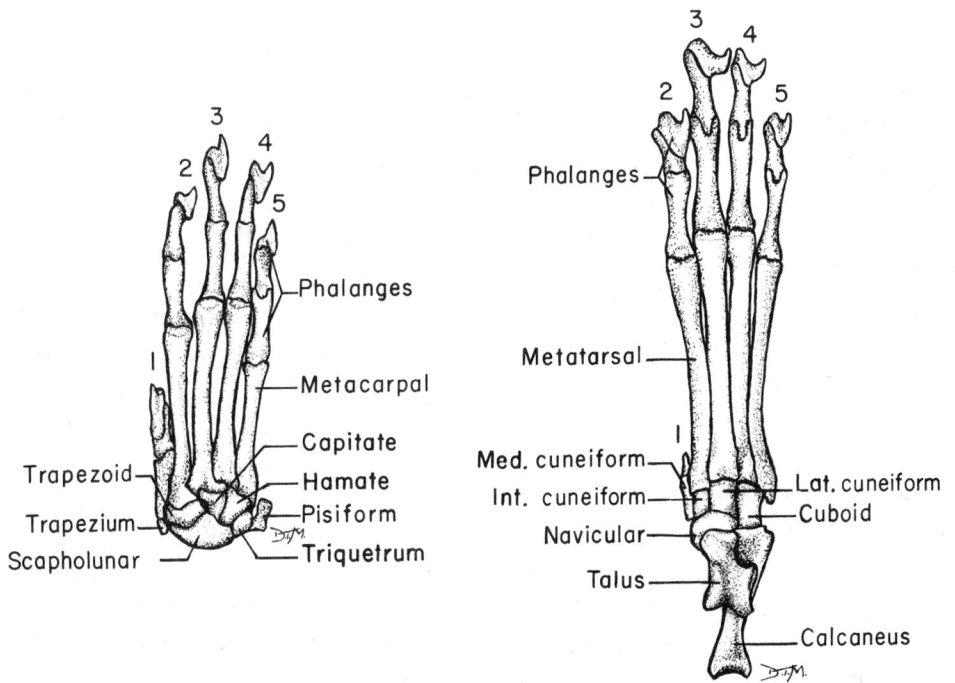

Figure 4–14. Dorsal views of the right manus (left drawing) and pes (right drawing) of the cat.

Pelvic Girdle and Appendage

The ilium, ischium and pubis on each side of the pelvic girdle have fused together in adult mammals to form an **os coxae,** but they can be seen clearly in young specimens (Fig. 5–5, *D*). The **ilium** extends dorsally from the **acetabulum,** or socket for the hip joint, to the sacrum. Its dorsalmost border is its **crest.** Notice that the ilium inclines cranially and unites with more sacral vertebrae than does the ilium in primitive tetrapods. The **ischium** surrounds all but the cranial portion, and some of the medial side, of the large opening (**obturator foramen**) in the ventral portion of the girdle. The enlarged caudo-lateral portion of the ischium is referred to as its **tuberosity.** The **pubis** lies cranial to the fenestra and completes its medial wall. Although the pubis enters the cranial portion of the acetabulum in most tetrapods, it is separated from it in the cat by a small **acetabular bone** (Fig. 5–5, *D*) of unknown phylogenetic significance. The pubes and ischia of opposite sides are united by **symphyses,** so that the pelvic girdle and sacrum form a complete ring, or **pelvic canal,** through which internal organs must pass to reach the anus and urogenital apertures.

Since the leg has rotated beneath the body, the thigh bone, or **femur,** articulates with the acetabulum by a **head** that projects from the medial side of the proximal end of the bone (Fig. 5–15). A large, lateral **greater trochanter** and a small, caudal **lesser trochanter** can also be seen on the proximal end. These processes are for muscle attachments. A depression called the **trochanteric fossa** is situated medial to the greater trochanter. The distal end of the femur has a smooth articular surface over which the **patella,** or kneecap (a sesamoid bone), glides. Posterior to this are smooth **lateral** and **medial condyles** for articulation with the tibia. The rough areas above each condyle are **epicondyles,** and the depression between the two condyles is the **intercondyloid fossa.**

The **tibia** is the larger and more medial of the two shank bones. Its proximal end has a pair of **condyles** for articulation with the femur, and an anterior, oblong bump (the **tuberosity**) for the attachment of the patellar ligament. Its shaft has a relatively sharp anterior margin which continues from the tuberosity. The distal end of the tibia, which articulates with the ankle, is prolonged on the medial side as a process called the **medial malleolus.** The **fibula** is a very slender bone. Notice that it does not enter the knee joint but does serve to strengthen the ankle laterally. Its distal end has a small, pulleylike process known as the **lateral malleolus.** Tendons pass posterior to this process.

Examine the foot (**pes**). The ankle, or **tarsus,** of the cat is typical of that of mammals for it consists of seven **tarsal bones** (Fig. 5–14). The one which articulates with the tibia and fibula is the **talus,** which appears to be homologous to the tibiale, intermedium and one centrale. The main joint in the mammal ankle is between this bone and the leg, rather than mesotarsal as in certain reptiles and birds. The large, posteriorly projecting heel bone is

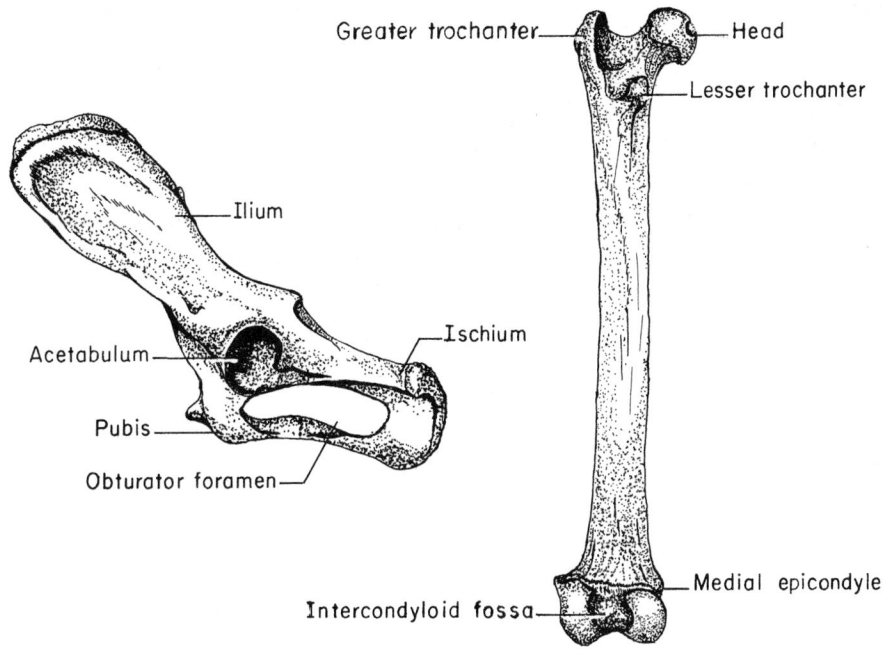

PELVIS   FEMUR

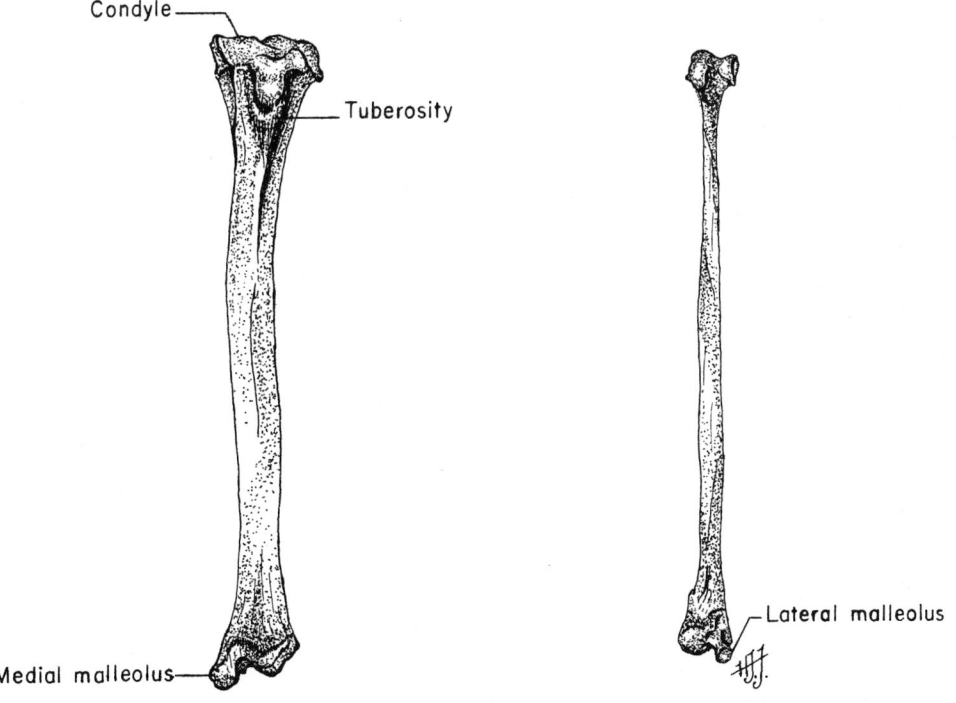

TIBIA   FIBULA

Figure 5–15. Pelvic bones from the left side of the cat.

called the **calcaneus** (fibulare). A **navicular** (centrale) lies just distal to the talus, and a row of four bones lies distal to the navicular and calcaneus. There are, from medial to lateral, the **medial cuneiform** (distal tarsal 1), **intermediate cuneiform** (distal tarsal 2), **lateral cuneiform** (distal tarsal 3), and **cuboid** (distal tarsal 4). Five elongated **metatarsals,** which occupy the sole of the foot, normally follow the tarsals, but in the cat the first toe is lost and its metatarsal is reduced to a small nubbin of bone articulated with the medial cuneiform. The bones in the free part of the digits are the **phalanges.** What is the phalangeal formula?

6

THE MUSCULAR SYSTEM

Groups of Muscles

Continuing on the general theme of the organ systems whose activities support and move the body, we will next consider the muscular system. In order to understand their evolution, the numerous muscles of vertebrates must be grouped in some way. Although the subdivisions of the muscular system according to histological structure (smooth vs. striated), or general type of innervation (involuntary vs. voluntary), are useful for certain types of work, the phylogenetically most natural method of subdivision is their mode of embryonic development in lower vertebrates (Fig. 6–1). Using this basis, there are two major groups of muscles that correlate with the two major divisions of the endoskeleton — the somatic and the visceral. **Somatic muscles** (parietal

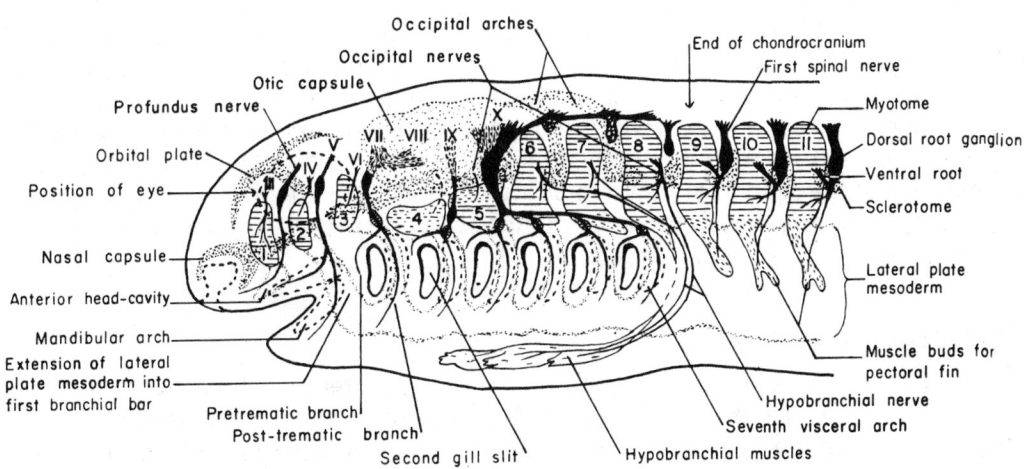

Figure 6–1. Diagram of the head of an embryo dogfish (*Scyllium*) to show the segmentation of the head and the origin of the cranial muscles. The myotomes, which give rise to the somatic muscles, are hatched and numbered with Arabic numerals. The lateral plate mesoderm, which gives rise to the visceral musculature, is shown in outline ventral to the posterior myotomes and the gill slits. This part of the mesoderm sends processes up into each of the interbranchial septa. The positions of the visceral arches are shown by broken lines. The nerves are shown in solid black and numbered with Roman numerals. The heavy stippling dorsally represents cartilage of the chondrocranium and vertebrae. (Slightly modified after Goodrich: On the development of the segments of the head of *Scyllium*, Quarterly Journal of Microscopical Science, Vol. 63.)

muscles) develop from embryonic myotomes,[11] and are associated with the "outer tube" of the body. They are the more conspicuous muscles of the body and the appendages. All are striated and voluntary. **Visceral muscles** arise from the embryonic lateral plate, or hypomere, and are associated with the "inner tube" of the body. They are the muscles of the gut, including the wall of the pharynx and its visceral arches. Although we generally think of these muscles as being smooth and involuntary, those of the visceral arches are striated and many are voluntary.

Both the somatic and visceral muscles can be further subdivided as shown in Table 4. The somatic muscles are broken down into axial and appendicular groups. The **axial muscles,** as the name implies, are the muscles in the axis of the body—the extrinsic muscles of the eyeball, the epibranchial and hypobranchial muscles which in fish lie, respectively, dorsal and ventral to the gill region, and the muscles of the trunk and tail. In gnathostomes, the hypobranchial and trunk muscles are further subdivided. The hypobranchial musculature is divided at the level of the hyoid into prehyoid and posthyoid groups; the trunk musculature into epaxial muscles lying dorsal and lateral to the vertebral column, and into more ventral hypaxial muscles. The innervation of these, and of other groups, is shown in Table 4.

The **appendicular muscles** develop embryologically from mesenchyme that lies within the limb bud. In the lower vertebrates, at least, this mesenchyme is derived early in development from myotomic buds. Thus the appendicular muscles are closely related to axial muscles, but are sufficiently distinct to warrant separate treatment. Since the mesenchyme in the limb bud becomes divided into dorsal and ventral premuscular masses before giving rise to the muscles, the appendicular muscles may be further subdivided into dorsal and ventral groups. Appendicular muscles lie within the appendage, or on the girdle, or may grow from the appendage back onto the trunk. The term, as here used, does not include those muscles that start their development in some other part of the body and then grow over to the girdle.

The visceral muscles are subdivided into two categories: (1) the **branchiomeric,** or branchial, muscles primitively associated with the visceral skeleton, and (2) the remaining muscles of the gut tube and associated structures. Only the branchiomeric group of visceral muscles will be considered. They, in turn, are grouped according to the arch with which they become associated—mandibular muscles, hyoid muscles, muscles of the third arch, etc. Branchiomeric muscles are innervated by the visceral motor fibers in the cranial nerves associated with the various visceral arches. The trigeminal nerve (V) is the nerve of the first arch and its muscles; the facial (VII) supplies the muscles of the second arch; the glossopharyngeal (IX), the third; and the vagus (X) and spinal accessory (XI), a derivative of the vagus present only in amniotes, supply the musculature of the remaining arches.

Muscle Terminology

In describing the individual muscles, a number of terms will be used, and these may appropriately be defined at this time. The ends of muscles are attached to skeletal elements, or to connective tissue septa, and the intermediate part of the muscle (its belly) is free. One end of a muscle tends to be fairly stationary during contraction, and the other end moves the bone to which it is attached as the muscle shortens. The more stable end is the **origin** of the muscle; the opposite end is the **insertion.** In the case of limb muscles, the origin is proximal and the insertion distal. A muscle may have two or more points of origin, or two or more points of insertion. If the multiple points of attachment are segmentally arranged, they are called **slips.** The multiple origins of a muscle which are not segmented are sometimes called **heads.**

---

[11]See footnote 2, p. 7.

The attachment of a muscle to a bone is by its investing connective tissue, not by the actual muscle fibers. If the muscle fibers come very close to the bone, and the muscle appears to attach to it, we speak of a "fleshy" attachment; if a narrow band of connective tissue extends from the muscle to the bone, we speak of a **tendon;** if the muscle attaches by a broad, thin sheet of connective tissue, we speak of an **aponeurosis.**

The connective tissue of a muscle, besides implementing the attachment, forms a covering about the muscle and spreads into the muscle, where it surrounds bundles of muscle fibers and even invests the individual muscle cells. The connective tissue covering of a muscle is a part of the **deep fascia**—a dense layer of connective tissue that forms a sheath for the individual muscles and also may hold groups of muscles together. Deep fascia is to be distinguished from **superficial fascia,** which is the layer of loose connective tissue beneath the skin. Superficial fascia generally contains fat; deep fascia does not.

Sometimes a transverse septum of connective tissue, called a **tendinous intersection,** is found in the middle of a muscle. Such an intersection often represents either a persistent myoseptum or the point at which two originally separate muscles have united.

Action of Muscles

Most muscles act in antagonistic groups; that is, the action of one muscle, or group of muscles, is offset by an antagonistic muscle or group. Various sets of terms describe these antagonistic actions. Unfortunately some of these have been used in different ways by different authors. The usage I shall follow is one proposed for quadruped mammals by Gray (1944) and slightly modified by Barclay (1946) to apply to all tetrapods except human beings. Barclay defines **protraction** and **retraction** of the humerus or femur as "movements causing the distal ends of these bones to move respectively forwards and backwards longitudinally." **Adduction** and **abduction** are defined as moving a segment respectively toward or away from a point of reference; for limb movements the reference is the midventral line of the body. These terms can also be applied to movements of the scapula. The terms flexion and extension are here limited to movements of the distal parts of the appendage and certain movements of the trunk. **Flexion** is the movement of a distal segment of the limb toward the next proximal segment, as in the approximation of the antebrachium and brachium, or the hand and antebrachium. Flexion also describes the bending of the head or trunk toward the ventral surface. **Extension** is movement in the opposite directions. Lateral bending of the trunk is called lateral flexion. The terms flexion and extension have also been used to describe movements of the entire limb at the shoulder and hip, but their usage at the shoulder has been inconsistent. Some authors equate protraction and flexion, but others, especially human anatomists, use the term extension for a forward movement of the arm. It seems best to drop the terms flexion and extension here and use protraction and retraction. **Rotation** can be illustrated by the movement of the radius on the ulna, or the axis on the atlas. In rotation of the radius on the ulna, special terms are often used. Rotation of the forearm to a position in which the palm of the hand faces ventrally, or toward the ground, is called **pronation;** the opposite rotation, which brings the palm up, is **supination.**

Muscles rarely contract alone, but usually contract in groups. The action of one muscle may have a synergistic effect upon another and modify its direction or strength of pull.

Names of Muscles

Muscle names are often confusing to the student, but they can be very helpful, for they describe one or more distinguishing features of a muscle: its shape (trapezius);

(Text continued on page 116)

Table 4. Major Muscles of Vertebrates

Showing the major muscles of the vertebrates studied in the laboratory, and the groups to which they belong. The main pattern of innervation and the probable homologies are also shown.

	SQUALUS	NECTURUS	MAMMAL
		A. Somatic Muscles	
		I. AXIAL MUSCLES	
		1. Extrinsic Ocular Muscles	
From First Somite (Oculomotor Nerve)	Dorsal rectus / Ventral rectus / Medial rectus / Ventral oblique	Dorsal rectus / Ventral rectus / Medial rectus / Ventral oblique	Levator palpebrae superioris / Dorsal rectus / Ventral rectus / Medial rectus / Ventral oblique
Second Somite (Trochlear Nerve)	Dorsal oblique	Dorsal oblique	Dorsal oblique
Third Somite (Abducens Nerve)	Lateral rectus	Lateral rectus / Retractor bulbi	Lateral rectus / Retractor bulbi
		2. Epibranchial Muscles (Dorsal Rami of Occipital and Anterior Spinal Nerves)	
	Epaxial portion of the myomeres in gill region	Anterior part of the dorsalis trunci	Anterior part of the epaxial muscles
		3. Hypobranchial Muscles (Ventral Rami of Spino-occipital or, in Amniotes, Hypoglossal Nerve and Cervical Plexus)	
Prehyoid Muscles	Coracomandibular	Genioglossus / Geniohyoid	Lingualis proprius / Genioglossus / Hyoglossus / Styloglossus / Geniohyoid
Posthyoid Muscles	Rectus cervicis (or Coracoarcual + Coracohyoid) / Coracobranchials	Rectus cervicis / Omoarcuals / Pectori-scapularis	Sternohyoid / Sternothyroid / Thyrohyoid

4. Axial Muscles of the Trunk (Spinal Nerves)

Epaxial (Dorsal Rami)	Interspinalis	Interspinalis, Intertransversarii	
	Dorsalis trunci	Occipitals, Multifidus, Spinalis, Semispinalis	Transverso-spinalis
		Longissimus	Longissimus
		Splenius, Iliocostalis	Iliocostalis
Hypaxial (Ventral Rami)	Subvertebralis	Longus colli, Psoas minor, Quadratus lumborum, Omotransversarius	Subvertebral
	Levator scapulae (Opercularis), Thoraci-scapularis	Serratus ventralis (part), Serratus ventralis (part), Rhomboideus, Rhomboideus capitis, Serratus dorsalis	
	External oblique	Scalenus, Rectus thoracis, External oblique, External intercostals	Lateral
	Internal oblique	Internal oblique, Internal intercostals, Transversus abdominis, Transversus thoracis	
	Transversus	Diaphragm muscles	
	Rectus abdominis	Rectus abdominis	Ventral

II. APPENDICULAR MUSCLES (Ventral Rami of Spinal Nerves)

1. Pectoral Muscles

Dorsal Group	**Abductor (Extensor)**	Latissimus dorsi	Cutaneous trunci (part), Latissimus dorsi, Teres major
		Subcoracoscapularis	Subscapularis
		Scapular deltoid	Scapulodeltoid, Acromiodeltoid
		Procoraco-humeralis	Cleidobrachialis, Teres minor
		Triceps	Triceps brachii, Tensor fasciae antebrachii
		Forearm extensors	Forearm extensors

Table 4 continued on following page

Table 4. Major Muscles of Vertebrates (Continued)

SQUALUS	NECTURUS	MAMMAL
Ventral Group Adductor (Flexor)	Pectoralis	Cutaneous trunci (part) Pectoralis complex
	Supracoracoideus	Supraspinatus Infraspinatus
	Coracoradialis	Biceps brachii (part) Biceps brachii (part)
	Humeroante-brachialis	Brachialis
	Coracobrachialis	Coracobrachialis
	Forearm flexors	Forearm flexors
2. Pelvic Muscles		
Dorsal Group Abductor (Extensor)	Iliotibialis	Sartorius Iliacus Psoas major Pectineus Vasti?
	Puboischiofemoralis internus	Rectus femoris Gluteus superficialis Tensor fasciae latae?
	Ilioextensorius Iliofibularis	Gluteus medius Gluteus profundis
	Iliofemoralis	
	Shank extensors	Shank extensors
	Puboischiofemoralis externus (Adductor femoris. In Necturus this is not clearly separated from the preceding)	Obturator externus Quadratus femoris
	Pubotibialis Ischiofemoralis	Adductor brevis et longus
Ventral Group Adductor (Flexor)		Adductor magnus Obturator internus Gemelli
	Caudofemoralis	Crurococcygeus (absent in some mammals) Piriformis
	Puboischiotibialis	Gracilis
		Semimembranosus Semitendinosus
	Ischioflexorius	Biceps femoris? Abductor cruris caudalis? (absent in some mammals)
	Shank flexors	Shank flexors

B. *Visceral Muscles*

I. BRANCHIOMERIC MUSCLES

Mandibular Muscles (Trigeminal Nerve)

- Adductor mandibulae, Levator palatoquadrati, Spiracularis, Preorbitalis → Levator mandibulae (3 parts) → Masseter, Temporalis, Pterygoids, Tensor veli palati, Tensor tympani, Mylohyoid, Anterior digastric
- Intermandibularis → Intermandibularis

Hyoid Muscles (Facial Nerve)

- Levator hyomandibulae, Dorsal constrictor → Depressor mandibulae, Branchiohyoideus → Stapedius; Platysma and facial muscles (part)
- Ventral constrictor, Interhyoideus → Interhyoideus, Sphincter colli → Platysma and facial muscles (part); Posterior digastric, Stylohyoid

Branchiomeric Muscles of Remaining Arches (Glossopharyngeal, Vagus and, in Amniotes, Spinal Accessory)

- Cucullaris → Cucullaris, Levatores arcuum → Trapezius complex, Sternocleidomastoid complex
- Interarcuals
- Branchial adductors → Dilatator laryngis
- Superficial constrictors and Interbranchials → Subarcuals, Transversi ventrales, Depressores arcuum → Intrinsic muscles of the larynx and certain muscles of the pharynx

location (intermandibularis); location and shape (dorsal rectus); action (adductor mandibulae); attachments in certain species (coracohyoid).

The Study of Muscles

As methods of body support changed during the transition from fish to tetrapod, and movements of the body and its parts became more complex during the evolution of vertebrates, the muscles became more numerous. It is not possible, in a course of this scope, to study all of them. One of two approaches may be taken. One could examine all the superficial muscles of the body, and omit the deeper ones; or one could examine certain muscle groups in more detail and omit other groups. I prefer the latter treatment for it gives a fairly complete evolutionary picture of at least certain groups. In this connection the details of the appendicular muscles of the distal portion of the limbs have been omitted except for those of mammals, as have those of the tail and perineum. Other groups are described with reasonable completeness. Insofar as possible, the muscles are described by natural groups. This will permit a further selection by the instructor of the muscles to be studied if time is short.

A few remarks concerning the dissection of muscles are appropriate. Insofar as possible, confine your dissection of the muscles to one side of the body and cut open the body, when you do, on the opposite side. Care must be exercised so as not to injure the underlying muscles when the skin is removed. It is best to try to peel the skin off by tearing underlying connective tissue with your fingers or with forceps. If you must cut with a scalpel, cut toward the underside of the skin, not toward the muscles. After the skin has been removed, the muscles must be carefully separated from each other. This involves cleaning off the overlying connective tissue with forceps until you can see the direction of the muscle fibers. Ordinarily, the fibers of one muscle are held together by a sheath of connective tissue, and all run in the same general direction between common attachments. The fibers of an adjacent muscle will be bound together by a different sheath and will have a different direction and attachments. This will give you a clue as to where to separate one from the other. Separate the muscles by picking away, or tearing, the connective tissue between them with forceps, watching the fiber direction as you do so. Do not try to cut muscles apart. If the muscles separate as units, you are doing it correctly; but if you are exposing small bundles of muscle fibers, you are probably separating the parts of a single muscle. It is best to expose and separate the muscles of a given region before attempting to identify them. It will be necessary sometimes to cut through a superficial muscle to expose deeper ones. In such cases, you usually should cut across the belly of a muscle at right angles to its fibers, and turn back (reflect) its ends, rather than detach its origin or insertion. The dissection will be more meaningful if you have a skeleton before you on which to visualize the points of attachment of the muscles.

FISHES

Although advanced in some respects when compared with such agnathous forms as *Petromyzon,* the muscles of *Squalus* are a good example of the condition of the musculature in primitive vertebrates. The natural groups of muscles can easily be recognized and studied in the adult, for the groups have not lost their identity, as they have to some extent in higher vertebrates, through the migration of muscles.

Axial Muscles

(A) TYPICAL MYOMERES OF THE TRUNK AND TAIL

The bulk of the musculature of fishes belongs to the axial group of somatic muscles, and the most conspicuous of these are the muscles of the trunk and tail. Remove a wide strip of skin from the posterior portion of the tail, and another strip from the front of the trunk between the pectoral and anterior dorsal fins (Fig. 6-2). These strips should extend from the middorsal to the midventral lines of the body. Try not to injure the underlying muscles when taking off the skin.

Notice that the trunk musculature consists of muscle segments, or **myomeres,** which develop from the embryonic myotomes, and that the segments are separated from each other by connective tissue septa, the **myosepta.** The myomeres are bent in a complex zigzag fashion and each is divided into dorsal and ventral portions by a longitudinal connective tissue septum (the **horizontal skeletogenous septum**), which lies deep to the lateral line. The dorsal portions of the myomeres constitute the **epaxial musculature;** the ventral, the **hypaxial musculature.** This division of the trunk musculature is found in all gnathostomes, but is absent in the Agnatha. In addition to their divisions into epaxial and hypaxial masses, the myomeres can be divided, at the apexes of the zigzags, into **longitudinal bundles.** The muscle fibers within the two, somewhat darker bundles adjacent to the horizontal septum, extend longitudinally, but those in the others are somewhat oblique.

Each myomere is a complex entity, as can be seen in Figure 6-2. The internal part of each V-like fold of a myomere is cone shaped and extends further cranially or caudally than the fold does at the body surface. An analogous arrangement can be made by cutting a V-shaped notch out of the top half of several conical paper cups, and then fitting them together. The cone-within-a-cone nature of the folds can be seen if you make a cross section of the tail. In such a section, the overlapping V-shaped folds of several myomeres appear as a series of nearly concentric rings (Fig. 6-6, p. 000). As described on page 000, the relationships of the myomeres to the vertebral column, and to the skeletogenous septa, also show in this view. In the abdominal region, the ventral band of connective tissue which separates the myomeres of opposite sides is called the **linea alba.**

The myomeres are the main locomotor organs of a fish, for they cause the lateral undulations of the trunk and tail whereby fish swim. Metachronal contractions are initiated anteriorly and pass toward the tail, affecting first one side of the body and then the other. These waves increase in amplitude as they move caudad. The large caudal fin, in addition to giving a lateral thrust, dampens the lateral displacement of the body and reduces the sinuosity of the fish's course through the water. As the body moves from side to side, its inclination to the vertical plane changes so that the thrusts it delivers against the water are directed downward and backward, not just laterally. The effect is very similar to the thrust given by a skulling oar.

How contraction of the myomeres brings about flexure of the body has been considered by Willemse (1959). As he points out, the fibers do not act directly on the

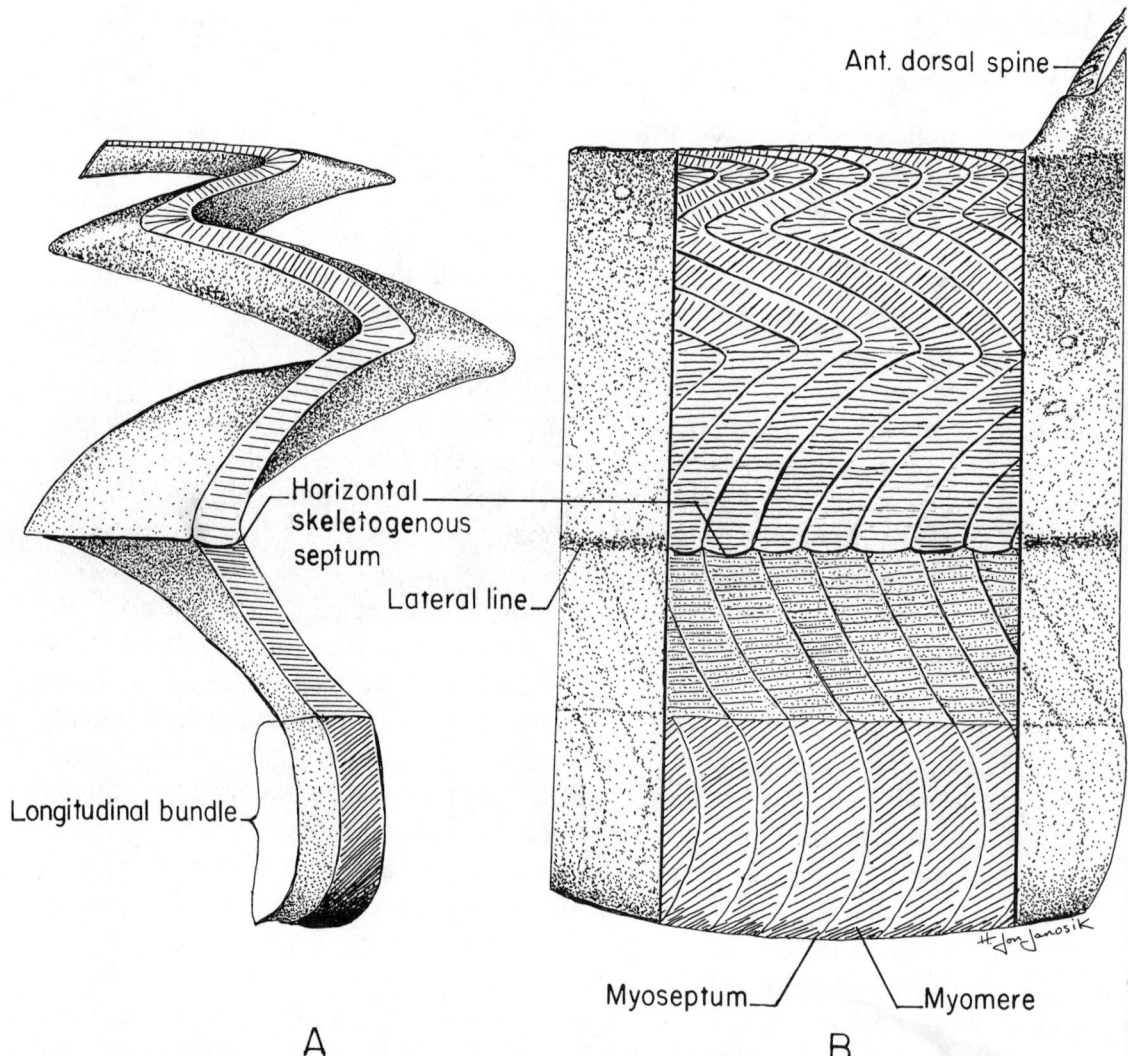

Figure 6-2. Trunk muscles of the dogfish. *A.* Diagram of a lateral view of an individual myomere; cranial is toward the left; *B,* lateral view of a group of myomeres. (*A* modified after Coles.)

vertebrae by way of the myosepta, as many investigators had thought. Rather, the contraction of the myomeres on one side of the body, being resisted by the incompressible vertebral column, simply causes a bending of the body to this side. The situation is analogous to the bending, upon cooling, of a strip composed of two different metals lying side by side. The shortening of one metal at a faster rate will be resisted by the other, and the strip will bend.

As Nursall has summarized (1962), the complex folding of the myomeres increases their fore and aft extension. This allows relatively short fibers to exert their effect over a considerable distance. Most of the muscle fibers attach diagonally onto the myosepta; therefore, when they are activated they pull the myosepta closer together (the myomeres contract) but do so by assuming a more vertical angle with respect to the myosepta rather than by shortening themselves. The situation is analogous to an oblique parallelogram in which its two sides can be brought closer together by changing the angle and not the length of its ends. Myomere contraction occurs with a minimum of bulging; this is important in the fish, which must maintain a streamlined body surface.

The darker (red) fibers that comprise certain of the longitudinal bundles are particularly rich in myohemoglobin. Although they contract less rapidly than the white fibers, they do not fatigue as rapidly.

(B) EPIBRANCHIAL MUSCLES

The axial musculature is interrupted in the head by the gill region, but extends forward above and below the gills as the **epibranchial** and **hypobranchial** musculature, respectively. Remove a strip of skin dorsal to the gill region, and the epibranchial muscles can be seen as an extension of the epaxial musculature to the chondrocranium (Fig. 6-4). Notice that the segmentation persists in this region. Certain of the deeper parts of this musculature, which are not easily seen, attach onto the tops of the branchial arches.

(C) HYPOBRANCHIAL MUSCLES

To see the hypobranchial muscles, which represent a forward extension of the hypaxial musculature, remove the skin on the underside of the head from the pectoral region forward to the lower jaw. It will be necessary in this case to skin both sides of the body. A broad sheet of transverse muscle fibers lies caudal to the jaws. This is part of the branchiomeric musculature, and will be studied presently. At this time, cut through it near the midventral line, and reflect it. Be careful not to injure a narrow, midventral, longitudinal muscle that lies just beneath the transverse layer.

The narrow midventral muscle (embryologically paired) thus exposed is called the **coracomandibular** (Fig. 6-3). It arises from the jaw and inserts caudally onto the surface of other muscles. It represents the **prehyoid** portion of the hypobranchial musculature, but in fishes this muscle commonly extends caudal to the hyoid, often as far as the coracoid. Cut through the caudal attachment of the coracomandibular and pull it forward. The dark, or pink, mass beneath its front end is the **thyroid gland.**

The paired muscles now exposed belong to the **posthyoid** portion of the hypobranchial musculature. The entire complex, which arises on the pectoral girdle and inserts on the hyoid arch, may simply be called the **rectus cervicis.** But it is sometimes considered to be two pairs of muscles, for the fibers of the caudal half converge toward the midline, while those of the cranial half extend longitudinally to the hyoid arch. The caudal pair is called the **coracoarcuals;** the cranial, the **coracohyoids.**

Make a cut from the ventral edge of the last gill slit to the caudal end of the rectus cervicis, and then dissect forward, gradually separating the branchial region from the hypobranchial muscles. Be careful not to injure blood vessels unduly. This will expose a deep group of muscles that arise from the coracoid, from the wall of the pericardial cavity beneath the posterior part of the rectus cervicis, and from the dorsal surface of the rectus cervicis itself. These muscles insert onto the ventral ends of the branchial arches and are called the **coracobranchials** (Fig. 6-3). They can also be seen in sections of the head (Fig. 9-9, p. 251, and Fig. 10-10, p. 294).

All the hypobranchial muscles support the floor of the pharynx and help to open the mouth and expand the gill pouches.

(D) EXTRINSIC MUSCLES OF THE EYE

The segments rostral to the epibranchial muscles disappear during embryonic development except for the most rostral three, and these give rise to the extrinsic muscles of the eye. These muscles extend from the wall of the orbit to the surface of the eyeball, and are responsible for the movements of the eyeball. They will be considered when the eye is studied (p. 173).

Appendicular Muscles

(A) MUSCLES OF THE PECTORAL FIN

The appendicular group of somatic muscles is small and simple in fishes, for the paired fins are used simply to increase stability and maneuverability,

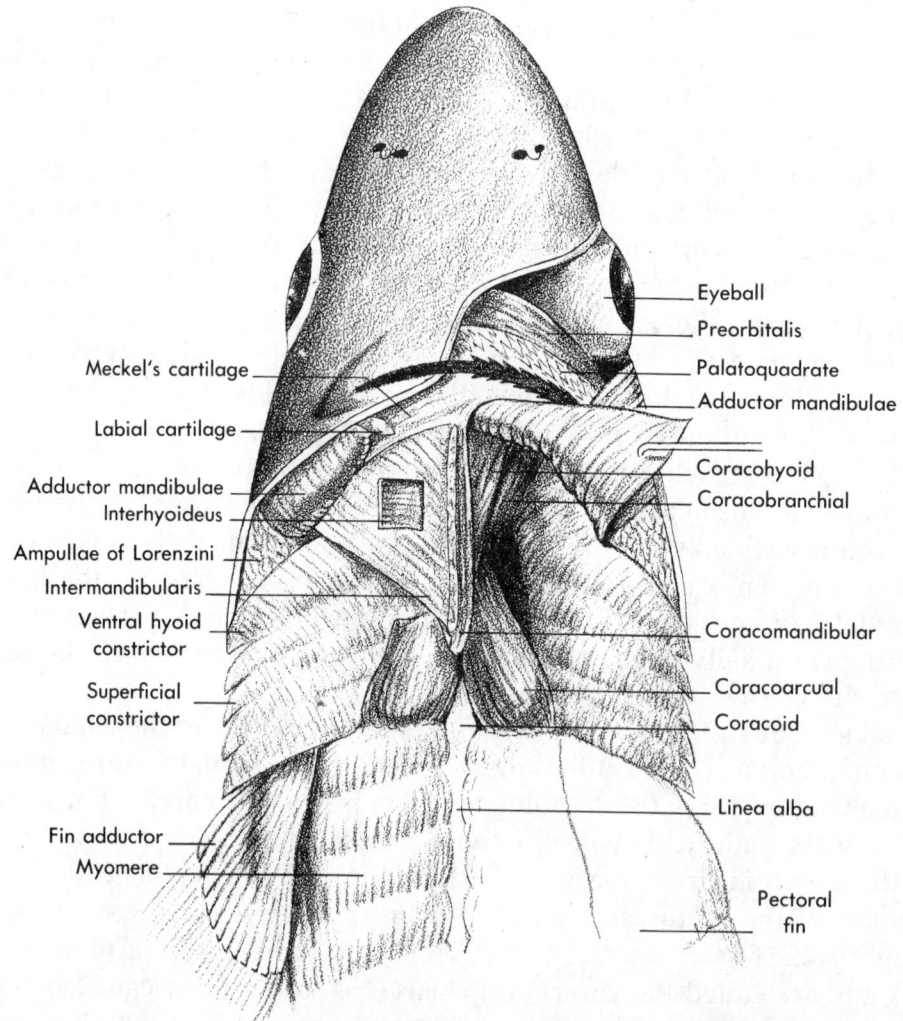

Meckel's cartilage

Labial cartilage

Adductor mandibulae
Interhyoideus

Ampullae of Lorenzini

Intermandibularis

Ventral hyoid
constrictor

Superficial
constrictor

Fin adductor
Myomere

Eyeball
Preorbitalis
Palatoquadrate
Adductor mandibulae
Coracohyoid
Coracobranchial

Coracomandibular

Coracoarcual
Coracoid

Linea alba

Pectoral
fin

Figure 6–3. Ventral view of the head muscles of *Squalus*.

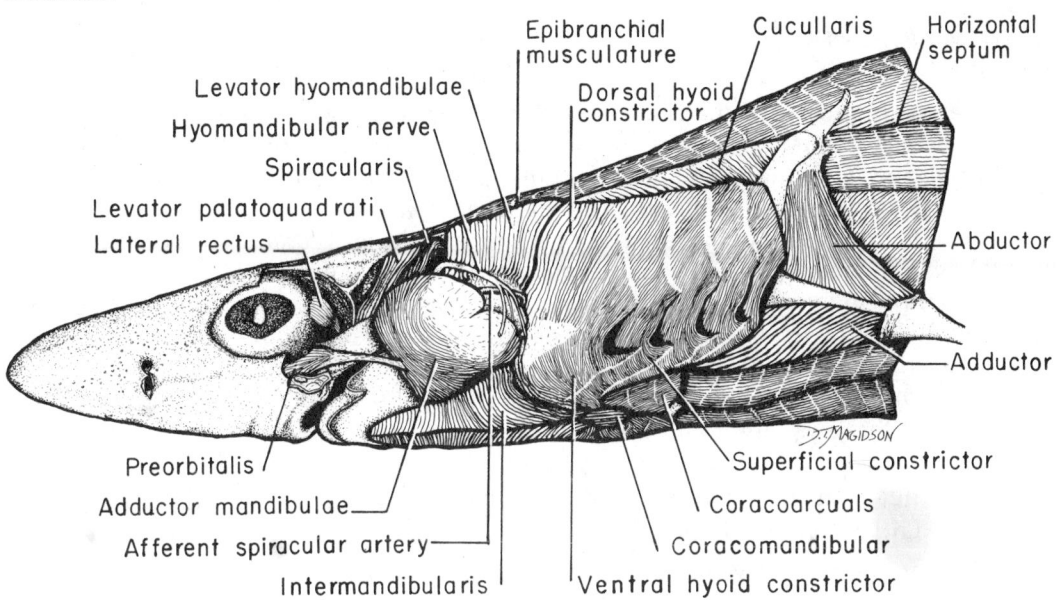

Figure 6–4. Lateroventral view of the muscles of the pectoral appendage and head of *Squalus*.

not for propulsive thrusts. Skin the base of the pectoral fin on its dorsal, ventral and cranial surfaces. The appendicular muscles consist of a single dorsal and a single ventral mass, each of which arises from the pectoral girdle and inserts by slips onto either the dorsal or ventral surface of the pterygiophores of the fin as far distal as the radial cartilages. The dorsal mass (**abductor** or extensor) elevates the fin and pulls it caudally (Fig. 6–4). The ventral mass (**adductor** or flexor), which also spreads onto the cranial surface of the fin, depresses the fin and pulls it forward.

(B) MUSCLES OF THE PELVIC FIN

Similarly, skin the base of the pelvic fin. In a female the muscles are uncomplicated by the presence of copulatory structures. The dorsal mass (**abductor**) arises from the surface of the posterior trunk myomeres, from the iliac process, and from the metapterygium. It inserts by slips onto the radial cartilages. Most of the ventral mass (**adductor**) is divided into proximal and distal portions. The former arises from the puboischiadic bar and inserts on the metapterygium; the latter arises from the metapterygium and inserts on the radials.

The appendicular muscles of the male are fundamentally the same, but portions of both the dorsal and ventral mass extend onto the clasper as distinct little muscles. Also, the adductor cannot be seen until one reflects a long muscular sac (the siphon, Fig. 11–3, p. 335) that lies between it and the skin. Do not remove the siphon completely, as it will be studied in connection with the reproductive system.

Branchiomeric Muscles

The basic pattern of the branchiomeric musculature is illustrated in Figure 6–5. As can be seen, most of the embryonic muscle plate of a given arch forms an **interbranchial** and **superficial constrictor,** but parts of the muscle plate separate as a **levator, interarcual,** and **adductor.** The interbranchials, superficial constrictors, interarcuals, and adductors pull the parts of a branchial arch together and compress the branchial pouches, thus aiding in expelling water. The branchial pouches are expanded and take in water through the elastic recoil of the branchial skeleton aided by the contraction of the levators and the coracobranchials (hypobranchial group) already described. As would be expected, this basic pattern is modified considerably in the region of the mandibular and hyoid arches since the muscles of these arches move and support the jaws, and control water flow through the spiracle.

It is necessary to remove the skin from the side of the gill region, jaws, spiracle, and from the back and underside of the eye. Cut close to the skin and do not remove any muscular tissue. In the spiracular region, remove the skin, but do not otherwise injure the anterior surface of the spiracular valve (p. 00). After the skin is removed, carefully clean the area by cutting and picking away connective tissue. Be particularly meticulous in the vicinity of the jaws, spiracle, and eye.

Study the muscles, beginning with those of the mandibular arch. The large mass of muscle at the angle of the jaws is the **adductor mandibulae** (Fig. 6–4). It arises from the posterior part of the palatoquadrate (note the large adductor mandibulae process in this region on a skeleton) and inserts on the mandibular cartilage. Dorsal to this, and in front of the spiracle, is another muscular mass that arises from the side of the otic capsule and inserts on the palatoquadrate. Careful dissection reveals that this mass is divided into an anterior **levator palatoquadrati** and a posterior **spiracularis.** The latter is on the cranial wall of the spiracular valve. Lift the eye, and a **preorbitalis** will be seen ventral to it. This muscle arises from the underside of the chondrocranium and inserts by a tendon on the mandibular cartilage. The last of the mandibular muscles is the ventral **intermandibularis** which was cut through during the dissection of the hypobranchial muscles. It originates from the mandibular cartilage and the fascia over the adductor mandibulae, and extends diagonally caudally and laterally to insert on a midventral tendinous intersection.

The intermandibularis overlies a deeper transverse sheet of muscle, the **interhyoideus.** Separate the two and note that the fibers of the interhyoideus are more nearly transverse and originate from the ceratohyal. Caudally, the interhyoideus is continuous with the **ventral hyoid constrictor.** Dorsally, the hyoid musculature is represented by a **dorsal hyoid constrictor,** which lies above and behind the angle of the jaw, and a more cranial **levator hyomandibulae.** The latter lies caudal to the spiracle and dorsal to the hyomandibular. It arises from the surface of the epibranchial musculature and otic capsule, and inserts primarily on the hyomandibular. A few fibers, however, extend onto the palatoquadrate.

The musculature of the remaining visceral arches (branchial arches), except for the last, is much the same. Note the four **superficial constrictors** that lie above and below the last four gill slits. The fibers of the superficial constrictors attach to connective tissue intersections and incline toward the gill slits. Open all the branchial pouches by cutting through the superficial constrictors along a line parallel to an intersection. Open them as wide as you can, without cutting through the margins of the internal gill slits (the openings communicating with the pharynx). The interbranchial septa, thus mobilized, support the gill lamellae. Remove the skin and gill lamellae from the anterior face of one septum, and it can be seen that the septum is composed largely of circularly arranged muscle fibers. This muscle, which is a deep part of the constrictor musculature, is called the **interbranchial** (Figs. 6–5 and 9–11). There are also only four of these, for the last branchial arch lacks an inter-branchial septum and associated muscles. Cut completely across the inter-branchial septum you have started to dissect in the frontal plane, the corresponding branchial arch, and into the pharynx. Examine the cut surface medial to the branchial arch, and you will see a cross section of the **branchial adductor,** a short muscle that extends from the epibranchial to the cerato-branchial. There are five of these, one for each branchial arch.

The **levators** of the five branchial arches have united to form a triangular muscle, the **cucullaris** (partly homologous to the trapezius of higher verte-brates, and sometimes called by that name), which lies dorsal to the superficial constrictors. Separate the cucullaris from the superficial constrictors. Ob-serve that it arises from the surface of the epibranchial musculature, and passes caudally to insert on the pectoral girdle and last branchial arch. Completely separate the cucullaris and the levator hyomandibulae from the

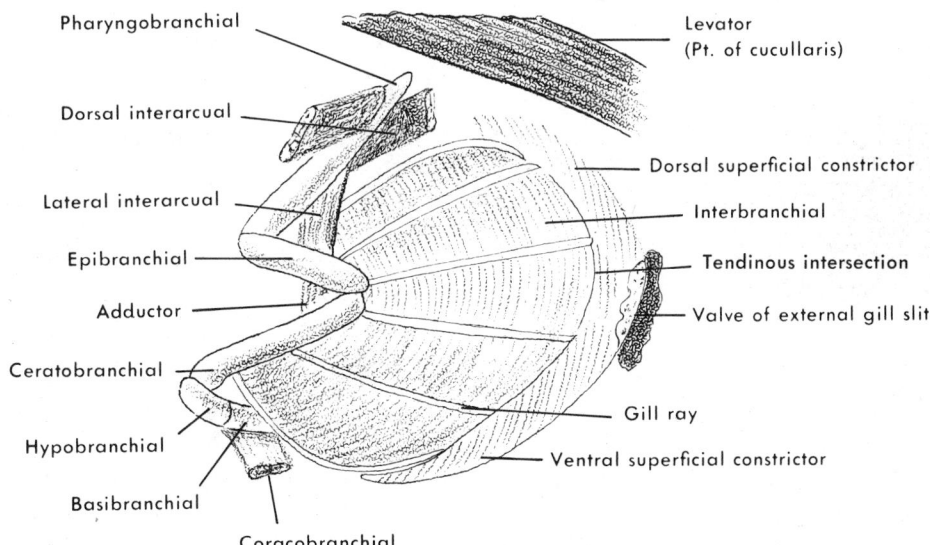

Pharyngobranchial

Levator
(Pt. of cucullaris)

Dorsal interarcual

Lateral interarcual

Dorsal superficial constrictor

Interbranchial

Epibranchial

Tendinous intersection

Adductor

Valve of external gill slit

Ceratobranchial

Hypobranchial

Gill ray

Basibranchial

Ventral superficial constrictor

Coracobranchial

Figure 6–5. A representative visceral arch of *Squalus* and associated muscles. All are branchiomeric except for the coracobranchial, which belongs to the hypobranchial group. Cranial is toward the left.

epibranchial muscles, and push the branchial area ventrally. The large cavity exposed is the anterior cardinal sinus, a part of the venous system. Clean out the coagulated blood, if necessary, and examine the tops of the branchial arches. Some of the epibranchial muscles may be cut away if the arches cannot be seen clearly. Small muscles will be seen extending from the epibranchials to the pharyngobranchials, and also between the pharyngobranchials. These are the **lateral** and **dorsal interarcuals,** respectively.

PRIMITIVE TETRAPODS

Many changes occur in the muscular system during the evolution from fish through tetrapods. These are correlated primarily with changes in the mode of support and locomotion, with the development of head movements independent of those of the trunk, with changes in the method of respiration, and with the development of a mobile tongue. As regards locomotion, the primitive dorsal and ventral appendicular muscle masses differentiate into a number of separate muscles, for the appendages become the organs of propulsive thrust, and their movements more complex. The appendicular muscles, however, retain their primitive dorsal and ventral groupings (Table 4). Concomitantly, the trunk and tail muscles become reduced and lose their segmentation, for lateral undulations play a less important role in locomotion. This is especially true for the hypaxial muscles. The epaxial muscles are retained, and differentiate further in connection with the movement of the vertebral column and head. In tetrapods that raise themselves off the ground, the epaxial musculature plays an important role in bracing and tying the elements of the vertebral girder together and in supporting the body against gravity. Some of the hypaxial musculature that persists forms thin layers on the ventrolateral portion of the body wall, some becomes associated with the ventral surface of the vertebral column (Fig. 6–6), and, especially in the pectoral region, some attaches onto the limb girdles.

With the shift from gill to pulmonary respiration, the more caudal visceral arches and most of their muscles are greatly reduced. But the cucullaris becomes associated with the pectoral girdle, and it is retained and elaborated in this connection. The cranial visceral arches contribute to the jaws, auditory ossicles, and hyoid apparatus. Their muscles have a comparable history to a large extent, except for the hyoid musculature. Most of the hyoid musculature moves to a superficial position, and forms, in mammals, the **platysma** and **facial muscles.** The hyoid apparatus and newly evolved tongue are moved instead by the hypobranchial muscles.

Necturus exemplifies the primitive tetrapod condition very well, for the changes just described are in an early stage. It must be remembered, however, that *Necturus* is a permanent larva that retains external gills, and hence retains also more of the branchial apparatus and muscles than would be found in a true adult tetrapod.

Axial Muscles

(A) Muscles of the Trunk

Confine your study of the muscles of the *Necturus* to one side of the body, preferably the side that has not been cut open to inject the circulatory system. Remove, from the middle of the trunk, a strip of skin that is about

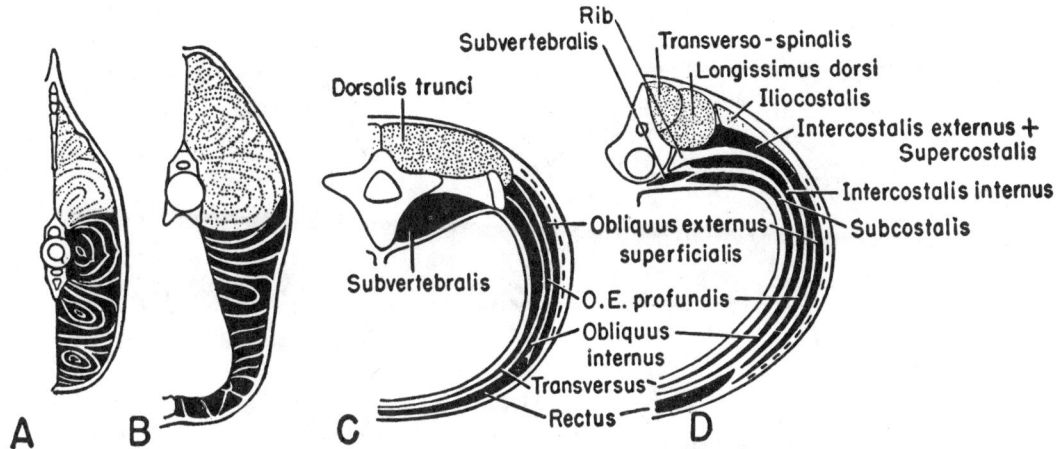

Figure 6–6. Diagrammatic cross sections to show the divisions of the trunk musculature in *A*, a shark tail; *B*, the shark trunk; *C*, a urodele; *D*, a lizard. The epaxial muscles are stippled; the hypaxial are black. In *D*, the dorsal labels of the hypaxial muscles pertain to a region in which ribs are present; the more ventral labels are those of the corresponding abdominal muscles. (From Romer, The Vertebral Body. Mainly after Nishi.)

two inches long and extends from the middorsal to the midventral line. Clean off any extraneous connective tissue, and note that the musculature still consists of **myomeres,** separated by **myosepta,** and divided into **epaxial** and **hypaxial** portions. The myomeres, however, are not as complexly folded as they are in fishes.

Most of the epaxial portions of the myomeres, including those in the epibranchial region, form a dorsal, longitudinal bundle of muscle called the **dorsalis trunci** (Figs. 6–6, *C*, and 6–7). Although the epaxial musculature is fairly uncomplicated in amphibians, the deeper portions of it are already beginning to become specialized. Make a deep transverse cut through the dorsalis trunci, take a firm grip with a pair of forceps on a bundle of muscle fibers beside the middorsal line, and pull the fibers caudad. This will expose the tops of the vertebral arches, and a series of short **interspinalis** muscles. Each interspinalis arises on the edge of a caudal vertebral zygapophysis of one vertebra, and inserts on the arch of the vertebra behind it.

Examine the hypaxial musculature and note that most of its superficial fibers incline posteriorly and ventrally. These constitute the **external oblique muscle.** The fibers next to the midventral line, however, extend longitudinally, and form an incipient **rectus abdominis.** Since *Necturus* is an incompletely metamorphosed species, the rectus abdominis is not as completely differentiated as it is in adult amphibians. Carefully cut into a myomere, reflecting the fibers of the external oblique as you do so. Presently, you will see a second layer, the **internal oblique,** whose fibers extend cranially and ventrally at nearly right angles to those of the external oblique. Uncover the internal oblique in several myomeres. Its ventral fibers form a part of the rectus. Now cut into the internal oblique in one segment and reflect its fibers. You will soon see a third layer, the **transversus,** whose fibers incline more vertically than those of overlying layers. Cut through this muscle and you reach the

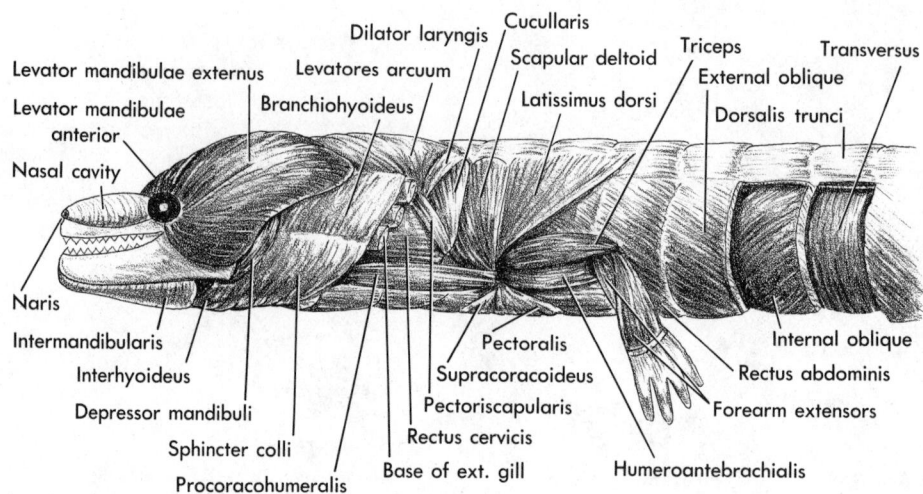

Figure 6–7. Lateral view of the trunk, pectoral, and head musculature of *Necturus*.

parietal peritoneum and body cavity. Another hypaxial muscle is the **sub-vertebralis,** which lies on the sides of the vertebral bodies ventral to the transverse processes (Fig. 6–6, *C*). It extends from vertebra to vertebra, attaching to the centra and transverse processes. It may be seen now by making a small incision through the body wall slightly to one side of the mid-ventral line, and pushing the viscera aside; or observation may be postponed until the abdominal viscera are studied. Several other hypaxial muscles attach onto the girdles and will be seen later.

(B) HYPOBRANCHIAL MUSCLES

To approach the hypobranchial muscles, remove the skin from the entire underside of the head and neck as far caudally as the pectoral region. Care-fully cut through the thin transverse sheet of branchiomeric muscles that lies under the head, and turn it to the sides. Cut slightly to one side of the mid-ventral line. A pair of midventral, longitudinal muscles will be seen extending posteriorly from the symphysis of the jaw. Their caudal end inserts on the posterior portion of the hyoid apparatus (basibranchial 2). These muscles, the **geniohyoids,** represent the bulk of the prehyoid portion of the hypo-branchial musculature (Fig. 6–8). In primitive tetrapods, deep fibers derived from the geniohyoid begin to spread into the newly evolved tongue. *Necturus* has one such deep layer, **genioglossus,** which can be seen most clearly after cutting through the caudal attachment of the geniohyoid and pulling it forward. The genioglossus appears as a thin layer of more or less longitudinal fibers that arise on the chin, extend caudally in the floor of the mouth, and insert on a transverse fold of mucous membrane at the base of the tongue. Its lateral portions are best developed.

Most of the posthyoid hypobranchial musculature is represented by a **rectus cervicis,** a wide band of muscle extending caudad in the neck from the hyoid apparatus toward the pectoral girdle. It is continuous caudally with the

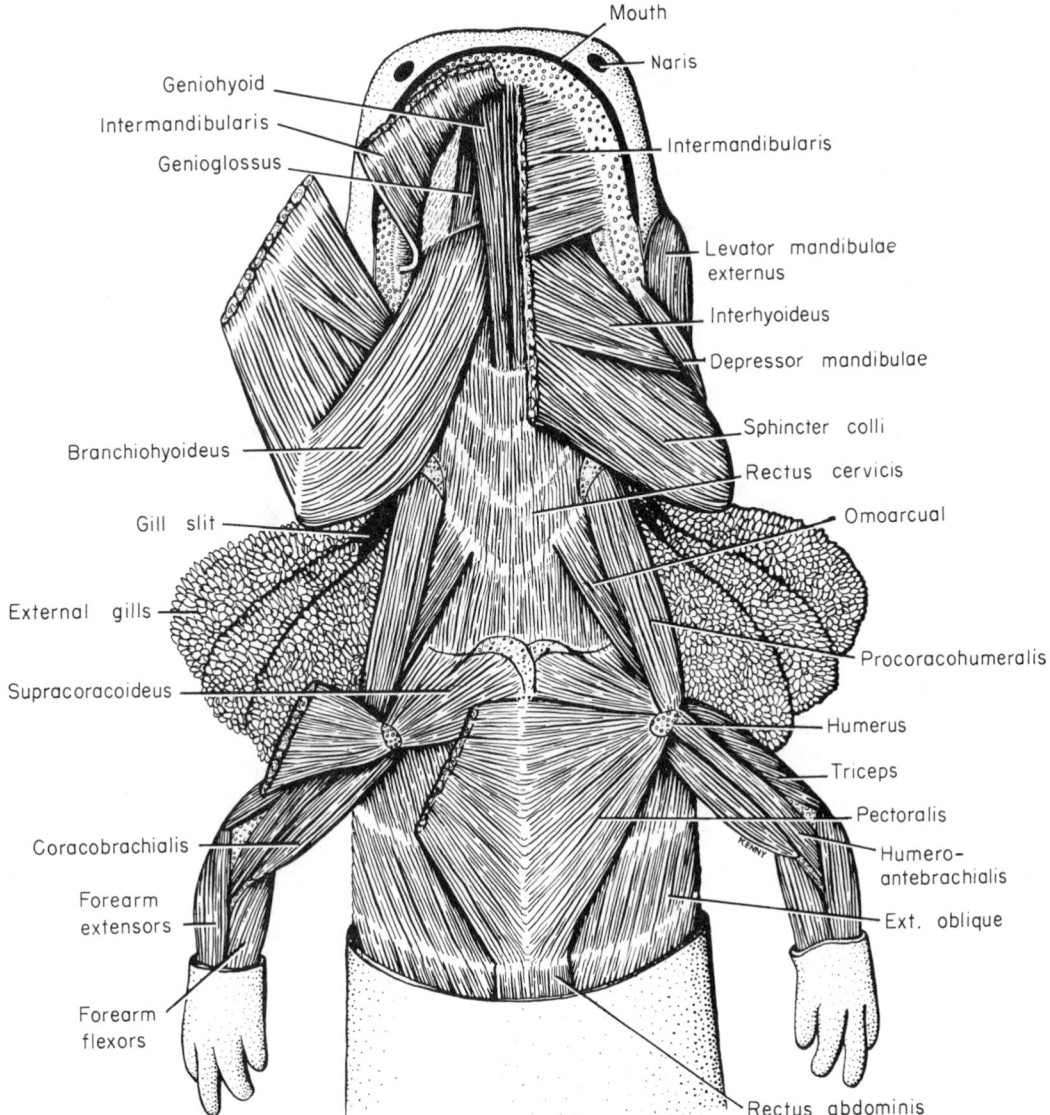

Figure 6–8. Ventral view of the head and pectoral muscles of *Necturus*.

rectus abdominis. Note that it retains traces of segmentation in the form of tendinous intersections. When the pectoral region is dissected, slips of the rectus cervicis that are called **omoarcuals** will be seen attaching onto the procoracoid process of the girdle. A **pectoriscapularis,** which is a derivative of the rectus cervicis, will also be seen at that time.

Appendicular Muscles

(A) PECTORAL MUSCLES

Cut off the external gills on one side and skin the pectoral appendage and the lateral and ventral surfaces of the trunk in its vicinity. A large fan-shaped muscle, the **pectoralis,** will be seen on the ventral surface (Fig. 6–8). It arises from the linea alba and hypaxial musculature, and inserts on the proximal end of the humerus. It lies caudal to the middle of the coracoid region. The ventral surface of the coracoid itself is covered primarily by a smaller fan-shaped muscle, the **supracoracoideus** that also inserts proximally on the humerus. The posterior part of the supracoracoideus lies beneath the anterior part of the pectoralis. By carefully picking away connective tissue and watching the direction of the muscle fibers, you will be able to separate the two. To see the supracoracoideus more clearly, cut across the pectoralis and reflect its ends. Still another fan-shaped muscle, the **coracoradialis** lies deep to the supracoracoideus, but do not look for it until you find its tendon of insertion passing along the brachium (p. 129). A **procoracohumeralis** will be found cranial and lateral to the supracoracoideus. It lies on the ventrolateral surface of the procoracoid process, from which it arises, and inserts proximally on the humerus. Some slips of the rectus cervicis (a hypobranchial muscle) attach onto the medial surface of the procoracoid process. These slips are sometimes called the **omoarcuals.**

Examine the lateral surface of the girdle (Fig. 6–7). The most caudal of a series of muscles that converge toward the shoulder joint is a large, fan-shaped **latissimus dorsi.** This arises from the surface of the dorsalis trunci. Next anterior is the **scapular deltoid,** which arises from the lateral surface of the scapula and suprascapular cartilage. Both of these insert on the proximal end of the humerus. The small muscle just cranial to the scapula is the **cucullaris,** a muscle belonging to the branchiomeric series, but inserting on the base of the scapula. The still narrower muscle cranial to the cucullaris is the **pectoriscapularis,** a hypobranchial muscle attaching onto the scapula.

Reflect the origin of the latissimus dorsi and examine the posterodorsal end of the scapula. A band of muscle will be seen extending from the hypaxial musculature forward to insert on the medial side of the scapula. This muscle, the **thoraciscapularis,** belongs to the hypaxial group. Another, and narrower, band of muscle may also be seen extending forward from the front of the dorsal end of the scapula. A better view of it will be obtained by cutting and reflecting the cucullaris. This muscle, which is also a hypaxial derivative, is called the **levator scapulae.** Its anterior end attaches to the occipital region of the skull. It is this muscle which becomes the opercularis in terrestrial urodeles (p. 187). Cut through the belly of the pectoralis, if this has not been done, and also through the latissimus dorsi. Pull the arm forward so you can get at the medial side of the girdle. (The sternal cartilages referred to on page 56 can now be seen.) Clean away blood vessels and nerves dorsal to the glenoid region, and you will find a small muscle, the **subcoracoscapularis,** arising from the caudal edge of the scapula and inserting on the humerus. It passes between two heads of the triceps.

The upper arm is covered for the most part by three muscles. A **humero-antebrachialis,** arising from the humerus and inserting on the proximal end of the radius, is located on the ventrolateral surface. A **coracobrachialis** (sometimes divided into **brevis** and **longus**) is located on the ventromedial surface. It arises from the coracoid and inserts on the shaft and distal end of the humerus. The tendon of the **coracoradialis,** described earlier, can be found between the coracobrachialis and humeroantebrachialis. After you find it, trace it medially to the belly of the muscle. A **triceps** covers the dorsal surface. Although the triceps arises from the girdle and humerus by three heads, only two of these are readily distinguishable—a medial **coracoid head** located on the medial side of the arm dorsal to the coracobrachialis, and a dorsolateral **humeral head** separated from the coracoid head by the subcoracoscapularis. All the heads converge and unite, forming a common tendon that passes over the elbow to insert on the proximal end of the ulna.

Clean the fascia from the surface of the forearm, and note that the muscles are aggregated into two groups—the **extensors** and the **flexors.** The extensors lie on the dorsal surface of the forearm, and arise from the lateral surface of the distal end of the humerus. The flexors have the opposite relationships. They lie on the ventral surface, and arise from the medial surface of the distal end of the humerus.

By studying Table 4, you should learn which of these muscles represent dorsal appendicular muscles and which ventral. In general, the ventral appendicular muscles protract and adduct the limb and flex its distal segments, while the dorsal muscles retract, abduct, and extend. Exceptions occur at the cranial end of the dorsal series where some of the muscles (the deltoid group) protract, and at the caudal end of the ventral series where some (the posterior fibers of the pectoralis) retract. In addition to moving the limbs, the appendicular muscles also serve as braces that can fix the bones at a joint and support the body. In general, the protractors are innervated by nerve fibers that leave the spinal cord more anteriorly than those to the retractors so that, as a wave of activity passes caudad along the spinal cord, the limb is first moved forward and toward the ground and then backward and away from the ground. In this way the limb of ancestral tetrapods probably aided locomotion as a "swimming wave" of contraction passed caudad along the myomeres.

(B) PELVIC MUSCLES

Skin one of the pelvic appendages and the adjacent trunk and tail. Also skin the cloacal region and remove, on one side, the large **cloacal gland** that lies between the skin and muscles. Clean the area, and note the ilium on the lateral surface dorsal to the base of the appendage. The hypaxial musculature attaches to it. Other parts of the girdle are covered by muscle.

Study the ventral surface of the girdle and thigh first. One of the most distinct muscles, and hence one that will serve as a good landmark, is the **pubotibialis**—a narrow band on the ventral surface of the thigh that arises from the lateral edge of the pubic cartilage and inserts on the proximal end of the tibia (Fig. 6–9). The ventral surface of the girdle, and of the thigh medial to the pubotibialis, is covered by a large triangular mass of muscle. The cranial two thirds of this represents a **puboischiofemoralis externus;** the caudal third, a **puboischiotibialis.** The line of separation between them can best be detected near the distal end of the thigh deep to the pubotibialis, and

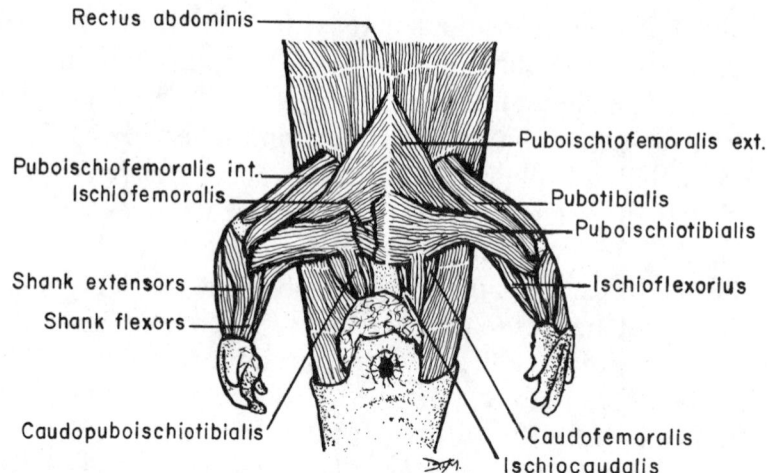

Figure 6–9. Ventral view of the pelvic muscles of *Necturus*.

then followed toward the midventral line of the body. Note that the front of the puboischiotibialis overlaps the back of the puboischiofemoralis externus. The puboischiofemoralis externus, in turn, overlaps the attachment of the rectus abdominis on the front of the pubic cartilage. These two muscles arise from the ventral surface of the girdle. The puboischiofemoralis externus inserts on the femur; the puboischiotibialis, on the proximal end of the tibia. An **ischioflexorius** lies along the medioventral surface of the thigh posterior to the puboischiotibialis, and at first appears to be a part of this muscle. The two can be separated most easily at the proximal end of the shank, and from the dorsal side. Try to follow the muscles medially. The ischioflexorius inserts on the fascia of the distal end of the shank. It arises from the posterior end of the ischium. It may be noticed that the muscle is divided into distal and proximal bellies separated by a short tendon.

Dissect in the area lateral to the cloaca, and you will find two longitudinal muscles. The narrower one beside the cloaca is the **ischiocaudalis.** It inserts on the caudal vertebrae and arises from the caudal end of the ischium. The more lateral muscle, the **caudopuboischiotibialis,** has a similar insertion, but arises from the puboischiotibialis. Both these muscles are hypaxial muscles that attach in the pelvic region. They serve to flex the tail. Dissect dorsal to the caudopuboischiotibialis and you will find a third longitudinal muscle. This is the **caudofemoralis,** an appendicular muscle arising from the caudal vertebrae and inserting on the proximal end of the femur.

Cut across the puboischiotibialis and reflect its ends. The small triangular-shaped muscle lying posterior to the puboischiofemoralis externus is the **ischiofemoralis.** It arises from the ischium and inserts on the proximal end of the femur. The slender muscle, arising from the underside of the femur near the insertion of the puboischiofemoralis externus and continuing across the underside of the knee joint to the fibula, is the **popliteus.** It is really a shank, rather than a thigh, muscle. (An **adductor femoris,** another deep muscle found cranial to the origin of the puboischiofemoralis externus in some urodeles, is absent as such in *Necturus*.)

The large muscle lying dorsal to the pubotibialis along the preaxial edge of the thigh is the **puboischiofemoralis internus.** Its origin is from the internal surface of the pubic cartilage and ischium. To get at it, one must cut through the pubic attachment of the rectus abdominis, and dissect deeply. The muscle inserts along most of the femur.

The dorsal surface of the thigh is occupied by a pair of muscles that arise from the base of the ilium, extend over the knee as a tendon, and insert on the tibia (Fig. 6–10). The cranial one is the **iliotibialis;** the caudal one is the **ilioextensorius.** Often these two appear as one muscle, for the separation between them may be obscure. Caudal to these, along the postaxial surface of the thigh, is a very slender **iliofibularis.** It, too, arises from the base of the ilium, but it diverges from the others and passes to insert on the fibula. Dissect away blood vessels and nerves posterior to the origin of the iliofibularis, and you will find a small, deep, triangular muscle that arises from the base of the ilium and inserts on the caudal edge of the femur. This is the **iliofemoralis.**

As with the forearm muscles, the muscles of the shank fall into extensor and flexor groups. The **extensors** lie on the dorsal surface of the shank; the **flexors,** on the ventral. The extensors arise for the most part from the dorsal surface of the distal end of the femur; the flexors, from the ventral.

Branchiomeric Muscles

Skin the underside of the head and the top and one side of the head. Be particularly careful above and below the stumps of the external gills. Note again the transverse sheet of muscle on the underside of the head that was cut and reflected during the study of the hyobranchial muscles. Approximately the rostral half of this represents an **intermandibularis,** a mandibular arch muscle; the caudal half an **interhyoideus,** a hyoid muscle (Fig. 6–8). Both insert on a median tendinous intersection, but they are distinct in their origins. The intermandibularis arises from the mandible; the interhyoideus, from the hyoid arch medial to the angle of the jaw and from the surface of a large muscle posterior to the angle of the jaw. The latter portion of the interhyoideus is sometimes called the **sphincter colli.**

The rest of the mandibular arch musculature is represented by a levator mandibulae complex that arises from the top and sides of the skull, inserts on the mandible, and serves to close the jaws (Fig. 6–7). A **levator mandibulae anterior** lies on the top of the head lateral to the middorsal line; a **levator mandibulae externus,** posterior to the eye.

The rest of the hyoid musculature is represented by the large muscle mass lying between the levator mandibulae complex and the gills. Close examination will reveal that this mass is divided into two muscles which arise in common from the first branchial arch in front of the gills. The more cranial one, the **depressor mandibulae,** inserts on the caudal end of the man-

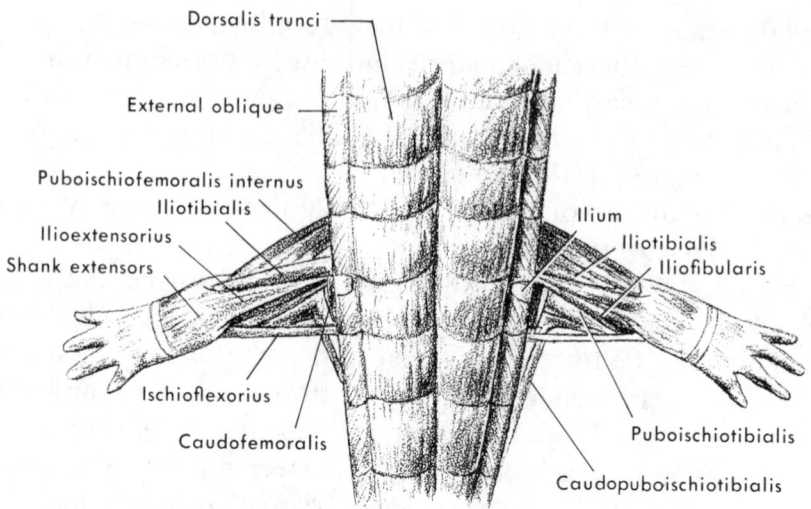

Figure 6–10. Dorsal view of the pelvic muscles of *Necturus*.

dible, and is the major muscle involved in opening the jaws. The larger and more caudal one, the **branchiohyoideus,** continues forward beneath the interhyoideus and intermandibularis to insert along the ventral portion of the hyoid arch (the ceratohyal).

The remaining branchiomeric muscles belong to the branchial arches, although some have acquired other attachments. A series of **levatores arcuum** will be found dorsal to the bases of the external gills. Caudal to these, and crossing the origin of the cucullaris, you will see a **dilatator laryngis.** The dilatator laryngis looks like a levator, but note that it curves in toward the base of the neck (on its way to the larynx) rather than attaching onto the gills. The **cucullaris,** observed during the dissection of the shoulder region, completes the dorsal branchial muscles. The ventral branchial muscles can be found by dissecting deep between the branchiohyoid and rectus cervicis. Pull the branchiohyoid forward, and the rectus cervicis caudally. A series of **subarcuals** will be seen—a longitudinal one extending from the ceratohyal to the first branchial arch; another longitudinal one extending from the first and second to the third branchial arch; and an oblique one medial to this and extending between the first and second branchial arches. The oblique fibers posterior to the subarcuals, and deep to the rectus cervicis, constitute the **transversi ventrales** (Fig. 10–12, p. 301). These fibers extend from the third branchial arch on one side to that on the other. The remaining ventral muscles are a series of three **depressores arcuum** that extend from the ventral portion of the three branchial arches to the bases of the three external gills. They are small, and may have been removed in skinning.

MAMMALS

The changes that were seen beginning in the muscles of primitive tetrapods continue in the evolution toward mammals. (Review the general remarks on tetrapod muscle evolution on page 124.) Many of the groups of muscles have lost their clear definition in adult mammals, for some of the muscles have migrated from their original positions. This makes it more difficult to study the muscles by natural groups, but it can still be done, if a few concessions are made to topography.

Skinning and Cutaneous Muscles

If you are studying a mink, the skin will have been removed, but it will be necessary to skin a cat or rabbit. Lay your specimen on its belly, and make a middorsal incision through the skin that extends from the back of the head to the base of the tail. Make additional incisions from this cut around the neck; around the tail, anus, and external genitals; down the lateral surface of each leg, and around the wrists and ankles. Skin is to be left for the time being on the head, tail and perineum, and feet.

Beginning on the back, gradually separate the skin from the underlying muscles by tearing through the superficial fascia with a pair of blunt forceps. As you separate the skin from the trunk, notice the fine, parallel, brown lines that adhere to its undersurface. They represent the **cutaneous trunci,** the largest of several cutaneous muscles that move the skin. The cutaneous trunci arises from the surface of certain appendicular muscles of the shoulder (latissimus dorsi and pectoralis), and from a midventral band of connective tissue (**linea alba**), fans out over most of the trunk, and inserts on the underside of the skin. It should be removed with the skin, except for that portion attached to the shoulder muscles posterior to the armpit. Several smaller cutaneous muscles, derived from the posterior musculature, become associated with the posterior part of the cutaneous trunci. They may not be noticed.

Much of the top and sides of the neck is covered by another cutaneous muscle, the **platysma,** which is derived from the hyoid musculature. As the platysma spreads over the face it breaks up into many, small cutaneous muscles associated with the lips, nose, eyes, and ears. They are collectively known as the **facial muscles.** You may see them later when the head is skinned.

As you continue to skin your specimen, you will come upon narrow tough cords passing to the skin. These are **cutaneous blood vessels** and **nerves,** and must be cut. Note that they tend to be segmentally arranged along the trunk. If your specimen is a pregnant or lactating female, the **mammary glands** will appear as a pair of large, longitudinal, glandular masses along the ventral side of the abdomen and thorax. They should be removed with the skin.

After the specimen is skinned, clean away the excess fat and superficial fascia on the side that is to be studied, but do not clean an area thoroughly until it is being studied. If your specimen is a male, be particularly careful in removing the wad of fat in the groin, for it contains on each side the proximal part of the **cremasteric pouch**—a part of the scrotum containing blood vessels and the sperm duct extending between the abdomen and scrotal skin. First find this pouch. It is large in the rabbit, but rather small in the cat and mink. Clean away connective tissue deep in the groin, or inguinal region, so as to find the actual boundary between the thigh and abdomen. In the rabbit, a shiny, white **inguinal ligament** will be seen in this region, extending from the pubis to the ilium.

Caudal Trunk Muscles

All the axial muscles cannot be studied at the same time, for many of them are located deep to the shoulder muscles. Those located on the trunk between the pectoral and pelvic appendages will be examined now, and the more cranial ones will be considered after the appendages have been studied (p. 160).

(A) HYPAXIAL MUSCLES

Continue to clean off the surface of the trunk between the pectoral and pelvic appendages. The wide sheet of tough, white fascia covering the lumbar region on the back is the **thoracolumbar fascia.** The wide sheet of muscle that runs cranially and ventrally from the anterior part of this fascia and disappears in the armpit is the latissimus dorsi (an appendicular muscle) (Fig. 6–12, p. 137). The large triangular muscle that covers the underside of the chest is the pectoralis (another appendicular muscle). The borders of the latissimus dorsi and pectoralis appear to run together posterior to the armpit. Separate the two in this region by removing the cutaneous trunci and carefully trace their edges forward. Remove the fat and loose connective tissue from beneath them. The hypaxial trunk muscles can now be studied.

As in other tetrapods (Fig. 6–6, C, D) the abdominal wall is composed of three layers of muscle, plus a paired longitudinal muscle along the midventral line. All serve to compress the abdomen. The **external oblique** forms the outermost layer (Fig. 6–11). This muscle arises by slips from the surface of a number of posterior ribs and from the thoracolumbar fascia. Part of its origin lies beneath the caudal edge of the latissimus dorsi. Its fibers then extend obliquely caudally and ventrally to insert by an aponeurosis along the length of the linea alba. In the rabbit, some of the fibers insert on the inguinal ligament. Using a sharp scalpel, or a razor blade, make a cut about 5 centimeters long through the external oblique at right angles to the direction of its fibers. Do not cut deeply, and as you cut watch for a deeper layer of muscle having a different fiber direction. When you reach the deeper layer, reflect part of the external oblique.

The **internal oblique** lies beneath the external oblique. It is most apparent high up on the side of the abdomen near its main origin from the

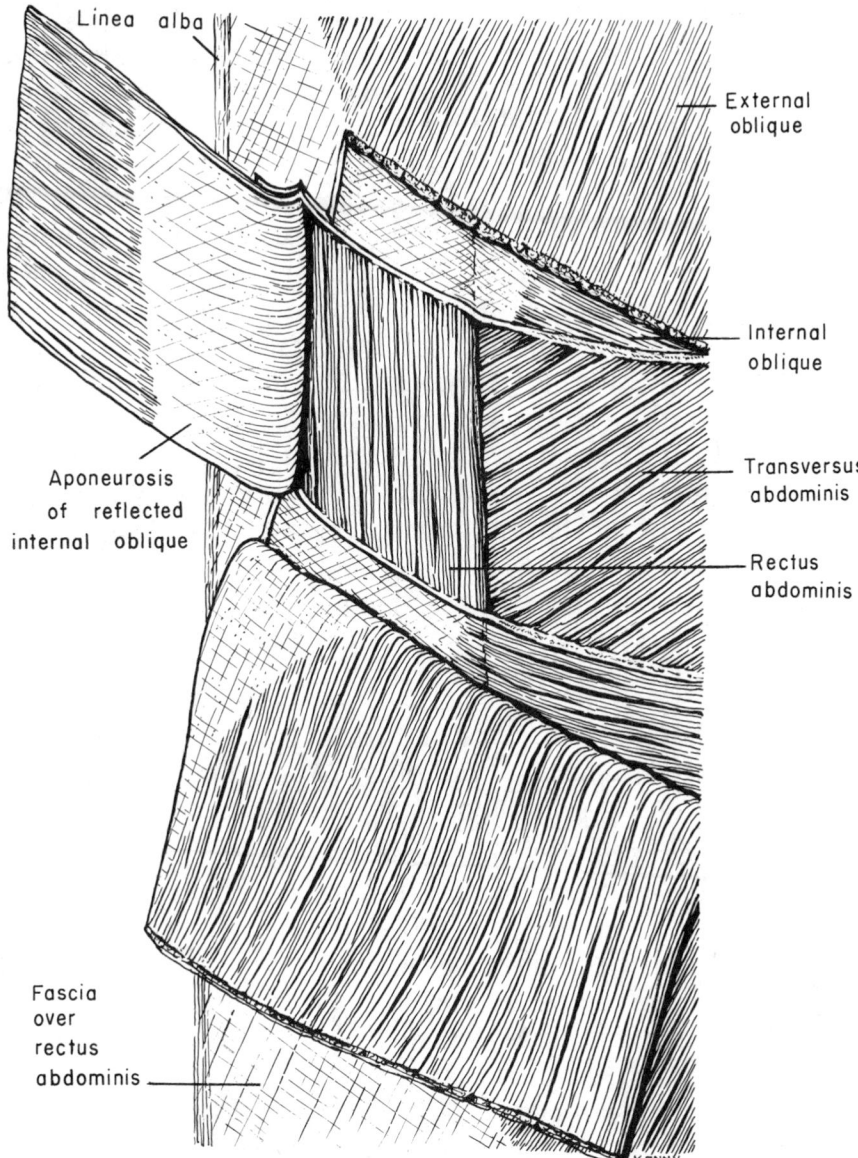

Linea alba

External oblique

Internal oblique

Transversus abdominis

Rectus abdominis

Aponeurosis of reflected internal oblique

Fascia over rectus abdominis

KENNY

Figure 6–11. Ventral view of the abdominal muscle layers on the left side of a cat. Part of the external oblique and internal oblique have been reflected.

thoracolumbar fascia. Its fibers extend ventrally and slightly cranially at right angles to the fibers of the external oblique and soon lead into a wide aponeurosis that inserts along the linea alba. Only the dorsal half of the muscle is fleshy. The ventral half is represented by an aponeurosis, but this may not be apparent at first, for one can see the third muscle layer, the transversus abdominis, through this aponeurosis.

To get at the **transversus abdominis,** make a longitudinal cut through the fleshy portion of the internal oblique, and reflect a part of this muscle. The transversus abdominis arises primarily from the medial surface of the more posterior ribs, and from the transverse processes of the lumbar vertebrae.

The latter portion of the origin lies in a furrow deep to the epaxial muscles. Note that this furrow has the same location as the horizontal skeletogenous septum of fishes. The fibers of the muscle extend ventrally, and slightly caudally, to insert along the linea alba by a narrow aponeurosis. Separate some of the fibers of the transversus, and you will expose the **parietal peritoneum** lining the abdominal cavity.

The reflection of the internal oblique also exposes a longitudinal band of muscle lying lateral to the midventral line. This is the **rectus abdominis.** It arises from the pubis and passes cranially to insert on the cranial costal cartilages and sternum. For much of its course, it lies between the aponeurosis of the internal oblique and the transversus. Transverse tendinous intersections can sometimes be seen in it.

In addition to the muscular layers of the abdominal wall, the posterior hypaxial musculature includes a subvertebral group (Fig. 6–6) that lies ventral to the lumbar and posterior thoracic vertebrae. This group, which includes the **quadratus lumborum** and **psoas minor,** is associated with certain pelvic muscles and is described in connection with the hind leg (p. 157).

(B) Epaxial Muscles

Lift up the thoracolumbar fascia with a pair of forceps, and make a longitudinal incision through it about one-half centimeter to one side of the middorsal line. Extend the incision from the latissimus dorsi to the sacral region, and reflect the superficial layer of the fascia. A deeper layer of this fascia will now be seen encasing the epaxial muscles. Make a longitudinal cut through it about one centimeter from the mid-dorsal line. The narrow band of muscle beside the spinous processes of the vertebrae is the **multifidus.** The wider lateral band of muscle represents a caudal union of the longissimus and iliocostalis. Together they constitute the **erector spinae.** In the cat and mink, a part of the fascia covering the caudal part of the erector spinae dips into it and subdivides it, but these subdivisions do not correspond with the subdivisions on the cranial part of the trunk. Ignore them until the cranial part of the epaxial musculature is studied.

Pectoral Muscles

Most of the muscles of the pectoral region are, of course, appendicular muscles, but a number of axial and several branchiomeric muscles have become associated with the girdle and appendage.

(A) Pectoralis Group

The large triangular muscle complex covering the chest is the **pectoralis.** It arises from the sternum and passes to insert primarily along the humerus. Its major actions are to pull the humerus toward the chest (adduction) and posteriorly (retraction), but it also helps in transferring body weight from the trunk to the pectoral girdle and appendage (Fig. 6–17, p. 144). Clean the surface of the muscle enough so that you can see the direction of its fibers.

A cleidobrachialis muscle covers the front of the shoulder cranial to the pectoral complex (Figs. 6–12 and 6–13). It will be considered later, but should be mobilized at this time because part of the pectoralis is located deep to it. Lift up the cleidobrachialis and clean away connective tissue from beneath it. The clavicle is imbedded on the underside of the medial portion of this muscle. Cut through the connective tissue that binds the clavicle to the manubrium so that you can push the cleidobrachialis forward.

The pectoralis of quadrupeds is divided into a pectoralis superficialis and a pectoralis profundus. These two parts are homologous to the pectoralis major and the pectoralis minor of human beings respectively, but the profundus is the larger muscle in quadrupeds. Each part may be further subdivided.

The **pectoralis superficialis** arises from approximately the cranial one third of the sternum. Its fibers extend more or less laterally to insert on the humerus (Figs. 6–12 and 6–13). The insertion in the cat (but not in the other

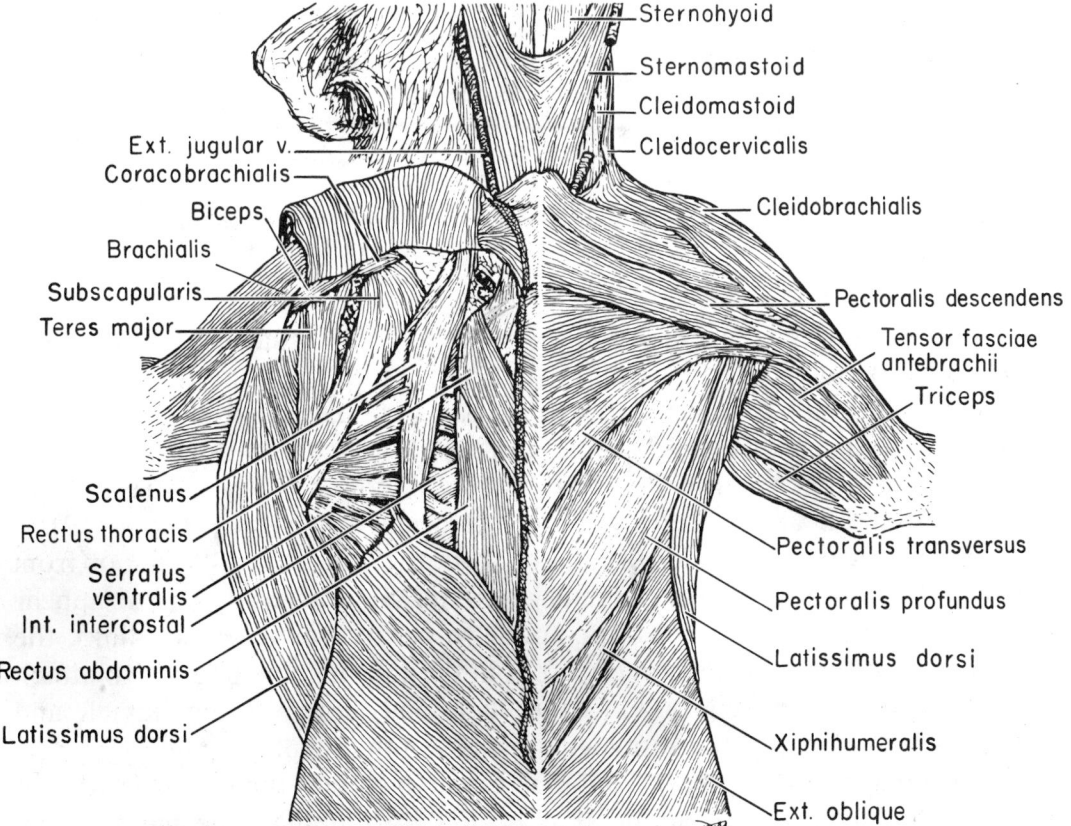

Figure 6–12. Ventral view of the muscles in the pectoral region of a cat. Superficial muscles are shown on the right side of the drawing. The pectoralis has been reflected on the left side in order to show certain deeper muscles.

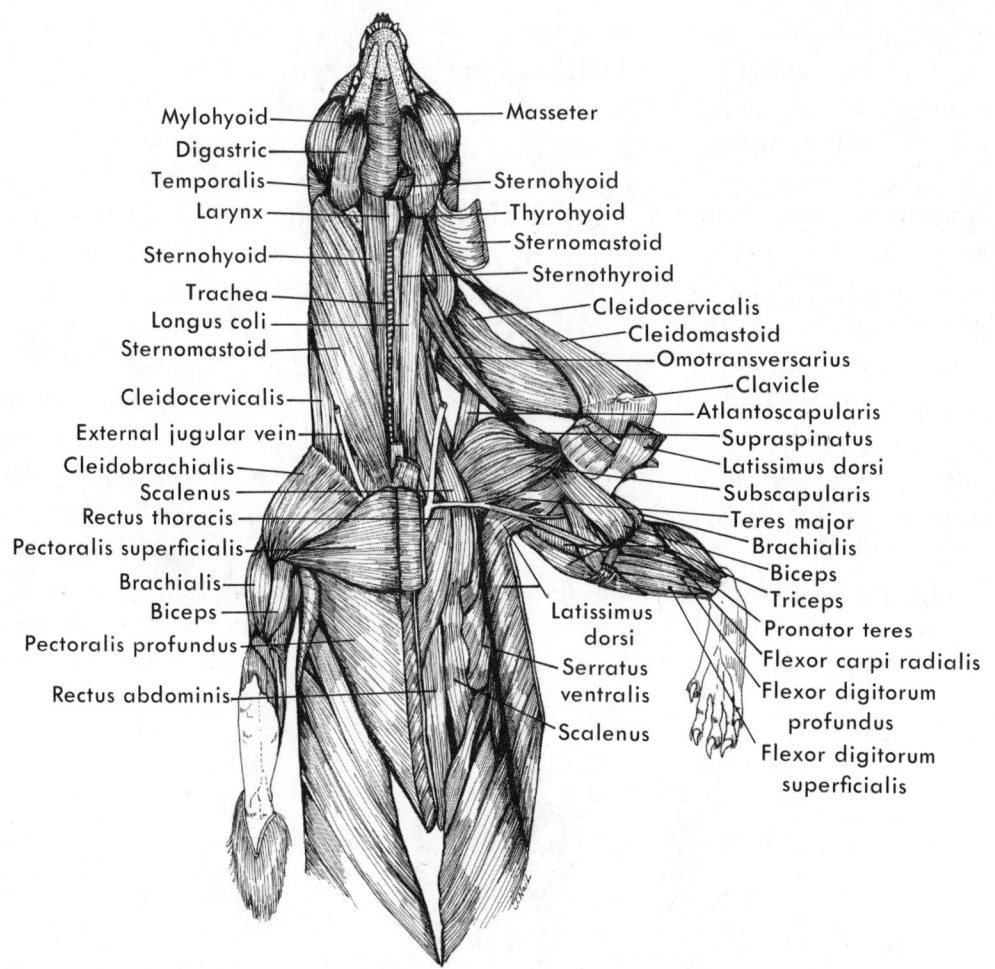

Figure 6–13. Ventral view of the pectoral, neck, and head muscles of a mink. Superficial muscles are shown on the left side of the drawing; deeper ones, on the right side.

mammals considered) extends to the distal end of the humerus and onto the antebrachium. In the cat and rabbit (but not mink), the superficial pectoral can be divided into two parts. A narrow band of very superficial fibers, the **pectoralis descendens,** extends from the front end of the sternum to the middle of the humerus (rabbit) or to the antebrachium (cat). The rest of the pectoralis superficialis is known as the **pectoralis transversus** (Fig. 6–12).

The **pectoralis profundus** originates from the caudal two thirds, or from the full length, of the sternum. Its fibers extend craniolaterally and disappear beneath those of the superficial pectoral. In carnivores (cat and mink), the insertion is confined to the humerus, but in the rabbit some fibers also attach onto the clavicle, and a group of deep fibers pass dorsal to the clavicle and sweep over the front of the shoulder to insert on the scapular spine. This portion will be seen later. The scapular fibers help to pull the dorsal border of the scapular forward, an action that occurs in hopping. The most caudal fibers of the cat's pectoralis profundus form a distinct, narrow band known as the **xiphihumeralis.**

(B) Trapezius and Sternocleidomastoid Group

The muscles belonging to the trapezius and sternocleidomastoid group are branchiomeric muscles that have become associated with the pectoral girdle, for they evolved from the cucullaris of lower vertebrates. The human trapezius is a large, undivided muscle, but it is subdivided in the mammals being considered here. Most parts act on the scapula, pulling it toward the middorsal line (abduction), cranially (protraction) and caudally (retraction). In the cat and mink, the most cranial part of the trapezius (cleidocervicalis) inserts on the clavicle and, together with the cleidobrachialis, protracts the arm. The sternocleidomastoid complex of muscles acts primarily upon the head, turning it to the side and flexing it. If the head is fixed, parts of the complex may act on the clavicle. Clean off connective tissue from the ventral, lateral, and dorsal surfaces of the neck and from the dorsal part of the shoulder. You may have to remove more skin from the back of the head. Do not injure the large external jugular vein located superficially on the ventrolateral surface of the neck.

All of the species considered have a **thoracic trapezius,** a thin sheet of muscle covering the cranial part of the latissimus dorsi, from which it should be separated (Figs. 6–14, 6–15, and 6–16). From their origin on the middorsal line of the thorax, the fibers of the thoracic trapezius converge to insert on the dorsal part of the scapular spine.

A **cervical trapezius** lies cranial to the thoracic trapezius in all of the mammals being considered. It arises from the middorsal line of the front of the thorax and neck, and from an aponeurosis that interconnects the left and right cervical trapezius. Its origin extends as far forward as the nuchal crest of the skull in the rabbit, but only to the base of the neck in the carnivores. Fibers of the cervical trapezius converge to insert on the ventral portion of the scapular spine and its metacromion process.

The cat and mink have a third component to the trapezius, the **cleidocervicalis** (clavotrapezius) (Figs. 6–14 and 6–15). From their origin on the nuchal crest and middorsal line of the cranial part of the neck, the fibers of the cleidocervicalis extend caudally and ventrally to merge, at the level of the clavicle, with those of the cleidobrachialis, the muscle seen earlier covering the humeral insertion of the pectoralis.

The **cleidobrachialis,** which is a part of the deltoid group, continues caudally from the clavicle in the carnivores and rabbit to insert on the humerus (mink and rabbit) or ulna (cat). The cleidobrachialis together with the carnivore cleidocervicalis constitute the **brachiocephalicus.** The cleidobrachialis and cleidocervicalis are distinct muscles in vertebrates with a well-developed clavicle. They tend to merge as the clavicle becomes reduced in carnivores, but careful dissection may reveal a tendinous intersection between them.

In all of the mammals under consideration, a **sternomastoid** arises from the manubrium and extends cranially and dorsally to insert on the mastoid

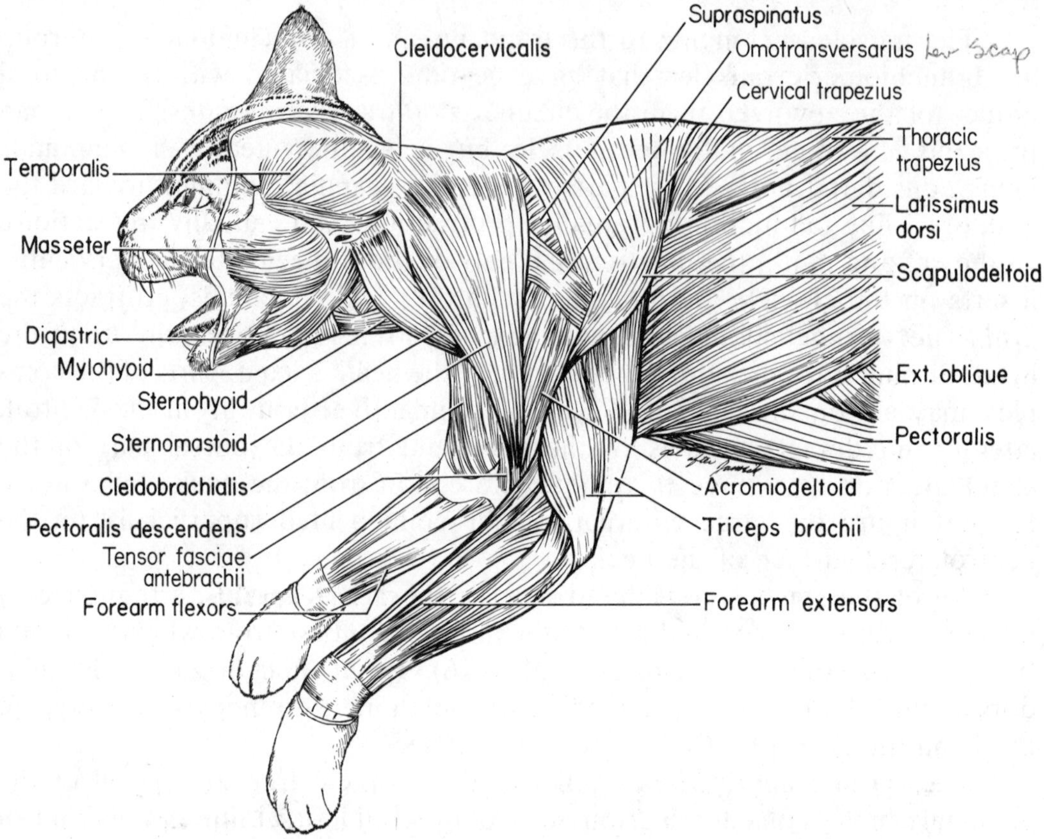

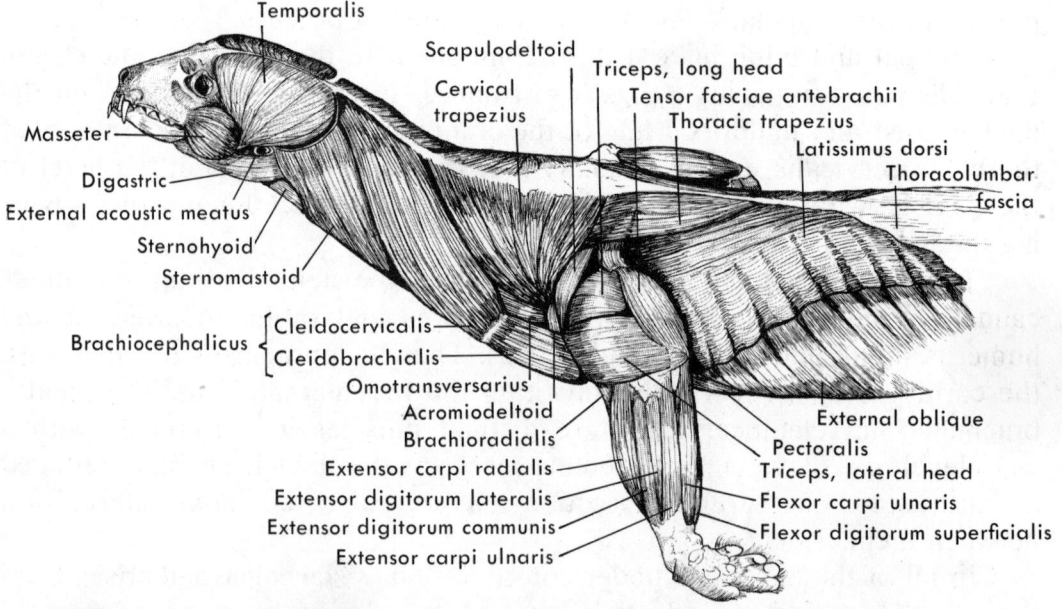

Cleidocervicalis

Supraspinatus

Omotransversarius *lev Scap*

Cervical trapezius

Thoracic trapezius

Latissimus dorsi

Scapulodeltoid

Temporalis

Masseter

Ext. oblique

Pectoralis

Digastric

Mylohyoid

Sternohyoid

Sternomastoid

Cleidobrachialis

Pectoralis descendens

Tensor fasciae antebrachii

Forearm flexors

Acromiodeltoid

Triceps brachii

Forearm extensors

Figure 6–14. Lateral view of the pectoral, neck, and head muscles of the cat.

Temporalis

Scapulodeltoid

Triceps, long head

Cervical trapezius

Tensor fasciae antebrachii

Thoracic trapezius

Latissimus dorsi

Thoracolumbar fascia

Masseter

Digastric

External acoustic meatus

Sternohyoid

Sternomastoid

Brachiocephalicus { Cleidocervicalis

Cleidobrachialis

Omotransversarius

Acromiodeltoid

Brachioradialis

Extensor carpi radialis

Extensor digitorum lateralis

Extensor digitorum communis

Extensor carpi ulnaris

External oblique

Pectoralis

Triceps, lateral head

Flexor carpi ulnaris

Flexor digitorum superficialis

Figure 6–15. Lateral view of the pectoral, neck, and head muscles of the mink.

region of the skull (Figs. 6–14, 6–15, and 6–16). As it extends forward, the muscle passes deep to the large external jugular vein. Muscle tissue superficial to the vein should be removed; it is either platysma or, in the rabbit, a specialized part of the platysma extending from the manubrium to the ear base (**depressor conchae posterior**). The sternomastoid of the rabbit is a narrow band, but it is wide in the carnivores and parallels the cranioventral border of the cleidocervicalis. Part of the insertion spreads onto the occipital region of the skull in carnivores, and some authors distinguish these fibers as a sterno-occipitalis. Left and right sternomastoids of the cat merge near the sternum (Fig. 6–12) and should be cut apart. *Deep*

Cat, mink, and rabbit have a **cleidomastoid** extending from the clavicle (where it joins other muscles attaching on the clavicle) to the mastoid region of the skull (Figs. 6–12, 6–13, and 6–16). Much of it lies deep to the wide sternomastoid in carnivores, but it lies dorsal to the sternomastoid of the rabbit.

In the rabbit only (Fig. 6–16), a **cleido-occipital** arises from the clavicle lateral and dorsal to the origin of the cleidomastoid. It extends forward deep to both the cleidomastoid and sternomastoid to insert on the basioccipital region of the skull.

(C) REMAINING SUPERFICIAL MUSCLES OF THE SHOULDER

The band of muscle whose caudal end can be seen on the side of the neck between the cleidocervicalis and cervical trapezius (carnivores) or between the sternocleidomastoid complex and trapezius (rabbit) is the **omotrans-** *lev Scap* **versarius** (Figs. 16–14, 16–15, and 16–16). It begins on the metacromion process of the scapula ventral to the insertions of the cervical trapezius, and extends forward to insert primarily on the transverse process of the atlas. In carnivores, much of it lies deep to the cleidocervicalis. The omotransversarius is a hypaxial muscle that helps to pull the scapula forward (protraction).

The deltoid lies caudal and ventral to the insertion of the omotransversarius. In the mammals being considered, it is subdivided into three parts. The **cleidobrachialis** (clavodeltoid) has been observed arising from the clavicle. It inserts on the humeral body in the mink and rabbit, and on the ulna in common with the brachialis (see the following section) in the cat. The **acromiodeltoid** lies dorsal and caudal to the cleidobrachialis. It arises from the acromion near the attachment of the omotransversarius, and inserts on the proximal portion of the humeral body. The **scapulodeltoid** lies along the caudal border of the scapula. It arises from the scapular spine and, in the rabbit, from the surface of the muscle covering the lower half of the scapula (infraspinatus). It inserts on the proximal end of the humerus. The muscle is thinner in the rabbit, and passes ventral to the large metacromion. But if the metacromion is broken and raised, the essential relationship will be seen to be the same as in the carnivores. The cleidobrachialis protracts the arm, but the caudal parts of the complex are retractors and abductors of the humerus.

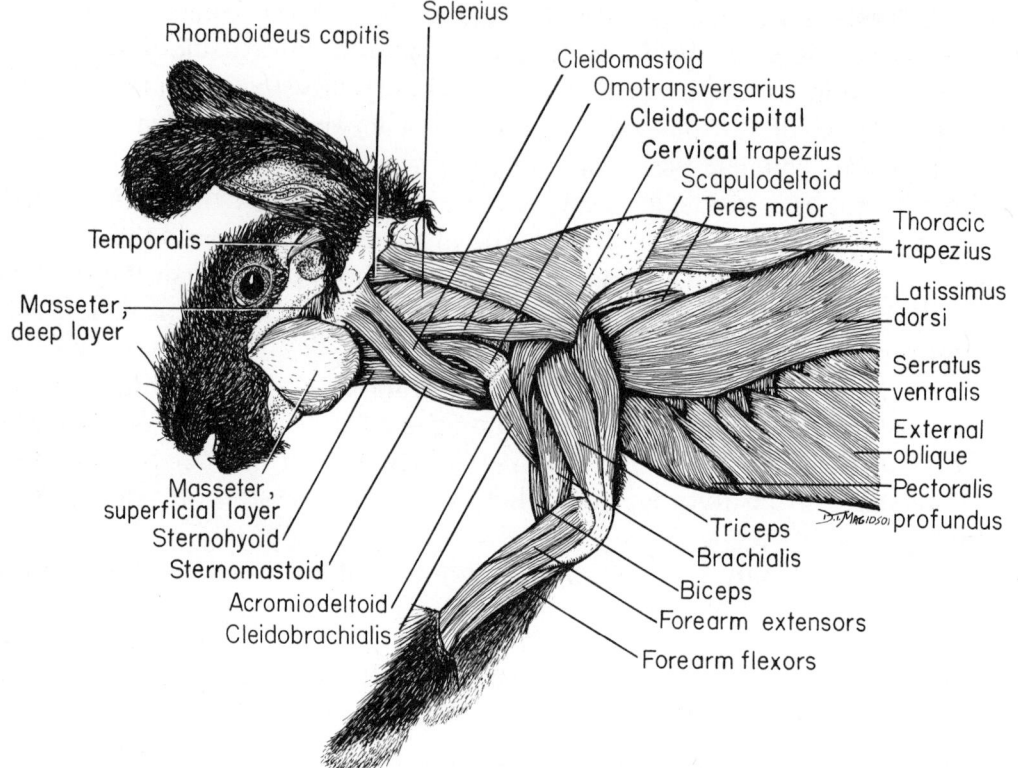

Figure 6–16. Lateral view of the pectoral, neck, and head muscles of the rabbit.

The **latissimus dorsi** has been observed on the side of the trunk caudal to the arm. It arises from the thoracolumbar fascia, and from the spinous processes of the last thoracic vertebrae. From here it passes forward and ventrally to insert on the proximal end of the humerus in common with the teres major. In carnivores, a small part of the muscle often forms a tendon that inserts with the pectoralis. The latissimus dorsi retracts the humerus.

(D) DEEPER MUSCLES OF THE SHOULDER

Cut across the center of the latissimus dorsi at right angles to its fibers, and also across the thoracic and cervical trapezius. Reflect the ends of these muscles, and clean out the fat and loose connective tissue from beneath them so as to expose the deeper muscles of the shoulder. In the rabbit, the pectoralis profundus can be seen sweeping over the front of the shoulder and scapula to insert on the scapular spine. It should be cut near its insertion and reflected.

The supraspinous fossa of the scapula is occupied by the **supraspinatus.** The muscle inserts on the greater tuberculum of the humerus, and protracts the humerus. The infraspinous fossa is occupied by the **infraspinatus.** Most of the muscle is covered by the scapulodeltoid in the rabbit. The infraspinatus inserts on the greater tuberculum of the humerus and rotates this bone outward.

A **teres major** arises from the posterior border of the scapula caudal and ventral to the infraspinatus. Part of the tensor fasciae antebrachii arises from its surface in the mink. The teres major passes forward to insert on the proximal end of the humerus in common with the latissimus dorsi. Its action is to rotate the humerus inward and to retract it.

Dissect deeply between the cranial part of the infraspinatus and the long head of the triceps (the large muscle on the posterior surface of the brachium). You will eventually come upon a very small triangular muscle arising by a tendon from the cranial part of the ventral border of the scapula and inserting on the greater tuberculum of the humerus. This is the **teres minor**; it helps the infraspinatus rotate the humerus outward.

Examine the scapular region from a dorsal view. The large muscle that arises from the tops of the posterior cervical and anterior thoracic vertebrae, and inserts along the dorsal border of the scapula, is the **rhomboideus.** It forms a continuous sheet of muscle bundles in carnivores, but in the rabbit it is clearly divided into a cranial **rhomboideus cervicis** and a caudal **rhomboideus thoracis.** The most cranial bundle in all cases extends more rostrally than the others to its origin from the back of the skull and is called the **rhomboideus capitis.** The rhomboideus is a hypaxial muscle that pulls the scapula toward the vertebrae, helps to hold it in place, and assists in its protraction and retraction.

Muscles that extend from the dorsal border of the scapula to the ribs and to the transverse processes of the cervical vertebrae constitute the serratus ventralis complex. Most lie deep to the rhomboideus, but in the mink a conspicuous slip (**atlantoscapularis**) attaches to the dorsal part of the scapular spine, crosses the dorsal part of the supraspinatus, and extends cranially and ventrally to attach onto the transverse process of the atlas (Fig. 6–13). Cut across the rhomboideus, pull the top of the scapula laterally, and clean away fat and loose connective tissue from the area exposed. The large fan-shaped muscle that you see is the **serratus ventralis** proper (Fig. 6–17). It is divided into a **serratus ventralis cervicis** and a **serratus ventralis thoracis** in the rabbit. Also approach the serratus ventralis ventrally by cutting through and reflecting the pectoralis. (The group of nerves passing into the arm dorsal to the pectoralis belong to the brachial plexus.) The serratus ventralis arises by a number of slips from the ribs just dorsal to the junction of ribs and costal cartilages, and from the transverse processes of the posterior cervical vertebrae. It inserts on the dorsal border of the scapula ventral to the insertion of the rhomboideus. The serratus ventralis is a hypaxial muscle. Together with the pectoralis, it forms a muscular sling that transfers much of the weight of the body to the pectoral girdle and appendage. The serratus ventralis is the major component in the sling.

Clean out the area between the serratus ventralis and the medial surface of the scapula. The subscapular fossa is occupied by a large **subscapularis,** which passes to insert on the lesser tuberculum of the humerus. This muscle pulls the humerus medially (adduction).

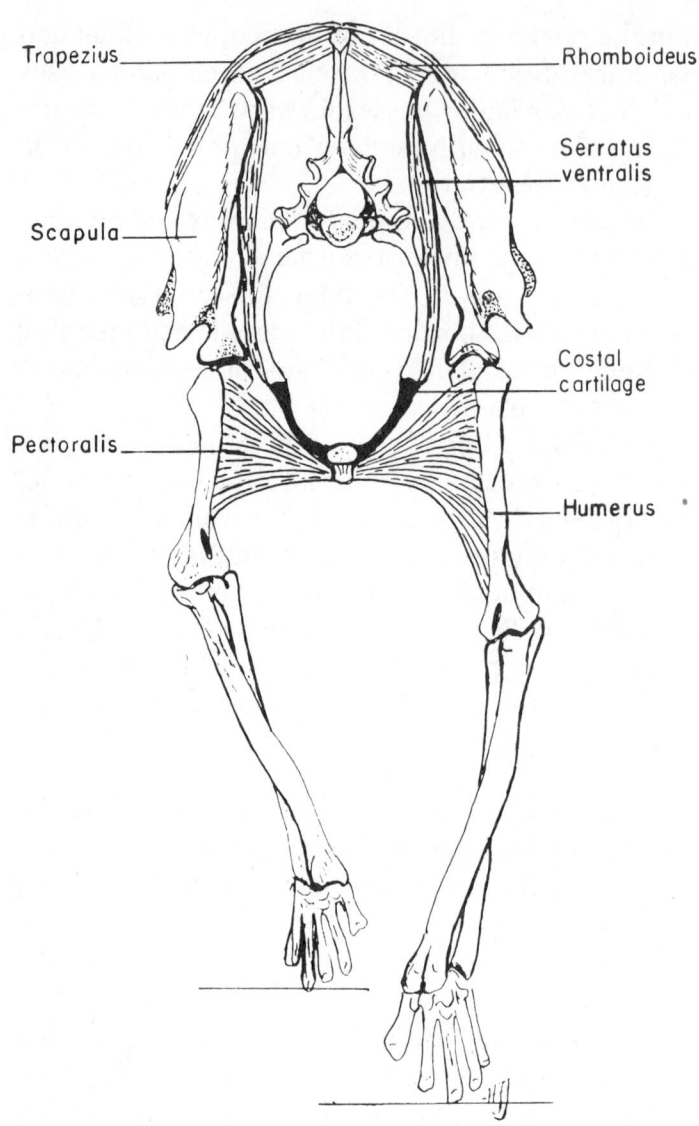

Trapezius

Rhomboideus

Serratus ventralis

Scapula

Costal cartilage

Pectoralis

Humerus

Figure 6–17. Diagrammatic cranial view of a section through the thorax of a cat at the level of the pectoral girdle, to show muscular connections between the trunk and appendage.

(E) MUSCLES OF THE BRACHIUM

Clean the muscles of the brachium and separate them from each other. The large muscle that covers the caudal surface and most of the medial and lateral surface of the humerus is the triceps brachii. A **tensor fasciae antebrachii** (dorsoepitrochlearis) is closely associated with the medial (in the mink, dorsal) surface of the triceps, and should be studied first. In the cat and rabbit, the tensor fasciae antebrachii arises primarily from the lateral surface of the latissimus dorsi and extends distally along the median surface of the arm to insert on the tendon of the triceps and on the antebrachial fascia. This part is present in the mink as well, but the most conspicuous part of the mink's tensor fasciae antebrachii is a powerful slip that arises from the surface of the teres major and extends distally along the caudal surface of the arm to join the tendon of the triceps (Fig. 6–15).

Cut through the origin of the tensor fasciae antebrachii and reflect the muscle. The **triceps brachii** is now exposed clearly (Figs. 6–14, 6–15, and 6–16). It has three main heads. The **long head,** located on the caudal surface of the humerus, is the largest. Note that it arises from the scapula caudal to the glenoid cavity. A large **lateral head** arises from the proximal end of the humerus and covers much of the lateral surface of this bone. It is quite distinct in the carnivores, but is partly united with the long head in the rabbit. A small **medial head** can be found on the medial surface of the humerus deep to several nerves and blood vessels. It arises from most of the shaft of the humerus. All of the heads of the triceps insert in common on the olecranon of the ulna.

A small **anconeus** arises from the distal portion of the lateral surface of the humerus. To see it, cut through and reflect the distal end of the lateral head of the triceps. The triceps, tensor fasciae antebrachii, and anconeus extend the forearm.

The craniolateral surface of the humerus is covered by the **brachialis.** It arises from the shaft of the humerus and inserts on the proximal end of the ulna (carnivores), or on the ulna and radius in common with the biceps (rabbit, Fig. 6–13).

The **biceps brachii** lies on the anteromedial surface of the humerus (Figs. 6–12 and 6–13). To see it clearly, cut through and reflect the pectoralis near its insertion on the humerus. (The mass of nerves passing to the arm dorsal to the pectoralis belong to the brachial plexus.) The biceps of the cat, mink, and rabbit has a single origin by a tendon from the edge of the glenoid cavity. This tendon lies in the intertubercular groove between the greater and lesser tubercles. The muscle forms a prominent belly beneath the humerus, and inserts by a tendon on the radial tuberosity (carnivores), or on both the radius and ulna (rabbit). Biceps and brachialis are the major flexors of the forearm. The biceps also assists in supination of the forearm.

Clean off connective tissue from the medial side of the shoulder joint, and you will find a short band of muscle arising from the coracoid process and inserting on the proximal end of the humerus. This is the **coracobrachialis** (Fig. 6–12). It helps to pull the humerus toward the body (adduction). In some individuals, a long portion of the coracobrachialis passes to the distal end of the humeral shaft. It is absent in the mink.

(F) MUSCLES OF THE FOREARM

Before the muscles of the forearm and hand can be studied it is necessary to remove the very extensive **antebrachial fascia.** The deeper part of this fascia is continuous with the tendons of the tensor fasciae antebrachii, triceps and (in the cat) pectoralis descendens. It forms a dense fibrous sheet which dips down between many of the muscles and also attaches onto the ulna and radius. At the level of the wrist, part of this fascia forms ligaments which encircle the wrist and hold the muscle tendons in place. A band of dense fibers, the **extensor retinaculum,** bridges the tendon grooves on the dorsal surface of the radius, and a comparable **flexor retinaculum** lies on the palmar side of the wrist. The antebrachial fascia continues into the hand and on the palmar side helps to form fibrous sheaths, the **vaginal ligaments,** through which the flexor tendons of the fingers run. After the fascia has been removed, separate the major muscles before attempting to identify them.

The muscles of the forearm and hand can be sorted into an extensor group located on the craniolateral surface of the forearm and the back of the hand, and a flexor group located on the caudomedial surface of the forearm and palm of the hand. In the elbow region, the insertion of the biceps, brachialis, and (in the cat) cleidobrachialis pass between these groups at the cranial border of the arm, and the ulna and insertion of the triceps separate them on the caudal border.

Most of the extensors arise from or near the lateral epicondyle of the humerus. Superficially, at the level of the elbow (Fig. 6–18), one can recognize from cranial to caudal a brachioradialis (absent in the rabbit), an extensor carpi radialis complex, an extensor digitorum communis, an extensor digitorum lateralis, and an extensor carpi ulnaris. The **brachioradialis** arises more

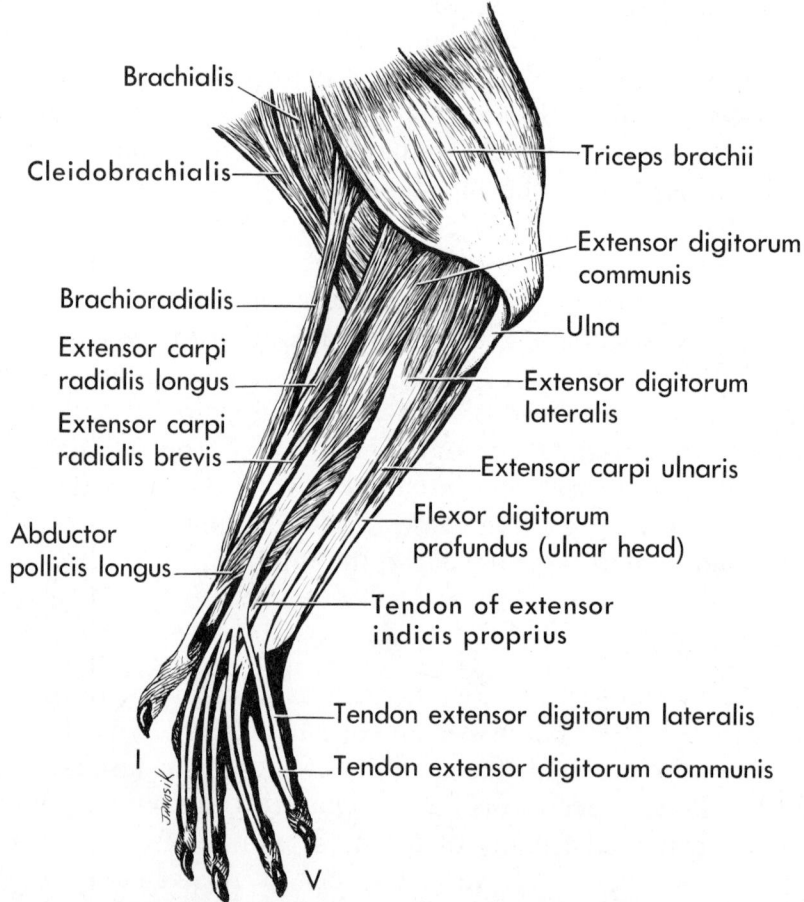

Figure 6–18. Lateral view of the extensor muscles of the forearm of the cat after removal of the antebrachial fascia and extensor retinaculum.

proximally from the humerus than the others and inserts on the styloid process of the radius. The extensor carpi radialis complex can be divided in the cat and rabbit (but not in the mink) into a more superficial and cranial **extensor carpi radialis longus** and a deeper and more caudal **extensor carpi radialis brevis.** The tendons of both pass deep to the tendon of the abductor pollicis longus (see below) to insert upon the bases of the second and third metacarpals, respectively. The **extensor carpi ulnaris** has a comparable position on the ulnar border of the forearm, and its broad tendon inserts upon the base of the fifth metatarsal. All these muscles act to extend the forearm and hand, and the brachioradialis also assists the supinator (see below) in rotating the radius in such a way that the palm is turned dorsally (supination).

The long digital extensors are more complex. The **extensor digitorum communis** breaks up into four tendons which pass along the dorsal surface of digits 2 to 5. The tendons are bound to each phalanx of the digits by connective tissue, but the most conspicuous attachment is on the terminal phalanges. The tendon of the **extensor digitorum lateralis** also passes through a groove on the radius and divides at about the level of the wrist into two (rabbit), or three or four (carnivores), parts which pass down the dorsal surface of the lateral digits. These tendons at first lie somewhat on the ulnar side of those of the extensor digitorum communis but eventually unite with those of the latter muscle. These two muscles are extensors of digits 2 to 5.

Muscles that lie deep to the preceding ones on the extensor surface of the forearm can be seen by cutting and reflecting the two long digital extensors and the carpi ulnaris. A narrow **extensor digit II,** which probably also incorporates the human extensor pollicis longus, arises deep to the extensor carpi ulnaris from the proximal three-fourths of the lateral surface of the ulna, and it inserts by a tendon that goes to the middle phalanx of the second digit (Fig. 6–18). Often part of the tendon passes to the thumb. The muscle assists in the extension of these digits. A powerful **abductor pollicis longus,** which includes the human extensor pollicis brevis, arises from much of the lateral surface of the ulna and adjacent parts of the radius. Its fibers converge, go deep to the tendon of the extensor digitorum communis, and form a tendon that goes superficial to the extensor carpi radialis tendons (Fig. 6–18) to insert upon the radial side of the first metacarpal. This muscle extends and abducts the thumb. Finally, in the cat but not in the rabbit, a **supinator** passes obliquely across the radius deep to the belly of the extensor digitorum communis and proximal to the abductor pollicis longus. It arises from the lateral epicondyle of the humerus and elbow ligaments and inserts upon the radius. Its diagonal fibers enable it to act as a powerful supinator of the hand.

Most of the flexor muscles arise from or near the medial epicondyle of the humerus. The superficial muscles are, from cranial to caudal, the pronator teres, flexor carpi radialis, flexor digitorum superficialis, and the flexor carpi ulnaris (Fig. 6–19). The **pronator teres** passes diagonally from the medial epicondyle to the medial border of the radius, and its action (pronation) rotates the forearm in such a way that the palm faces the ground.

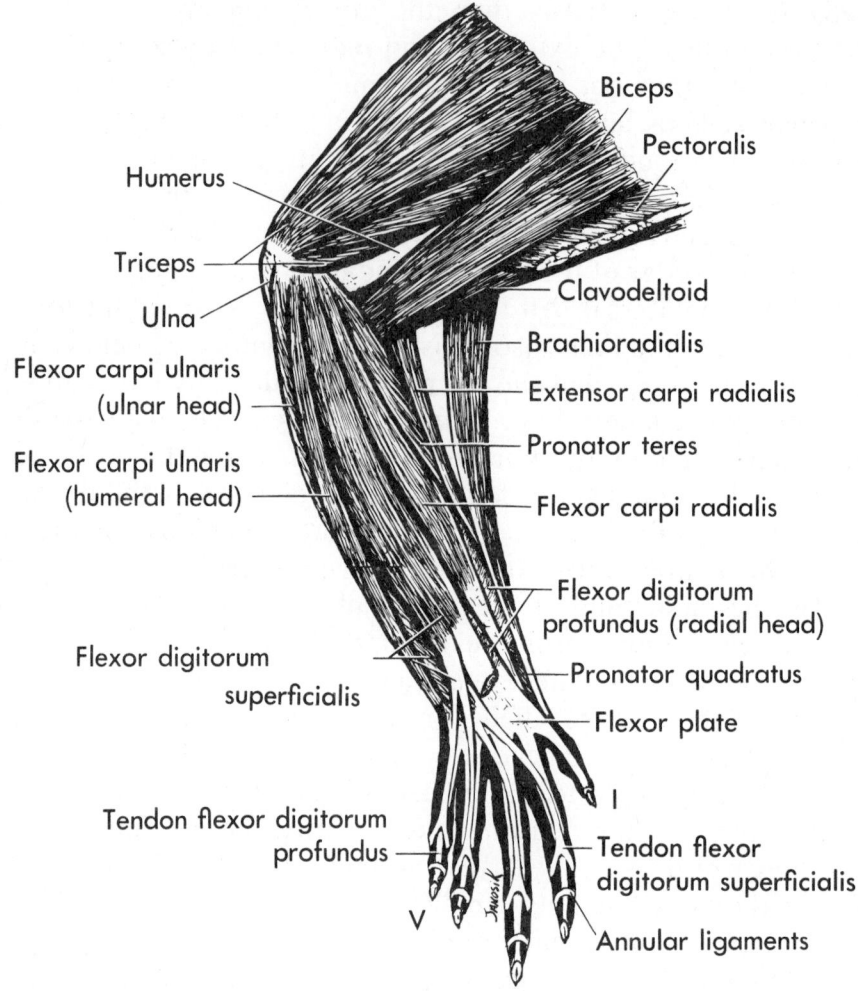

Figure 6–19. Medial view of the flexor muscles of the forearm of the cat after removal of the antebrachial fascia, flexor retinaculum, and palmar aponeurosis.

The **flexor carpi radialis** extends down the radial border of the forearm and forms a long tendon which passes deep to other tendons in the hand to insert on the proximal ends of the second and third metacarpals. The **flexor carpi ulnaris** has a comparable position on the ulnar border of the forearm. It arises by two heads, one from the medial epicondyle and one from the surface of the ulna; it inserts upon the pisiform bone. The chief action of these two muscles is to flex the hand at the wrist.

The **flexor digitorum superficialis,** widest of the superficial muscles, lies between the two carpal flexors. It arises partly from the medial epicondyle of the humerus and partly from the surface of a deeper muscle (flexor digitorum profundus). At the level of the wrist it forms a tendon that passes deep to the flexor retinaculum. Part of this tendon has cutaneous attachments to the foot pads, but most divides into four tendons that go to the bases of the second through fifth fingers. Here each tendon splits and attaches onto the side of the middle phalanx of the digit. The tendons of the flexor digitorum profundus pass through the openings formed by the splitting of the superficialis tendons. This muscle flexes the digits near their middle.

Cut and reflect the flexor carpi radialis, flexor digitorum superficialis and flexor carpi ulnaris. The muscle complex deep to them is the **flexor digitorum profundus**. It consists of five heads, all of which converge at the wrist to form a powerful flexor tendon. The first, or ulnar head, arises from most of the length of the outer border of the ulna, and part of it is visible from the extensor side of the forearm (Fig. 6–18); the second, third and fourth heads arise more or less in common from the medial epicondyle deep to the origin of the flexor digitorum superficialis and extensor carpi ulnaris; the fifth or radial head (Fig. 6–19) is the deepest and arises from the middle third of the radius, the interosseous ligament stretching between radius and ulna and from the adjacent parts of the ulna. Their common tendon extends into the palm as a powerful flexor plate and then breaks up into five strong flexor tendons which run through ligamentous sheaths, beneath annular ligaments and finally insert on the terminal phalanges of the digits. The tendons on the second to last finger perforate the tendons of the flexor digitorum superficialis. This muscle flexes all segments of the digits. Certain small intrinsic muscles in the hand (the **lumbricales**) arise superficially from the flexor plate. Other intrinsic hand muscles are not described.

Separate the tendon of the flexor carpi radialis, which goes deep to the flexor plate, and the radial head of the flexor digitorum profundus. The very deeply situated muscle which you see in the cat and mink is the **pronator quadratus** (Fig. 6–19). Rabbits lack this muscle along with the supinator on the extensor surface of the forearm. Its fibers pass diagonally from their origin on the distal third of the ulna to their insertion on the outer border of the radius. It assists in hand pronation.

Pelvic Muscles

Although the thigh muscles of all mammals are essentially the same, there are some distinct differences in relationships between certain muscles of carnivores and rabbits. Frequent reference should be made to the figures. Clear the fat and superficial fascia from the surfaces of the pelvic region and thigh as a preliminary to the dissection. Include the large pads of fat lateral to the tail base and in the depression (**popliteal fossa**) behind the knee. Start to separate the more obvious muscles.

(A) LATERAL THIGH AND ADJACENT MUSCLES

The most cranial muscle on the thigh of carnivores is the **sartorius** (Figs. 6–20 and 6–21). It is a band extending from the crest and ventral border of the ilium (its origin) to the patella and medial side of the thigh (its insertion). Most of the muscle lies on the medial surface of the thigh. In the

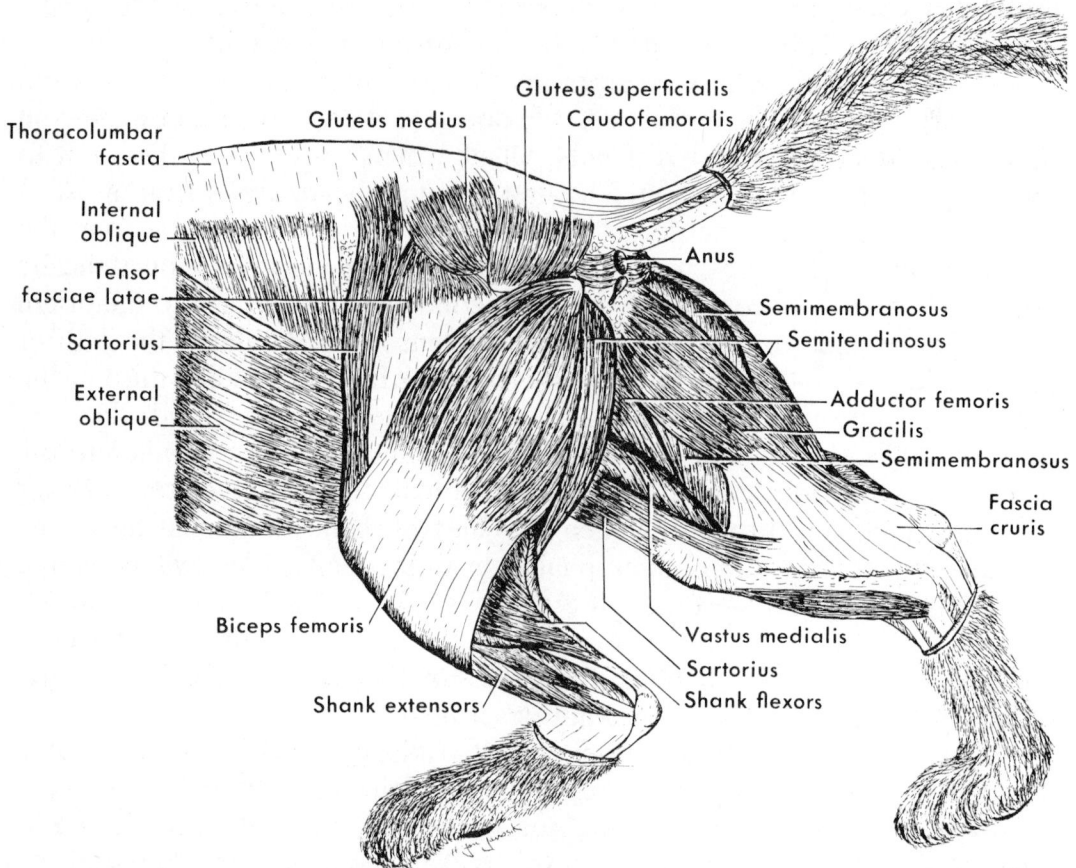

Figure 6–20. Superficial pelvic and thigh muscles of the cat. Lateral muscles can be seen on the left leg and medial muscles on the right leg.

rabbit, the muscle is entirely on the medial surface (Fig. 6–25). It originates from the inguinal ligament and inserts on the medial tibial condyle. It is fused with the gracilis distally. In all of these mammals, the sartorius adducts and rotates the femur and helps to extend the shank.

A tough, white fascia, the **fascia lata,** lies on the lateral surface of the thigh caudal to the sartorius in carnivores, or on the craniolateral surface of the thigh in the rabbit. The **tensor fasciae latae** arises from the ventral border of the ilium and from the surface of adjacent muscles, and inserts in the fascia lata. This muscle lies on the lateral surface of the thigh in carnivores (Figs. 6–20 and 6–21), but it extends onto the medial side in the rabbit (Figs. 6–22 and 6–25).

The lateral surface of the thigh caudal to the fascia lata is covered by a very broad **biceps femoris** (Figs. 6–20, 6–21, and 6–22). It has a narrow origin from the tuberosity of the ischium, and then fans out to insert by a broad aponeurosis on the patella and much of the tibial shaft. The biceps femoris forms the lateral wall of the popliteal fossa. Its action is to flex the shank and abduct the thigh.

A more slender **semitendinosus** lies caudal to the origin of the biceps

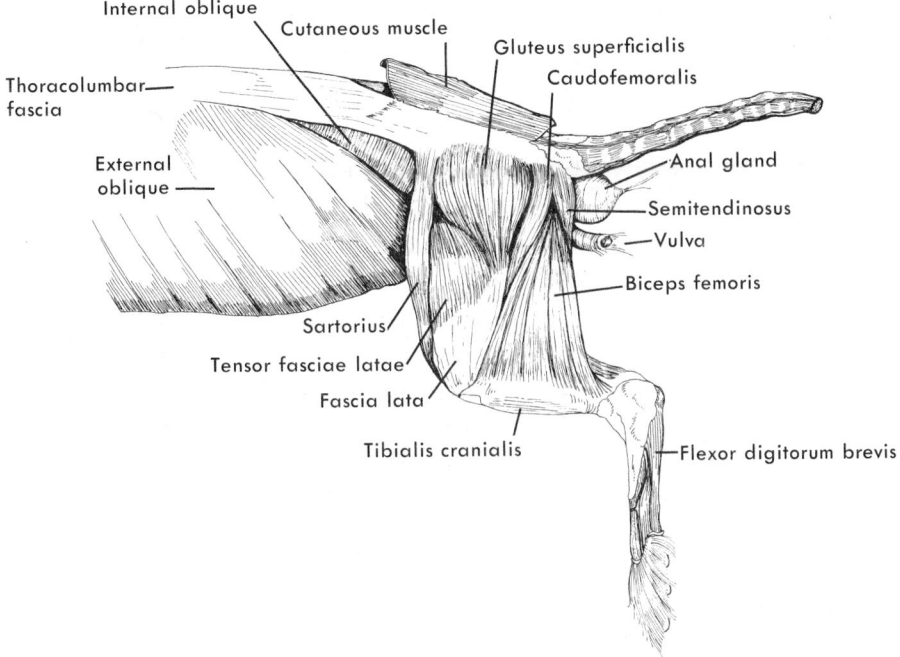

Figure 6–21. Lateral view of the superficial pelvic and thigh muscles of the mink.

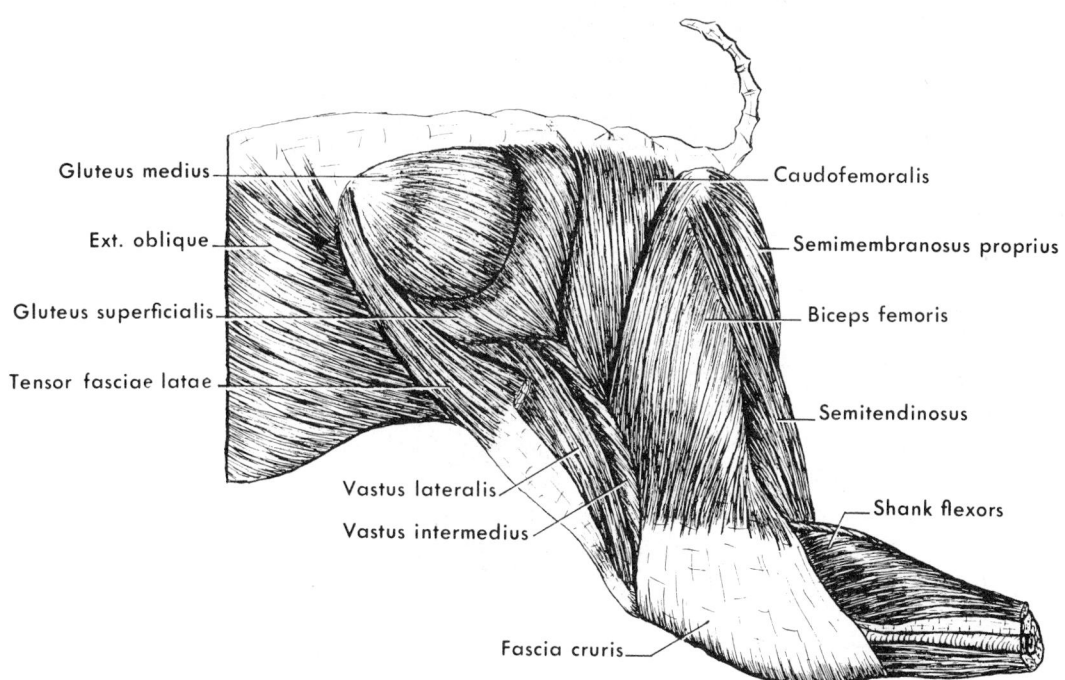

Figure 6–22. Lateral view of the pelvic and thigh muscles of the rabbit. Part of the fascia lata has been removed.

femoris and contributes to the medial wall of the popliteal fossa. It also arises from the ischial tuberosity, but in the rabbit a part of it arises from the fascia overlying the biceps femoris. It inserts on the medial side of the proximal end of the tibia. In the rabbit, part of its insertion joins the calcaneus tendon of several of the shank flexors (see below). Its main action is flexion of the shank.

The band of muscle cranial and dorsal to the origin of the biceps femoris is the **caudofemoralis** (coccygeofemoralis). It arises from the more posterior sacral and anterior caudal vertebrae. It passes beneath the cranial border of the biceps and forms a narrow tendon that inserts upon the distal end of the femur (mink), or on the patella in common with part of the insertion of the biceps (cat and rabbit). It abducts and helps retract the thigh.

Cut across the biceps and caudofemoralis near their origins, being careful not to cut a very slender muscle that lies beneath them. This band, which will be seen on reflecting these muscles, is the **abductor cruris caudalis** (tenuissimus) (Fig. 6–23). It arises from a sacral or caudal vertebra, inserts

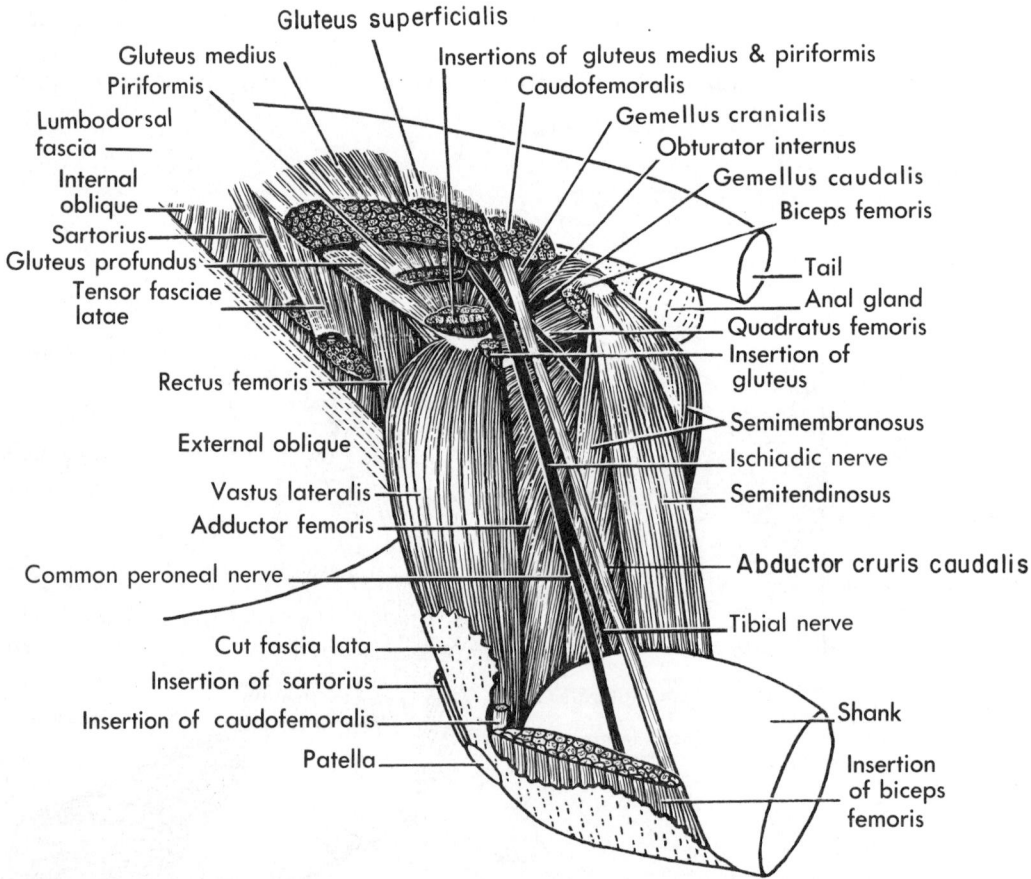

Figure 6–23. Lateral view of the deep muscles of the pelvis and thigh of a cat. The sartorius, tensor fasciae latae, biceps femoris, gluteus medius, gluteus maximus, caudofemoralis, and piriformis have been largely cut away to expose the deeper muscles. However, the origins and insertions of these muscles are shown as points of reference.

on the tibia and adjacent crural fascia with the biceps, and assists the biceps in thigh abduction and shank flexion. The abductor cruris caudalis is closely associated with the biceps and may have evolved from it.

(B) Gluteal Complex and Deeper Pelvic Muscles

The muscle mass covering the dorsolateral surface of the sacrum between the caudofemoralis and sartorius (carnivores), or between the caudofemoralis and tensor fasciae latae (rabbit), is the gluteus complex (Figs. 6–20, 6–21, and 6–22). Carefully remove overlying connective tissue and fascia, and start to separate the layers of the complex. The most superficial part is the gluteus superficialis; the next deeper layer, the gluteus medius. In the cat and rabbit the superficialis is also somewhat caudal and distal to the medius, but the medius is completely covered by the superficialis in the mink. (The gluteus superficialis is homologous to the human gluteus maximus, but it is usually not so large as the medius in quadrupeds.)

The **gluteus superficialis** arises from the sacral fascia and from the spinous processes of sacral and anterior caudal vertebrae. In the rabbit, part also arises from the ventral border of the ilium. Fibers converge to insert on the greater trochanter of the femur (carnivores), or, in the rabbit, on a third trochanter which lies on the lateral surface of the femur slightly distal to the greater trochanter.

Cut through the belly of the gluteus superficialis and reflect its ends. The **gluteus medius,** which lies partly (cat, rabbit) or entirely (mink) deep to the superficialis, arises from the crest and lateral surface of the ilium and adjacent vertebrae. It inserts upon the greater trochanter. Both gluteus superficialis and medius act primarily as thigh abductors.

A thin **piriformis** lies deep to the caudal portion of the gluteus medius, from which it cannot always be separated easily (Fig. 6–23). It arises from the last sacral and first caudal vertebrae and inserts with the medius on the greater trochanter. It too abducts the thigh.

Cut through and reflect the piriformis if this was not done with the reflection of the gluteus medius. The prominent **ischiadic nerve** lies deep to the piriformis. A cranial **gluteus profundus** (gluteus minimus of human anatomy) and a more caudal **gemellus cranialis** lie deep to the gluteus medius and piriformis. The ischiadic nerve crosses the latter muscle. These two muscles are more or less united, and are somewhat difficult to separate except at their origins. The gluteus profundus arises from the lateral surface of the ilium; the gemellus cranialis, from the dorsal borders of the ilium and ischium. They insert in common on the greater trochanter and rotate and abduct the thigh.

If you are dissecting the cat, look deep to the cranioventral border of the gluteus profundus (Fig. 6–23). The short muscle observed is the **articularis coxae.** It arises from part of the lateral iliac surface, inserts upon the proximal end of the femur, and helps to flex the thigh. This muscle is absent in the mink and rabbit.

The narrow band of muscle located just caudal to the gemellus cranialis in all of the mammals being studied is a part of the **obturator internus** (Fig. 6–23). The obturator internus arises on the inside of the pelvis from the borders of the obturator foramen and passes over the dorsal rim of the ischium (where it can now be seen) to its insertion in the trochanteric fossa. It is a thigh abductor.

A **gemellus caudalis** lies caudal and partly deep to the obturator internus. It arises from the lateral surface of the ischium cranial to the origin of the caudofemoralis, and inserts with the obturator internus in the trochanteric fossa. It is a thigh abductor and retractor.

The **quadratus femoris** is the rather thick band of muscle distal to the gemellus caudalis. It arises from the ischial tuberosity deep to the origin of the biceps femoris, and passes to insert on the femur at the bases of the greater and lesser trochanters. It is primarily a thigh retractor.

Cut through and reflect the gemellus caudalis and quadratus femoris. An **obturator externus** lies deep to them. It arises on the lateral surface of the pelvis from the borders of the obturator foramen and inserts deep in the trochanteric fossa. Its action is thigh rotation and retraction.

(C) QUADRICEPS FEMORIS COMPLEX

The front portion of the thigh of mammals is covered by a group of four muscles which insert in common on the patella and **patella tendon,** which, in

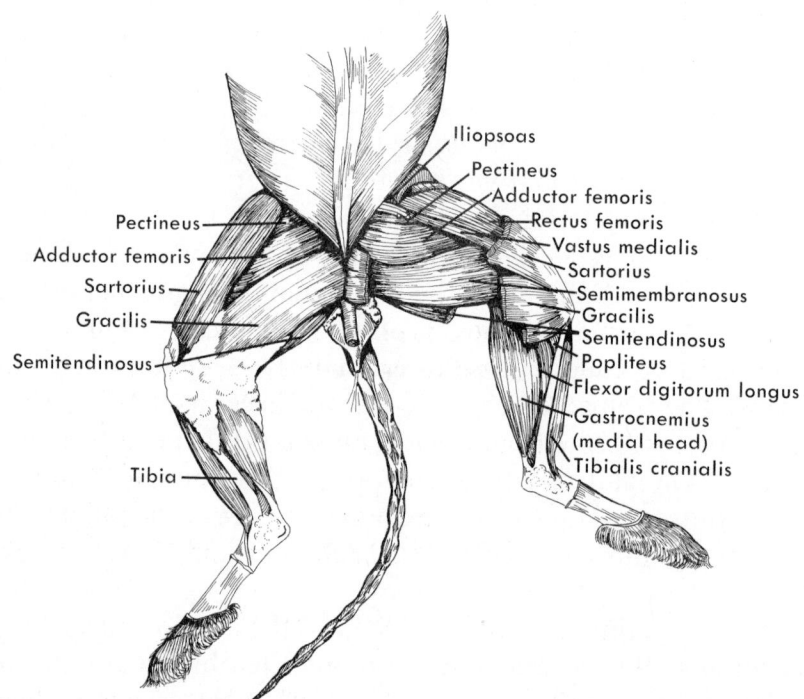

Figure 6–24. Medial thigh muscles of the mink. Superficial muscles are shown on the left side of the drawing; deeper ones, on the right side.

turn, attaches to the tuberosity of the tibia. The whole complex is often referred to as the **quadriceps femoris,** and it is the primary shank extensor. The patella permits the common tendon of these muscles to slide easily across the knee joint.

Cut through the tensor fasciae latae (and in carnivores, the sartorius) at right angles to their fibers and reflect their ends. Recall that the tensor fasciae latae of the rabbit extends onto the medial side of the thigh and is partly united with the vastus medialis (Fig. 6–25). Separate the tensor fasciae latae from adjacent muscles, and clean the area exposed. The large muscle on the craniolateral surface of the thigh, which was largely covered by the tensor fasciae latae, is the **vastus lateralis** (Figs. 6–22 and 6–23). It arises from the greater trochanter and adjacent parts of the femoral body. A **rectus femoris** lies on the cranial thigh surface medial to the vastus lateralis (Figs. 6–23, 6–24, and 6–25). Since it arises from the ilium just cranial to the acetabulum, it also acts across the hip joint and helps to protract the thigh. A **vastus medialis** lies on the medial surface of the thigh caudal to the rectus femoris. It arises from the femoral body (Figs. 6–20, 6–24, and 6–25). The **vastus intermedius** of the carnivores can be found by dissecting deeply between the vastus lateralis and rectus femoris. It arises from the femur lateral

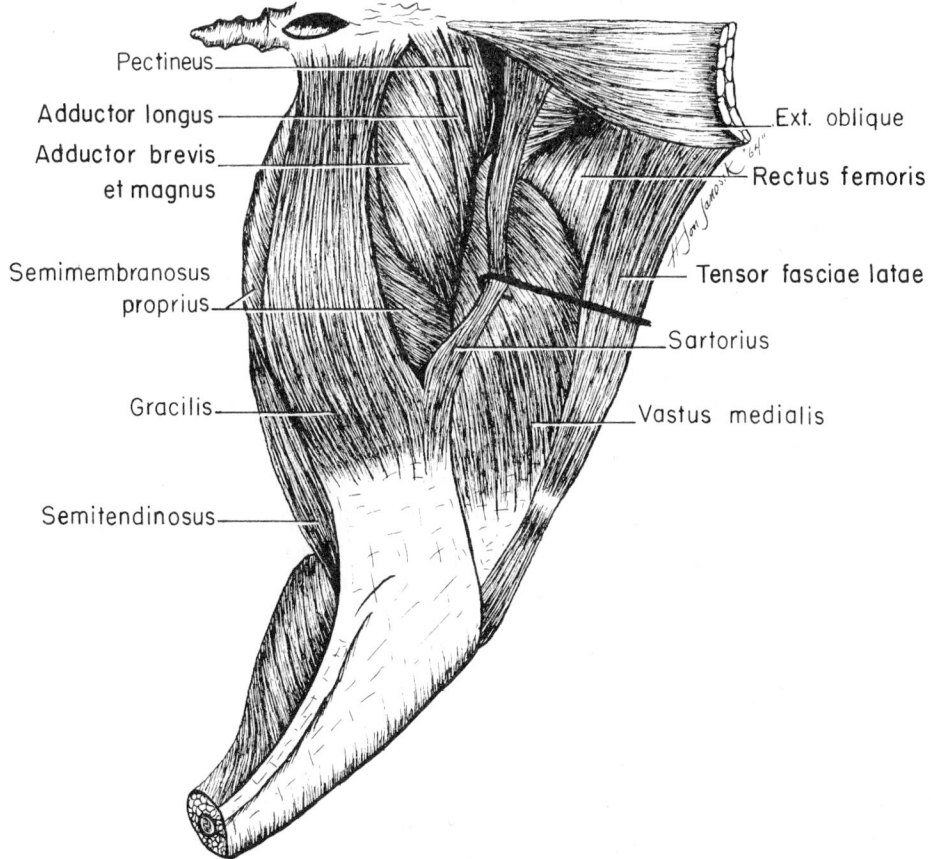

Figure 6–25.   Medial view of the thigh muscles of the rabbit.

to the origin of the vastus medialis. It has a similar origin and relationships in the rabbit, but is larger, so part of the muscle can be seen on the lateral surface of the thigh between the vastus lateralis and caudofemoralis (Fig. 6–22).

(D) CAUDOMEDIAL THIGH MUSCLES

The medial surface of the thigh caudal to the sartorius and quadriceps femoris is largely covered by the **gracilis** (Figs. 6–20, 6–24, and 6–25), a broad, thin muscle that arises from the pubic and ischial symphyses. It inserts by an aponeurosis onto the tibia and crural fascia. The distal part of the rabbit's sartorius unites with it. The gracilis adducts and retracts the thigh, and flexes the crus.

Cut through the gracilis and reflect its ends. Notice again the semi-tendinosus seen earlier from the lateral surface. It inserts on the tibia and, in the rabbit, part of it joins the calcaneous tendon. It forms much of the medial wall of the popliteal fossa. The broad, thick muscle lying deep to the gracilis and cranial to the semitendinosus is the **semimembranosus.** It arises from the tuberosity and caudal border of the ischium, and inserts upon the medial epicondyle of the femur and the adjacent part of the tibia. It is a retractor of the thigh and flexor of the shank. The semimembranosus of the rabbit can be divided into two parts: the **semimembranosus proprius,** which you have exposed, and a **semimembranosus accessorius** imbedded within the proprius. In order to see the latter, carefully cut through the belly of the proprius until you come to a darker, cylindrical portion. This is the accessorius. Its origin is limited to the ischial tuberosity and its insertion to the medial condyle of the tibia.

A small, triangular-shaped **pectineus** arises from the cranial border of the pubis just caudal to the point where the femoral blood vessels emerge from the body wall. It inserts on the femoral body beside the origin of the vastus medialis. It is a thigh adductor.

An adductor group of muscles lie between the pectineus and semi-membranosus, and take their origin from the cranial border of the pubis and from the pubic and ischial symphyses. They insert along the femoral body and adduct the thigh. The group is subdivided somewhat differently in different mammals. In the rabbit, the slender, cranial component is an **adductor longus,** and the rest constitutes an **adductor brevis et magnus.** In the carnivores being studied, the complex is more or less a unit and may be called the **adductor femoris.**

(E) ILIOPSOAS COMPLEX AND ADJACENT MUSCLES

Note the thick bundle of muscle that emerges from the body wall medial to the origin of the rectus femoris. This is the iliopsoas complex. It may be studied now, or it may be postponed until the abdominal viscera have been dissected. If the former choice is made, trace the bundle forward by cutting through the muscle layers of the abdominal wall. The complex lies in a retroperitoneal position. The main part of the bundle represents the **psoas major** (Fig. 10–25, p. 319). This portion arises primarily from the bodies of the last two or three thoracic vertebrae and from the bodies of all the lumbar vertebrae. It inserts on the lesser trochanter and protracts and rotates the thigh.

Lateral and slightly dorsal to its extreme caudal portion you will see a group of fibers arising from the ventral border of the ilium plus (in the rabbit) adjacent vertebrae. These fibers represent the **iliacus.** They insert and act in common with the psoas major. Psoas major and iliacus are more intimately united in the cat and mink than in the rabbit; together they may be called the **iliopsoas.**

The thin muscle medial to the psoas major is the **psoas minor.** It arises (carnivores) from the bodies of the caudal thoracic and cranial lumbar vertebrae, or (rabbit) from the bodies of the caudal lumbar vertebrae. It inserts on the pelvic girdle near the origin of the pectineus. The psoas minor is one of the subvertebral hypaxial muscles referred to on page 136, rather than an appendicular muscle. Its action is flexion of the back.

Lift up the lateral border of the psoas major near its middle. The thin muscle that lies on the ventral surface of the transverse processes is the **quadratus lumborum,** another subvertebral hypaxial muscle. It arises primarily from the bodies and transverse processes of the lumbar and last several thoracic vertebrae. It extends further forward in the rabbit, and some of its fibers also spring from the ribs. It inserts on the ilium cranial to the origin of the iliacus. Its action is lateral flexion of the vertebral column.

(E) MUSCLES OF THE SHANK

The shank is covered by a tough **fascia cruris,** which is partly united with the tendons of certain thigh muscles, including the biceps femoris and gracilis. Remove the fascia and reflect thigh muscles inserting on the shank, but try to leave their tendons intact. Separate the more obvious shank muscles. They are very similar in cat, mink, and rabbit.

As was the case in the forearm and hand, the muscles of the shank and foot fall into extensor and flexor groups, but the groups are not quite so clearly separated as in the forearm. The extensors lie on the craniolateral surface of the shank; the flexors, on the caudomedial surface. They are separated by an exposed strip of the tibia on the medial side (Fig. 6–27); the position of the fibula indicates their separation laterally, although the fibula is not exposed at the surface (Fig. 6–26).

The large calf muscle on the caudal surface of the shank is a functional unit composed of the gastrocnemius, flexor digitorum superficialis and soleus (Figs. 6–26 and 6–27). The lateral head of the **gastrocnemius** arises from the lateral epicondyle of the femur, lateral surface of the patella and adjacent parts of the tibia and, in the cat, by a small slip from the crural fascia. Its medial

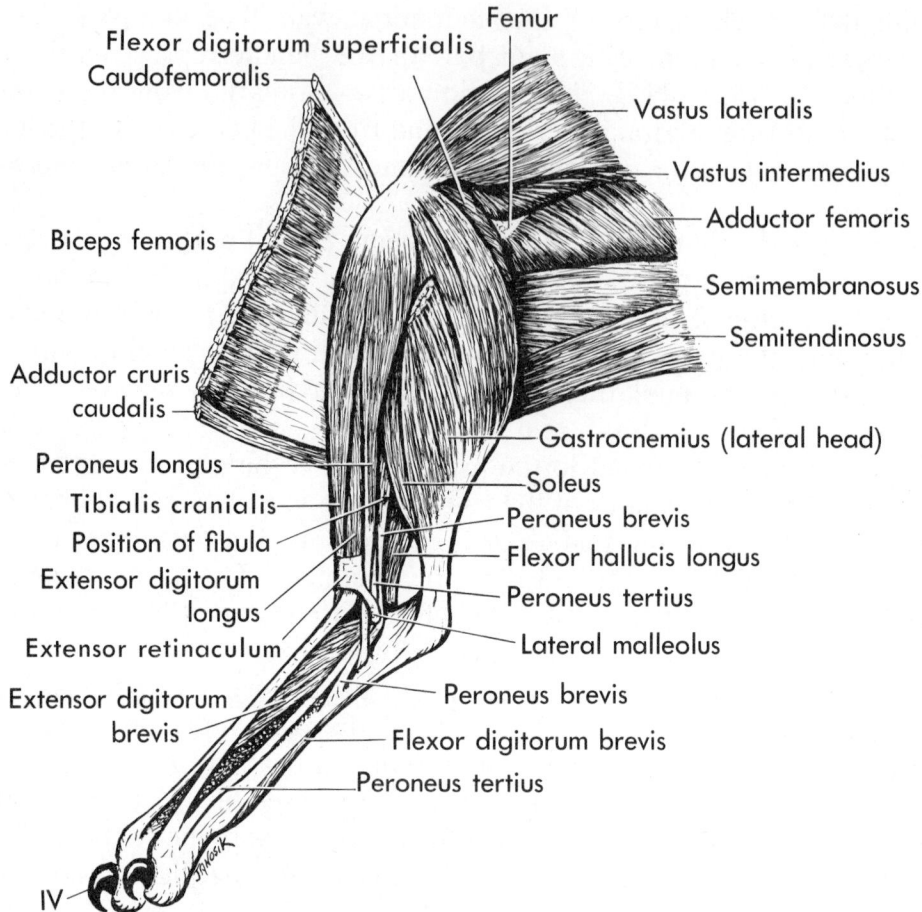

Figure 6–26. Lateral view of the shank muscles of the cat after reflection of the biceps femoris and removal of the crural fascia.

head arises from the medial epicondyle of the femur. A **flexor digitorum superficialis** (plantaris) is situated between the two heads of the gastrocnemius, where it takes its origin deep to the lateral head of the gastrocnemius from the lateral epicondyle of the femur and adjacent part of the patella. The fleshy part of the **soleus** lies deep to the distal portion of the lateral head of the gastrocnemius (Fig. 6–26). It arises from the proximal one third of the fibula in the cat, but by a narrow tendon from the proximal end of the fibula in the mink and rabbit. All converge to form a large common tendon, the **calcaneus tendon** (Achilles tendon), which inserts upon the calcaneus. As mentioned earlier, part of the tendon of the semitendinosus joins this one in the rabbit. From an evolutionary viewpoint these muscles belong to the ventral limb musculature or flexors, and their action upon the foot is one of plantar flexion. This action is often called extension in human anatomy. These muscles are particularly important in thrusting the foot upon the ground and raising the body, hence their large size.

The remaining four flexor muscles of the shank are best seen from the medial side where they lie between the tibia and the group just described

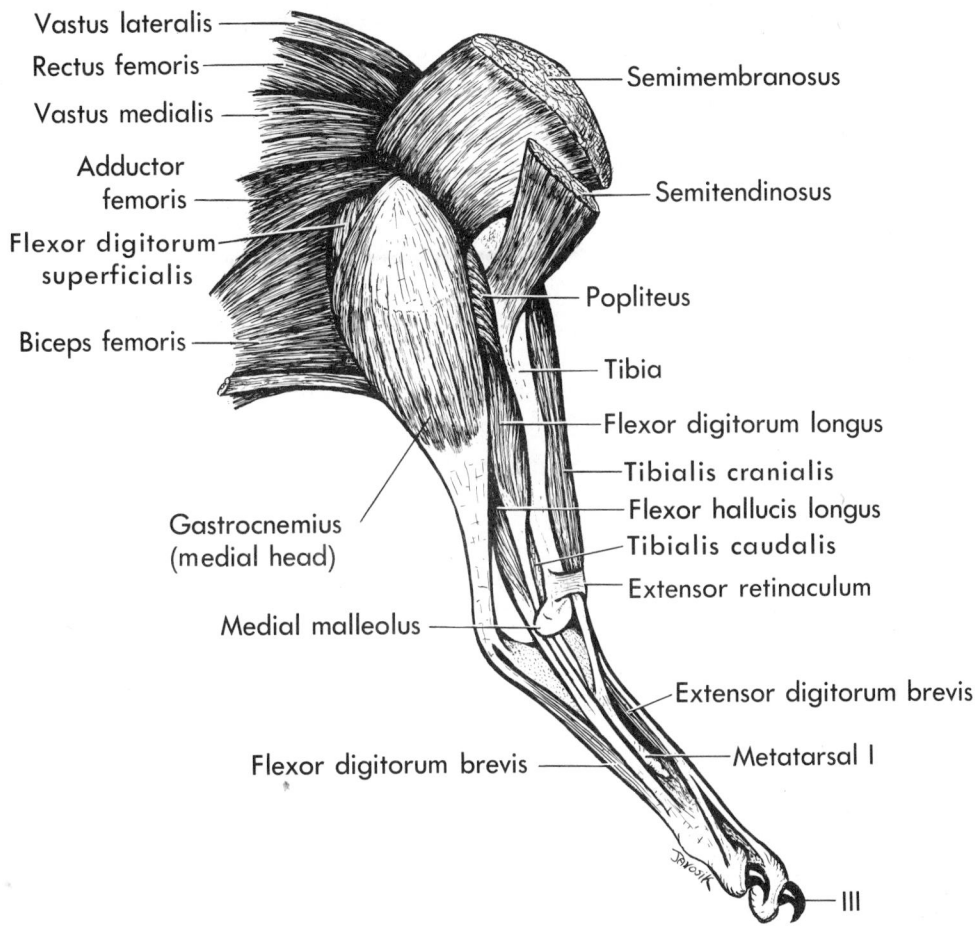

Vastus lateralis
Rectus femoris
Vastus medialis
Adductor femoris
Flexor digitorum superficialis
Biceps femoris
Gastrocnemius (medial head)
Medial malleolus
Flexor digitorum brevis

Semimembranosus
Semitendinosus
Popliteus
Tibia
Flexor digitorum longus
Tibialis cranialis
Flexor hallucis longus
Tibialis caudalis
Extensor retinaculum
Extensor digitorum brevis
Metatarsal I
III

Figure 6–27. Medial view of the shank muscles of the cat after reflection of the tensor fascia lata, gracilis, semitendinosus, and semimembranosus, and removal of the crural fascia.

(Fig. 6–27). A triangular-shaped **popliteus** arises by a narrow tendon from the lateral epicondyle of the femur, extends toward the medial side of the shank, passing caudal to the knee joint, and fans out to insert upon the proximal one third of the tibia. It helps to flex the shank and also rotates it slightly, turning the foot toward the midventral line.

The **flexor digitorum longus** arises from the head of the fibula and the shaft of the tibia deep to the insertion of the popliteus. The **flexor hallucis longus** arises from the caudal surface of much of the rest of the fibula and tibia. Each muscle forms a powerful tendon which passes caudal to the medial malleolus before uniting to form a broad tendon plate which covers much of the sole of the foot deep to an intrinsic foot muscle, the **flexor digitorum brevis.** At the level of the toes this plate breaks up into four tendons which extend down the flexor side of the digits to the terminal phalanges. These two muscles flex the toes and assist in plantar flexion of the whole foot.

Last of the flexors is a small **tibialis caudalis,** which lies between the flexores digitorum longus and hallucis longus. Its origin from the fibula and tibia extends proximally to the insertion of the popliteus. Its tendon of insertion, too, passes caudal to the medial malleolus and inserts upon certain distal tarsals. It assists the other muscles in plantar flexion of the foot.

The dorsal, or extensor, musculature of the shank is much less massive than the flexors for it is involved primarily in the recovery movements of the limb. Most cranial of the extensors is the **tibialis cranialis** (Figs. 6–26 and 6–27). It arises from about the proximal one third of the fibula and adjacent parts of the tibia, forms a long tendon which crosses the cranial surface of the tibia, goes beneath a band of connective tissue near the ankle joint, the **extensor retinaculum,** and finally inserts upon the first metatarsal. Together with other muscles in this group it extends the foot—a motion often called dorsiflexion.

The remaining extensors all lie on the lateral side of the shank (Fig. 6–26). An **extensor digitorum longus** is located caudal and deep to the tibialis cranialis, which partly covers it. It arises from the lateral epicondyle of the femur by a tendon which traverses the knee joint capsule. After passing beneath the extensor retinaculum near the ankle joint, the muscle breaks up into four tendons which pass down the dorsum of digits 2 to 5, finally inserting on the terminal phalanges. Its attachments on the digits are closely associated with those of intrinsic foot muscles, including the **extensor digitorum brevis.** The extensor digitorum longus extends the digits and assists in dorsiflexion of the foot.

A peroneus complex lies caudal to the extensor digitorum longus and takes its origin from the full length of the fibula. The complex is subdivided into three components which, although they arise from different parts of the fibula, are most distinct at their insertion (Fig. 6–26). The tendon of the **peroneus longus** passes through a groove on the surface of the lateral malleolus and then runs through a diagonal groove deep in the sole of the foot attaching onto metatarsals 2 to 4. Its primary action is abduction and eversion of the foot. The tendons of the **peroneus tertius** and **peroneus brevis** pass through a groove on the caudal border of the lateral malleolus. That of the tertius continues down the dorsum of the fifth digit, finally uniting with the extensor tendon of the extensor digitorum longus. The much stouter tendon of the peroneus brevis inserts onto the lateral side of the fifth metatarsal.

Cranial Trunk Muscles

The caudal trunk muscles were described before the appendages were dissected, and other trunk muscles were seen during the dissection of the shoulder. Now that the appendages have been examined, it is possible to resume the study of the cranial trunk muscles.

(A) HYPAXIAL MUSCLES

All the trunk muscles that become associated with the pectoral girdle belong to the hypaxial group. They are the **omotransversarius, rhomboideus, rhomboideus capitis,** and **serratus ventralis.** Find these again.

Lay your specimen of the cat, mink, or rabbit on its back, reflect the pectoralis, and examine the muscles on the ventrolateral portion of the thoracic wall. The **rectus abdominis** will be seen passing forward to its insertion on the sternum and anterior costal cartilages (Figs. 6–12; 6–13). The thoracic wall is composed of three layers of muscle comparable to those of the abdominal wall. Observe that the outermost layer, the **external intercostals,** consists of fibers that pass from one rib caudally and ventrally to the next caudal rib. This layer does not extend all the way to the midventral line. Cut through, and reflect, an external intercostal and you will find an **internal intercostal.** Its fibers extend cranially and ventrally. The third layer, **transversus thoracis,** is incomplete and found only near the midventral line. To see it, lift up the rectus abdominis, and cut through and reflect the ventral portion of an internal intercostal. The transversus thoracis arises from the dorsal surface of the sternum and is inserted by a number of slips into the costal cartilages. A better view of the muscle will be had when the thorax is opened (p. 268).

In addition to these layers, other muscles are associated with the thoracic wall. The diagonal muscle that arises near the middle of the sternum, and crosses the cranial end of the rectus abdominis to insert on the first rib, is the **rectus thoracis** (Fig. 6–12). Dorsal to it is a fan-shaped muscle complex which extends between the cervical vertebrae and the ribs. This is the **scalenus.** It arises from the transverse processes of most of the cervical vertebrae (carnivores), or the last four (rabbit), and has multiple insertions on various ribs. In the cat and mink one portion of its insertion extends as far caudad as the ninth or tenth rib. In the rabbit much of the insertion passes between the cranial and caudal halves of the serratus ventralis.

Turn your specimen on its side, reflect the latissimus dorsi, and pull the top of the scapula away from the trunk. Medial to the scapula and serratus ventralis, you will see a number of short muscular slips that arise from the thoracolumbar fascia and insert on the dorsal portion of the ribs. These constitute the **serratus dorsalis** (Fig. 6–28). This muscle can be seen better by making a longitudinal incision through the thoracolumbar fascia and reflecting it. It can be divided into cranial and caudal parts.

All these thoracic muscles, together with the muscular diaphragm and the abdominal muscles, are concerned with respiration. The muscular movements of respiration are very complex, but the major movements during inspiration are a contraction of the diaphragm and a forward movement of the ribs through the action of such muscles as the external intercostals, scalenus, and the cranial portion of the serratus dorsalis. During expiration these muscles relax; the diaphragm is pushed cranially by the contraction of abdominal muscles and the pressure of the abdominal viscera; and the ribs move caudally by their elastic recoil and by the action of such muscles as the internal intercostals, transversus thoracis, and the caudal part of the serratus dorsalis.

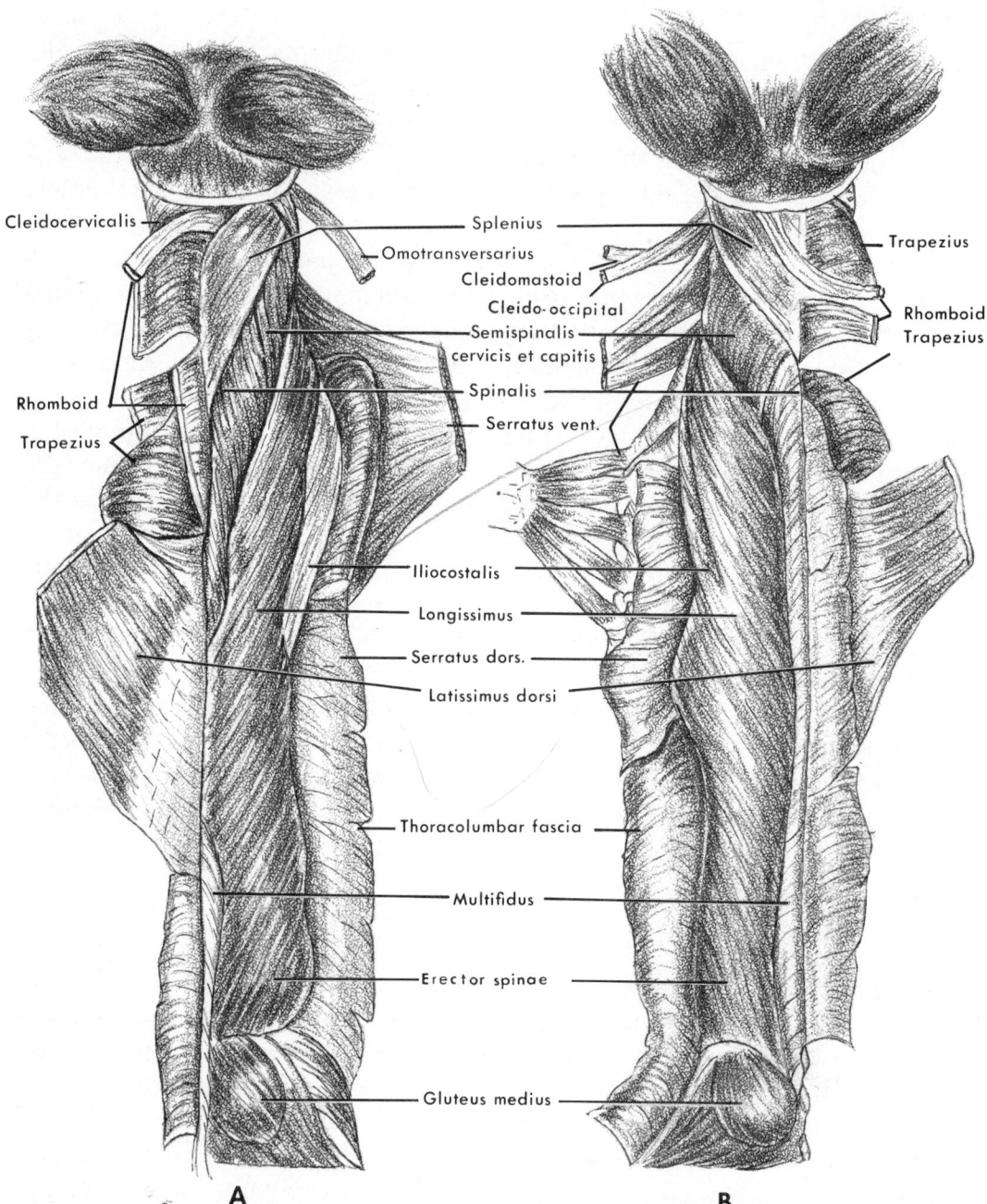

Cleidocervicalis

Rhomboid

Trapezius

Splenius

Omotransversarius

Cleidomastoid

Cleido-occipital

Semispinalis

cervicis et capitis

Spinalis

Serratus vent.

Iliocostalis

Longissimus

Serratus dors.

Latissimus dorsi

Thoracolumbar fascia

Multifidus

Erector spinae

Gluteus medius

Trapezius

Rhomboid

Trapezius

A   B

Figure 6–28. Dorsal views of the epaxial musculature of the cat *(A)* and rabbit *(B)*. The overlying muscles have been reflected, but their attachments are shown.

The subvertebral portion of the cranial hypaxial musculature is represented primarily by the **longus colli.** This muscle appears as a band in the neck lying ventral and medial to the origin of the scalenus. It arises from the bodies of the first six thoracic vertebrae, and as it passes forward it receives other slips of origin from the transverse processes and bodies of the cervical vertebrae. Portions of it insert on the bodies and transverse processes of each of the cervical vertebrae. Its action is flexion and lateral flexion of the neck.

(B) Epaxial Muscles

The most superficial of the cranial epaxial muscles is the **splenius** (Fig. 6–28). It is a thin but broad triangular muscular sheet that covers the back of the neck deep to the trapezius and the cranial portions of the rhomboideus and serratus dorsalis. The splenius arises from the middorsal line of the neck and passes forward and laterally to insert on the occipital region of the skull (nuchal crest) and transverse process of the atlas. Each splenius individually acts as a lateral flexor of the head; together they elevate the head (extension).

Much of the thoracolumbar fascia was reflected during the study of the caudal epaxial muscles and the serratus dorsalis. Complete the reflection of this fascia. The epaxial mass should now be well exposed. Find the multifidus and erector spinae posteriorly (p. 136) and trace them forward.

The **multifidus** can be seen most clearly in the lumbar region. It consists of bundles of muscle fibers that extend from the mamillary processes, transverse processes, and zygapophyses of caudally lying vertebrae to the spinous processes of more cranial ones. Most bundles cross two vertebrae between their origin and insertion. More cranial parts of the multifidus lie deep to the spinalis and will not be seen.

The **spinalis** lies lateral to the spinous processes of the thoracic and the more posterior cervical vertebrae. In the cat and mink, most of it arises from the fascia covering the erector spinae, and for this reason the spinalis is sometimes considered to be a division of the erector spinae. Deeper parts arise from the dorsal surface of the vertebrae. The spinalis inserts on the spinous processes of the vertebrae. The spinalis of the rabbit is similar except that it is not as intimately associated with the erector spinae. Fibers of both the multifidus and spinalis extend diagonally craniomedially.

As the **erector spinae** continues forward from its origin on the iliac crest and dorsal surfaces of the more caudal trunk vertebrae, it splits into a **longissimus** lying lateral to the spinalis and a more lateral **iliocostalis.** Fibers of these muscles extend diagonally craniolaterally; those of the iliocostalis insert on the ribs, those of the longissimus insert chiefly on the transverse processes of thoracic and cervical vertebrae.

The group of muscle bundles lying deep to the splenius, and arising from the vertebrae between the cranial ends of the spinalis and longissimus, constitute the **semispinalis cervicis et capitis.** They insert on the back of the skull.

All of these epaxial muscles are extensors and lateral flexors of the back, neck, and head.

As shown in Figure 6–6, the epaxial muscles of reptiles fall into three groups— transversospinalis, longissimus, and iliocostalis. The epaxial group becomes further complicated in mammals, but the reptilian subdivisions can still be recognized. The transversospinalis system of reptiles is represented by several deep muscles (interspinalis, intertransversarii, occipitals) and by the multifidus spinae, spinalis, and semispinalis; the longissimus, by the longissimus and splenius; and the iliocostalis, by the iliocostalis.

Hypobranchial Muscles

The hypobranchial muscles are an axial group located on the ventral side of the neck and throat. All move the larynx, hyoid apparatus, and tongue. The group is utilized in swallowing. Complete the skinning of this region in your specimen as far forward as the chin. Clean away the loose connective tissue and fat. You may turn back, but do not destroy, prominent glands, ducts, blood vessels, and nerves.

(A) POSTHYOID MUSCLES

Find the sternomastoid muscles (p. 139) and push them laterally. The thin, midventral band of muscle that covers the windpipe (trachea) and extends from the cranial end of the sternum to the hyoid is the **sternohyoid** (Fig. 6–29). Actually there is a pair of sternohyoids, but they are generally fused. Their origin is the sternum; their insertion, the hyoid.

Carefully separate the sternohyoid from another band of muscle that lies dorsal and lateral to it. This band, the **sternothyroid,** has a similar origin but passes forward to insert on the thyroid cartilage of the larynx. The larynx, or Adam's apple, lies caudal to the hyoid. Thus the sternothyroid is not as long

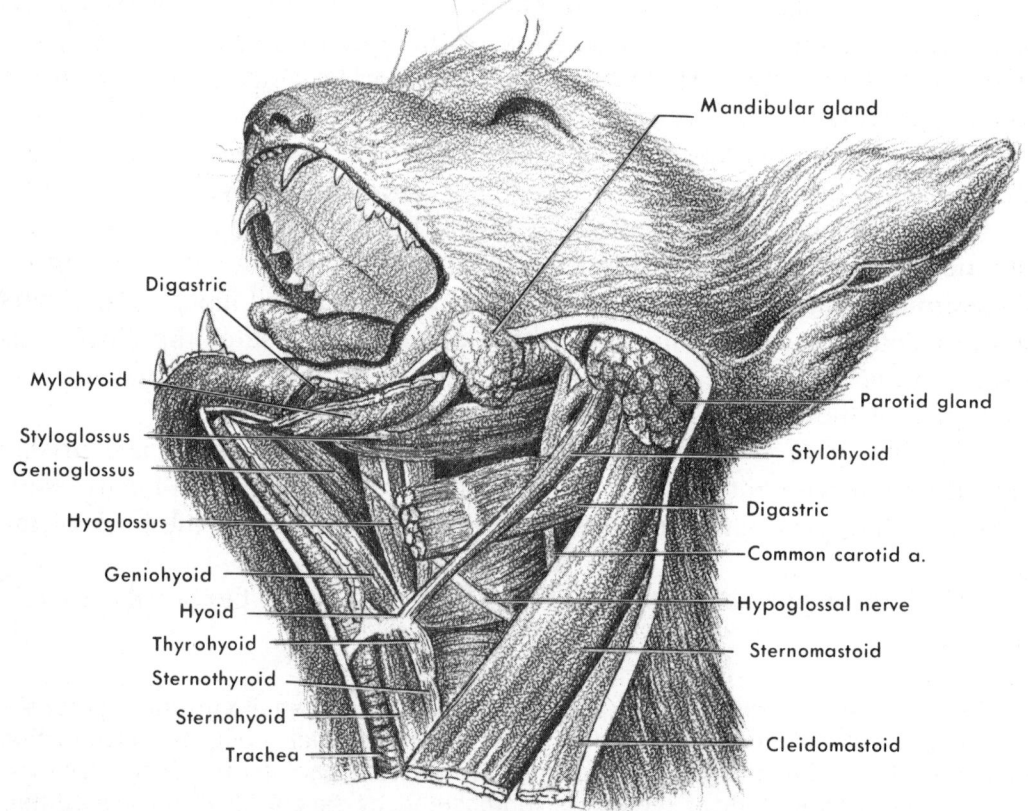

Figure 6–29. Lateroventral view of the musculature in the cranial part of the neck and the floor of the mouth in the cat.

a muscle as the sternohyoid. A short band of muscle, the **thyrohyoid,** lies on the lateral surface of the larynx. It arises at the point of insertion of the sternothyroid and passes forward to insert on the hyoid. Sternothyroid and thyrohyoid appear as one band unless their attachments on the larynx are carefully exposed.

(B) PREHYOID MUSCLES

Two branchiomeric muscles must be studied and reflected in order to get at the prehyoid muscles. Note the stout band of muscle that is attached to the ventral border of the mandible. This is the **digastric.** It has a fleshy origin in the cat and mink from the paracondyloid and mastoid processes of the skull. In the rabbit its origin is by a tendon from just the paracondyloid process. In all three animals it inserts along the ventral border of the lower jaw and acts to open the jaw. Disconnect its insertion from the jaw on one side and reflect the muscle but do not break the narrow ribbon of muscle (stylohyoid, Fig. 6–29) that extends across it. The sheet of more or less transverse fibers that lies between and deep to the insertions of the digastric muscles of opposite sides of the body is the **mylohyoid.** It arises from the mandible and inserts on a median tendinous intersection and on the hyoid. It acts to raise the floor of the mouth and to pull the hyoid forward. Make a longitudinal incision through the muscle and reflect it on one side. It is not very thick.

The longitudinal muscles that lie deep to the mylohyoid constitute the prehyoid portion of the hypobranchial musculature. It may be necessary to cut through the mandibular symphysis and spread the two halves of the lower jaw apart to see these muscles clearly. The midventral band of muscle (really a pair of muscles that have united) is the **geniohyoid** (Fig. 6–29). It arises from the front of the mandible and inserts on the hyoid. Cut across the geniohyoid and reflect its ends. The muscle that arises from the hyoid lateral, and deep, to the insertion of the geniohyoid is the **hyoglossus.** It passes forward into the tongue. Pull the tip of the tongue and you will note that the muscle is moved. The band of muscle that arises from the chin deep to the origin of the geniohyoid is the **genioglossus.** It passes caudally into the tongue, lying medial to the hyoglossus. The band of muscle that arises from the mastoid process at the base of the skull and passes forward into the tongue is the **styloglossus.** It lies lateral to the rostral portion of the hyoglossus. The glossus muscles, together with intrinsic muscle fibers within the tongue (**lingualis proprius**), form the substance of the tongue and manipulate this organ.

Branchiomeric Muscles

(A) MANDIBULAR MUSCLES

The **mylohyoid** and the rostroventral half of the **digastric,** which were seen during the dissection of the hypobranchial muscles, are branchiomeric muscles of the mandibular arch. To see other mandibular arch muscles, skin the top of the head and the cheek region on one side. The auricle should also be cut off. The platysma, facial muscles (both belonging to the hyoid group) and loose connective tissue must be removed, but be careful not to injure glands, nerves and vessels in this region. Special care should be exercised in skinning and cleaning the cheek, for the duct of one of the salivary glands crosses the cheek just beneath the facial muscles. Find the zygomatic arch. The powerful muscle that lies ventral to the arch is the **masseter.** It arises from the arch and inserts in the massenteric fossa and adjacent parts of the mandible (Figs. 6–14, 6–15, and 6–16). Deep and superficial layers can be recognized in most mammals, but they are particularly distinct in the rabbit, where the deep layer arises from the caudal part of the zygomatic arch and extends rostrally and ventrally nearly at right angles to the superficial layer.

Another mandibular muscle, **temporalis,** lies dorsal to the zygomatic arch. In carnivores it is a sizable muscle and fills the large temporal fossa. It arises primarily from the surface of the cranium, but some fibers spring from the top of the zygomatic arch. It passes deep to the zygomatic arch and inserts on the coronoid process of the mandible.

In the rabbit the temporal fossa and muscle are very small. To see the muscle, cut through a powerful facial muscle that passes from the top of the skull caudally to the base of the auricle. The **temporalis** extends from a point above the base of the auricle to the back of the orbit (Fig. 6–16). Its insertion is by a tendon that passes down the caudal wall of the orbit to the coronoid process of the mandible.

A **pterygoid** muscle, extending from the skull base to the medial side of the lower jaw, cannot be seen at this time. It will be noticed when the mouth and pharynx are opened (Fig. 9–17, p. 263).

Jaw mechanics are quite different in a cat and rabbit (Fig. 6–30). The jaw joint of a cat, as is characteristic of carnivores, is in line with the tooth row, so that the upper and lower jaws come together in the manner of a pair of scissors. The shape of the condyle and mandibular fossa are such that only a hinge action is permitted. However, the lower jaw as a whole can move slightly to the left or right side so that the carnassial teeth can engage more intimately on one side or the other. In the rabbit, as is characteristic of gnawing and herbivorous mammals, the jaw joint is situated well dorsal to the tooth row so that all of the teeth of the upper and lower jaw come together simultaneously. The shape of the condyle and mandibular fossa permits the lower jaw to move back and forth and from side to side, actions which occur in gnawing and grinding.

The adductor muscles of the jaws must exert a strong force for closing the jaws, and also balance forces at the jaw joint so that there is little tendency for the jaws to become disarticulated. Both masseter and temporal muscles are large in carnivores, but a low condyle and a high coronoid process give the temporal muscle a somewhat longer moment arm (the perpendicular distance between the line of action of the muscle and the jaw joint), and hence a greater mechanical advantage, than the masseter. Both exert strong forces for closing the jaws. The direction of pull of the temporal muscle also resists any forward pull on the canine that may occur when a carnivore is seizing prey, and the pull of the masseter helps to resist any tendency for the jaw

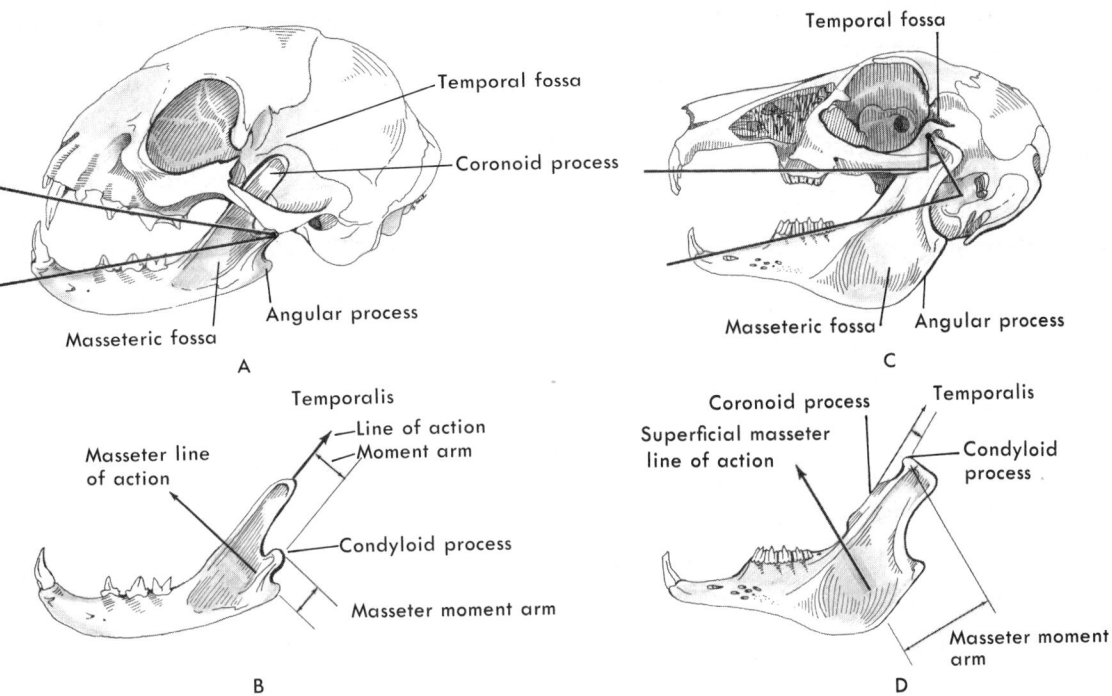

Figure 6–30. Jaw mechanics in a cat (*A* and *B*) and rabbit (*C* and *D*). *A* and *C,* jaw closure; *B* and *D,* lines of action and moment arms of the masseter and temporalis muscles.

condyle to slip ventrally and caudally out of the mandibular fossa, which could occur if the temporal muscle alone were used in cutting up the food. The pterygoids of carnivores are modest-sized muscles whose actions are similar to those of the masseters.

The masseter of herbivores is a very large muscle divided into superficial and deep layers that are oriented nearly at right angles to each other. The superficial layer provides a particularly strong force for closing the jaws because it has a very long moment arm. Its tendency to pull the lower jaw forward out of the mandibular fossa is opposed by the deep layer. These two layers, and a large pterygoid, provide the forces needed to close the jaws, move them back and forth and from side to side in a grinding action, and balance forces at the jaw joint. The temporal muscle is not an important muscle in herbivores, and it, together with its points of attachment on the skull and jaws, is small.

(B) HYOID MUSCLES

The major hyoid arch muscles are the **platysma, facial muscles,** caudal half of the **digastric** and **stylohyoid.** The first three have been seen. The stylohyoid is a small ribbon of muscle that lies lateral (carnivores) or caudal (rabbit) to the caudal portion of the digastric (Fig. 6–29). In the carnivores it arises from the stylohyal bone and inserts on the body of the hyoid. In the rabbit it arises from the base of the skull and inserts onto the greater horn of the hyoid. It acts on the hyoid.

Another hyoid arch muscle, which is not feasible to find at this time, is the stapedius. It lies within the middle ear and acts on the stapes. The muscle is described more fully on page 188.

Considerable variation is seen in the digastric of different mammals. In many species, human beings among them, the rostral and caudal halves of the muscles form distinct bellies which are interconnected by a prominent, round tendon. This is also the situation during one of the embryonic stages of the rabbit, but, as development proceeds, the posterior belly is reduced to a tendon. In other mammals the two halves of the digastric are united by a tendinous intersection. Such an intersection is present embryologically in carnivores, but it disappears in the adult.

(C) CAUDAL BRANCHIOMERIC MUSCLES

Much of the caudal branchiomeric musculature is lost during the course of evolution, but some becomes associated with the pectoral girdle as the **trapezius** and **sternocleidomastoid** complexes. These were described in connection with the shoulder region. Some of the remaining caudal branchiomeric musculature forms the intrinsic muscles of the larynx (e.g., **thyroarytenoid, cricoarytenoid, cricothyroid;** (see Fig. 9–18, p. 265), and some is contributed to the wall of the pharynx. Certain of the intrinsic muscles of the larynx can be seen on the ventral surface of the larynx deep to the cranial end of the sternohyoid.

7

THE SENSE ORGANS

Although the sense organs and nervous system are concerned with the integration of the activities of all parts of the body, they may appropriately be considered at this time, for the most conspicuous effector organs are the muscles described in the previous chapter. The sense organs, the central nervous system, and the basic pattern of distribution of the peripheral nerves will be the topic of this and the following chapter. If separate heads cannot be provided for the study of *Squalus,* and separate brains for the mammal, this unit of work should be postponed until the end of the course.

Irritability

Irritability, that is, the ability to receive sensations and respond to stimuli, is a basic property of protoplasm. All the aspects of irritability are combined in the individual cells of the more primitive organisms, but in higher animals there is a division of labor. Certain cells receive, others transmit, and still others respond to the stimuli. The receptive cells are the sense organs, or **receptors;** the transmitting cells are the neurons of the **nervous system;** and the responding cells are the **effectors** (muscles, glands, cilia, etc.).

Classification of Sense Organs

The stimuli causing the sensation of pain may be received by free nerve endings, but stimuli producing other sensations in vertebrates are received by very specific sense organs. These range from microscopic sensory corpuscles to such large and complex organs as the eye. As with the skeleton and muscles, this array of sense organs, and the sensory nerves which lead from them, may be divided into two groups— somatic and visceral. **Somatic sensory organs** lie in the "outer tube" of the body—the skin (exteroceptors of the physiologist) and the somatic muscles (proprioceptors of the physiologist). Occasionally proprioceptive organs, which are the organs of muscle sense, are found in branchiomeric muscles. **Visceral sensory organs** (interoceptors) are associated with the "inner tube" of the body, i.e., the viscera. Only those sense organs that can be seen grossly will be considered.

THE LATERAL LINE SYSTEM

Fishes and larval amphibians have a unique sensory system, known as the lateral line system, by which currents and other water movements can be detected. Experiments on a variety of fishes have shown that the lateral line system enables them to localize objects at a distance, even in turbid water where visibility would be reduced, either by the disturbances produced by a moving object or by reflected water waves set up by their own movements. The system has been called one of "distant touch." In addition, the system enables a fish to detect its own body movements. This could be important, for fish lack the proprioceptive muscle spindles found in terrestrial vertebrates. The actual receptive organs in the system are groups of mechanoreceptive cells termed **neuromasts.** Hairlike processes from each cell in a neuromast protrude into a cap of gelatinous material, the **cupula,** that is bent by movements of the water. The distribution of the neuromasts varies. Most lie in canals, one of which extends along the side of the trunk; the others ramify over the head. These **lateral line canals** are usually beneath the skin but open to the surface by pores. In a few fishes the canals take the form of open grooves. Other neuromasts are in isolated, but often linearly arranged, pits (**pit organs**). The neuromasts have a spontaneous activity, but the level of this activity can be altered by water movement along the flank. In a ray, perfusion of the lateral line canal from the head toward the tail increases the frequency of discharge of certain units and decreases that of others. Perfusion in the opposite direction has opposite effects. A special group of somatic sensory neurons, called the **lateralis neurons,** lead from the neuromasts and enter the brain through the seventh, ninth and tenth cranial nerves.

Cartilaginous fishes also have an unusual sensory system known as the **ampullae of Lorenzini.** The system appears to be a modification of the lateral line system because the receptors are neuromastlike cells supplied by lateralis neurons. The receptors lie in little swellings (ampullae) at the base of jelly-filled tubes that open on the body surface. The tubes lie just beneath the skin and parallel to it. They vary in length and taken all together extend in all directions, i.e., some groups extend rostrally, some caudally, some medially, and so forth. Although the ampullae do respond to tactile stimuli and to temperature and salinity changes, their primary function is believed to be electroreception. The skin has such a high electrical resistance that a small voltage change in the external environment does not affect deeper tissues directly through it. The ampullae, however, act as electrical capacitors. They can hold a voltage change for a moment and transmit it without a voltage drop through the receptive cells. The length and direction of a tube determines the stimulus intensity at the receptive cells. Tubes directly in line with the voltage gradient would be most affected. This would enable the fish to detect the source of the voltage. The ampullae have been shown to be very sensitive; some can detect a two microvolt change (Murray in Cahn, 1967). Skates have been trained to give conditioned responses to the electrical potentials emanating from the respiratory muscles of a buried flatfish up to 16 centimeters away. The system may enable fish possessing it to detect hidden prey. It is also sensitive enough so that a skate could detect the weak electric currents set up as it moves through the earth's magnetic field.

As stated, the lateral line system is limited to fishes and larval amphibians. It is completely lost during the evolution of reptiles and is never reacquired, even in such aquatic tetrapods as the Cetacea. Parts of the system have been noted during the study of the external features of *Petromyzon* (p. 17), *Squalus* (p. 34), and *Necturus* (p. 37). The system will be considered in detail only in *Squalus.*

PIT ORGANS

Pit organs occur on various parts of the head of the dogfish but cannot be distinguished grossly from other lateral line organs.

AMPULLAE OF LORENZINI

Examine your specimen of a large head of *Squalus*. The pores on the snout through which a jellylike substance extrudes when the area is squeezed are the openings of the ampullae of Lorenzini. Note their distribution. They are very abundant on the ventral surface of the snout and form two, large, V-shaped groups on the dorsal surface of the snout (Fig. 7–1). Other smaller groups are located caudal to the jaws. Make a deep V-shaped cut through the skin in one of the dorsal snout patches. The apex of the V should be directed caudally. Pull the skin flap and underlying tissue forward and examine its underside. Note that each pore leads into a long, jelly-filled tube which, in this region, extends rostrally to a round swelling, the ampulla proper. Small nerve twigs attach onto each ampulla.

LATERAL LINE CANALS

Make a fairly long transverse cut through the skin, caudal and medial to the spiracle. Examine the cut surface. The hole that you can see in the deeper layers of the skin is a cross section of the **lateral line canal** proper. This canal extends the length of the trunk and tail. Trace it forward by carefully slicing off the superficial layers of the skin with a sharp scalpel. The object is to make a horizontal section of the canal, taking the top half off and leaving the bottom half on the specimen. Keep your eye on the canal as you expose it to be sure you are cutting at the right level. Pores leading into the canal can be seen on the underside of the skin you remove. Farther forward on the head the lateral line canal leads into others (Fig. 7–1), which should be uncovered in the same manner.

A **supratemporal canal** crosses the top of the head caudal to the endolymphatic pores. An **infraorbital canal** passes ventrally caudal to the eye, zigzags beneath the eye, and then extends forward to the tip of the snout in a somewhat meandering fashion, passing medial to the nostril. A **hyomandibular canal** extends caudally from the bottom of the zigzag of the infraorbital canal. A **supraorbital canal** passes forward dorsal to the eye and onto the snout. It then turns on itself, and extends caudally to connect with the infraorbital canal. The latter portion of the supraorbital canal passes just dorsal to the nostril. A short **mandibular canal** is not connected with the others. It appears as a row of pores overlying the mandible caudal to the labial pocket. Make a cut through the skin at right angles to the row of pores, examine the cut surface, and verify that a canal is present.

THE EYEBALL AND ASSOCIATED STRUCTURES

The eyes are somatic sensory organs. Two types are found in vertebrates, the conventional pair of **lateral eyes** and a **median eye** on the top of the head. Median eyes, either **pineal** or **parietal**, develop as outgrowths from the diencephalic region of the

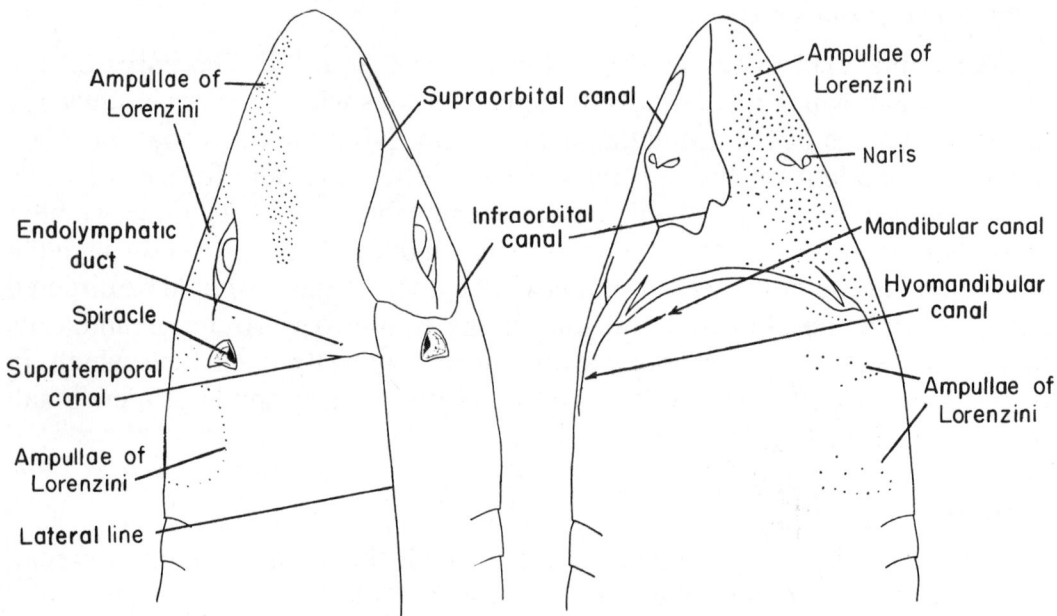

Figure 7-1. Dorsal and ventral views of the lateral line canals, and ampullae of Lorenzini, of the dogfish, *Squalus suckleyi.* (From Daniel, The Elasmobranch Fishes, University of California Press.)

brain (Fig. 7-2, *A*). They are not visual but simply light receptive. As one ascends the evolutionary scale, the parietal eye becomes adapted primarily for light reception. It is well developed in many reptiles, in which it detects the amount of solar radiation and initiates behavioral responses by which the animal regulates its body temperature. The parietal eye is lost in birds and mammals, which have evolved physiological mechanisms of thermoregulation. The pineal eye, or epiphysis, tends to lose its photo-receptive function and become glandular, forming in mammals the well-developed pineal gland (p. 218). Secretion of the pineal gland is affected by light received by the parietal eye or in other ways.

It is with the lateral eyes that you will be concerned at this time. Although the eye-ball differs in its method of accommodation and in its adaptive details, its basic anatomy is much the same in all vertebrates. Evolutionary tendencies, however, can be seen in associated structures. For example, the surface of the fish eye is bathed in water which keeps it moist and clean. Tetrapods have evolved movable eyelids of various types, as well as tear glands and ducts that protect, cleanse, and moisten the eye surface.

Fishes

Study the eye of *Squalus* on the same side of the head on which you dissected the lateral line. *Squalus,* like other cartilaginous fishes, has an upper and lower **eyelid** formed of skin folds, but they are immovable. (Most fishes lack eyelids altogether.) Note that the epidermis on the inner surface of the eyelids reflects onto and over the surface of the eyeball. This layer is the **conjunctiva.**

Remove the upper eyelid and free the eye from the lower lid by cutting through the conjunctiva. The **eyeball** *(bulbus oculi)* lies in a socket, the **orbit,**

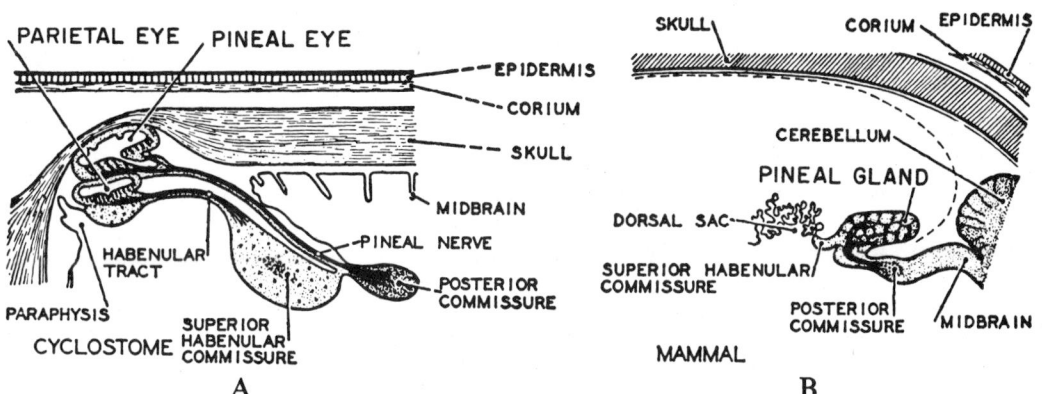

Figure 7–2. Sagittal diagrams of *A,* the pineal eye of a lamprey; *B,* the pineal gland of a mammal. It will be noted that the lamprey also has a parietal eye, but that this does not reach the surface. A parietal eye connects with the superior habenular commissure, a pineal eye with the posterior commissure. (From Rand, The Chordates, Blakiston Company. After Studnicka.)

on the side of the chondrocranium. Cut away the cartilage that forms the roof of the orbit (supraorbital crest, antorbital process, postorbital process), but do not cut into the otic capsule. A mass of gelatinous connective tissue surrounds and helps to support the eye. It must be picked away.

A group of ribbon-shaped muscles passes from the medial wall of the orbit to the eyeball. These are the extrinsic muscles of the eye, and they are responsible for the various movements of the whole eyeball. As explained on page 120, these muscles belong to the axial subdivision of the somatic muscles. Although they are derived from three myotomes (Table 4), they fall grossly into an oblique and a rectus group. Push the eyeball caudally and note the two muscles that arise from the rostromedial corner of the orbit and insert on the eye (Fig. 7–3). The one that inserts on the dorsal surface is the **dorsal (superior) oblique;** the one that inserts on the ventral surface, the **ventral (inferior) oblique.** The muscles that arise from the caudomedial corner are all recti. Three can be seen in a dorsal view. The one that inserts on the top of the eyeball adjacent to the insertion of the dorsal oblique is the **dorsal (superior) rectus;** the one that lies medial to the eye is the **medial rectus;** the one that lies caudal to the eye is the **lateral rectus.** Lift up the eye and look at its ventral surface. A **ventral (inferior) rectus** passes to insert on the ventral surface of the eyeball beside the insertion of the ventral oblique.

Most of the other strands that are seen passing to the muscles and through the orbit are nerves and will be considered later; at this time find the **optic nerve.** It is the large nerve passing from the eyeball caudal to the ventral oblique.

Cut across the extrinsic muscles of the eye near their insertions, trying not to injure the nerves going to them. The stalk of cartilage that will be seen passing to the back of the eyeball between the four rectus muscles is the **optic pedicle.** It is shaped like a golf tee and helps to support the eyeball. Disconnect it from the eyeball and also free the small nerve that crosses the

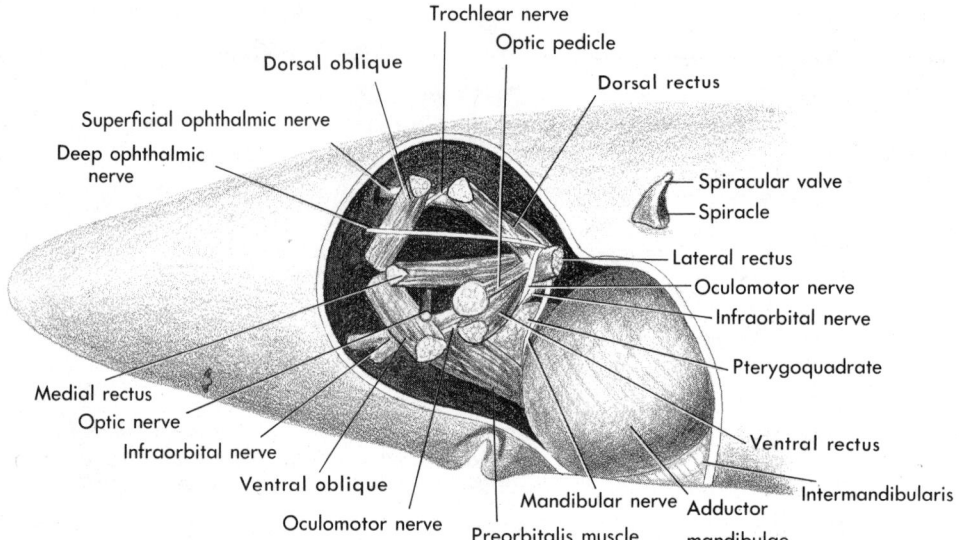

Figure 7–3. Lateral view of orbit of *Squalus* after removal of the eyeball.

medial surface of the eye. Cut the optic nerve and remove the eye. Some **conjunctiva** will cling to the front of the eye.

The outermost layer of the eyeball proper is a tough **fibrous tunic,** the anterior portion of which is modified to form the transparent **cornea,** through which you can see an opening, the **pupil,** surrounded by the pigmented iris. Conjunctiva and cornea are fused.

Submerge the eyeball in a dish of water and cut off its dorsal third. Note the large, spherical **lens.** Try not to pull it away from surrounding tissues. The three layers or coats of which the eyeball is composed can be seen at the back of the eye. As mentioned, the outermost layer is the fibrous tunic. Its front portion is modified as the cornea, the rest of it constitutes the **sclera** (Fig. 7–4). Much of the dogfish's sclera is cartilaginous, and this provides a great deal of support for the eyeball. The pigmented layer internal to the sclera is the **vascular tunic,** most of which forms a **choroid.** The **iris** is the modified anterior portion of the vascular tunic, and the **pupil** is a hole through the iris. The whitish layer internal to the choroid is the **retina.** It is an incomplete layer that disappears near the lens. The point at which the optic nerve connects with retina is the **optic disc,** sometimes called the blind spot because there are no receptive cells here.

Carefully pull the lens away from the iris. A ring of black material will probably adhere to it. This black material is in the form of small radiate folds and represents the **ciliary body** — a modification of the vascular tunic located near the base of the iris. A gelatinous material (the **zonule**), which will not be seen, extends between the ciliary body and the lens. The lens is supported by the ciliary body, zonule, a middorsal suspensory ligament (which cannot be seen grossly) and the **vitreous body.** The vitreous body is the gelatinous material that fills the large cavity (**chamber of the vitreous body**) between the lens and the retina. Other cavities, which are filled with a watery

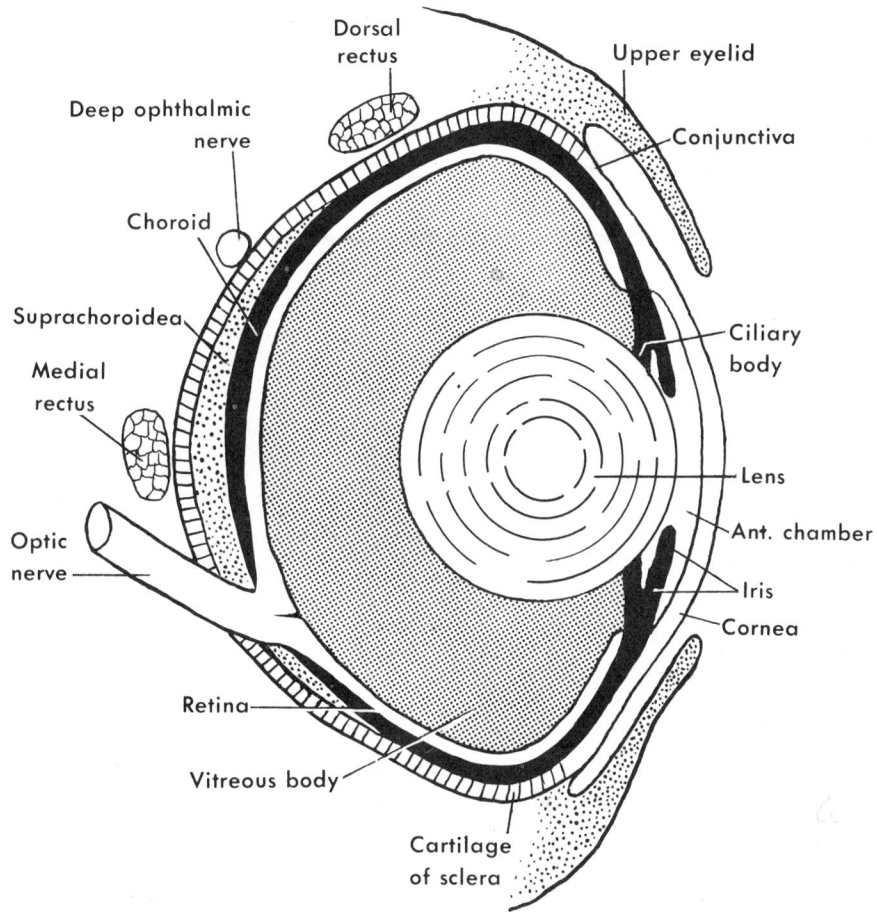

Figure 7–4. Vertical section through the right eyeball of *Squalus* at the level of the optic nerve. Viewed from behind.

aqueous humor, are located in front of the lens. The very small cavity that lies between the lens and the iris is the **posterior chamber;** the larger one between the iris and the cornea, the **anterior chamber.**

Cut across the back of the eyeball where the optic pedicle attaches. The relatively thick layer of material between the sclera and choroid is known as the **suprachoroidea.** It is a vascular connective tissue that is found only in species with an optic pedicle.

Cut into the lens and observe that it is composed of layers of modified epithelial cells arranged concentrically like the skins of an onion.

The sclera is the supporting layer of the eyeball, and the presence of cartilage within it is not surprising, since the sclera develops in part from the optic capsule of the embryonic chondrocranium (p. 47). The **sclerotic bones** present in some large-eyed vertebrates ossify in this cartilage. The choroid performs several functions. It is vascular and helps to nourish the light-sensitive, avascular retina. Its pigment, along with pigment of the retina, prevents internal reflections of light that would otherwise blur the image. In addition, the choroid of many vertebrates, including the dogfish, is so constructed that it can reflect some light back onto the retina. Such a reflecting device, known as a **tapetum lucidum,** is found in vertebrates that live under conditions of low light intensity—certain fishes and nocturnal tetrapods. The elasmobranch tapetum depends on **guanine crystals** within certain choroid cells. It is among the most remarkable of vertebrate tapeta in that adjacent pigment cells have the ability to extend or retract their pigment across the guanine layer, thus adapting the eye to light or dark conditions (Fig. 7–5). Further adjustments to light and dark are, of course, made by the iris. The dogfish's iris is unusual in that the sphincter muscle fibers within it are independent effectors which contract upon direct stimulation by light. The radial muscle fibers in the iris, which dilate the pupil, are innervated by neurons in the oculomotor nerve.

Since the refractive index of the cornea and humors of the eye is nearly the same as that of water, they do not take part in the refraction of light rays. Nearly all the refraction in fishes occurs at the thick, spherical lens. In elasmobranchs the lens is held in such a position that distant objects are in focus on the retina. In bright light, acuity can be increased, regardless of the distance of the object, by the contractions of the iris until the pupil is a pinpoint. Additional accommodation for near objects is accomplished by the forward movement of the lens through the contraction of a small **protractor lentis** muscle that is located ventrally in the ciliary body. The pressure of the aqueous humor pushes the lens back to its resting position when this muscle relaxes.

Mammals

We may pass directly to the mammal eye, for that of the aquatic *Necturus* contributes little to an understanding of the evolution of this organ. The basic structure of the mammalian eyeball is much the same as in *Squalus*. Details differ, however. Ac-

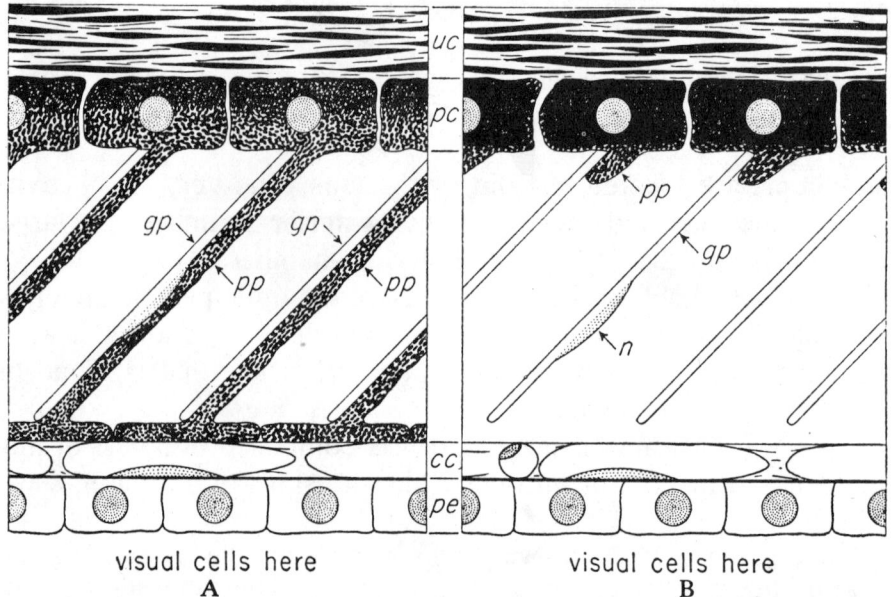

Figure 7–5. Diagrams of section through the choroid of the dogfish *Mustelus mustelus,* to show the tapetum lucidum adapted for *(A)* light, and *(B)* dark conditions. In the light-adapted eye, processes of the pigment cells cover the guanine plates; in the dark-adapted eye they are withdrawn and allow the guanine to reflect light back to the visual cells of the retina. Abbreviations: *cc,* choroid capillaries; *gp,* guanine plates; *n,* nucleus of guanine cell; *pc,* migratory pigment cells of the choroid; *pe,* pigmented layer of the retina (in this case devoid of pigment); *pp,* pigment processes; *uc,* unmodified pigment cells of the choroid. (From Walls, The Vertebrate Eye, Cranbrook Institute of Science. After Franz.)

commodation is accomplished primarily by changing the shape of a biconvex lens. When the eye is at rest, distant objects are in focus. To accommodate for near objects, the curvature of the front of the lens is increased through the action of muscles in the ciliary body.

Examine the eye of your specimen on the side that was used for the study of the muscles. Movable **upper** and **lower eyelids** (*palpebrae*) are present. The slitlike opening between them is called the **palpebral fissure.** The corners of the eye where the lids unite are the **ocular angles.** Cut through the lateral ocular angle and pull the lids apart. The **nictitating membrane** (*semilunar fold*) can now be seen clearly. It is attached at the medial ocular angle, but its lateral edge can spread over the surface of the eye if the eyeball is retracted slightly. In man the nictitating membrane is reduced to a small, semilunar fold that can be seen covering the medial corner of the eye. Examine the edge of each lid three or four millimeters from the medial ocular angle with a hand lens. If you are fortunate, you will see on each a minute opening (**lacrimal punctum**) that leads into a lacrimal duct. If you cannot find them in your specimen, look for one on the human eye by pulling down the lower lid and examining its edge near the most medial eyelash.

Cut off the upper and lower lids, leaving a bit of skin around the medial ocular angle. As you remove the lids, note that the **conjunctiva** on the underside of the lids reflects over the **cornea.** If the cornea is not too opaque, the **pupil** and **iris** can be seen. A facial muscle, the **orbicularis oculi,** encircles the eyelids and will be cut off with them. It closes the lids.

Free the eyeball and associated glands from the bony rims of the orbit by picking away connective tissue. Do not dissect deeply in the region of the medial ocular angle, and try not to destroy a loop of connective tissue attached to the rostrodorsal wall of the orbit. One of the ocular muscles passes through the latter. Using bone scissors, cut away the zygomatic arch beneath the eye, the postorbital processes and the crest of bone above the orbit (supraorbital arch). Push the eye rostrally. The dark glandular mass that lies on the caudodorsal surface of the eyeball is the **lacrimal gland.** It is larger in the rabbit than in the cat or mink and extends ventral to the eyeball (Fig. 7–6). Its secretions, the tears, enter near the lateral ocular angle, bathe the surface of the cornea, and pass into the lacrimal duct through the **lacrimal puncta.** The lacrimal duct enters the nose through a canal in the **lacrimal bone.** Find the nasolacrimal canal on a skull (p. 85). You may be able to find the lacrimal duct by dissecting ventral and rostral to the medial ocular angle. First find the rostral border of the orbit and the position of the nasolacrimal canal. Do not injure a muscle that is attached to the orbital wall near the lacrimal bone.

A second tear gland, the **gland of the nictitating membrane** (Harderian gland), lies just rostral to the nictitating membrane. It is not large in carnivores but is quite large in the rabbit, and part of the rabbit's gland extends ventrally, going deep to the ventral oblique muscle, to appear on the underside of the eye.

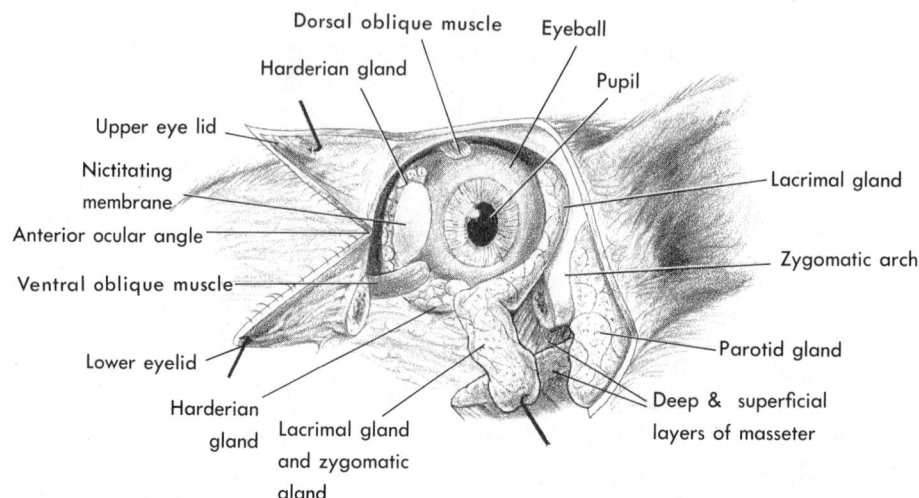

Dorsal oblique muscle Eyeball

Harderian gland

Pupil

Upper eye lid

Nictitating
membrane

Lacrimal gland

Anterior ocular angle

Zygomatic arch

Ventral oblique muscle

Parotid gland

Lower eyelid

Deep & superficial
layers of masseter

Harderian
gland   Lacrimal gland
and zygomatic
gland

Figure 7–6. Dissection of the lacrimal apparatus of the rabbit's eye. Anterior is toward the left.

The cat, mink, and rabbit have a **zygomatic gland** located in the floor of the orbit ventral to the eye. It is difficult to separate it from the ventral portion of the lacrimal gland in the rabbit. The zygomatic gland is not a part of the lacrimal apparatus; rather, it secretes saliva that is discharged into the mouth at a point near the last upper tooth.

Remove the glands and connective tissue from around the eye of your cat or rabbit and expose the extrinsic muscles of the eyeball. The pattern of the muscles is much the same as in fishes, except for two additional muscles. A **levator palpebrae superioris** arises from the medial wall of the orbit dorsal to the optic foramen and inserts on the upper eyelid which it raises. Its lateral end may be found unattached, for its insertion was probably cut in removing the eyelids. The rest of these muscles move the entire eyeball. Two oblique muscles pass to the rostral wall of the eyeball. The **dorsal** (superior) **oblique** arises from the wall of the orbit slightly rostral to the optic foramen and goes through a connective tissue pulley (the **trochlea**), which is attached to the rostrodorsal wall of the orbit, before inserting on the eye. The **ventral** (inferior) **oblique** arises from the maxillary or lacrimal bone. Four recti arise from the margins of the optic foramen and pass to the caudal portion of the eyeball. A **dorsal** (superior) **rectus** inserts on the dorsal surface of the eyeball; a **ventral** (inferior) **rectus,** on the ventral surface; a **medial rectus** on the medial surface; and a **lateral rectus** on the lateral or posterior surface. A **retractor bulbi,** which can be divided into four parts, passes to the eye deep to the recti. The derivation of the extrinsic ocular muscles are shown in Table 4.

The structure of the mammalian **eyeball** *(bulbus oculi)* is similar in basic structure to that of a fish. Directions for its dissection are included here for those who wish to make a comparison or prefer to study the mammal's eye instead of the fish's. It can be seen most clearly in a large eye such as that of a cow or sheep, although your specimen's eye may be used. Carefully clean up the surface of the eyeball by removing the extrinsic ocular muscles and associated fat. Notice that the optic nerve attaches somewhat excentrically, toward the rostroventral portion of the eyeball. Leave a stump of it attached. Some conjunctiva will adhere to the surface of the cornea.

Open the eyeball by carefully cutting a small window through its dorsal surface. Observe that its wall consists of the same three layers as in a fish's eye: an outer **fibrous tunic;** a middle dark layer, the **vascular tunic;** and the inner, whitish **retina** (Fig. 7–7). The fibrous tunic is a dense, supporting connective tissue. Approximately the medial two thirds of it form the opaque **sclera;** the lateral one third, the transparent **cornea** through which light enters the eye. The vascular tunic is rich in blood vessels; most of this layer is a **choroid** that lies behind the retina and helps to nourish it. The portion of the retina which you see is the nervous layer which contains the receptive rods and cones on its choroid surface. Embryonically there is a pigmented layer to the retina, but this becomes associated with the choroid in the adult. The pigment reduces light reflections within the eye.

With a pair of fine scissors extend a cut from the window which you made completely around the equator of the eyeball, thereby separating the eyeball into an anterior half containing the lens and cornea and a posterior half containing most of the retina. Cut through all the layers. The jellylike mass filling the eyeball between the lens and retina is the **vitreous body,** a medium which helps to support the lens and also helps to refract light entering the eye. The space in which it lies is the **chamber of the vitreous body.** Keep the vitreous body with the anterior half of the eyeball. Submerge both halves in a dish of water.

Examine the posterior half (Fig. 7–7, *A*). The retina has probably become partly detached from the choroid but can be floated back into its normal position. Note the round spot (**optic disc**) at which the optic nerve attaches to the retina. This region is devoid of rods and cones, and hence is often called the blind spot. Remove the retina and observe that an extensive section of the choroid dorsal to the optic disc is quite iridescent. This is the **tapetum lucidum,** an area which reflects some of the light passing through the

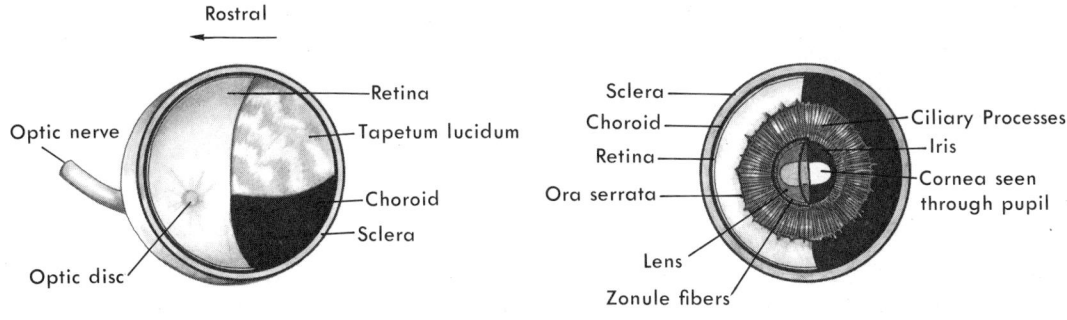

Figure 7–7. Dissection of a cow eye. *A,* Posterior half; *B,* anterior half. A portion of the retina has been removed in each half to show the vascular tunic.

retina back onto the rods and cones and, hence, facilitates the animal seeing in dim light. This tapetum is dependent on a compact layer of endothelial cells in the choroid and not on guanine granules. Many mammals, including human beings, lack a tapetum.

Carefully remove the vitreous body from the anterior half of the eyeball; notice that the white, nervous layer of the retina does not extend far into this half of the eyeball. The line of demarkation between it and the dark choroid (plus embryonic pigment layer of retina) is the **ora serrata** (Fig. 7–7, *B*). The portion of the vascular tunic which you see extending from the ora serrata toward the **lens** is the **ciliary body.** The portion of it next to the lens has a pleated appearance; the individual folds are the **ciliary processes.** While observing the area with a dissecting microscope, carefully stretch the region between the lens and the ciliary processes. Many delicate **zonule fibers** will be seen passing from the ciliary body to the equator of the lens.

Refraction and accommodation differ in mammals and fishes. Intraocular pressures cause the wall of the eyeball to bulge outward, and this force is transmitted via the zonule fibers to the elastic lens, which is consequently under tension and somewhat flattened. Under these circumstances the lens has its minimum thickness and refractive powers, so the eye is focused on distant objects. The major refraction of light is caused by the cornea in a terrestrial vertebrate because of the sharp contrast between its index of refraction and that of air; the role of the lens is more analogous to the fine adjustment of a microscope. Accommodation for a close object requires a greater bending of light rays. Muscle fibers in the base of the ciliary body contract and bring the base of the ciliary body a bit closer to the lens. This releases the tension on the zonule fibers and permits the lens to bulge and increase its thickness.

Carefully remove the lens and notice that the vascular tunic continues in front of it to form the **iris.** The **pupil,** of course, is the opening through the iris. Its diameter, and the amount of light it permits to pass, is regulated by circular and radial muscle fibers within the iris. The space between the lens and iris is the **posterior chamber;** that between the iris and cornea, the **anterior chamber.** Both are filled with a watery **aqueous humor** produced by the ciliary processes. This liquid maintains the intraocular pressure. Excess liquid is drained off by a microscopic **scleral venous sinus** (canal of Schlemm), which encircles the eye between the base of the cornea and the iris. If one makes a vertical cut through the anterior half of the eyeball and examines it under a dissecting scope, this canal and the ciliary muscles can sometimes be seen.

THE NOSE

In fishes the nose typically consists of a pair of **olfactory sacs,** each of which connects to the surface by a pair of external nostrils (**nares**), through which water carrying odoriferous particles circulates. In sarcopterygian fishes and tetrapods, each olfactory sac, now usually called a **nasal cavity,** has but one naris, but each opens into the mouth through an internal nostril (**choana**). Thus the nose serves as an air passage as well as retaining its original olfactory function. In the evolution through tetrapods, the olfactory and respiratory roles of the nasal cavities become segregated to some extent, the olfactory epithelium becoming restricted for the most part to the dorsal parts of the cavities. But in many tetrapods a part of the olfactory epithelium remains in the ventral part of the cavity where it forms the **vomeronasal organ** (Jacobson's organ). This organ apparently detects the odor of food within the mouth; it sometimes has a separate connection with the mouth (p. 85). In mammals the respiratory passages are prolonged through the evolution of a bony hard palate (p. 77) and a fleshy soft palate.

In order for a substance to stimulate the olfactory epithelium it must be in solution. This is no problem for the fish, but it is for a terrestrial animal. Tetrapods have met the problem by the evolution of glands whose secretions keep the epithelium moist. The secretions of these glands, and the mucosa of the nasal passages as a whole, also condition the air that passes to the lungs by moistening, cleansing and, in birds and mammals, warming it. In birds and mammals the mucosal surface is increased through the evolution of **conchae** or turbinate bones (p. 83).

The receptive olfactory cells of the nasal organ are of a unique type in that the cells not only receive the stimuli but also have long processes that conduct the impulses back to the brain. This type of cell, which is believed to be very primitive, is called a neurosensory cell. Although related to the visceral sensation of taste, olfaction is considered to be somatic sensory.

Fishes

Study the nose of *Squalus*. Note that each **naris** is divided into two openings (Fig. 7–8, *A*) by a superficial flap of skin and a deeper ridge. The lateral opening, which also faces rostrally to a slight extent, is the incurrent aperture through which a current of water enters the olfactory sac; the medial one, the excurrent aperture. A thin, flaplike valve along its caudal margin prevents water from entering the excurrent aperture. Remove the skin and other tissue from around the **olfactory sac** on the side of the head used for the dissection of the eye. The nasal capsule and antorbital process of the chondro-cranium must also be cut away. You can then see that the olfactory organ is a round sac having no connection with the mouth. An **olfactory tract** of the brain extends to the caudomedial surface of the sac and there expands into an **olfactory bulb.** Cut open the sac and note how its internal surface is increased by many septalike folds (**olfactory lamellae**) which, in turn, bear minute secondary folds (Fig. 7–8, *B*).

Primitive Tetrapods

Note the pair of widely separated **nares** on your specimen of *Necturus*. Remove the skin between the naris and the eye on the side of the head used

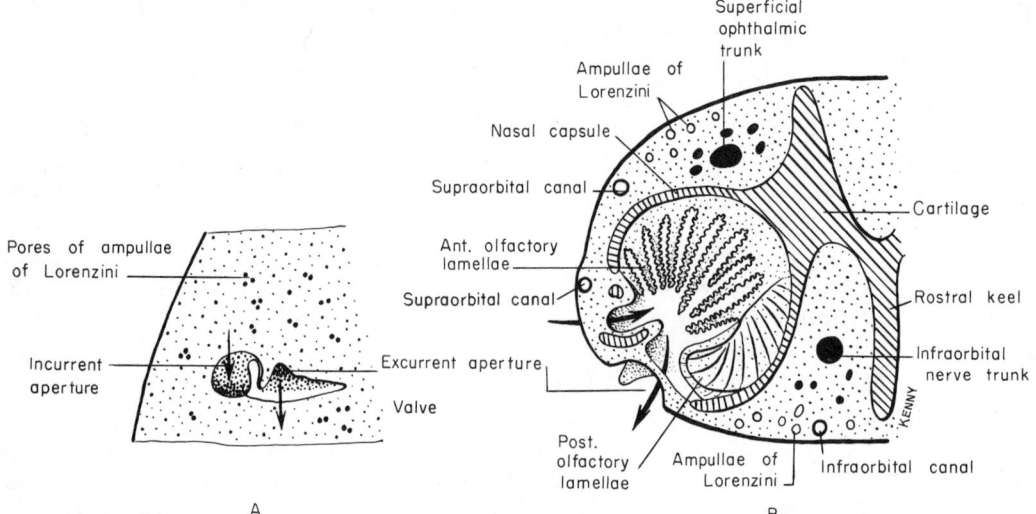

Figure 7–8. *A,* Ventral view of the right naris of *Squalus;* anterior is toward the top of the page; *B,* diagram of the left olfactory sac as seen in a cross section taken just posterior to the naris. Viewed from behind. Arrows indicate the direction of the water current.

for the dissection of the muscles. The long, tubular **nasal cavity** will be seen after picking away surrounding connective tissue (Fig. 6–7, p. 126). Open the mouth, cutting through the angle of the jaw on the side being studied, and find the slitlike **choana** through which the nasal cavity enters the mouth. It lies lateral to the most caudal, and shortest, row of teeth (pterygoid teeth). Cut open the nasal cavity and find the **olfactory lamellae** within it. What is their function? A vomeronasal organ is not developed in *Necturus* but is present in metamorphosed amphibians.

Mammals

The nose of mammals is to be studied on sagittal sections of the head cut in such a way that one half shows the nasal septum, and the other the inside of the nasal cavity. The nose should be studied from demonstrated preparations, unless this unit of work has been postponed to the end of the course, or unless heads from specimens of a previous year's class have been saved for the purpose. This work should also be correlated with the description of the sagittal section of the skull on page 82. The following account is based on the cat but is applicable to many other mammals.

The **nares,** which are close together in mammals, lead into paired **nasal cavities.** The nasal cavities occupy the area of the skull rostral to the cribriform plate of the ethmoid bone and dorsal to the hard palate. On the larger section it can be seen that they are completely separated from each other by a **nasal septum.** The ventral portion of the septum is formed by the **vomer bone,** the posterodorsal portion by the **perpendicular plate of the ethmoid,** and the rest by **cartilage.**

On the smaller section it can be seen that each nasal cavity is filled to a large extent by three folded **conchae** or turbinate bones (Figs. 4–26 and 9–16). These are, of course, covered with the nasal mucosa. The **ventral concha** (maxilloturbinate) is represented by a simple fold that extends from the dorsal edge of the naris caudally and ventrally to about the middle of the hard palate. The nasolacrimal duct from the eye enters lateral to the ventral concha. It is best seen on a skull in which this concha has been removed. The **dorsal concha** (nasoturbinate) is represented by a single longitudinal fold in the dorsal part of the nasal cavity lying deep to a median, perpendicular plate of the nasal bone. The area between and caudal to these two conchae is filled by the highly folded **middle concha** (ethmoturbinate).

Air passages lead from the naris between the conchae. The most prominent of these is the **ventral meatus,** which lies between the ventral and middle conchae on the one hand and the hard palate on the other. It opens by the **choana** into the nasopharynx. The choanae are located at the caudal border of the hard palate. The nasopharynx is separated from the mouth cavity by the fleshy soft palate. A **dorsal meatus** extends from the dorsal part of the naris to the caudal portion of the middle concha. A **common meatus** lies between the nasal septum and the conchae, and other air passages lie between the complex scrolls of the conchae. Most of the olfactory epithelium is limited to the more caudal parts of the conchae. **Olfactory nerves** may be seen, with a hand lens, passing through the cribriform foramina of the cribriform plate to the olfactory bulb of the brain.

Paired **vomeronasal organs** are present in the cat. The entrance to one can be seen on the roof of the mouth just caudal to the first incisor tooth. From here an **incisive duct** leads through the palatine fissure (p. 85) to the organ. The vomeronasal organ can be found by carefully dissecting away the rostroventral portion of the nasal septum. It appears as a cul-de-sac with a cartilaginous wall lying on the hard palate and against the nasal septum. It extends approximately one centimeter caudad to the incisor teeth.

THE EAR

The ear of vertebrates is closely related to the lateral line system. Indeed, the inner, receptive part of the ear is often regarded as a deeply set portion of this system. Among the features supporting this homology are the common mode of embryonic formation, similarity of the receptive "hair cells" of the ear to neuromasts, and the close relationship of the statoacoustic nerve to nerves from the lateral line.

The inner ear consists of a series of thin-walled canals and sacs filled with a fluid known as the **endolymph.** These canals and sacs are collectively called the **membranous labyrinth.** They may lie within a single chamber in the otic capsule, but often they are imbedded within a series of parallel canals and chambers within the capsule known as the **osseous labyrinth.** The membranous labyrinth and osseous labyrinth are separated from each other by spaces filled with fluid and crisscrossed by minute strands of connective tissue. This fluid is the **perilymph.**

Only an internal ear is present in fishes (Fig. 7–9). As in higher vertebrates, it is an organ of equilibrium, and in many teleosts the lower part of it (sacculus and lagena) has also been shown to be sensitive to sound waves that reach it by passing from the water through adjacent tissues, or by way of special sound concentrating mechanisms such as the swim bladder and weberian ossicles. Evidence that elasmobranchs can also hear is twofold. First, they show behavior responses to low frequency sound vibrations in the water even if the lateral line and major cutaneous nerves are cut, but they lose these responses when the statoacoustic nerve is cut. Second, nerve impulses have been recorded from the nerve twigs coming from the sacculus and utriculus of a ray while it was being subjected to sound stimulations up to 120 cycles per second. It has not been possible to record similar impulses from the lagena twig. In terrestrial vertebrates a part of the inner ear is definitely specialized for the reception of sound waves.

Whereas sound waves pass easily from water into the tissues of a fish, which are mostly water, they do not pass easily from air into water. Tetrapods sensitive to air-borne sound vibrations have evolved a special mechanism that receives such vibrations and increases their pressure amplitude sufficiently to overcome the inertia in the

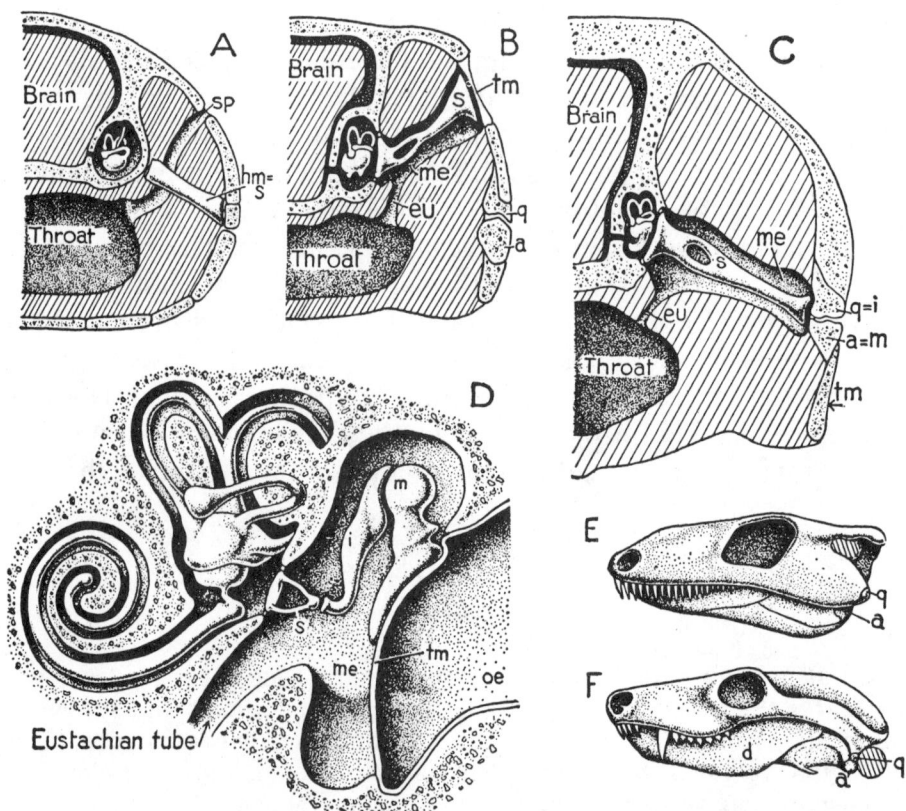

Figure 7–9. Diagrams to show the orthodox hypothesis of the evolution of the middle ear and auditory ossicles. The cross sections are through the otic region of *A*, a fish; *B*, a primitive amphibian; *C*, a primitive reptile; and *D*, a mammal. The lateral views are of *E*, a primitive amphibian, and *F*, a mammal-like reptile. They show in particular the shift of the eardrum (hatched) from the otic notch to a point behind the jaw articulation. Abbreviations: *a*, articular; *d*, dentary; *eu*, auditory tube; *hm*, hyomandibular; *i*, incus; *m*, malleus; *me*, middle ear, or tympanic, cavity; *oe*, external acoustic meatus; *q*, quadrate; *s*, stapes; *sp*, spiracle; *tm*, tympanum, or eardrum. (From Romer, Man and the Vertebrates, University of Chicago Press.)

liquids of the inner ear. In most living amphibians and reptiles, sound waves impinge upon the ear drum, or **tympanum**, located on the body surface or at the base of a canal, the **external acoustic meatus.** They are transmitted across a **tympanic cavity** (homologous to the spiracular pouch) by the **stapes** (homologous to the hyomandibular). The foot plate of the stapes fits into a **fenestra vestibuli,** or oval window, on the side of the otic capsule, and a specialized part of the perilymph transmits vibrations from there to the receptive part of the membranous labyrinth. The difference in size between the large tympanum and small fenestra ovalis increases the pressure amplitude. Vibrations are finally released from the inner ear through a **fenestra cochlea,** or round window.

The orthodox view of ear evolution (Fig. 7–9) assumes that labyrinthodonts and cotylosaurs had an ear of this type. This has been retained in most of their descendants and has been further elaborated in mammals. The sound detecting portion of the mammalian inner ear is a long, spiral **cochlea;** two additional ossicles, **malleus** and **incus,** appear in the middle ear as a corollary of a new jaw joint and the inward movement of the articular and quadrate (p. 77); and an **auricle** develops about the entrance of the external acoustic meatus. But some investigators question the presence of a tympanum in early mammal-like reptiles and their cotylosaur and labyrinthodont ancestors. They believe that these early tetrapods could detect vibrations only by bone conduction, especially through the lower jaw. According to this view, an ear with a tympanum has evolved independently in those amphibians that have it (frogs), in most living reptile groups, and in mammals. Animals such as *Necturus* have not lost an air-sensitive ear, but evolved from ancestors that never had one.

Fishes

Dissect the ear of *Squalus* on the side of the head used for the study of other sense organs. First uncover the otic capsule by removing the skin and muscles from its dorsal, lateral and caudal surfaces. The spiracle and adjacent parts of the mandibular and hyoid arch may also be cut away. Note the duct leading from one of the endolymphatic pores in the skin to the endolymphatic fossa on the top of the chondrocranium. It ultimately connects with the sacculus of the membranous labyrinth. This duct, usually called the **endolymphatic duct,** is not strictly homologous to the endolymphatic duct of higher vertebrates, for it represents a retention of the pathway along which the ear invaginated from the surface during embryonic development. The endolymphatic duct of higher vertebrates is a secondary evagination from the inner ear toward the surface and is not the original invagination pathway.

To expose the **membranous labyrinth** of the inner ear you must carefully shave away the surrounding cartilage of the otic capsule, beginning on the dorsal and lateral surfaces and gradually working ventrally. You can generally see the various canals and chambers through the cartilage shortly before you reach them. Use special care as you dissect the cartilage from around the parts of the membranous labyrinth and try not to break them. You will first expose the three **semicircular canals — anterior vertical, posterior vertical,** and **lateral,** or horizontal. As you continue to dissect you will come upon the **sacculus** lying in a large cavity medial to the lateral canal. The sacculus is generally partly collapsed, so that the dorsolateral wall of the cavity in which it lies can be dissected away without injuring it. Continue to remove cartilage from around the sacculus and semicircular canals, tracing the latter to their points of attachment on chambers called utriculi. Each end of each canal attaches to one of two utriculi. The anterior and lateral canals attach onto the **anterior utriculus,** the posterior canal onto the **posterior utriculus.** The utriculi connect with the sacculus by inconspicuous openings. The ventral end of each canal bears a round swelling, the **ampulla,** containing a sensory patch known as a **crista.** Branches of the **statoacoustic (vestibulocochlear) nerve** can be seen leaving each ampulla, but the cristae may not be apparent. Dissect away as much of the cartilage as possible from the ventral side of the sacculus and observe the short extension or bulge from its caudoventral portion. This is the **lagena.** The sacculus and lagena contain large sensory patches called **maculae.** These sensory patches are overlaid by a mass of calcareous concretions and sand grains that form an **otolith.** Remove as much cartilage as possible from the medial side of the membranous laby-

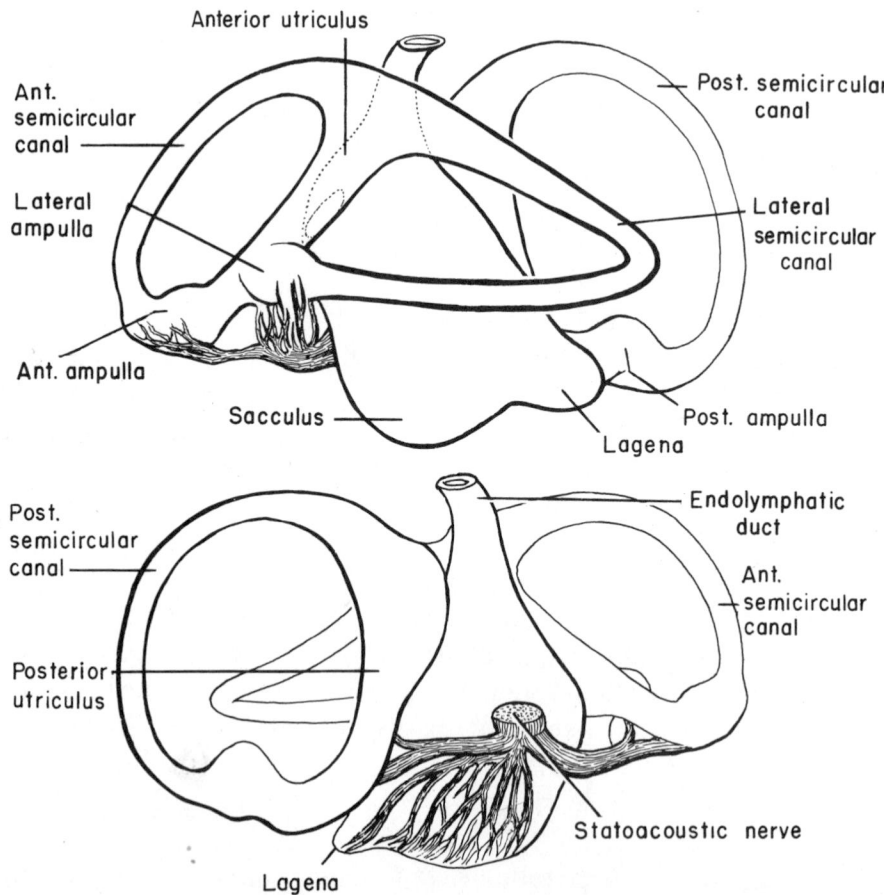

Figure 7–10. Left ear (membranous labyrinth) of the selachian, *Heptanchus maculatus.* Lateral view above, medial view below. (Redrawn from Daniel, The Elasmobranch Fishes, University of California Press.)

rinth, cut the nerves coming from the sensory patches, and gently lift out the ear and float it in a dish of water. It is well to leave a bit of cartilage on the medial and ventral surfaces. Review the parts of the ear mentioned and compare your dissection with Figure 7–10.

The basic structure of the membranous labyrinth can be seen better in the cartilaginous fishes than in any other vertebrates, for the labyrinth is relatively large and can be freed more easily from cartilage than from bone. The elasmobranch ear is atypical, however, in having two utriculi. Usually all semicircular canals connect into one. Also the connection of the endolymphatic duct to the surface is lost in the adults of other vertebrates.

Primitive Tetrapods

The structure of the ear of a primitive tetrapod should be studied on demonstration dissections of a bullfrog, since the urodele ear lacks the **tympanum** and middle ear cavity found in most terrestrial vertebrates. The large external tympanum is easily seen. Remove this on one side and the **tympanic cavity** or middle ear cavity will be exposed. Open the mouth and note that the tympanic cavity communicates with the back of the mouth cavity by the **auditory** (Eustachian) **tube.** A long, rodlike **stapes** crosses the middle ear cavity.

Expose the **otic capsule,** which contains the inner ear, by removing the overlying skin and muscle. Also remove the musculature lying caudal to the middle ear. Cut through the back of the middle ear cavity and trace the stapes to the otic capsule. Its inner end is associated with the knoblike, specialized **operculum** (homologous to the urodele structure of the same name, page 62). Remove the operculum and the inner end of the stapes and you will see the **oval window** into which they fit. Vibrations enter the inner ear at this point and are released through a fenestra, which, in living amphibians, opens into the cranial cavity dorsal to the acoustic nerve. It may be found by opening the cranial cavity and removing the brain. This fenestra is analogous to the **round window,** which opens into the tympanic cavity of certain reptiles and mammals and is usually given this name.

The otic opercular bone of terrestrial frogs and salamanders has no homologue in other vertebrates. An opercularis muscle extends caudad from it to insert on the pectoral girdle. This gave rise to the notion that salamanders, who have no tympanic membrane, perceive groundborne vibrations by way of the pectoral limb, girdle, and opercular apparatus. Lombard and Straughan (1974) have found that the opercular apparatus is functional in frogs, who do have a tympanic membrane, so the situation is more complex than was first thought. Inactivation of the apparatus in frogs significantly lowers their sensitivity to tones below 1000 Hz. Salamanders hear only low tones: inactivation of their opercular apparatus also decreases their sensitivity, and renders them essentially incapable of sound detection. The method of sound perception in salamanders is still not clear, but appears not to be through the limbs as was previously thought.

Mammals

The external ear can be seen easily on your mammal specimen. It consists of the external ear flap, the **auricle,** and a canal with a cartilaginous wall leading inward to the skull (**external acoustic meatus**). The **tympanum** lies at the base of the meatus between the external and middle ears. It can be found on the side of the head on which the pinna was removed by cutting away as much of the external acoustic meatus as possible and shining a light into the remainder. Note that the tympanum is set at an angle; its rostral portion extends more medially than its caudal portion. The opaque line seen through the membrane is the handle of the malleus.

The rest of the ear is difficult to dissect and should be observed on demonstration preparations. These can be prepared from the sagittal sections of the head used for the study of the nose. The following account is based on the cat but is applicable to many other mammals. Remove the muscles and other tissue from around the tympanic bulla except at its rostromedial corner. The middle ear, or **tympanic cavity,** lies within the bulla and opens into the nasopharynx by the **auditory tube** (Eustachian tube). The opening of the tube appears as a slit in the lateral wall of the nasopharynx (Fig. 9–16). Careful dissection between this slit and the rostromedial corner of the bulla will reveal the tube, part of whose wall is cartilaginous and part bony.

The rest of the dissection will be much easier to do if the specimen is first decalcified by placing it in a weak solution (0.06) of nitric acid for a few days. Break away the caudomedial portion of the bulla and also the mastoid and paracondyloid processes and adjacent parts of the nuchal crest. This exposes the caudomedial chamber of the tympanic cavity. Note that it is largely separated from a smaller rostrolateral chamber by a more or less vertical plate of bone. A hole through the dorsolateral portion of this plate passes between the two chambers of the middle ear cavity. The **fenestra cochlea,** or round window, can be seen through this hole. Carefully break away all this plate of bone and open up the rostrolateral chamber. The handle of the **malleus** can be seen on the inside of the tympanum, and it will be noted that the fenestra cochlea is situated on a round promontory of bone. A fingerlike process of cartilage extends from the caudolateral wall of the tympanic cavity between these two structures. The tiny nerve that runs along it and leaves its tip is the **chorda tympani,** a branch of the facial nerve going to the taste buds of the tongue and certain salivary glands. Break away bone from the rostromedial corner of the bulla and find the entrance of the auditory tube. The other auditory ossicles (**incus** and **stapes**), and the **fenestra vestibuli,** or oval window, in which the stapes fits, lie dorsal to the fenestra cochlea. To see them one must remove a piece of bone caudal and dorsal to the external acoustic meatus without injuring the plate of bone supporting the tympanic membrane. You will also notice two small muscles passing to certain of the ossicles. A **stapedius** arises from the medial wall of the tympanic cavity caudad to the fenestra cochlea and inserts on the stapes. A **tensor tympani** arises from the medial wall rostral to the fenestra vestibuli and inserts on the malleus. The groups to which these muscles belong are indicated in Table 4. They adjust the auditory ossicles and tympanum to the intensity of the sound waves. Their contraction, for example, reduces the amplitude of the vibration of the ossicles and protects the delicate structures of the inner ear from movements resulting from loud noises.

The inner ear lies within the petrosal portion (otic capsule) of the temporal bone. Portions of it may be noted by removing the brain and chipping away pieces of the petrosal, but it cannot be dissected satisfactorily by this method. The **internal acoustic meatus** for the vestibulocochlear and facial nerves lies in the cranial cavity on the caudomedial surface of the petrosal.

8

THE NERVOUS SYSTEM

As explained in the introduction to the chapter on sense organs, the nervous system integrates, or coordinates, the actions of the various parts of the body so that all will function in harmony with one another and with the external environment. But it should be pointed out that the nervous system is not the only system involved in body coordination. The endocrine glands also play an important role. Nervous coordination is implemented by nerve impulses that travel along the nerve cells or neurons. It is characterized as being rapid and specific. Endocrine coordination, on the other hand, is by way of the secretions of hormones that are carried by the circulatory system. It tends to be slower and often one hormone may affect many different parts of the body. Since the endocrine glands are widely scattered, they will not be discussed in one place but as they are observed. The nervous system will be considered at this time, as it is the last of the group of organ systems that deals directly with the general function of body support and movement.

Neurons

The morphological and functional unit of the nervous system is the **neuron,** just as the muscle cell is of the muscular system. A typical neuron has a **cell body,** which includes the nucleus and most of the cell's cytoplasmic organelles. Extending from the cell body are cytoplasmic processes: functionally **dendrites** carry information to the cell body, and an **axon** carries information away from the cell body. In most cases the axon is by far the longer process; in large vertebrates it may extend many feet. The one exception is in the case of primary sensory neurons carrying information toward the spinal cord or brain. In this case the dendrite is very long, extending from the skin or sense organ to the cell body which lies just outside the brain or cord; the axon may be much shorter. The term **nerve fiber** refers to a long process, usually an axon. Protective cells surround all nerve fibers, and in many cases contain a fatty substance, **myelin,** which forms many layers of wrapping around the fibers. Aggregates of nerve cell bodies within the brain are known as **nuclei;** aggregates of cell bodies outside the brain and spinal cord, as **ganglia. Tracts** are groups of fibers running together within the brain and cord. **Nerves** are groups of fibers outside the brain and cord.

Divisions of the Nervous System

Grossly, the nervous system can be divided into a central and a peripheral portion. The **central nervous system** consists of the **brain** and **spinal cord** (Fig. 8–1). Both cord

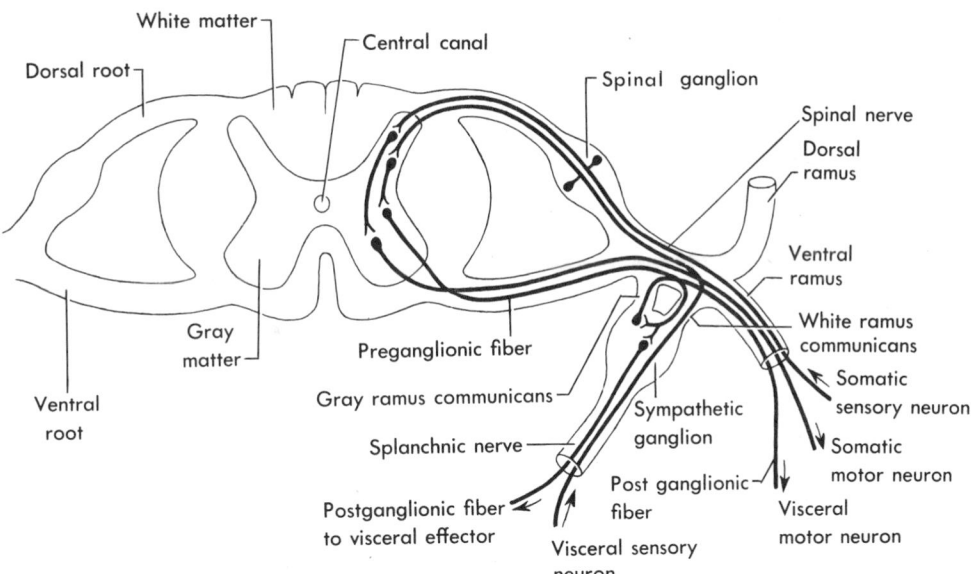

Figure 8–1. Diagram of a cross section through the spinal cord and a pair of spinal nerves to show the organization and functional components of the nervous system. Two internuncial neurons are shown in the gray matter between the sensory and motor neurons. (From Villee, Walker, and Barnes, General Zoology.)

and brain are hollow, for they contain a **central canal,** which expands to form **ventricles** in certain regions of the brain. You will recall that a single, dorsal, tubular nerve cord is one of the diagnostic chordate characteristics. The distribution of the neurons within the central nervous system is such that we can speak of **gray** and **white matter.** The gray matter contains the cell bodies of neurons and unmyelinated fibers; the white, myelinated fiber tracts. Most of the gray matter is centrally located. In the cord it forms a continuous column that, in higher vertebrates, has the appearance of an H, or a butterfly, in cross section. In the brainstem, the gray matter tends to break up into distinct nuclei, but the nuclei have a relationship to each other similar to the relationship between the parts of the gray matter in the cord. Some gray matter migrates to the surface parts of the brain and forms a conspicuous **cortex** in the higher vertebrates.

The **peripheral nervous system** consists of all the neural structures—ganglia and nerves—lying outside the spinal cord and brain. **Spinal nerves** are segmentally arranged, and each connects to the central nervous system by a **dorsal** and **ventral root** (Fig. 8–1). The dorsal root bears a **ganglion** containing the cell bodies of sensory neurons. More distally the spinal nerve breaks up into branches, or rami, going to various parts of the body—a **dorsal ramus** to the epaxial region, a **ventral ramus** to the hypaxial region, and often one or more **communicating rami** with visceral connections. Although most of the **cranial nerves** may have had this pattern at one stage in their evolution, their segmentation and organization into dorsal and ventral roots are not so apparent in living vertebrates as those in the spinal nerves.

The **autonomic** portion of the peripheral nervous system consists of motor neurons going to the visceral organs, glands, and smooth muscle generally. These leave the cord or brain in certain of the spinal and cranial nerves. Some of the fibers of the autonomic system remain in the spinal and cranial nerves, but some leave to travel in special branches to the organs in question. Thus, the autonomic system, while clearly definable functionally, is not completely separated from the rest of the peripheral nervous system morphologically.

A unique feature of the autonomic system is that there is a peripheral relay (Fig.

8–1). A **preganglionic neuron,** having its cell body in the gray matter of the cord, passes out through the ventral root of a spinal nerve and goes to an autonomic ganglion. There it synapses with a **postganglionic neuron** whose axon extends to the visceral organ. In contrast, only one neuron is involved in the innervation of the branchiomeric and somatic musculature.

Another unique feature of the autonomic nervous system is its subdivision, in the higher vertebrates at least, into **sympathetic** and **parasympathetic** portions. A given visceral organ receives both types of fibers. One activates the organ; the other inhibits it. Sympathetic innervation increases the activity of the heart, slows down digestive processes and has other effects that help the body to adjust to conditions of stress. Parasympathetic innervation has the opposite effect. The sympathetic fibers of the autonomic system leave through the thoracic and anterior lumbar spinal nerves; the parasympathetic, through certain cranial and sacral nerves. The peripheral relay of the sympathetic fibers is in a ganglion at some distance from the organ being supplied, so the postganglionic fiber is quite long; the parasympathetic relay, on the other hand, is in or very near the organ being supplied, thus its postganglionic fibers are relatively short. Many of the sympathetic ganglia lie against the back of the body cavity, lateral and ventral to the vertebral column. Those on each side of the body are interconnected by visceral fibers to form a chain known as the **sympathetic cord.**

Functional Components and Their Interrelations

A nerve may contain both **sensory (afferent)** and **motor (efferent) neurons.** These are mixed in the distal parts of a nerve, but they tend to segregate near and within the central nervous system. In a typical spinal nerve of a higher vertebrate, for example, the cell bodies of the motor neurons lie within the ventral portions of the gray matter of the cord and their fibers generally pass out through the ventral root of the spinal nerve (Fig. 8–1). In very primitive vertebrates, however, the autonomic fibers leave through the dorsal root. The sensory neurons approach the cord through the dorsal root of the spinal nerve, and their cell bodies are located in the dorsal root ganglion. The sensory neurons ultimately enter the dorsal portions of the gray matter, but they may travel for some distance in the white matter of the cord before doing so.

Besides distinguishing between sensory and motor neurons, it is possible to distinguish between sensory neurons coming from somatic receptors and those coming from visceral receptors. The same is true for motor neurons. Thus, there are four major types of neurons within the peripheral nerve—**somatic sensory, visceral sensory, visceral motor,** and **somatic motor.** These four types of neurons also correlate with areas of the gray matter in that they begin, or end, in definite regions, or columns. Somatic sensory neurons end in the most dorsal part of the gray matter (Fig. 8–1), which constitutes a **somatic sensory column;** visceral sensory neurons, just ventral to them in a **visceral sensory column.** The cell bodies of the visceral motor neurons constitute a **visceral motor column** located in the lateral part of the gray matter; those of the somatic motor neurons constitute a **somatic motor column** located in the most ventral part of the gray matter. Certain of these four categories may be further subdivided into special and general types. For example, a distinction can be made between the visceral motor fibers to the viscera (autonomic system) and those to the branchiomeric musculature. The former are **general visceral motor fibers,** and the latter **special visceral motor fibers.**

As explained, somatic and visceral sensory neurons ultimately enter their respective columns of the gray matter. These primary sensory axons typically send branches ascending and descending along the spinal cord white matter. This arrangement enhances the formation of multiple synaptic connections responsible for both simple and complex reflexes, the latter always involving **internuncial neurons** between the sensory input and motor output. These internuncial neurons may pass through the gray matter to a motor neuron (a three neuron reflex mechanism), or they may ascend in the white matter to the brain. Here they may form reflex connections or be relayed to still other parts of the brain. Internuncial neurons that ascend to the brain are called **afferent internuncial neurons,** and at some point during their ascent they usually cross (decussate) to the side of the central nervous system opposite to the one on which they originated. Impulses that originate in the brain descend through the white matter to the motor neurons on other internuncial neurons. These, too, usually decussate at some point during their descent.

Development of the Brain

As might be expected, the brain is the most complex part of the nervous system. An understanding of the various regions of which it is formed is best gained from a consideration of its development. As shown in Figure 8–2, the brain arises as an en-

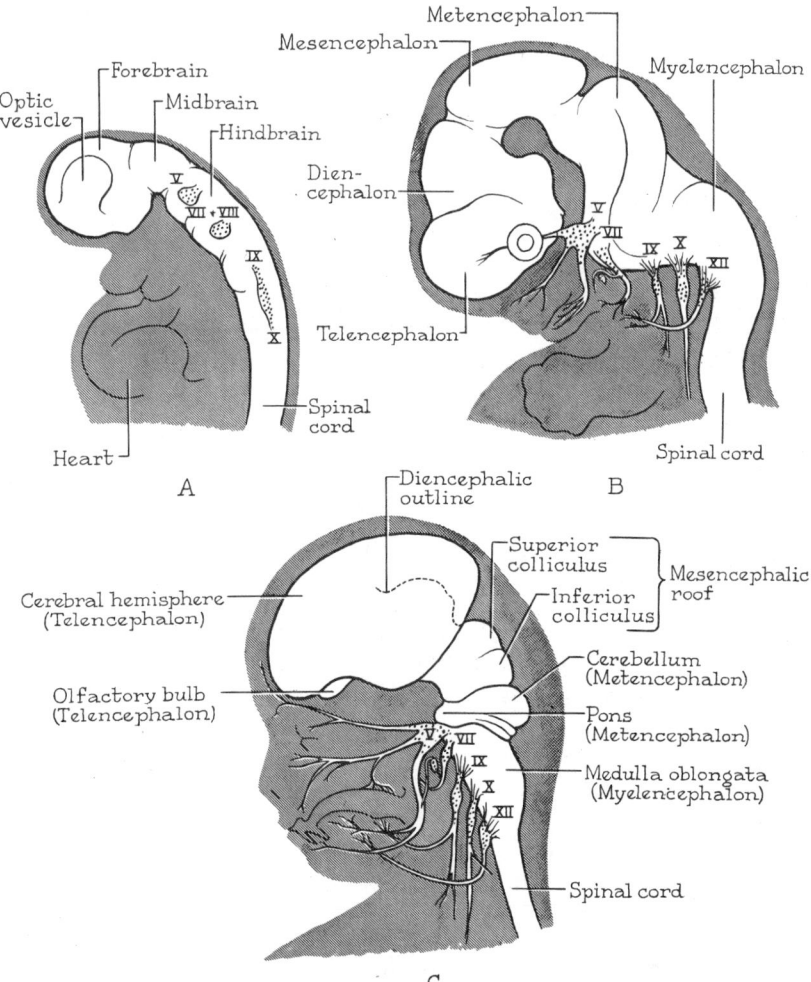

Figure 8–2. Diagrams of three stages in the development of the human brain to show the differentiation of the principal brain regions. Certain cranial nerves are identified by Roman numerals. (From Villee, Walker, and Barnes, General Zoology. Modified after Patten.)

largement of the cranial end of the neural tube. Very soon it becomes divided by certain constrictions into three vesicles, or regions: an anterior forebrain, or **prosencephalon;** a middle midbrain, or **mesencephalon;** and a posterior hindbrain, or **rhombencephalon.**

The mesencephalon does not divide further, but the other two regions do. The prosencephalon divides into a rostral **telencephalon** and a caudal **diencephalon;** the rhombencephalon into a rostral **metencephalon** and a caudal **myelencephalon.** Each of these five regions further differentiates. The telencephalon gives rise to the cerebral hemispheres and olfactory bulbs; the diencephalon, to the thalamus, hypothalamus, epithalamus and optic vesicles; the mesencephalon, to the tectum (roof of the midbrain); the metencephalon, to the cerebellum; the myelencephalon, to the medulla oblongata.

FISHES

The nervous system of the dogfish should be studied carefully, for the basic structure of the nervous system can be seen exceptionally well. Not only can the system be easily exposed by the removal of cartilage, but it is in a morphologically primitive and generalized stage. The nervous system of *Squalus* is a good prototype for that of all vertebrates.

Dorsal Surface of the Brain

The dissection of the nervous system should be done on the large head of *Squalus* that was used for the study of the sense organs. The cranial nerves should be studied primarily on the intact side of the head, for certain of them may have been destroyed during the dissection of the sense organs. Remove the skin and underlying tissue from the dorsal surface of the chondrocranium and from around the eye. Be careful not to cut a large dorsal nerve (**superficial ophthalmic nerve**) that lies rostral to the orbit and lateral to the rostrum. Cut away the cartilaginous roof of the cranial cavity. As you do so, look into the rostral part of the cavity and you may see a delicate, threadlike stalk extending from a depressed area on the top of the brain (diencephalon) to the epiphyseal foramen in the roof of the cranial cavity. This is the **epiphysis,** a vestige of the pineal eye of more primitive vertebrates (Fig. 8–3). Also cut away the supraorbital crest and as much of the lateral walls of the cranial cavity as is possible without injuring nerves. Much of the ear on the intact side will have to be cut away. Be particularly careful not to break the small **trochlear nerve** that leaves the brain dorsally and passes to the dorsal oblique muscle. *Have to break on one side*

The brain should now be well exposed. Its surface is covered with a delicate, vascular connective tissue, the **meninx primitiva.** Strands of connective tissue pass from the meninx to the connective tissue lining the cranial cavity (the **endochondrium**). In life, cerebrospinal fluid lies in the apparently empty **perimeningeal space** between the brain and the wall of the cranial cavity.

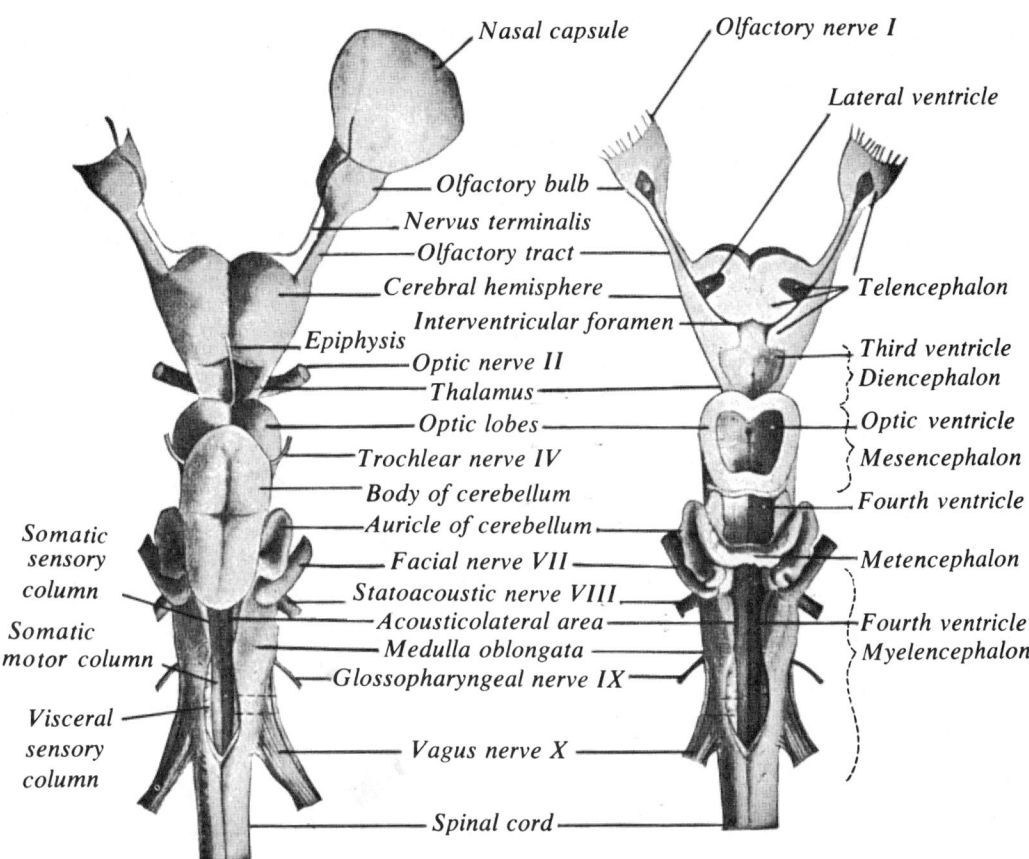

Figure 8–3. Dorsal views of the dogfish brain. The ventricles have been opened in the right figure. (From Ranson and Clark, Anatomy of the Nervous System.)

The paired **olfactory bulbs** form the most cranial part of the brain. They are the lateral enlargements in contact with the olfactory sacs, and they receive the primary olfactory neurons coming from the olfactory epithelium. Secondary olfactory neurons originate in the bulbs and form the **olfactory tracts** that extend caudally to the **cerebral hemispheres.** These neurons terminate in a ventrolateral portion of the hemisphere that is essentially homologous to the mammalian piriform lobe.

The **diencephalon** is the depressed area, often with a dark roof, situated caudal to the cerebral hemispheres. The roof of the diencephalon is known as the **epithalamus,** its lateral walls as the **thalamus,** and its floor (which will be seen later) as the **hypothalamus.** The **epiphysis** may be seen attaching to the caudal part of the epithalamus. Most of the roof is very thin and membrane-like, forming a **tela choroidea.** The tela choroidea consists of only the ependymal epithelium that lines the central nervous system and the meninx that covers it. Carefully remove this part of the tela choroidea and, while doing so, note that part of it extends onto the caudal surface of the cerebral hemispheres. This part of the roof actually does not belong to the diencephalon but forms a thin-walled sac, the **paraphysis,** which is considered to be a part of the telencephalon (Fig. 8–4). A prominent fold extends from the tela choroidea down into the large cavity (**third ventricle**) within the dien-

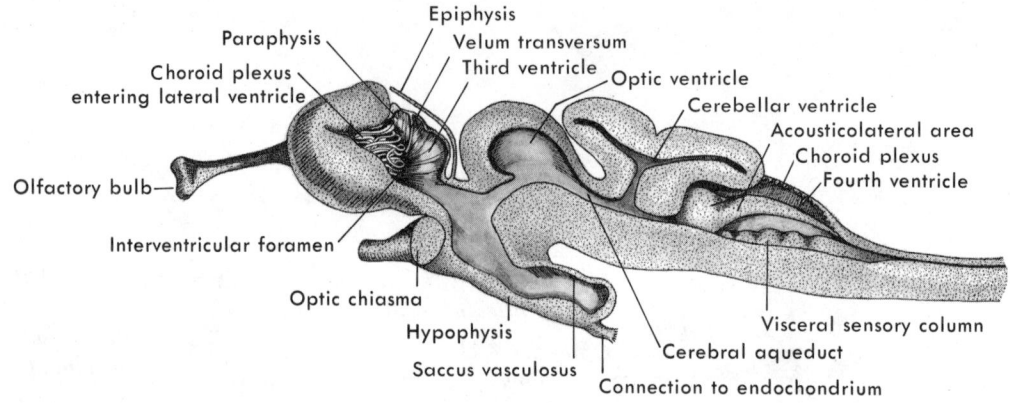

Figure 8–4. Sagittal section of the brain of *Squalus.*

cephalon. This fold, the **velum transversum,** represents the rostral end of the diencephalon. Vascular tufts extend forward from the velum transversum into a **lateral ventricle** within each cerebral hemisphere. A tuft in the lateral ventricle constitutes the **choroid plexus of the lateral ventricle.** Choroid plexuses secrete the cerebrospinal fluid into the ventricles. Some of the fluid escapes from certain regions of the brain and circulates around the outside of the central nervous system. The small transverse fold of nervous tissue at the very caudal part of the epithalamus is the **habenular region.**

The habenula lies just rostral, and slightly ventral, to the large pair of **optic lobes.** The optic lobes develop in the roof, or **tectum,** of the **mesencephalon** and are the only part of the mesencephalon apparent in a dorsal view.

The **metencephalon** lies caudal to the mesencephalon and consists dorsally of the **cerebellum.** The **body of the cerebellum** is the large, median, oval mass whose cranial end overhangs the optic lobes. Note that it is partially subdivided into four parts by a longitudinal and transverse groove. The pair of earlike flaps that lie on either side of the caudal part of the body of the cerebellum are the **auricular lobes of the cerebellum** (Fig. 8–3). The ventral part of the metencephalon contributes to the medulla oblongata in fishes.

The **myelencephalon** lies caudal to the metencephalon and forms the greater part of the **medulla oblongata** — the elongated region of the brain that is continuous caudally with the spinal cord. Most of the roof of the medulla is a thin tela choroidea which covers a large brain cavity, the **fourth ventricle.** Carefully remove it and also its forward extension from the auricles of the cerebellum. Vascular folds, the **choroid plexus of the fourth ventricle,** can be seen extending into the ventricle. Lift up the caudal end of the cerebellum and notice that the auricles are continuous with each other.

The columns of gray matter that continue from the cord into the brain can be seen in the ventral and lateral walls of the fourth ventricle. The pair of midventral, longitudinal folds that lie on the floor of the ventricle are the **somatic motor columns.** They contain the cell bodies of somatic motor neurons. There is a deep, longitudinal groove lateral to each somatic motor column. The lateral wall of either groove constitutes the **visceral motor column,** a column containing the cell bodies of visceral motor neurons. A longitudinal row of small bumps lies dorsal to the visceral motor column. This is the **visceral sensory column** (Fig. 8–5), and it receives and relays impulses from the visceral sensory neurons. The longitudinal groove between the visceral motor and visceral sensory columns is the **sulcus limitans,** a landmark separating the ventral motor from the dorsal sensory portion of the cord and brainstem. (The brainstem is the brain minus the cerebrum and cerebellum.) The dorsolateral rim of the fourth ventricle constitutes the **somatic sensory column,** and it receives and relays impulses from the somatic sensory neurons. The relationship between these columns can be visualized particularly well in a cross section which you may make after the brain and nerves have been studied. The rostral part of the somatic sensory column is enlarged. This portion, known as the **acousticolateral area** (Fig. 8–3), receives the neurons from the ear and lateral line organs. It is continuous rostrally with the auricles of the cerebellum. Although it cannot be seen in a gross dissection, these sensory and motor columns continue forward through the metencephalon and into the mesencephalon. Their rostral ends, however, become discontinuous, forming discrete patches, or nuclei, of gray matter.

Cranial and Occipital Nerves

The cranial and occipital nerves must be considered before the ventral and internal parts of the brain can be examined. More of the lateral wall of cranium will have to be removed as the nerves are studied.

(A) NERVUS TERMINALIS

Fishes are usually described as having 10 pairs of cranial nerves, and these are both named and numbered. However, an additional rostral nerve has been left out of this numbering system. The **nervus terminalis** (Fig. 8–6)

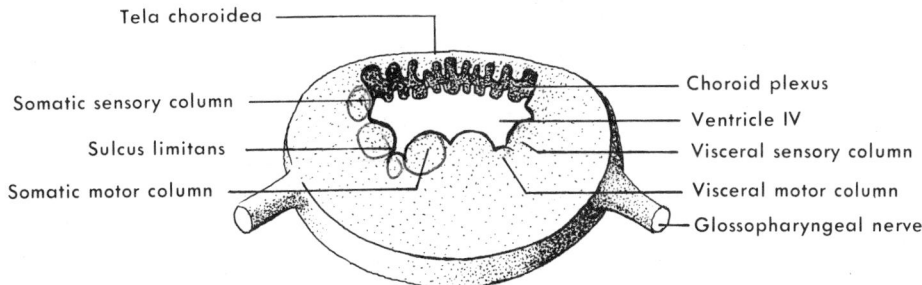

Figure 8–5. Cross section through the medulla oblongata of *Squalus* made at the level of the glossopharyngeal nerve.

is a tiny nerve that lies along the medial surface of the olfactory tract and extends between the olfactory sac and cerebral hemisphere. It is seen best at the medial angle formed by the junction of the olfactory tract with the cerebral hemisphere, for it separates slightly from the olfactory tract in this region. Although the terminalis is found in all vertebrates, except cyclostomes and birds, its function is uncertain. The consensus is that it carries both somatic sensory fibers (cutaneous, not olfactory fibers) from the nasal area, and visceral motor fibers of the autonomic system. The latter are probably vasomotor. Some investigators have reported the presence of ganglia along the nerve.

(B) OLFACTORY NERVE

The **olfactory nerve** (I) carries olfactory impulses (somatic sensory) from the olfactory sac to the olfactory bulb. Since the sac and bulb are adjacent to each other in the dogfish, the olfactory nerve is not compact but consists of a number of minute groups of neurons passing between these structures. These may be seen by making a section through the olfactory sac and bulb. The olfactory neurons are of the neurosensory type (p. 181).

(C) OPTIC NERVE

The **optic nerve** (II) brings in optic impulses (somatic sensory) from the eye. Find it in the orbit and trace it medially. It is a thick nerve. Push the brain away from the cranial wall and note that the optic nerve attaches to the ventral surface of the diencephalon. Since the retina of the eye develops embryologically from an outgrowth of the brain, the optic nerve is really a brain tract rather than a true peripheral nerve.

(D) OCULOMOTOR NERVE

The **oculomotor nerve** (III) carries somatic motor impulses to most of the extrinsic ocular muscles, receives proprioceptive impulses from these muscles (somatic sensory), and carries autonomic fibers (visceral motor) to the eye. To see it, mobilize the eye on the intact side of the head in the manner described in connection with the dissection of the eye (p. 172), remove the gelatinous connective tissue lying in the orbit, and look on the ventral surface of the eyeball. The branch of the oculomotor going to the **ventral oblique** muscle will be apparent (Fig. 7–3, p. 174). Follow it caudally and medially. It passes ventral to the ventral rectus, and at the caudal margin of this muscle crosses a small, often whitish, blood vessel. This vessel, an artery, follows the margin of the ventral rectus and enters the eyeball. The autonomic fibers of the oculomotor form a small **ciliary nerve** that travels along this vessel, but these fibers can seldom be seen grossly. The branch of the oculomotor to the **ventral rectus** lies between the ventral rectus and the small vessel. Turn your specimen over, and pick up the oculomotor nerve from the dorsal side. It extends dorsal to the origin of the dorsal rectus and enters the cranial cavity. Just before it enters, it gives off one branch to the **dorsal rectus** and another to the **medial rectus.** (Do not confuse the oculomotor with another nerve of the same size, the profundus, which crosses the base of the oculomotor and extends along the medial surface of the eyeball.) Push the brain away from the cranial wall and note the attachment of the oculomotor on the ventral surface of the mesencephalon.

(E) TROCHLEAR NERVE

The **trochlear nerve** (IV) has been noted crossing an optic lobe. Lift up the cerebellum and note where it attaches on the brain. The trochlear passes through the cranial wall, goes ventral to, or perforates, the large superficial ophthalmic nerve and extends to the **dorsal oblique** muscle. Like the oculomotor, it is primarily a somatic motor nerve, but it carries a few proprioceptive fibers (somatic sensory).

(F) ABDUCENS NERVE

Skip the fifth nerve for a moment and consider the **abducens** (VI), which carries somatic motor fibers to the **lateral rectus** and returns proprioceptive (somatic sensory) impulses from this muscle. It can be seen on the ventral surface of the lateral rectus. Its attachment on the ventral surface of the medulla will be seen later when the brain is removed.

(G) TRIGEMINAL NERVE

Return to the **trigeminal nerve** (V), which is the nerve of the mandibular arch and the general cutaneous sensory nerve of the head (Fig. 8-6). The trigeminal attaches more or less in common with the facial (VII) and stato-acoustic (VIII) nerves on the dorsolateral surface of the medulla just caudal to the auricles of the cerebellum. It is difficult to separate these nerves at their attachments on the brain, but they can be sorted out to some extent by their peripheral branches. The trigeminal has four branches in fishes. A **superficial ophthalmic branch,** together with a comparable branch of the facial nerve, forms the large **superficial ophthalmic nerve** that has been noted passing through the dorsal region of the orbit and along the lateral surface of the rostrum (Fig. 7-3, p. 174). A **deep ophthalmic branch** (or **profundus nerve**) enters the orbit dorsal to the oculomotor, adheres to the connective tissue on the medial surface (back) of the eyeball, and leaves the front of the orbit through a small foramen to join the superficial ophthalmic nerve. The deep ophthalmic should be traced on the side of the head in which the eyeball has been left. Both ophthalmic branches of the trigeminal return somatic sensory impulses from general cutaneous sense organs (not lateral line organs) in the skin on the top and side of the head. In addition, the deep ophthalmic has several minute and inconspicuous branches to the eyeball.

These are considered to be homologous to the long ciliary nerve of mammals and, for the most part, return sensory fibers from parts of the eye other than the retina. In some selachians there is evidence that they also carry a few autonomic fibers to the eye (Norris and Hughes, 1920).

A **mandibular branch** of the trigeminal can be found by dissecting away the connective tissue on the caudal wall of the orbit. It is a fairly thick nerve that lies caudal to the lateral rectus. The mandibular carries special visceral motor fibers to the branchiomeric muscles of the first visceral arch and returns some somatic sensory fibers from general cutaneous sense organs in the skin overlying the lower jaw.

The **maxillary branch** of the trigeminal, together with the buccal branch of the facial, forms the large infraorbital nerve that extends rostrally across the floor of the orbit. The **infraorbital nerve** is fully as wide as any of the ocular muscles and is easily confused with a muscle. It divides near the rostral border of the orbit and is distributed to the skin overlying the upper jaw and the underside of the rostrum. The maxillary portion of this nerve returns somatic sensory fibers from general cutaneous sense organs in this region.

Cut away enough cartilage and connective tissue from the caudomedial corner of the orbit to be able to see where all of the branches of the trigeminal come together, and again note the attachment of the trigeminal to the medulla. The main part of the trigeminal bears a slight enlargement, the **semilunar ganglion,** that contains the cell bodies of the sensory neurons; however, it is unlikely that you can distinguish this ganglion.

(H) FACIAL NERVE

The **facial nerve** (VII) is the nerve of the hyoid arch, spiracle, and the cranial lateral line organs. As mentioned above, its **superficial ophthalmic** and **buccal branches** contribute to the superficial ophthalmic and infraorbital nerves, respectively. They return somatic sensory impulses from the lateral line organs. The attachment of certain fibers from these trunks to the ampullae of Lorenzini can be seen. Cut away the skin adjacent to the caudoventral corner of the spiracle on the intact side of the head, and pick away the underlying connective tissue. The large nerve that will be seen is the **hyomandibular nerve,** a branch of the facial (Fig. 6–4, p. 121). Follow it peripherally, noting that it is distributed to the hyoid muscles (special visceral motor fibers), skin (somatic sensory fibers from lateral line organs), and lining of the mouth (visceral sensory fibers of both general and taste nature). Follow it medially to its union with the other branches and its attachment on the brain, which, as stated, is more or less in common with that of the trigeminal and statoacoustic nerves. You will have to cut away some of the spiracle, surrounding muscles and otic capsule as you go. About one centimeter from the brain the hyomandibular bears a slight enlargement, the **geniculate ganglion,** that contains the cell bodies of sensory neurons. Another, and smaller, branch of the facial leaves from the rostroventral surface of this ganglion. This is the **palatine nerve,** and it returns visceral sensory neurons from the mouth lining.

(I) STATOACOUSTIC NERVE

A part of the **statoacoustic** *(vestibulocochlear)* **nerve** (VIII)[12] may have been noted coming from the ampullae of the anterior vertical and lateral semicircular canals during the dissection of the hyomandibular nerve. Continue to cut away the otic capsule and note another, and longer, part of this nerve coming from the ampulla of the posterior vertical semicircular canal, the sacculus, and parts of the utriculus. The statoacoustic contains somatic sensory fibers from various parts of the inner ear.

(J) GLOSSOPHARYNGEAL NERVE

The **glossopharyngeal nerve** (IX) is the nerve of the third visceral arch and the first of the five definitive gill pouches. It can be seen crossing the floor of the otic capsule caudal to the sacculus. It passes ventral to the caudal branch of the statoacoustic and at first may be confused with this nerve. Cut away this part of the statoacoustic and find the attachment of the glosso-pharyngeal on the side of the medulla. Trace the glossopharyngeal laterally. This will be facilitated if you open the first gill pouch by cutting through the skin and muscle dorsal and ventral to the first external gill slit. Cut all the way to, but not through, the internal gill slit (the opening between the gill pouch and pharynx). As the nerve leaves the otic capsule, it bears an oval-shaped swelling, the **petrosal ganglion,** that contains the cell bodies of sensory neurons. Several branches leave from the petrosal ganglion. A large **posttrematic** passes down the caudal face of the first gill pouch to carry special visceral motor fibers to the branchial muscles and return visceral sensory fibers from this region. A smaller **pretrematic** branch, which is entirely visceral sensory, passes down the cranial face of the first pouch. A still smaller **pharyngeal** branch follows the pretrematic a short distance, then curves around a tendon near the dorsal edge of the internal gill slit and is distributed to the wall of the pharynx. It, too, is entirely visceral sensory. There is finally a small **dorsal** branch distributed to lateral line organs, and often to the skin in the supratemporal region, but it is impractical to find.

omit

(K) VAGUS NERVE

The **vagus** (X) is the nerve of the remaining visceral arches. Find its attachment on the dorsolateral surface of the caudal end of the medulla and follow it caudally out of the otic capsule. To see the rest of the nerve, you

---

[12]This nerve is called the vestibulocochlear nerve in mammals. Lower vertebrates lack a cochlea in their ear, so it is appropriate to retain the term statoacoustic for their eighth nerve.

must cut open the remaining gill pouches in the manner described for the first. If you cut as far as you should, you will cut into a large blood space, the **anterior cardinal sinus,** lying dorsal to the internal gill slits. This sinus must also be opened by a longitudinal incision. A large branch of the vagus (**visceral nerve**) lies beneath the connective tissue on the dorsomedial wall of the anterior cardinal sinus. It gives off four **branchial branches** that cross the floor of the sinus and are distributed to the remaining four visceral arches and pouches. Each branchial branch follows the pattern of the glossopharyngeal, having a sensory ganglion from which **posttrematic, pretrematic,** and **pharyngeal branches** arise. After giving off the branchial branches, the visceral nerve continues as the **intestino-accessory nerve.** It gives off a small branch, which probably will not be seen, to the cucullaris (Fig. 8–6) and then follows along the wall of the esophagus to the viscera. It also sends a branch to the pericardial cavity. The intestino-accessory nerve contains visceral motor fibers of the autonomic system, visceral sensory fibers, and special visceral motor fibers to the cucullaris.

Just before the visceral nerve enters the rostral end of the anterior cardinal sinus, the vagus gives off a **lateral** (or dorsal) nerve. This branch lies medial to the visceral nerve and extends caudally between the epaxial and hypaxial musculature. It receives somatic sensory fibers from the lateral line canal proper, and in some elasmobranchs a few general cutaneous fibers from the skin in the gill region. Cell bodies of these neurons lie in a ganglion near the proximal end of this branch.

(L) OCCIPITAL NERVES

Two or three **occipital nerves** lie caudal to the vagus. Although the occipital nerves are closely related to cranial nerves, and are destined to contribute to the formation of a tetrapod cranial nerve (the hypoglossal), convention excludes them from the cranial series because the caudal limit of the cranium is not the same in all fishes. Sometimes these nerves emerge from the skull and sometimes just posterior to the skull. The occipital nerves are serially homologous with ventral roots of spinal nerves. The dorsal roots of these nerves are believed to have become incorporated in the vagus (p. 206).

In *Squalus* the occipital nerves emerge from the back of the chondrocranium. Find the junction between cranium and vertebral column by cutting away the muscle and cartilage overlying this region. There will, of course, be an abrupt change in the width of the axial skeleton at this point. The nerve that emerges between the chondrocranium and first vertebra is the **first spinal nerve.** The small nerves that arise from the lateroventral surface of the neural tube between this point and the large root of the vagus are the occipital nerves. There are generally two such nerves; rarely, three. At first they may appear to be accessory, posterior roots of the vagus, for they pass through the cartilage and appear to join the vagus. Actually they run with the vagus for only a short distance and then leave to form the **hypobranchial nerve.** The hypobranchial nerve also receives contributions from the first two or three spinal nerves; it is the nerve that carries somatic motor fibers to the hypobranchial muscles. It also contains some somatic sensory fibers of a proprioceptive and general cutaneous nature. The general cutaneous fibers enter the spinal cord through the spinal nerves.

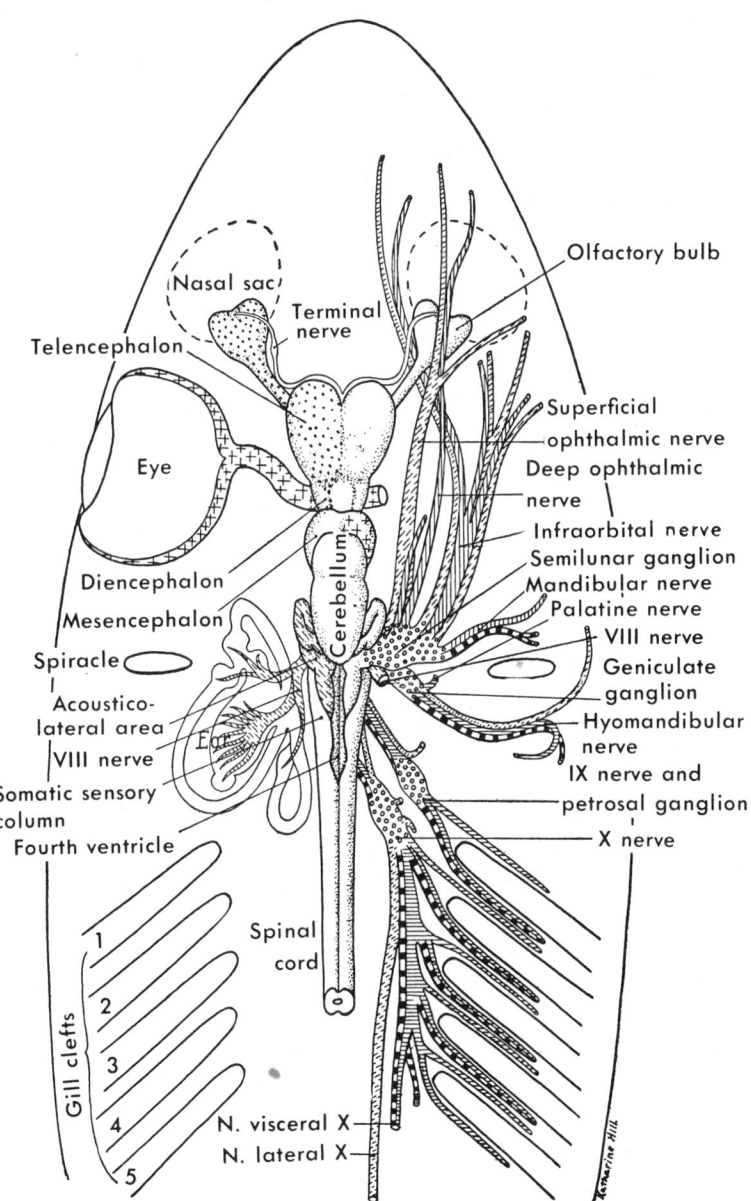

Figure 8–6. Diagram of the brain and most of the cranial nerves of the dogfish, showing functional components: olfactory, stipple; visual, crosses; acousticolateral, broken oblique lines; visceral sensory, horizontal lines; general cutaneous, vertical lines; visceral motor, black and white rectangles; ganglia, circles. (From Ranson and Clark, Anatomy of the Nervous System. After Herrick and Crosby.)

Table 5. Distribution and Components of the Cranial and Occipital Nerves of Selachians

The cranial and occipital nerves of selachians, together with their general distribution and functional components, are shown in this table. An X indicates the presence of the various components; and (X) that the component is found in some, but not all selachians.

NERVE	BRANCHES	DISTRIBUTION	SOMATIC SENSORY			VISCERAL SENSORY		VISCERAL MOTOR		SOMATIC MOTOR
			General (Cutaneous)	Special including Lateral Line (L)	Proprioceptive	General	Special (Taste)	General (Autonomic)	Special (Branchiomeric)	
Terminalis		Olfactory sac	X					X		
I. Olfactory		Olfactory epithelium		X						
II. Optic		Retina		X						
III. Oculomotor	4 Muscular Branches	Ventral oblique; ventral, dorsal and medial rectus			X					X
	Ciliary	Ciliary body of eye						X		
IV. Trochlear		Dorsal oblique			X					X
V. Trigeminal	Superficial Ophthalmic	Skin over top and side of head	X							
	Deep Ophthalmic (Profundus)	Skin over top and side of rostrum; sensory fibers from the eye and in some cases a few motor fibers to the eye	X					(X)		
	Mandibular (Post-trematic)	Mandibular muscles and skin over lower jaw	X						X	
	Maxillary	Skin over upper jaw and underside of rostrum	X							
VI. Abducens		Lateral rectus			X					X
VII. Facial	Superficial Ophthalmic	Lateral line organs over top and side of head	(X)	X(L)						
	Buccal	Lateral line organs over upper jaw and underside of rostrum		X(L)						

Table 5. *(Continued)*

Nerve	Branch	Distribution							
	Hyomandibular (Post-trematic)	Hyoid muscles; mouth lining; and lateral line organs near lower jaw		X(L)	X	X		X	
	Palatine (Pharyngeal)	Mouth lining			X	X			
VIII. Stato-acoustic (Vestibulo-cochlear)		Sensory patches of inner ear		X				X	
IX. Glossopharyngeal	Dorsal	Supratemporal lateral line organs; adjacent skin	(X)	X(L)					
	Pretrematic	Anterior wall of first typical gill pouch			X	X			
	Posttrematic	Posterior wall of first typical gill pouch			X	X		X	
	Pharyngeal	Pharyngeal lining			X	X			
X. Vagus	Lateral (Dorsal)	Most of lateral line canal; skin in dorso-lateral gill region	(X)	X(L)					
	Visceral	4 branchial branches to remaining pouches and then continues as the intestino-accessory branch to the cucullaris, heart and abdominal viscera. Each branchial branch has pretrematic, posttrematic, and pharyngeal branches. At least one posttrematic supplies ventral pit organs		X(L)	X	X	X	X	
Occipital Nerves (2-3 in *Squalus*)		Anterior epibranchial and hypobranchial musculature (the first 2-3 spinal nerves also supply the hypobranchial musculature in *Squalus*)		X					X

The separation of the occipitals from the vagus can most easily be seen by finding the hypobranchial nerve and tracing it medially. Completely free the visceral branch of the vagus from the wall of the anterior cardinal sinus. The hypobranchial nerve emerges from the musculature, crosses the visceral branch of the vagus near the level of its last branchial branch, and curves ventrally near the point where the visceral branch passes onto the esophagus.

A summary of the distribution and functional components of the cranial and related occipital nerves is presented in Table 5. This appears to be a confusing array, but it is felt that, at one stage in evolution, most of these nerves were related to head segments, as spinal nerves still are to trunk segments. Subsequent evolution has led to such a great modification of the head that this relationship is now obscure. However, a clue to the relationship can be obtained by an analysis of the components in the cranial and occipital nerves and a comparison between them and primitive spinal nerves.

Primitive spinal nerves differ from those of higher vertebrates in failing to have a union of the dorsal and ventral roots and in having visceral motor fibers traveling through the dorsal roots instead of the ventral. In view of this, the terminalis, trigeminal, facial, glossopharyngeal and vagus are considered by some investigators to be serially homologous to primitive dorsal roots of spinal nerves. The deep ophthalmic branch of the trigeminal also should be added to this series, for it is a separate nerve in agnathous vertebrates and in embryos of the dogfish. The oculomotor, trochlear, abducens and occipitals are considered to be ventral roots that have secondarily acquired proprioceptive fibers in addition to their somatic motor fibers. The olfactory, optic and statoacoustic are believed to be unique nerves that have developed in association with the special sense organs rather than with head segments.

Relating these dorsal and ventral nerves to each other, and to head segments, is still another problem. One clue to this has been found in the embryo of the dogfish (Scyllium), for in such a form there is a complete series of myotomes in the early stages of development (Fig. 6–1, p. 109). Eight segments are found in the cranial region. The first three form the extrinsic eye muscles, one or more of those adjacent to the otic capsule degenerate, and the more caudal contribute to the epibranchial and hypobranchial muscles. Some of the segments caudal to the cranium also contribute to these muscles in certain vertebrates. The nerves that are believed to be related to these segments can best be expressed in tabular form (Table 6). It will be noted in this table that the terminalis has no segment; but the presence of such a nerve may indicate that a segment was present rostral to the ocular myotomes at a very early stage of vertebrate evolution. The other nerves can be related to myotomes, but it should be emphasized that such a scheme is hypothetical. This hypothesis, however, helps to bring order to a seemingly chaotic pattern of nerves.

It is also apparent from this analysis that most of the dorsal nerves have shifted their orientation from the myotomes to the visceral arches and, thus, have become essentially branchiomeric nerves. The most generalized branchiomeric nerve is believed to be the glossopharyngeal (Fig. 8–7). As was observed, the main branch of this nerve is its posttrematic branch, which extends down the caudal face of the first definitive gill slit, supplying visceral motor fibers to the musculature of the third visceral arch and visceral sensory fibers to this region. The nerve also has visceral sensory pretrematic and pharyngeal branches and a somatic sensory dorsal branch. The vagus, facial and, to some extent, the trigeminal represent modifications of this basic pattern, as can be seen by comparing their branches with those of the glossopharyngeal in Table 5.

Table 6. Segmentation of the Cranial Nerves

MYOTOME	VENTRAL ROOT	DORSAL ROOT	VISCERAL ARCH
1	Oculomotor	Terminalis Deep ophthalmic	Trabeculae of chondrocranium
2	Trochlear	Trigeminal	Mandibular
3	Abducens	Facial	Hyoid
4		Glossopharyngeal	3rd
5	Occipitals, of which a variable number per- sist in the adult	Vagus	4th
6			5th
7			6th
8			7th

Ventral Surface of the Brain

It is now possible to return to the brain, remove it, and study its ventral surface. Before lifting the brain out, cut across the caudal end of the medulla, the olfactory tracts, and the roots of the cranial nerves. Try to leave stumps of the nerves attached to the brain. After you have cut across the trigeminal, facial and statoacoustic nerves, push the brain to one side and note the small abducens nerve leaving. It, too, must be cut. Lift up the caudal end of the brain and carefully pull it rostrally. You will soon see a part of the brain extending into a recess (the **sella turcica**) in the floor of the cranial cavity. It will probably be necessary to cut away some of the floor caudal to this recess to get the brain out intact.

After the brain has been removed, examine its ventral surface. Identify the attachments of the cranial nerves and the major regions of the brain. Several new structures can be seen in the region of the **hypothalamus** (floor of the diencephalon). The optic nerves attach at the rostral end of the hypothalamus and their fibers decussate at this point, forming an X-shaped structure known as the **optic chiasma.** In all vertebrates except mammals with

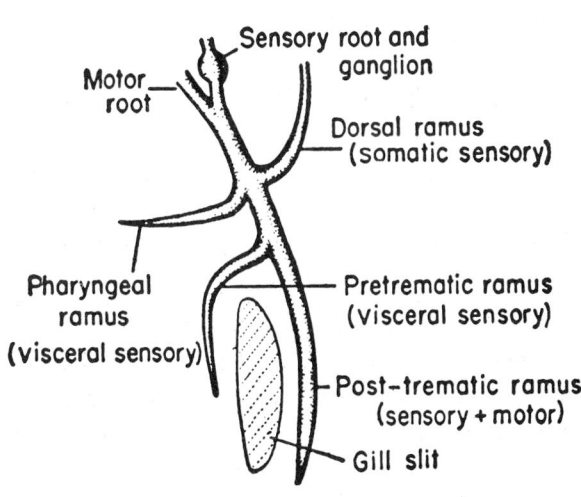

Figure 8–7. Diagram of a lateral view of a typical branchiomeric nerve (dorsal root cranial nerve). This diagram is based on the glossopharyngeal nerve, but the vagus, facial, and, to some extent, the trigeminal represent modifications of this basic pattern. (From Romer, The Vertebrate Body.)

stereoscopic vision, the decussation is complete, for all the fibers of the left optic nerves pass to the right side of the brain, and vice versa. After passing through the optic chiasma, the optic fibers form a band, the **optic tract,** that leads caudally and dorsally to the optic lobes. The meninx primitiva will have to be removed to distinguish the tract. Most of the rest of the hypothalamus consists of a large, caudally projecting **infundibulum.** Just caudal to the optic chiasma, the infundibulum bears a pair of prominent lateral lobes known as the **inferior lobes** of the infundibulum (Fig. 8–8). The **hypophysis,** or pituitary gland, lies partly between and partly caudal to the inferior lobes. The infundibulum finally forms a thin-walled, dark **saccus vasculosus** that lies caudal to the inferior lobes and dorsal to the hypophysis. The saccus vasculosus is well developed in deep sea fishes and, hence, is believed by some to be a pressure receptor; others, on cytological grounds, believe that it has some secretory function.

If the brain was carefully removed, the various lobes of the hypophysis can be recognized (Fig. 8–8). Two conspicuous lobes lie in the midventral line—an **anterior lobe** between the inferior lobes of the infundibulum, and an **intermediate lobe** caudal to this. A pair of **superior lobes** lie between the intermediate lobe and the saccus vasculosus.

The hypophysis is an endocrine gland of dual origin. The part of it that attaches intimately to the inferior lobes of the hypothalamus (a part of the cranial end of the intermediate lobe of the hypophysis) develops as an outgrowth of the diencephalon and, therefore, is homologous to the mammalian **neurohypophysis.** The rest of the hypophysis develops from an embryonic hypophyseal pouch, which is an invagination from the stomodeum, and hence is comparable to the **adenohypophysis.** Functions of the fish hypophysis are not fully known, but they are probably similar to those of a mammal. The neurohypophysis stores and releases hormones involved in water balance. These hormones are actually secreted by certain cells in the hypothalamus, and reach the neurohypophysis by travelling along processes of the secretory cells. The adenohypophysis produces a hormone that helps regulate growth, and others that interact with the thyroid, adrenal and gonad hormones in the regulation of the level of body metabolism, chromatophores, mineral metabolism, and reproduction.

Ventricles of the Brain

The central cavity that characterizes the neural tube of chordates expands to form large chambers, or ventricles, in the brain (Fig. 8–4). Some have been observed. To see the others cut off the roofs of the cerebral hemispheres, the optic lobes and the body of the cerebellum in the horizontal plane. **Lateral ventricles** lie in the paired cerebral hemispheres. Each connects with the **third ventricle** of the diencephalon by a narrow passage known as the **interventricular foramen** (foramen of Monro). The **third ventricle** connects with the **fourth ventricle** of the hindbrain by a narrow, ventral passage known as the **cerebral aqueduct** (aqueduct of Sylvius). **Optic ventricles** in the optic lobe and a **cerebellar ventricle** in the body of the cerebellum lie dorsal to the cerebral aqueduct.

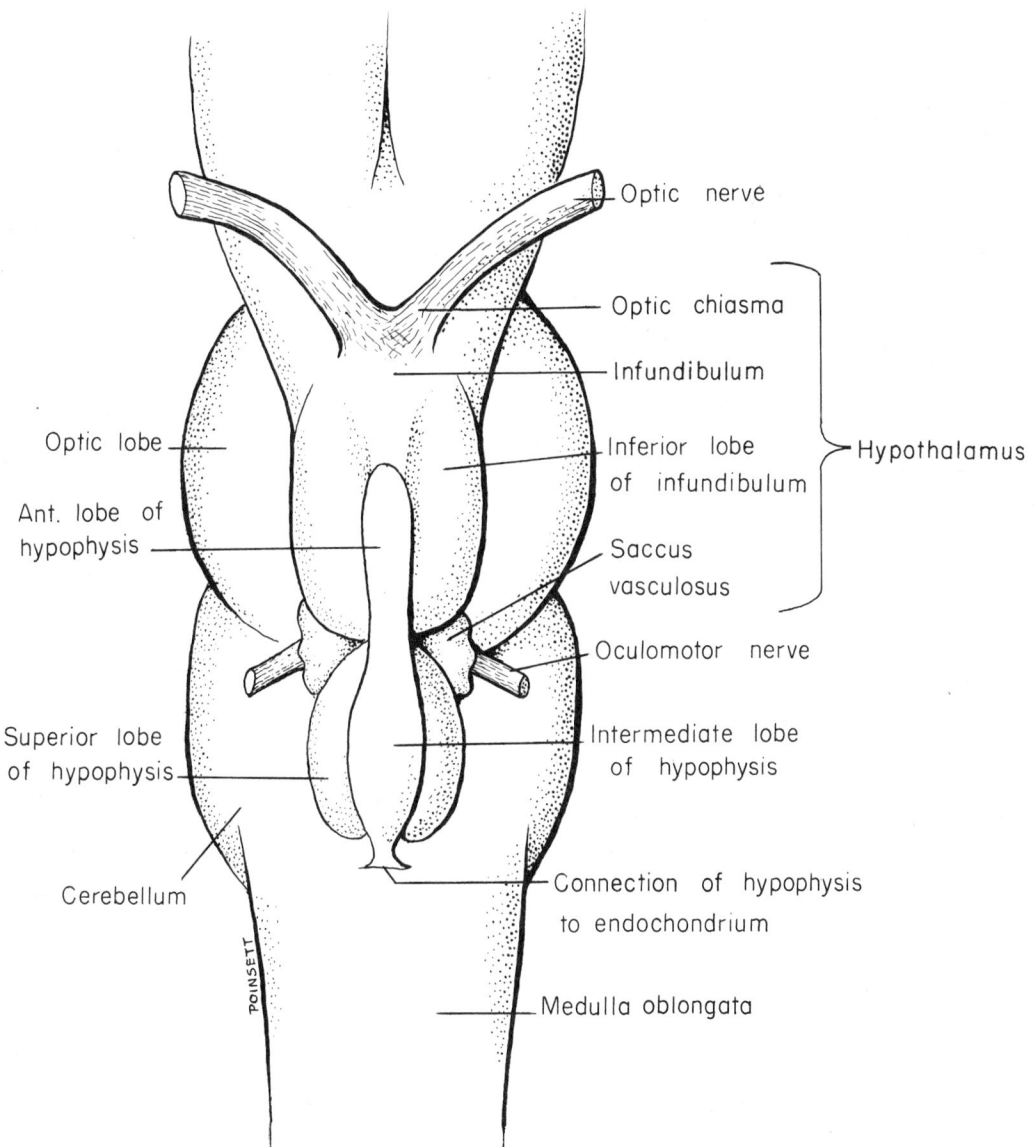

Figure 8–8. Ventral view of the hypophysis and adjacent parts of the brain of *Squalus*.

Although brain function cannot be considered in detail in a laboratory manual, the anatomy of the brain is more meaningful if one has some conception of the function of its various parts. As noted, the sensory and motor columns of the gray matter of the cord continue into the brain and extend through the medulla oblongata as far forward as the mesencephalon. Within the brain, the columns break up into discrete patches of gray matter (nuclei) rather than remaining continuous. The primary sensory neurons in the cranial nerves terminate in various sensory nuclei located in the dorsal part of the brainstem, e.g., taste fibers of the facial, glossopharyngeal, and vagus nerves in a gustatory nucleus (nucleus solitarius) in the myelencephalon, tactile fibers of the trigeminal nerve in a trigeminal sensory nucleus in the metencephalon. The primary sensory neurons, by way of internuncial neurons, make appropriate connections with motor neurons whose cell bodies form the motor nuclei, which are located in the ventral part of the brainstem. Motor nuclei of the facial, glossopharyngeal, vagus, and occipital nerves lie in the myelencephalon and initiate impulses to muscles involved in swallowing, respiratory movements, regulation of heart beat, and digestive tract movements. Motor neurons to jaw muscles start in a trigeminal motor nucleus located primarily in the metencephalon; the motor nuclei of the extrinsic ocular muscles lie in the mesencephalon.

The diencephalon has many functions. The habenular region in the epithalamus is an olfactory center. The thalamus relays impulses to and from the cerebral hemispheres and other parts of the brain. It is not large in fish. The hypothalamus is the largest part of the diencephalon in fish and contains important visceral and autonomic centers. Gustatory, olfactory, and general visceral sensory impulses are projected to this region, and efferent pathways lead to visceral motor nuclei and indirectly to the hypophysis. Body activity (rest or wakefulness), digestion, blood sugar level, water balance, and sexual activity are regulated here.

Certain parts of the brain have become elaborated, probably in relation to the sensory inflow from the three major distance receptors: nose, eye, and ear plus lateral line. Parts of the cerebral hemispheres receive the olfactory input; the optic lobes, the visual input; and the cerebellum, input from the ear and lateral line. These major areas to which the special senses are projected have to some extent "attracted" related sensory impulses and, thus, serve as important correlation centers. The various sensory data that are projected to these regions are integrated, and appropriate impulses leave on efferent pathways to the motor nuclei and columns. The brain, however, does not exert as much influence on the motor columns of the cord in the lower vertebrates as it does in the higher vertebrates.

The cerebellum, which is believed to have evolved as an elaboration of the acousticolateral area, is a center for muscular coordination. Its major afferent inflow is from the inner ear and the lateral line. It may initiate motor impulses related to the maintenance of body orientation and equilibrium, but much of its activity is to insure that motor directives beginning in other areas are carried out smoothly and with reference to the existing orientation of the body. The auricular lobes are related primarily to equilibrium and balance; the cerebellar body, to locomotor activity. The cerebellar body is large in the dogfish and other active fishes, but it is small in the lamprey and less active fishes.

The optic lobes receive projections not only from the eyes but from many other sense organs and sensory centers in the brain: olfactory, gustatory, acousticolateral and general cutaneous. Motor impulses initiated here pass to the floor of the mesencephalon and back to the motor nuclei and columns in the medulla and spinal cord. In short, the optic lobes of fishes are the master integration center and are analogous to the cerebral cortex of higher vertebrates.

The cerebral hemispheres of lower vertebrates, such as fishes, are not highly developed. In addition to receiving olfactory information, they receive a very limited projection of optic and other sensory information through the thalamus (Cohen and Duff, 1973). Little is known about their function in fishes, but it is clear that they do not have the dominant, integrating role that they have in the higher vertebrates. Efferent pathways lead back to the habenula, hypothalamus, and optic lobes.

The Spinal Cord and Spinal Nerves

The spinal cord lies in the vertebral canal of the vertebral column. To see it and the spinal nerves, remove the muscles overlying several centimeters of the vertebral column and carefully shave away the dorsal part of the vertebral arch. You will soon see the **spinal cord** and the **dorsal roots** of the spinal nerves through the cartilage. A **dorsal root ganglion** is present on each but is rather small. Each dorsal root passes through a foramen in the dorsal intercalary plate. A dorsal root lies slightly caudal to its corresponding ventral root, so to find the **ventral root** it is necessary to cut away the lateral wall of the vertebral arch cranial to the dorsal root. The ventral root arises from the cord by a fan-shaped group of rootlets and passes out of the vertebral canal through a foramen in the neural plate. The two roots unite in the dogfish in the musculature lateral to the vertebral column, but this union, and the subsequent splitting of the spinal nerve into dorsal, ventral and communicating rami, is difficult to find.

The components in the roots of the dogfish spinal nerves appear to be similar to those in the nerves of higher vertebrates—sensory neurons in the dorsal roots, motor in the ventral. Autonomic fibers present in the dorsal root of many fishes and amphibians have not been positively identified in selachians, but some may be present. Certainly most of the preganglionic visceral motor fibers of the autonomic system extend through the ventral roots to synapse with postganglionic fibers in a series of ganglia lying dorsal to the posterior cardinal sinus and kidneys. The autonomic system of the dogfish thus has cranial and spinal contributions, but the visceral organs apparently do not have the double innervation that they have in higher vertebrates. Also, autonomic fibers do not go to the skin in the dogfish.

PRIMITIVE TETRAPODS

The nervous system of such primitive tetrapods as the amphibians has progressed little beyond the condition seen in fishes. As a matter of fact, the cerebellum is not so well developed in living amphibians as in active fishes like *Squalus*. *Necturus* may be studied as an example of the amphibian condition, but it need not be studied in detail as the changes are not great.

Dorsal Surface of the Brain

Expose the dorsal surface of the cranium by removing the skin and muscles overlying it; then, carefully chip away the bone that forms its roof. The meninx primitiva of fishes is represented by two layers in primitive tetrapods—a tough outer **dura mater** and an inner, vascular **pia-arachnoid mater** applied to the surface of the brain. If the dura was not removed with the skull bones, it will have to be cut off in order to see the brain.

The rostral region of the brain is formed by the paired **cerebral hemispheres** and **olfactory bulbs** (Fig. 8–9). The olfactory bulbs lie rostral to the hemispheres but are not clearly separated from them. At most, only a slight lateral indentation may be seen between bulbs and hemispheres. Since the olfactory bulbs are adjacent to the rest of the telencephalon rather than to the olfactory sacs, there is no long, narrow olfactory tract as there was in the dogfish. The bands of nervous tissue that extend between the olfactory sacs and the olfactory bulbs are not the tracts, but the **olfactory nerves.** The small,

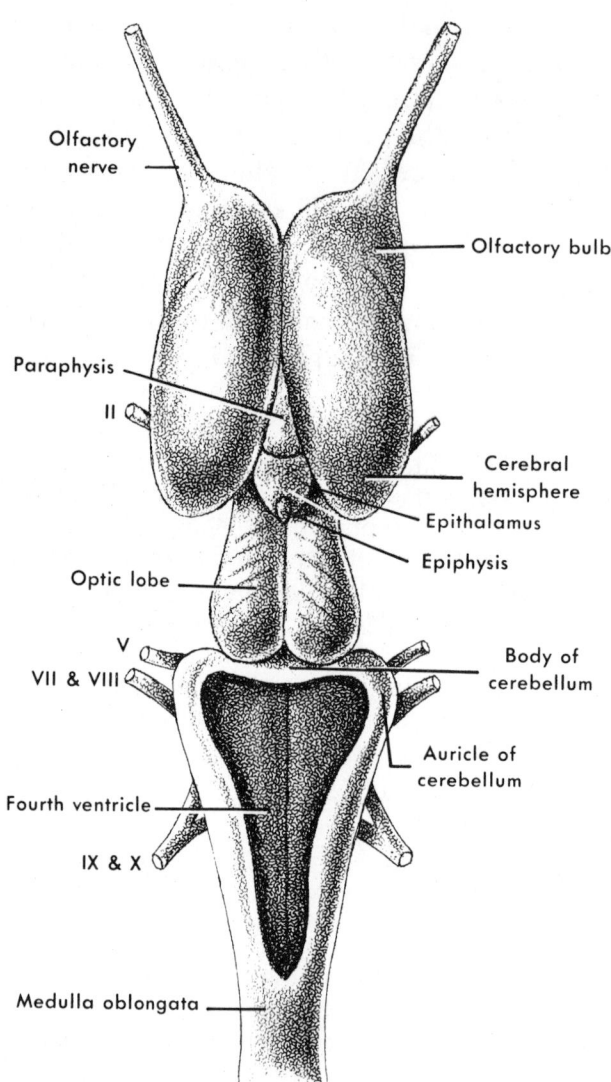

Figure 8–9. Dorsal view of the brain and cranial nerves of *Necturus*.

dark body that lies between the caudal ends of the cerebral hemispheres is the **paraphysis.** This is an evagination from the telencephalon.

The **epithalamic** region of the diencephalon appears as a small, triangular, light area posterior to the paraphysis. As usual, the **epiphysis** and **habenula** are present in this region, but they are hard to distinguish. From the tela choroidea in this region a **choroid plexus** has invaginated into the third ventricle, and has extended caudad into the cavity of the midbrain. It can be seen by cutting open the epithalamus.

The **optic lobes** lie caudal to the epithalamus. They are relatively small in *Necturus* and do not bulge dorsally to the extent that they do in *Squalus.* The line of separation between optic lobes and epithalamus is not sharp.

The **medulla oblongata** lies caudal to the optic lobes and can be recognized by the very dark tela choroidea that forms most of its roof. Remove the tela with its **choroid plexus** and note the heartshaped **fourth ventricle.**

As mentioned at the outset, the **cerebellum** is very small in amphibians. Its **body** consists of only a narrow, transverse band lying between the optic lobes and the medulla. This fold may be difficult to distinguish from the optic lobes. The **auricles** of the cerebellum are represented by the tissue that surrounds the anterolateral corners of the fourth ventricle.

Cranial and Occipital Nerves

The **olfactory nerve** (I) was observed coming into the brain from the olfactory sac. The **optic nerve** (II) from the eye crosses the cranial cavity and attaches to the brain at the rostral end of the diencephalic floor.

The nerves to the eye muscles — **oculomotor** (III), **trochlear** (IV) and **abducens** (VI) — are present but can be seen only under magnification. They have the same relationships as in *Squalus* and supply the same muscles. The abducens, however, also goes to the newly evolved **retractor bulbi** (Table 4).

The large trunk arising from the lateral surface of the rostral end of the medulla is the **trigeminal nerve** (V); the trunk caudal to it the common attachment of the **facial** (VII) and **statoacoustic** (VIII) **nerves.** The roots caudal to this trunk represent the origin of the **glossopharyngeal** (IX) and **vagus** (X). Glossopharyngeal and vagus run together until they leave the cranium, then they separate. The distribution and composition of these nerves is substantially the same as in *Squalus* but, of course, there would be some modification in the gill region. Lateral line fibers are present in *Necturus* but are lost in metamorphosed Amphibia.

Necturus, like other salamanders, has a hypobranchial nerve to the hypobranchial musculature. This nerve is formed primarily of fibers from the first spinal nerve, but it also receives contributions from the second spinal nerve and, in some salamanders, a small twig from the glossopharyngeal-vagus trunk. Certain of these contributions would obviously be homologous to the occipital nerves of *Squalus,* but just which ones is uncertain. It is of interest in this connection that the first, and sometimes the second, spinal nerve consists of only the ventral root.

Ventral Surface of the Brain

Cut across the caudal end of the medulla and across the cranial nerves and remove the brain. Try to lift the hypophysis from the sella turcica without breaking it off. Note the major regions of the brain and the stumps of the cranial nerves, and study the floor of the diencephalon (hypothalamus) in more detail. The optic nerves decussate to form an **optic chiasma** at the rostral end of the hypothalamus, but the chiasma is small and inconspicuous in *Necturus.* As in *Squalus,* the greater part of the hypothalamus consists of a large, posteriorly projecting **infundibulum** to which the hypophysis is attached. A small **saccus vasculosus** is present in the roof of the infundibulum but is difficult to see grossly.

MAMMALS

In the evolution through reptiles to mammals, numerous changes occur in the nervous system which are related in large measure to the increased activity and flexibility of response of mammals. A major change in the brain is the evolution of a **neopallium,** or **neocortex,** in the cerebral hemispheres to which sensory impulses are projected and where many motor impulses originate. Although a neopallium may have first appeared in mammal-like reptiles (Fig. 8–10, *D*), it has not yet been definitely demonstrated in any living reptiles. It is, however, present in all mammals, although its extent in different groups varies considerably. The neopallium, developing between the olfactory **paleopallium** and **archipallium,** forces them apart as it comes to occupy a greater part of the cerebrum. With the increasing dominance of the neopallium as the primary integrating structure, the tectum is left with relatively minor optic and auditory centers.

Another important change is the great enlargement of the cerebellum. This is correlated with the increased complexity of muscular movement, an increased projection of sensory data to this region, and an interconnection of cerebral hemispheres and cerebellum. To the original auricular lobes (vestibular in nature) and body (locomotor) are added a pair of hemispheres related to the more delicate and precise musculoskeletal movements.

Other changes in the brain tend to be correlated with the increased importance of the neopallium and the cerebellum. The thalamus enlarges as it becomes an important pathway and relay station between the cerebrum and other parts of the central nervous system. Other interconnections between the thalamus and the cerebrum suggest that these two parts of the brain may form a complex interacting mechanism, so that the thalamus appears to be more than just a relay station. Important centers develop in the ventrolateral portions of the midbrain (**red nucleus**) and metencephalon (**pons**) for the interconnection of the cerebellum and cerebrum. More and larger fiber tracts, both afferent and efferent, evolve in the mesencephalon and hindbrain because the cerebrum, the cerebellum and the brain in general exert more influence over the body than they did in lower vertebrates.

The basic pattern of the cranial nerves of mammals remains much the same as in fishes (Table 5), but a few changes have been superimposed upon this plan during the evolution to mammals. The loss of the lateral line system in lower tetrapods entails the loss of lateral line neurons from nerves that previously carried them (VII, IX, X). Except for the trigeminal outflow, mammals retain the autonomic fibers present in the cranial nerves of fishes and, in addition, acquire autonomic fibers in the facial and glossopharyngeal nerves. Those in the facial go to the tear and salivary glands; those in the glossopharyngeal, to the parotid gland. With the elaboration of the cucullaris to form the trapezius and sternocleidomastoid complex, we find that the special visceral motor fibers that supply these muscles separate from the vagus to form a new cranial nerve, the **accessory** (XI). This change occurs first in reptiles. Finally, the caudal limit of the cranium becomes fixed and the occipital nerves (plus, possibly, a rostral spinal nerve or two) form a definite cranial nerve, the **hypoglossal** (XII). Among living vertebrates this change is first seen in the reptiles, but it probably occurred earlier, for there is a foramen for such a nerve in the skulls of crossopterygians and labyrinthodonts. In modern amphibians the hypobranchial musculature is supplied by fibers that travel in the first spinal nerve.

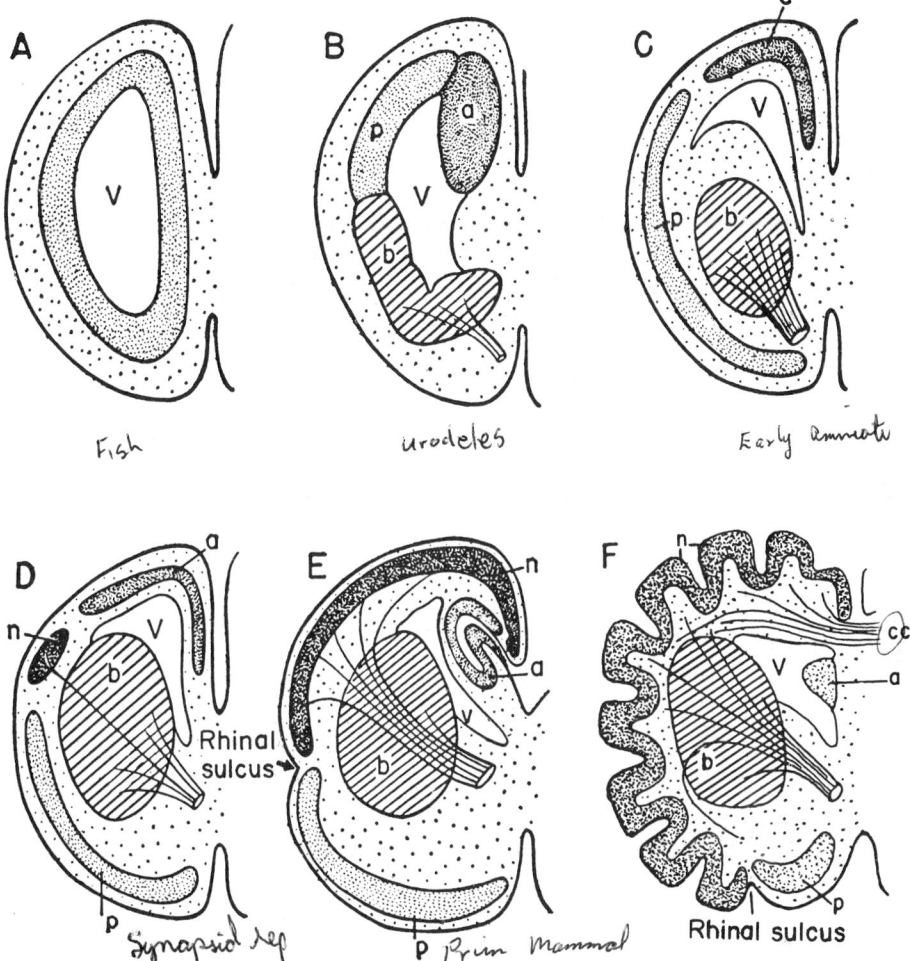

Fish urodeles Early amniote

Rhinal sulcus

n a

Synapsid rep Prim Mammal Rhinal sulcus

Figure 8–10. Diagrammatic cross sections of the left cerebral hemisphere, showing plausible stages in its evolution. *A,* Very primitive fish stage, in which the hemisphere may serve only as a central olfactory structure, and all the gray matter lines the ventricle, with very little, if any, differentiation of parts; *B,* a situation similar to that of today's living urodeles, probably a very generalized tetrapod condition, with differentiation of neural elements into three distinct groups; *C,* early amniote stage, in which pallial portions of gray matter migrate toward the surface but the corpus striatum remains internal and has increased fiber connections, other than those for olfaction; *D,* a synapsid reptile in which the neopallium is beginning to appear; *E,* a primitive mammal, possibly a stage reached in the mid-Cretaceous, in which neopallium becomes more important and pushes original pallial areas apart; *F,* a modern placental mammal in which neopallium is extremely developed and connected with the neopallium of the opposite side by way of the corpus callosum. Abbreviations: *a,* archipallium; *b,* corpus striatum; *cc,* corpus callosum; *n,* neopallium; *p,* paleopallium; *V,* lateral ventricle. Major fiber tracts are shown by groups of lines. The lines passing through the corpus striatum represent the internal capsule. (From Romer, The Vertebrate Body. *F* redrawn and slightly modified.)

Meninges

The mammalian brain and the stumps of the cranial nerves should be studied from isolated sheep brains. As explained, the peripheral distribution and composition of the nerves is, with a few exceptions, essentially the same as in fishes. The foramina through which they leave the cranium are listed in

Table 3 (p. 86), and certain of the nerves will be seen during later dissections. If it is not possible to study the brain from isolated specimens, it may be removed from your own specimen. But if this is to be done, postpone this study until the end of the course. To remove the brain, first make a sagittal section of the head and then carefully loosen the halves of the brain and pull them out. Leave on the tough membrane covering the brain (dura mater) as you take it out. The cranial nerves will, of course, have to be cut, but leave as long stumps as possible.

The tough outer membrane that covers the brain is known as the **dura mater.** Actually, it represents the dura of lower vertebrates united with the periosteum lining the cranial cavity. Carefully remove the dura in order to see the other membranes. As you do so note that it sends one extension down between the cerebrum and cerebellum and another between the two cerebral hemispheres. The former extension is called the **tentorium;** the latter, the **falx cerebri** (Fig. 9–16, p. 262). These membranes help to stabilize the brain in the cranial cavity and protect it from distortion during sudden rotational movements of the head. The pia-arachnoid of lower tetrapods has separated into two layers in mammals, but they are hard to distinguish grossly. The **pia mater** is the vascular layer that closely invests the surface of the brain. The **arachnoid** lies between the pia and dura and is most easily distinguished from the pia in the region overlying the grooves on the brain surface, for the arachnoid does not dip into them, whereas the pia does. In life the dura and arachnoid adhere to each other, but there is a definitive **subarachnoid space** between the arachnoid and pia which is crisscrossed by weblike strands of connective tissue—a feature that gives the name to the arachnoid. **Cerebrospinal fluid** circulates in the subarachnoid space, where it forms a liquid cushion around the central nervous system.

The liquid cushion around the central nervous system is very important, for it helps to protect and support the exceedingly soft and delicate nervous tissue. Cerebrospinal fluid gives the brain a great deal of buoyancy. It has been calculated, for example, that a human brain weighing 1500 g. outside the body has an effective weight of only 50 g. in situ. The cerebrospinal fluid also helps to provide the brain with carefully selected nutrients and other substances, for fewer substances leave the capillaries within the central nervous system than leave capillaries elsewhere in the body.

External Features of the Brain and the Stumps of the Cranial Nerves

(A) TELENCEPHALON

The paired cerebral hemispheres and the cerebellum are so large that little else is at first apparent in a dorsal view. Notice that the **cerebral hemispheres** are separated from each other by a deep longitudinal furrow which is known as the **longitudinal cerebral fissure.** Spread the dorsal parts of the cerebral hemispheres apart and note the thick transverse band of fibers that connects them. This is the **corpus callosum,** a neopallial commissure found only in placental mammals. The surface of each hemisphere is thrown into many irregular folds (the **gyri**), which are separated from each other by grooves (the **sulci**). A pair of **olfactory bulbs** project from the rostroventral portion of the cerebrum (Fig. 8–11). They can be seen best in a ventral view. The olfactory bulbs lie over the cribriform plate of the ethmoid and receive the groups of olfactory neurons from the nose. These neurons, which cannot be seen grossly, constitute the **olfactory nerve (I).** A whitish band, the **lateral olfactory tract,** extends at an angle caudally and laterally from each bulb. The ventral portion of the cerebrum, to which each olfactory tract leads, is known as the **piriform lobe.** It is separated laterally from the rest of the cerebrum by the **rhinal sulcus.**

The piriform lobe represents the **paleopallium** of lower vertebrates. The remaining olfactory portion of the cerebrum, **archipallium,** has been pushed internally and does not show on the surface. Thus, all the superficial part of the cerebrum that lies lateral and dorsal to the rhinal sulcus is **neopallium**—the major integrating region of the brain.

(B) DIENCEPHALON

The telencephalon has enlarged to such an extent that it has grown back over and covers the diencephalon and much of the mesencephalon. To see

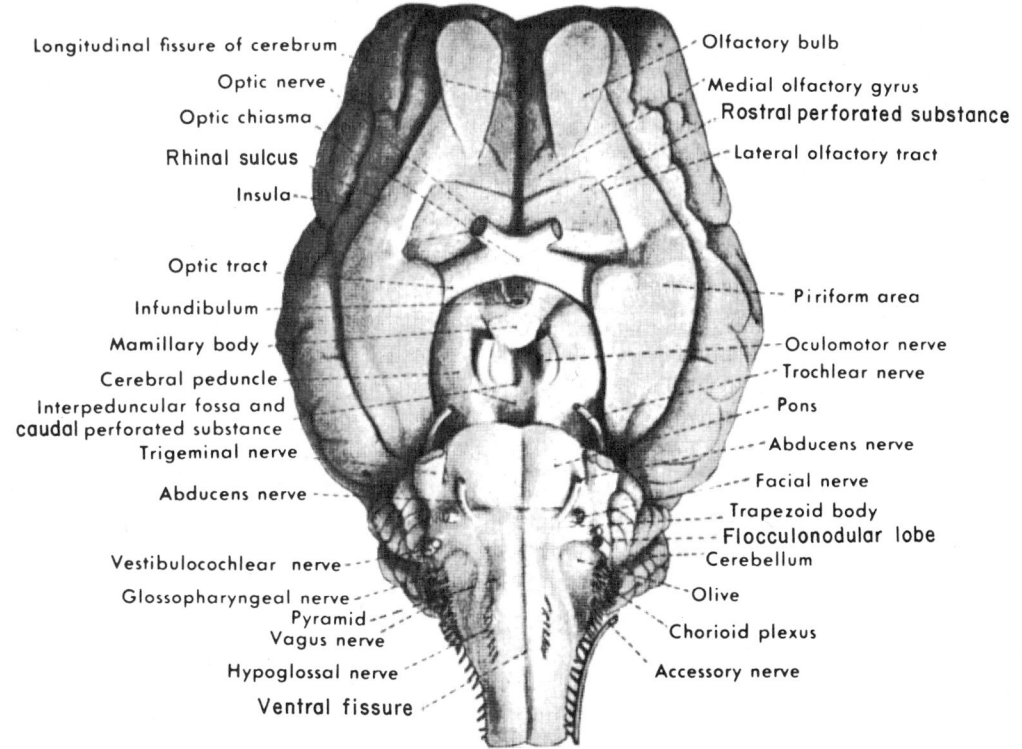

Figure 8–11. Ventral view of the sheep brain. (From Ranson and Clark, Anatomy of the Nervous System.)

the dorsal portion of the diencephalon (**epithalamus**), it is necessary to spread the cerebral hemispheres apart, cutting the corpus callosum. Pick away the tela choroidea and its choroid plexus forming part of the roof of the diencephalon. The longitudinal slit that is then exposed is the **third ventricle.** The knoblike **pineal gland** lies caudal to the ventricle. It has been implicated in the regulation of sexual development, for secretions of the gland inhibit gonadal development in some species. Extracts of the gland also cause retraction of the pigment in the chromatophores of lower vertebrates. Exposure to light decreases the activity of the gland. The narrow transverse band of tissue between the pineal body and the ventricle is the **habenular commissure,** and the tissue forming the posterolateral rim of the ventricle is the **habenula.**

Turn the brain over and examine the ventral surface of the diencephalon (**hypothalamus**). The **optic nerves** (II) undergo a partial decussation at the rostral border of the hypothalamus, forming the prominent **optic chiasma.** The rest of the hypothalamus is the oval area lying caudal to the optic chiasma. The **hypophysis** (pituitary gland) may still be suspended by its narrow stalk, the **infundibulum,** from the hypothalamus. If so, remove it in order to get a clearer view of the region. The cavity in the infundibulum represents an extension of the third ventricle. That portion of the hypothalamus adjacent to the attachment of the infundibulum is known as the **tuber cinereum.** A pair of rounded **mamillary bodies** forms the caudal end of the hypothalamus.

In order to see the **thalamus,** or lateral wall of the diencephalon, you must carefully pull one of the cerebral hemispheres forward and look beneath it. A better view will be had after the cerebrum has been dissected (p. 225), so you may wish to postpone a study of the thalamus until then. An **optic tract** leads from the optic chiasma to terminate in an enlargement of the thalamus known as the **lateral geniculate body** (Fig. 8–12). The meninges will have to be removed to see the area clearly. The smaller enlargement caudal to the lateral geniculate body is the **medial geniculate body,** and that portion of the thalamus lying dorsal to the geniculate bodies is known as the **pulvinar.**

The thalamus is an important relay station between the cerebral hemispheres and the rest of the brain. All sensory impulses, except for olfaction, pass through the dorsal portion of the thalamus on the way to the cerebrum. The medial geniculate body, for example, relays auditory impulses to the cerebrum. Most fibers in the optic tract terminate in the lateral geniculate body, from which impulses are relayed to the visual cortex. Some optic fibers go to the rostral colliculus, from which impulses go to the pulvinar and thence to the visual cortex. The ventral portion of the thalamus, which is not exposed on the surface, relays certain efferent impulses on the way back from the cerebral hemispheres. In addition, the thalamus may act as a subcortical center of integration, and its numerous interconnections with the cortex suggests that it plays a role in many cortical functions. The hypothalamus of mammals, as in lower vertebrates, is an important integrating center for many autonomic and visceral functions (p. 210). Temperature regulation is added to its activities in birds and mammals. The habenula continues to be an important olfactory center. Impulses reach it from the piriform lobe of the cerebrum and go out to nuclei in the floor of the mesencephalon.

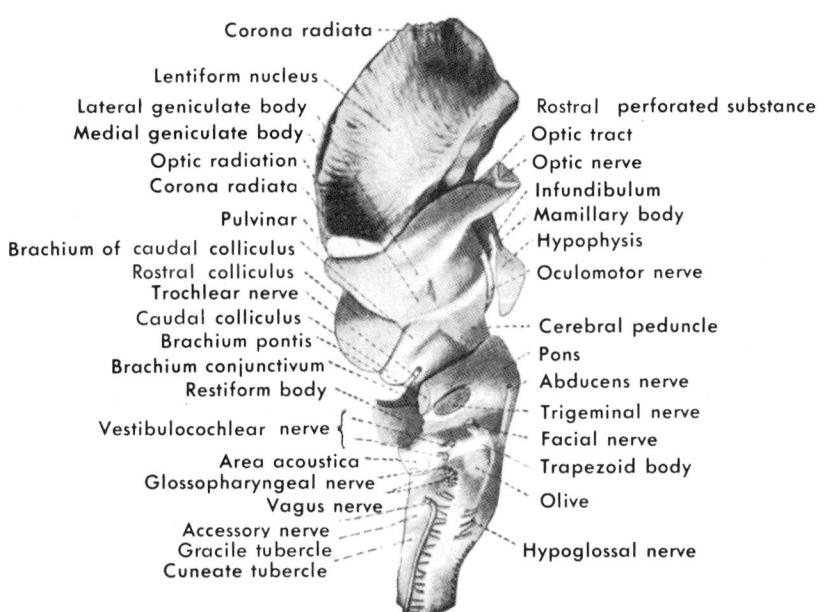

Figure 8–12. Lateral view of the sheep brain stem. (From Ranson and Clark, Anatomy of the Nervous System.)

(C) MESENCEPHALON

The roof, or **tectum,** of the mesencephalon can be seen by spreading the cerebrum and cerebellum apart. Four prominent, round swellings (the **corpora quadrigemina**) characterize this region. The larger, rostral pair are the **rostral** (superior) **colliculi,** the smaller, caudal pair, the **caudal** (inferior) **colliculi.** Note that the **trochlear nerves** (IV) arise slightly caudal to the caudal colliculi (Fig. 8–12).

A pair of **cerebral peduncles** lie along the ventrolateral surface of the mesencephalon. Each emerges from beneath the optic tract and is as wide as the distance from the medial geniculate body to the hypothalamus. An **oculomotor nerve** (III) arises from the surface of each peduncle. The depression between the two peduncles is the **interpeduncular fossa.** If you strip the meninges from this region, you may be able to see small holes through which blood vessels enter the brain. This region constitutes the **caudal perforated substance.** A comparable **rostral perforated substance** lies rostral to the optic chiasma (Fig. 8–11).

The evolution of the neopallium has robbed the tectum of its original importance as the major integrating area, but optic and auditory fibers are still projected to the rostral and caudal colliculi, which remain significant visual and auditory centers, respectively. The peduncles are large bundles of fibers that extend caudally from the cerebral hemispheres. Most efferent impulses from the cerebrum pass back through them.

(D) METENCEPHALON

The dorsal portion of the metencephalon forms the **cerebellum.** It will be noted that the surface area of the cerebellum is increased by numerous platelike folds (**folia**) separated from each other by **sulci.** The median part of the cerebellum, which has the appearance of a segmented worm bent nearly in a circle, is called the **vermis;** the lateral parts are the **hemispheres.** The lobe of each hemisphere that lies ventral to the main part of the hemisphere, and lateral to the region where the cerebellum attaches to the rest of the brain, is known as the **flocculonodular lobe.** These lobes are homologous to the auricular lobes of lower forms and receive vestibular impulses; most of the vermis is homologous to the body; and most of the hemispheres are new additions with cerebral connections.

The cerebellum is connected with other parts of the brain by three prominent fiber tracts, or peduncles (Figs. 8–12 and 8–13). The **brachium pontis,** or middle peduncle, lies medial to the rostral half of the flocculonodular lobe. You will have to dissect off this lobe on one side to see the brachium clearly. Note that the brachium pontis connects ventrally with a transverse band of fibers known as the **transverse fibers of the pons.** The tissue caudal and slightly medial to the brachium pontis constitutes the **restiform body,** or caudal peduncle. It also continues along the dorsolateral margin of the medulla. The **brachium conjunctivum,** or rostral peduncle, lies medial to the brachium pontis and can be seen by looking in the area between the cerebellum and caudal colliculi.

The ventral portion of the metencephalon has differentiated sufficiently from the medulla oblongata in mammals to be considered a distinct region — the **pons.** Grossly, the pons includes the region of transverse fibers bordered rostrally by the interpeduncular fossa, and caudally by the trapezoid body

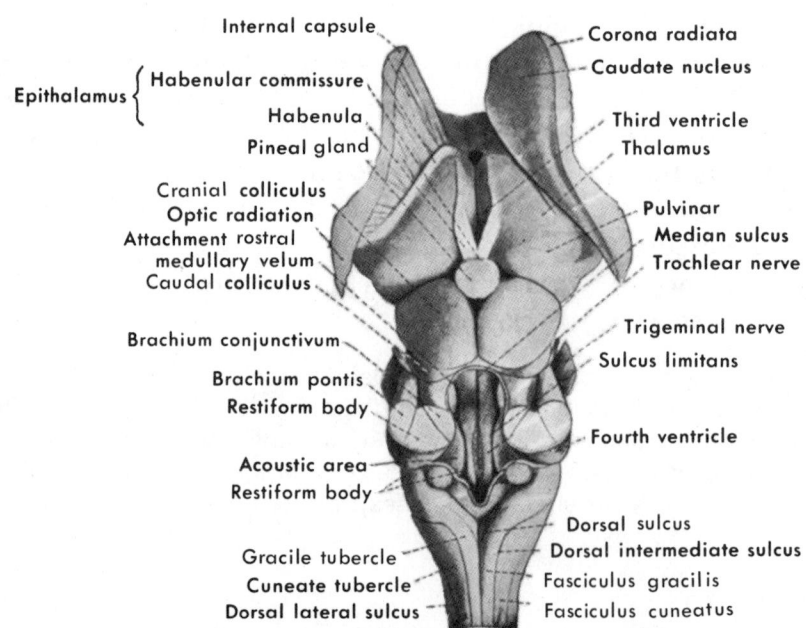

Figure 8–13. Dorsal view of the sheep brain stem. (From Ranson and Clark, Anatomy of the Nervous System.)

(part of the myelencephalon). The large trigeminal nerve (V) arises from the lateral portion of the pons and extends anteriorly across the base of the brachium pontis (Fig. 8–11).

The cerebellum is the center for equilibrium and motor coordination. It monitors the motor activity of the body and initiates corrective impulses. In this connection, it receives vestibular fibers from the inner ear, proprioceptive fibers from the muscles of the body, and a variety of impulses from the cerebrum. Fibers from the cerebrum are relayed in the pons, decussate in the transverse fibers of the pons and ascend to the cerebellum through the brachium pontis. Proprioceptive fibers come in through the restiform bodies. Vestibular fibers also enter in this region. Efferent fibers from the cerebellum pass out through the brachia conjunctiva to the ventral portion of the midbrain and metencephalon. Here they are relayed to the thalamus and to the motor nuclei and columns.

(E) MYELENCEPHALON

All the rest of the brain belongs to the myelencephalon, and forms the **medulla oblongata.** In order to see the parts of the medulla clearly, the meninges must be stripped off on at least one side, but the remaining cranial nerves should be identified before this is done. At the border between the pons and the medulla oblongata are the **abducens** (VI) and **facial** (VII) **nerves;** the abducens lies more medial and extends rostrally across the pons, and the facial passes out laterally with the eighth nerve. The stump of the **vestibulocochlear nerve** (VIII) lies dorsal to the seventh nerve and ventral to the caudal end of the flocculonodular lobe. The **glossopharyngeal** (IX) and **vagus** (X) **nerves** are represented by a number of fine rootlets caudal to, but in line with, the eighth nerve. Since one cannot trace these rootlets into the peripheral parts of the nerves, it is impossible to say more than that the rostral rootlets belong to the glossopharyngeal and the caudal ones to the vagus. The **accessory nerve** (XI) is the large, longitudinal nerve caudal to the vagus. It arises by a number of fine rootlets from the caudal end of the medulla and the rostral end of the spinal cord. The rostral end of the nerve is the end that leads out of the skull to the muscles that the nerve supplies. If the nerve also has a cut caudal end on your specimen, all its origin is not intact. The **hypoglossal nerve** (XII) is represented by the rootlets on the caudoventral portion of the medulla. If any of the cranial nerves cannot be found, refer to Figure 8–11.

Now strip off the meninges on one side of the medulla, and the **tela choroidea** with its **choroid plexus** that forms much of the roof of the medulla. Pull the cerebellum forward and note the large **fourth ventricle** extending forward into the metencephalon. The caudal part of the roof of the ventricle was formed by the tela choroidae, but the anterior part of the roof is formed by a thin layer of fibers termed the **rostral medullary velum,** which can be seen by pushing the cerebellum caudally. The trochlear nerve decussates in the velum.

Note the enlargement on the dorsal rim of the medulla just caudal to the point at which the restiform body turns into the cerebellum, and dorsal to the vestibulocochlear nerve. Observe that it extends medially to an oval-shaped enlargement in the ventrolateral part of the floor of the fourth ventricle. This enlargement constitutes the **area acoustica.**

Examine the dorsal surface of the caudal end of the medulla. (The medulla extends as far caudad as the **first spinal nerve.**) The prominent **dorsal sulcus** (Fig. 8–13) continues onto the medulla nearly to the fourth ventricle. A less distinct **dorsal intermediate sulcus** lies about one-half centimeter lateral to the preceding, and a **dorsal lateral sulcus** slightly lateral to the dorsal intermediate. These grooves outline two longitudinal fiber tracts, a dorsal **fasciculus gracilis** and a more lateral **fasciculus cuneatus.** The rostral end of the former tract expands slightly to form a structure called the **gracile tubercle.** The latter has a comparable enlargement known as the **cuneate tubercle.** The tubercles are the outward manifestations of nuclei. As already stated, the restiform body forms the dorsolateral rim of the medulla rostral to these tubercles. It then turns dorsally to enter the cerebellum.

Turn the brain over and examine the ventral surface of the medulla. The narrow, transverse band of fibers immediately caudal to the pons is the **trapezoid body.** Notice that its fibers can be followed dorsolaterally into the area acoustica. The midventral groove extending the length of the medulla oblongata is the **ventral fissure.** The longitudinal bands of tissue on either side of it that are approximately one centimeter wide are known as the **pyramids.** Note that some of the pyramidal fibers lie superficial to the trapezoid body. An area known as the **olive** lies lateral to the pyramids, caudal to the trapezoid body and ventral to the glossopharyngeal and vagus nerves.

The medulla is a transitional region between the rostral parts of the brain and the spinal cord. As in fishes (p. 210), most of its gray matter represents the forward continuation of the columns of the cord and the break-up of these into nuclei. Numerous reflex activities occur between these nuclei, and many important visceral activities are controlled here: respiratory movements, salivation, swallowing, rate of heart beat, and blood pressure. Most of the nuclei cannot be seen grossly, but some form bulges on the surface. The area acoustica is a region in which auditory fibers from the ear undergo various relays. Some of the relays extend auditory impulses through the trapezoid body. The olive represents another nucleus, but its function is uncertain. It has connections with the cerebellum via the restiform bodies and may be related to muscular coordination. Much of the white matter of the medulla represents fiber tracts passing through the region. The pyramids, for example, are a posterior continuation of some of the fibers that form the cerebral peduncles. The fasciculus gracilis and cuneatus are largely composed of ascending proprioceptive fibers. These relay in the nucleus gracilis and cuneatus before going to the thalamus (for relay to the cerebrum) or to the cerebellum via the restiform bodies.

Sagittal Section of the Brain

Cut the brain in half as close to the sagittal plane as possible. If you deviate from the plane, take the larger half and dissect away enough tissue to be able to see the median cavities of the brain clearly (Fig. 8–14). Many of the features just described can also be seen in this view. Note in particular the way in which the cerebral hemispheres extend back over the diencephalon and mesencephalon. The **corpus callosum** also shows particularly well. Its expanded rostral end is known as the **genu;** its expanded caudal end, as the **splenium;** and the thinner region between, as the **trunk of the corpus callosum.** A thin, vertical septum of tissue, **septum pellucidum,** lies ventral to the rostral part of the corpus callosum. It consists of two thin plates of gray matter. The lateral ventricle lies lateral to the septum and may be seen by breaking it. A band of fibers called the fornix lies caudal to the septum pellucidum. The **body of the fornix** begins near the splenium, and then the band curves forward and ventrally as the **column of the fornix.** It passes out of the plane of the section caudal to a small, round bundle of fibers which is a cross section of the **rostral** (anterior) **commissure,** an olfactory decussation. The thin ridge of tissue extending ventrally from the rostral commissure to the optic chiasma is the **lamina terminalis**—a landmark representing the rostral end at the embryonic neural tube. The cerebral hemispheres are lateral evaginations that extend forward from the lamina.

The third ventricle and diencephalon lie caudal to the column of the fornix, rostral commissure and lamina terminalis. Note that the **third ventricle** is very narrow but has a considerable dorsal–ventral and rostral–caudal extent. It is lined by a shiny epithelial membrane (the **ependymal epithelium**), as are all the cavities within the neural tube. The thalamus lies lateral to the

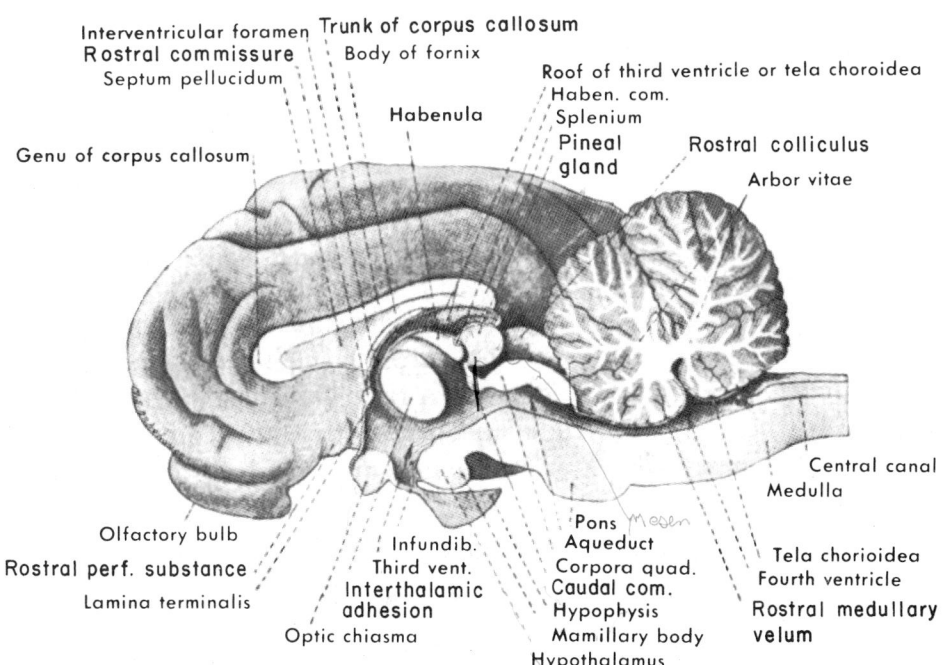

Figure 8–14. Sagittal section of the sheep brain. (From Ranson and Clark, Anatomy of the Nervous System.)

third ventricle, but a portion of it, the **interthalamic adhesion,** extends across the third ventricle and will appear as a dull, circular area not covered by the shiny ependyma. The **interventricular foramen** (foramen of Monro), through which each lateral ventricle communicates with the third ventricle, lies in the depression rostral to the massa intermedia. The hypothalamus lies ventral to the third ventricle and the epithalamus dorsal to it. Note again the **pineal gland,** the **habenular commissure** (which will show in cross section), and the **habenula.** These features show unusually well in the section. In addition, the epithalamus includes a **caudal** (posterior) **commissure.** This is the tissue ventral to the attachment of the pineal gland and anterior to the corpora quadrigemina.

Carefully dissect away tissue between the rostral commissure and the mamillary bodies. You will soon find a distinct band of fibers, the **post-commissural fornix,** which is continuous with the column of the fornix and leads to the mamillary bodies (Fig. 8–15). This is one of the main connections between the cerebrum and the hypothalamus. Dissecting just rostral to the rostral commissure, you will find fibers from the column of the fornix leading into the area of gray matter just under the septum pellucidum. This is the **septal region** and these fibers in front of the rostral commissure constitute the **precommissural fornix.**

A narrow **cerebral aqueduct** (aqueduct of Sylvius) leads through the mesencephalon to the fourth ventricle of the hindbrain. The cerebellum lies above the fourth ventricle. Note that most of its **gray matter** is in the form of a gray **cortex** over the surface of the folia, while the **white matter** is centrally located. The white matter, which represents fiber tracts extending between the cerebellum and other parts of the brain, has the appearance of a tree and is called the tree of life (**arbor vitae**).

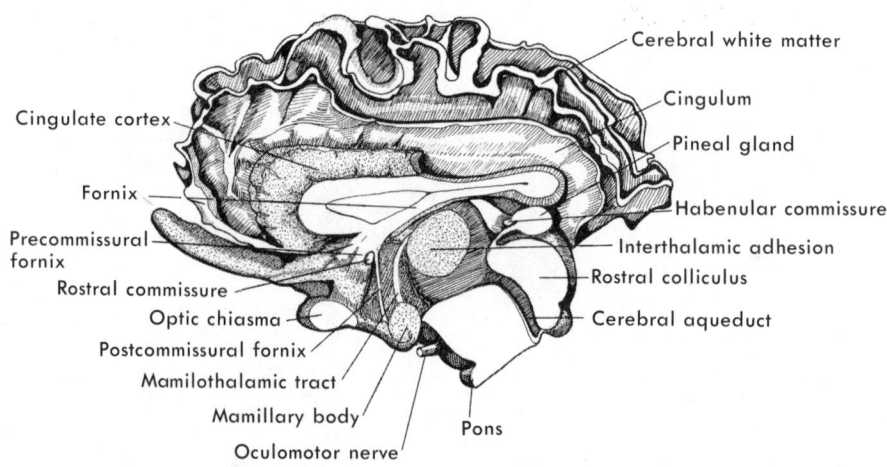

Figure 8–15. Dissection of a sagittal section of the cerebrum and diencephalon of a sheep brain.

Dissection of the Brain

By DOUGLAS B. WEBSTER, *Louisiana State University Medical Center*

So far your study of the brain has involved primarily the identification of external landmarks, which, though a vital first step, gives but little feeling of the structural and functional continuity of parts. After you have mastered these landmarks, therefore, a dissection is necessary to gain appreciation of the brain's structural organization.

(A) CEREBRAL CORTEX

Take one half of your sheep brain, preferably the half on which you have demonstrated the precommissural and postcommissural fornix. With a pair of fine forceps carefully remove the pia mater and arachnoid from the cerebral hemisphere and cerebellum. Notice that the pia mater and its blood vessels dip deep into each sulcus of the neocortex. To realize the extent of the **gray matter** of the neocortex compared to the underlying **white matter** (*i.e.,* fibers leading to, from and within the neocortex), either cut off a one- or two-centimeter slice of the cerebrum in the frontal plane or, if time permits, dissect off the gray matter. This dissection is best accomplished by scraping gently with a blunt instrument such as the handle of your scalpel or forceps or an orangewood manicure stick. Scrape deeply into each sulcus to remove all the gray matter, leaving only white. Do this for all the neocortex with the exception of the **insular cortex,** which will be removed later. The insular cortex is that part of the neocortex just above the rhinal sulcus at the level where the lateral olfactory tract meets the piriform lobe (Fig. 8–11).

When all this gray matter has been removed, notice the shape of the white matter deep to it. Especially notice how the fibers extend upward to form a huge core in each gyrus of the cerebrum (Fig. 8–15). Also notice the large mass of tissue you have removed; this gives you a better idea of just how much of the brain is neocortex.

The white matter you are now looking at is a tangle of heterogeneous fibers. Some are short **arcuate fibers** connecting one gyrus with an adjoining gyrus. Others are long **association fibers** connecting one gyrus with cells many gyri away. Still others are **commissural fibers,** passing through the **corpus callosum** to connect right and left hemispheres. Some are **projection fibers: thalamocortical fibers,** ascending from the thalamus to the cerebrum; and **corticobulbar** and **corticospinal fibers,** leading from cells in the neocortex down to other brain or spinal cord structures. These many fibers are not organized into easily dissected groups; only in histological sections, and especially through the use of degeneration studies, can they be definitely traced and identified.

(B) CORPUS STRIATUM

The deeper part of the cerebral hemispheres, especially anteriorly, has other groups of gray matter interspersed with fibers, and hence is known as the **corpus striatum.** The gray matter can be divided grossly into a lateral portion (the lentiform nucleus) and a medial portion (the caudate nucleus) in each hemisphere. Carefully scrape away the reserved gray matter of the insular cortex until you come to a very thin layer of white fibers—the **external capsule.** Break through this external capsule and you will find a large mass of gray material; this is the **lentiform nucleus.** After identifying the lentiform nucleus and determining its boundaries, scrape it away. You will find a very large mass of fibers running between the thalamus and the deep white matter below the neocortex. This is the **internal capsule,** which contains the axons leading from neurons of the thalamus up to the cerebral hemisphere as well as the axons leaving the neocortex to pass down to other brain and spinal cord structures (Fig. 8–16).

The medial nucleus of the basal ganglia is best approached from the medial side of the brain half. Again locate the **septum pellucidum.** Cut it open if this has not already been done and look into the **lateral ventricle.** In doing this, be careful not to injure the part of the brain between the septum and the olfactory bulb. Note the very large head of the **caudate nucleus** (the medial portion of the corpus striatum), forming the ventrolateral border of the rostral horn of the lateral ventricle (Fig. 8–17). After identifying the large caudate nucleus, carefully scrape away its surface and see fibers of the internal capsule from the medial side. The caudate nucleus and the lentiform nucleus are penetrated and separated by the internal capsule, which connects neocortex with the rest of the brain. After passing between these nuclei, the fibers of the internal capsule fan out to all parts of the neocortex in tracts called the **corona radiata** (Fig. 8–16).

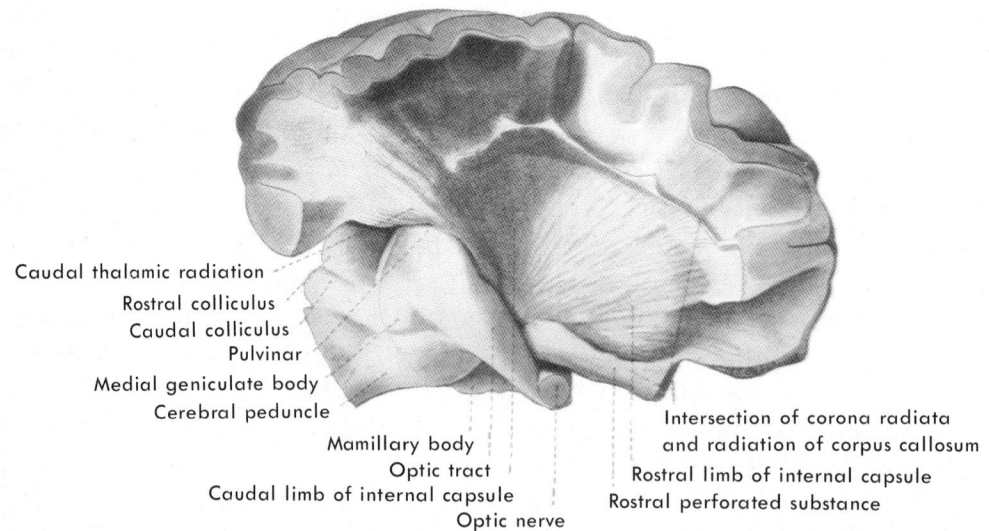

Caudal thalamic radiation
Rostral colliculus
Caudal colliculus
Pulvinar
Medial geniculate body
Cerebral peduncle
Mamillary body
Optic tract
Caudal limb of internal capsule
Optic nerve
Intersection of corona radiata and radiation of corpus callosum
Rostral limb of internal capsule
Rostral perforated substance

Figure 8–16. Lateral view of a dissection of the sheep brain to show the internal capsule. The lateral surface of the cerebrum and the lentiform nucleus have been removed. (From Ranson and Clark, Anatomy of the Nervous System.)

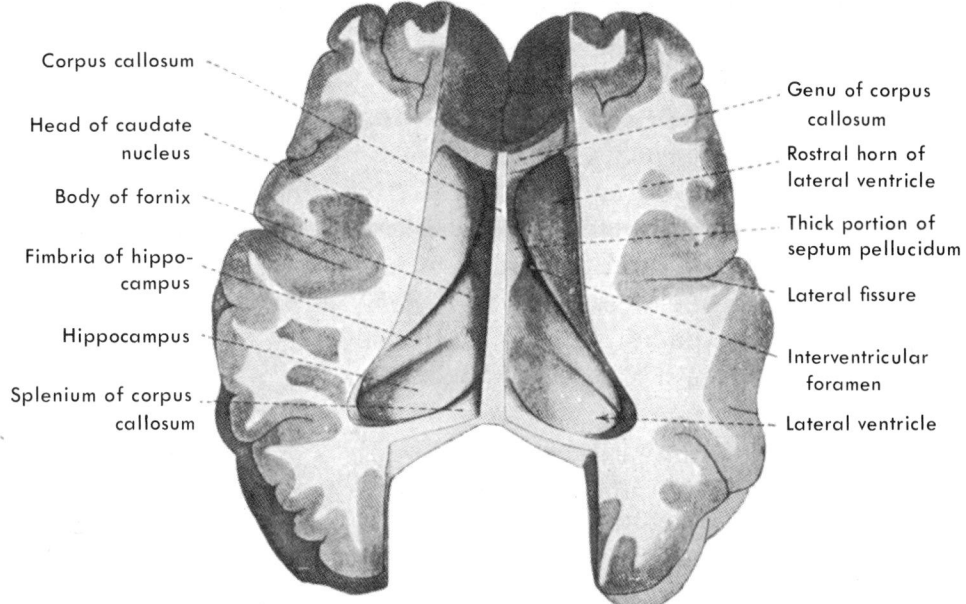

Figure 8–17. Dorsal view of a dissection of sheep cerebrum. (From Ranson and Clark, Anatomy of the Nervous System.)

(C) RHINENCEPHALON AND PAPEZ CIRCUIT

Now turn your attention to the rhinencephalon, or olfactory portions of the telencephalon. During the evolution of the mammalian neocortex, outgrowths from the primitive olfactory system eventually developed into the bulk of the cerebral hemispheres. In the process the olfactory portions themselves were displaced and now follow circuitous routes (see Fig. 8–10). The olfactory nerves terminate in the olfactory bulbs, and each olfactory bulb, in turn, connects with more caudal parts of the cerebrum by means of three olfactory tracts—the lateral, intermediate, and medial olfactory tracts.

The **lateral olfactory tract,** which you have already seen, is the largest and most easily demonstrated. Careful scraping of the superficial gray matter will enable you to follow this tract into the **piriform lobe** and a small part of the adjoining **neocortex.** The smaller and shorter **intermediate olfactory tract,** which lies on the medial side of the preceding one, may be traced into the **rostral perforated substance,** where it terminates primarily in a nuclear concentration called the **olfactory tubercle.** Although in the sheep the olfactory tubercle is not grossly distinguishable from the rest of the rostral perforated substance, in many mammals its large surface bulges from the rostral perforated substance. By scraping away gray matter from the rhinencephalon near the midline and caudal to the olfactory bulb you will uncover the **medial olfactory tract.** Part can be traced to the **septal area** just under the genu of the corpus callosum; part, into the **rostral commissure.** The olfactory bulb also connects with the mamillary bodies, the tegmentum (or floor) of the midbrain and the hippocampus, but these connections cannot be seen in gross dissection.

Prominent in the rhinencephalon and demonstrable by gross dissection is the **limbic lobe,** an important part of which is known as the **Papez circuit** (rhymes with "tapes"). In recent years the structures of the Papez circuit have come under a great deal of experimental scrutiny and they are now thought to be important in such diverse functions as emotional behavior and short-term memory.

Identify the **fornix** once again in the sagittal section, and follow its postcommissural portion to the **mamillary body.** Now carefully scrape away thalamic tissue just above the mamillary body and caudal to the postcommissural fornix to reveal a tract running rostrodorsally toward the rostral part of the thalamus. This is the **mamillothalamic tract.** Trace it up to the rostral part of the thalamus, where it ends in the **rostral thalamic nuclei;** it runs parallel to the postcommissural fornix.

The rostral thalamic nuclei send fibers by way of the internal capsule and corona radiata into the cerebral cortex, and particularly into that part of the cerebral cortex which lies just above the corpus callosum. This cortical area, which can be seen on the undissected brain half, is called the **cingulate cortex** (Fig. 8–15). It is possible to dissect this portion of the internal capsule, but since that would be destructive to other structures you will be seeing, this dissection should be saved for a later time. However, do identify the cingulate region where you have already removed cortex and notice there a distinct long association tract, called the **cingulum,** running rostrocaudally, which is composed of fibers from cells of the cingulate cortex. Follow it caudally, around the back end of the corpus callosum, and then see it dip deep into the hemisphere where it will connect with the hippocampus.

The **hippocampus** is thought by some to be the most ancient cortex of vertebrates and is thus given the name **archipallium.** It lies in the floor of the lateral ventricle (Fig. 8–17) behind the caudate nucleus and will be seen best by spreading open the lateral ventricle where you have already cut the septum pellucidum. (The irregular-shaped, dark-colored frill in the floor of the ventricle is the **choroid plexus of the lateral ventricle.** It has invaginated from the thin, ventral wall of the hemisphere.) The surface of the hippocampus is covered by a very thin layer of white matter, called the **alveus.** Deep to this thin tract is the gray matter of the hippocampus. The cingulum enters it caudally along its entire caudal border. The hippocampus extends lateroventrally, deep to the white matter of the neocortex, toward and almost reaching the piriform lobe. You may have to cut through the dorsolateral part of the hemisphere to trace it. The rostral border of the hippocampus lies free; along it is a tract of white matter called the **fimbria** of the hippocampus. Follow the fimbria of the hippocampus rostromedially and notice that it is continuous with and forms the body of the fornix.

You have now dissected the Papez circuit, going from the fornix by way of postcommissural fornix to the mamillary body, thence by way of the mamillothalamic tract to the rostral nucleus of the thalamus, thence by way of the internal capsule and corona radiata to the cingulate cortex, thence by way of cingulum back around to the hippocampus and, finally, by way of the fimbria, rejoining the fornix.

(D) CEREBELLUM

Cerebrum and cerebellum are also intimately connected to one another by circuit pathways, many of them demonstrable by dissection. Note how many of the lateral fibers of the internal capsule continue down into and form part of the **cerebral peduncle** and how these disappear into the pons. Within the pons, and you should verify this by careful scraping, is a great deal of gray material called the **pontine nuclei.** Many of the fibers coming down the cerebral peduncles terminate in synaptic connections with the cells of the pontine nuclei. Fibers from these cells of the pontine nuclei decussate—that is, cross the midline—in the pons, and then enter the cerebellum (Fig. 8-12) by way of the **brachium pontis.**

Follow the brachium pontis into the cerebellum by scraping away part of the cerebellar hemisphere. These fibers continue up through the white matter of the cerebellum—the **arbor vitae**—and terminate in the cerebellar cortex (Fig. 8-14). Fibers leaving the cerebellar cortex follow the arbor vitae down as far as the base of the cerebellar peduncles and terminate there in large nuclear groups which may be demonstrated by dissecting shallowly into the lateral aspect of the brachium pontis.

The largest and most easily demonstrated of these cerebellar nuclei are the **dentate nuclei.** Their cells send fibers out of the cerebellum by way of the **brachium conjunctivum** (Fig. 8-12). These fibers cross the midline in the midbrain below the cerebral aqueduct, and many of them pass forward to terminate in the thalamus. In gross dissection the brachium conjunctivum can be seen leaving the cerebellum and entering the brain stem, but the mesencephalic and diencephalic portions can be visualized only in microscopic sections.

The fibers end in the lateral part of the dorsal thalamus, synapsing on thalamic cells which, in turn, send their fibers up through the internal capsule and corona radiata to the neocortex. As we see, therefore, there are distinct channels for information transfer and sharing between cerebrum and cerebellum by way of the cerebral peduncle, pontine nuclei, brachium pontis, cerebellar cortex, arbor vitae, cerebellar nuclei, brachium conjunctivum, thalamus, internal capsule, corona radiata and neocortex.

(E) VISUAL SYSTEM

The cerebrum and cerebellum, the two main integrating centers of the brain, are concerned with information input and output—input being the information brought from each of the sense organs to the brain, and output the information sent out to skeletal muscles and other effectors. Since you have seen the structural connections between the cerebrum and cerebellum, it is now time to study the sensory channels leading to them and the motor pathways leaving them.

We have already examined the circuitous routes of the olfactory system, evolved as a by-product of the telencephalon's long phylogeny. Two other sensory systems — visual and auditory — lend themselves to gross dissections which will give you an idea of the structural apparatus for sending information up into the cerebral hemispheres.

The visual system is easily dissected, starting at the stumps of the two **optic nerves** where they enter the **optic chiasma.** Remember that the "optic nerve," strictly speaking, is not a true nerve but a brain tract containing axons of third order neurons from the ganglion cells of the retina. The optic nerve partially decussates in the optic chiasma: its fibers from the lateral half of the retina remain ipsilateral, while fibers from the medial half of the retina cross the midline.

Remove the caudolateral part of the cerebrum, including the hippocampus, but do so in such a way that you can place it back in position for further study. Remove the meninges from the lateral surface of the thalamus and mesencephalon. Follow the **optic tract** up the wall of the brain stem to the thalamic swelling known as the **lateral geniculate body** (Fig. 8–12). To see the gray matter of the lateral geniculate body, carefully lift the optic tract off the brainstem and peel it up and over the lateral geniculate body. You will notice many fibers penetrating the gray matter of the lateral geniculate body. Other, more superficial fibers pass over the lateral geniculate body and extend as the **brachium of the rostral colliculus** to terminate in the **rostral colliculus.** Still other fibers pass over the lateral geniculate body and enter the caudal thalamus in the region known as the **pulvinar,** whence they extend into the **pretectal region** between pulvinar and rostral colliculus. Careful dissection can reveal the **optic radiation,** the group of fibers that extend from the lateral geniculate body by way of the caudal part of the internal capsule and corona radiata to the **occipital neocortex.**

(F) Auditory System

Now turn your attention to the auditory system and identify again the area where the eighth cranial nerve enters the brain stem. Here you will find, just behind the restiform body, the tubercle of gray matter known as the **area acoustica,** composed of the **dorsal cochlear nucleus** and, deep to it, part of the **ventral cochlear nucleus** which also extends into the brain stem deep to the area acoustica (Fig. 8–12). The auditory portion of the eighth cranial nerve terminates in these cochlear nuclei.

Proceeding ventrally from the area acoustica, just behind the pons, identify again the **trapezoid body** as it passes around the ventral aspect of the brainstem and appears to join with its fellow on the other side just dorsal to the pyramids. Most, although by no means all, of the second-order auditory fibers leaving the cochlear nuclei cross the midline in the trapezoid body. Around the trapezoid body is a ventral nuclear group, unidentifiable by gross dissection, the **superior olivary complex**, one of the main termination points for second order auditory neurons.

From here the auditory system moves forward as a major tract, the **lateral lemniscus**, which you will first see emerging between the rostral edge of the brachium pontis and the brachium conjunctivum and then running up into the **caudal colliculus**. If you carefully peel off the brachium pontis it is possible to trace the lateral lemniscus all the way from the trapezoid body up into the caudal colliculus, where most or, more likely, all of its fibers terminate. From here, follow the **brachium of the caudal colliculus**, a prominent surface tract of fibers, as it crosses in a rostroventral direction down to a large thalamic prominence. This surface swelling is the thalamic relay station of the auditory system, the **medial geniculate body**; the **auditory radiation** arises here, then passes into the internal capsule, and then by way of the corona radiata terminates in the **temporal neocortex**.

(G) PYRAMIDAL SYSTEM

The major motor system leaving the cerebral hemisphere is best picked up at the cerebral peduncle, which contains **corticospinal fibers** as well as the **corticopontine fibers** described before. Identify the cerebral peduncle once again and then, by scraping, follow it rostrodorsally, up into the internal capsule and corona radiata. Its fibers begin in the motor cortex, perimotor cortical areas, and from areas of frontal and parietal cortex.

Now follow the cerebral peduncle caudally once again until it penetrates the pons. Further dissect the pons, demonstrating the gray as well as the white matter. Note that some of the fibers of the cerebral peduncle penetrate caudally directly through the pons and continue beyond it, forming the **pyramids**. Although much smaller than the cerebral peduncles, the pyramids contain the axons of the direct **corticospinal fibers** (Fig. 8–11). The pyramids continue caudally on both sides of the midventral line to the back part of the medulla oblongata, where most of their fibers decussate. Here you will see a slight bulge in the midventral surface of the brain stem. From this point on, the corticospinal fibers travel in different funiculi in different orders of mammals as they pass down the cord. However, they always terminate in the ventral horn cells of the spinal cord, and from these cells fibers go to the skeletal muscles, thus completing the output of motor information. The corticospinal fibers, with their cell bodies in the neocortex and their axons traveling all the way down to the ventral horn cells in the spinal cord, constitute the pyramidal system; this system, unique to mammals, is a major motor pathway for voluntary movements.

Spinal Cord and Spinal Nerves

(A) THE CORD AND SPINAL NERVES

The **spinal cord** (*medulla spinalis*) is a subcylindrical cord lying within the vertebral canal of the vertebral column. It is not uniform in diameter, for it bears **cervical** and **lumbosacral enlargements** from which nerves to the appendages arise, and the caudal end tapers as a fine **terminal filament** to end in the base of the tail. Only a section of the cord need be studied. To get at it, remove the epaxial muscles from your specimen of a mammal so as to expose several centimeters of the vertebral column. This should be done in the caudal thoracic region. With bone scissors, carefully cut across the pedicles of the vertebrae and remove the tops of the vertebral arches. The spinal cord will be seen lying in the vertebral canal. Continue chipping away bone and removing fat from around the cord until you have satisfactorily exposed it along with several roots of the spinal nerves.

The cord and roots are covered by the tough **dura mater**. Note that the dura is not fused with the periosteum lining the vertebral canal as it is in the cranial cavity. Leave the dura on for the present and examine the roots of the spinal nerves. At each segmental interval, there is a pair of dorsal and ventral roots (Fig. 8–1). Trace a dorsal and ventral root laterally on one side. They pass into the intervertebral foramen before uniting to form a **spinal nerve**. Just before uniting, the dorsal root bears a small round enlargement – the **spinal ganglion**. If you trace the spinal nerve laterally you may see it divide into a **dorsal ramus** to the epaxial regions of the body and a **ventral ramus** to the hypaxial regions. The small **communicating rami** to the sympathetic cord probably will not be seen.

The first spinal nerve emerges through a foramen in the vertebral arch of the atlas, and it is called the first cervical nerve. The eighth cervical nerve leaves just caudal to the seventh cervical vertebra. Thereafter, the spinal nerves carry the name and number of the vertebra caudal to which they emerge, e.g., the first thoracic nerve lies just behind the first thoracic vertebra. The number of spinal nerves varies with the number of vertebrae. Most quadrupeds have 8 cervical nerves, 12 to 14 thoracic nerves, 6 to 7 lumbar nerves, 3 to 4 sacral nerves, and 6 or more caudal nerves.

Slit open the dura at one end of the exposed area. This opens the **subdural space**. Observe that the roots of the spinal nerves do not have a simple attachment on the cord but unite by a spray of fine **rootlets**. Cut out a segment of the spinal cord and strip off the remaining meninges – **arachnoid** and **pia mater**. There is a deep ventral furrow on the cord known as the **ventral fissure**; a less distinct dorsal furrow, the **dorsal sulcus**; and a more prominent furrow slightly lateral to the middorsal line, the **dorsal lateral sulcus**. Note that the dorsal rootlets enter along the dorsal lateral sulcus. Recall that these same grooves extend onto the medulla oblongata.

Make a fresh cross section of the cord with some sharp instrument such as a razor blade. Examine the cut surface and compare it with Figure 8–1. Also look at demonstration slides if possible. The tiny **central canal** can generally be seen grossly, and if you are fortunate you may be able to distinguish the butterfly-shaped central **gray matter** from the peripheral **white matter.** As explained in the introduction to this chapter, the gray matter consists of unmyelinated fibers and the cell bodies of motor and internuncial neurons; the white, of ascending and descending myelinated fibers. That segment of the white matter that lies between the dorsal fissure and the dorsal lateral sulcus is called the **dorsal funiculus**; that segment between the dorsal lateral sulcus and the line of attachment of the ventral roots, the **lateral funiculus**; and that portion between the ventral roots and the ventral median fissure, the **ventral funiculus**.

(B) THE BRACHIAL PLEXUS

The dorsal rami of spinal nerves extend straight out into the epaxial region, but many of the ventral rami unite in a complex manner to form networks, or **plexuses**, before being distributed to the musculature and skin. This is especially true in the region of the appendages. In a typical mammal the anterior cervical nerves form a **cervical plexus** supplying the neck region; the posterior cervical and rostral thoracic nerves form a **brachial plexus** supplying the pectoral appendage; and the lumbar, sacral, and anterior caudal nerves form a **lumbosacral plexus** supplying the pelvic appendage.

The brachial plexus lies medial to the shoulder and rostral to the first rib; it should be approached from the ventral surface. If it is still intact on the side on which the muscles were dissected, study it there; otherwise, cut through the pectoralis complex of muscles on the other side. The dissection of the plexus involves the meticulous picking away of fat and connective tissue from around the nerves and the accompanying blood vessels. If you find it necessary to cut any of the larger vessels, do so in such a way that you will be able to appose the cut surfaces when you study the circulatory system. Clean off the nerves from a point as near to the vertebral column as you can reach to the point at which they disappear into the shoulder muscles and brachium.

The brachial plexus is formed by the union of the ventral rami of the sixth to eighth cervical and first thoracic nerves in the cat and mink. The fifth cervical also contributes to it in the rabbit. Considerable variation occurs in the details of the union of these nerves to form the plexus and in the origin of peripheral branches; but a common pattern for the mammals under consideration is shown in Figure 8–18. The ventral rami of the nerves that enter the plexus are referred to as the **roots** of the plexus. It will be noted that each root tends to split into two **divisions** and that the divisions of different nerves unite to form **trunks** from which peripheral nerves arise. It will also be noted that the splitting of roots into divisions, and the union of divisions to form trunks, occurs in such a way that there tends to be an early segregation of the nerves supplying the dorsal appendicular muscles from those supplying the ventral. The parts of the plexus going to the dorsal musculature have been stippled in the diagram. Many of the nerves are also cutaneous, being distributed to the skin, but only the major cutaneous branches will be described.

The most ventral nerves of the plexus are several **pectoral nerves**. They are small nerves which arise from the ventral divisions of the plexus, may or may not unite with each other, and pass to the pectoralis complex.

A large **suprascapular nerve** leaves the front of the plexus, where it

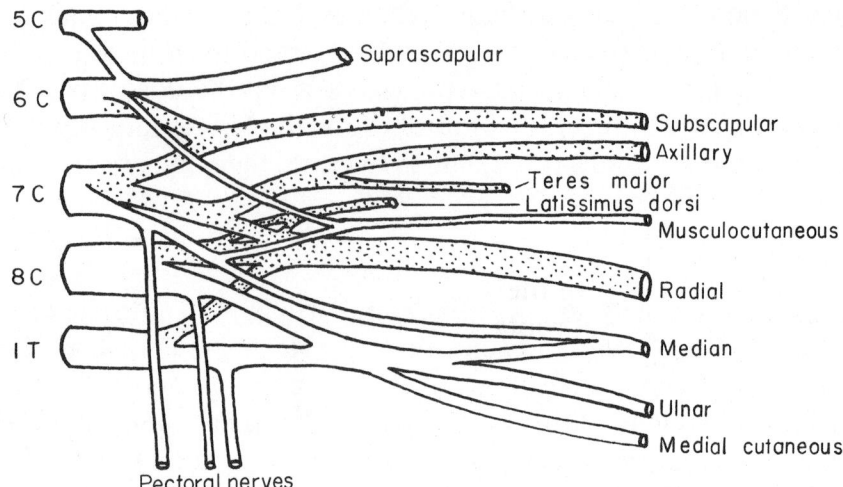

Figure 8–18. Diagram of the major parts of the mammalian brachial plexus based on the cat, mink, and rabbit, ventral view of the left side. 5 C, 6 C, 7 C, 8 C, and 1 T are ventral rami of the fifth to eighth cervical and first thoracic nerves forming the roots of the plexus. The fifth cervical nerve does not enter the plexus in the cat or mink. Most of the roots split into two divisions, and most of the divisions unite with divisions of adjacent nerves to form trunks that lead into the peripheral nerves, or give rise to the peripheral nerves. The divisions, cords, and peripheral nerves to the dorsal appendicular muscles have been stippled.

arises for the most part from the sixth cervical, and passes between the subscapular and supraspinatus muscles to supply the supraspinatus, infraspinatus, and some of the skin over the shoulder and brachium.

One or more **subscapular nerves** leave caudal to the suprascapular nerve and pass to the large subscapular muscle. They arise, for the most part, from the sixth and seventh cervicals.

A large **axillary nerve**, which lies caudal to the subscapular nerve, arises from the seventh cervical nerve. It passes through the proximal part of the brachium ventral to the long head of the triceps to supply the teres minor and deltoid complex.

Smaller nerves to the **latissimus dorsi** and **teres major** arise from the plexus near the origin of the axillary. Often the teres major nerve springs from the axillary. These are sometimes called subscapular nerves.

The large, deep nerve caudal to the axillary that is formed by the union of the dorsal divisions of the seventh and eighth cervicals and first thoracic is the **radial**. This is the largest nerve of the plexus. It passes between the triceps and the humerus to the lateral surface of the arm and thence down to the extensor portion of the forearm. It supplies the epitrochlearis, triceps, and forearm extensors. A large branch of the nerve is cutaneous.

All the above nerves go to dorsal appendicular muscles, except the pectoral nerves and the suprascapular, which supply parts of the ventral musculature. The remaining nerves innervate the rest of the ventral appendicular muscles. It will be noted that all the nerves to the ventral muscles arise from the ventral divisions of the plexus. A small **musculocutaneous** springs from the sixth and seventh cervicals, passes superficial to the radial nerve and enters the biceps. It generally branches before reaching the biceps. It supplies the biceps, coracobrachialis, brachialis and some of the skin over the forearm.

The ventral divisions of the seventh and eighth cervical and first thoracic combine to form two prominent nerves that run down the medial side of the brachium. The more cranial of these is the **median nerve**, the more caudal the **ulnar nerve.** They are distributed to the forearm flexors and skin of the hand. The median nerve passes through the supracondylar foramen of the humerus.

A small **medial cutaneous nerve**, which arises from the first thoracic, runs parallel with, and caudal to, the ulnar nerve. It supplies some of the skin over the forearm.

(C) THE LUMBOSACRAL PLEXUS

The **lumbosacral plexus**, which supplies the skin and muscles of the pelvis and hind leg, is located so deep within the abdominal and pelvic cavities that it cannot be dissected until the abdominal and pelvic viscera have been studied. If you are to dissect this plexus return to the following description after you have completed the urogenital system. The following description is based on the cat, but is applicable to the mink and rabbit.

The pelvic symphysis will have been cut. Push the two hind legs dorsally thereby spreading open the pelvic canal, and push the abdominal and pelvic viscera to one side. Also cut the external iliac artery and vein shortly before they pass through the abdominal wall, and reflect them. Identify the psoas minor muscle (p. 157), and cut and reflect it. A longitudinal cleft can be found on the lateroventral part of the psoas major near the point where the deep circumflex iliac vessels cross it. Separate the psoas major along this cleft into superficial (ventral) and deep (dorsal) portions. Notice the nerves of the lumbosacral plexus emerging through this cleft, and dissect away the superficial portion of the psoas major as you trace them medially to the intervertebral foramina through which they leave the vertebral column.

The lumbosacral plexus is formed by the ventral rami of seven spinal nerves (the fourth lumbar to third sacral). As is the case with the brachial plexus, some variation occurs in the way the seven roots of the plexus split to form divisions and the divisions unite to form the trunks from which the peripheral nerves arise. A common pattern is shown in Figure 8–19.

Lumbar nerve 4 splits soon after emerging from the intervertebral foramen into a **genitofemoral nerve** and a branch that passes caudad to join the divisions of the fifth lumbar nerve. The genitofemoral nerve continues caudad close to the external iliac vessels and passes through the body wall with the external pudendal vessels to supply the skin in the groin and on parts of the external genitalia. In a male small branches also go to the cremasteric muscle.

One division of **lumbar nerve 5** unites with the caudal branch of the fourth lumbar to form the **lateral femoral cutaneous nerve**, which extends laterally near the deep circumflex iliac vessels, perforates the body wall, and sup-

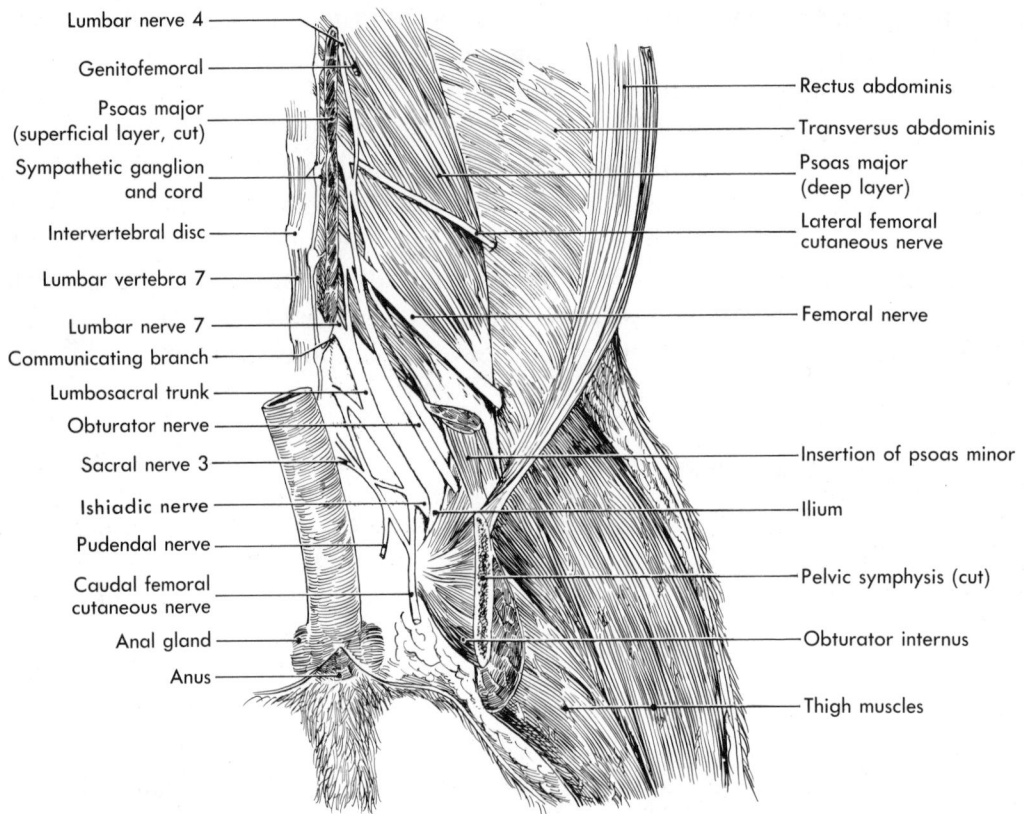

Figure 8–19. Ventral view of a dissection of the left lumbosacral plexus of a cat.

plies the skin over the lateral surface of the hip and thigh. The other division of lumbar 5 continues caudad to unite with the divisions of the sixth lumbar nerve. Branches to the psoas muscles may arise from this division or from the lateral femoral cutaneous.

Lumbar 6 is a large nerve. Its largest division unites with a branch of the caudal division of the fifth lumbar to form the large **femoral nerve**. After perforating the body wall with the femoral artery and vein, the femoral nerve innervates the quadriceps femoris and certain other extensor thigh muscles. It also gives rise to a prominent cutaneous branch (the **saphenous nerve**) that supplies the skin on the medial side of the thigh and shin.

A second division of lumbar 6 receives another division of the fifth lumbar to form the **obturator nerve**. This nerve extends caudolaterally, perforates the obturator internus near the brim of the pelvis, and goes through the obturator foramen of the pelvic girdle to supply primarily the gracilis, obturator externus, and adductors of the thigh. These muscles belong to the ventral appendicular group.

A third division of lumbar 6 continues caudad to join **lumbar 7** and form the large **lumbosacral trunk**. As the lumbosacral trunk continues caudolaterally it soon receives a division from sacral nerve 1 and a bit farther distally a contribution from sacral 2. The trunk leaves the pelvic canal by passing between the ilium and sacrum, and breaks up into several branches. **Gluteal nerves** supply the gluteal and other laterodorsal hip muscles, and a large **ischiadic (sciatic) nerve**, which is the main continuation of the trunk, continues down the lateral surface of the thigh (Fig. 6–23, p. 152), innervating the biceps femoris, semimembranosus, semitendinosus and other flexor muscles of the thigh. It bifurcates near the distal end of the thigh into **tibial** and **common peroneal** (fibular) **nerves**. These innervate the flexors and extensors of the shank respectively. Cutaneous branches of the ischiatic, tibial, and common peroneal nerves help to supply adjacent skin.

Sacral nerves 1, 2, and 3 also interconnect with each other to form a network from which several small nerves arise, the most conspicuous being the pudendal and caudal femoral cutaneous nerves. The **pudendal nerve** contains motor fibers to striated muscles in the anal region and sensory fibers coming from the anal region and from the penis and clitoris. The **caudal femoral cutaneous nerve** helps supply the skin in the anal area and adjacent parts of the thigh.

A portion of the **sympathetic cord** and **ganglia** can also be seen during this dissection lying on the ventral surface of the lumbar vertebrae (Fig. 8–19). Each ganglion receives a **communicating branch** from adjacent lumbar spinal nerves. Pelvic viscera receive their sympathetic innervation by minute branches from the pelvic extension of the sympathetic cord that follow the blood vessels to the organs. Parasympathetic innervation is from a **pelvic nerve** formed by very small branches from the sacral nerves. These nerves are seldom seen in dissections.

9

THE COELOM AND THE DIGESTIVE AND RESPIRATORY SYSTEMS

We now turn from the organ systems that support, move and integrate the body's activities to a group that sustains metabolism. The digestive system brings in the raw materials needed by the body; the respiratory system takes care of the essential gas exchanges; the circulatory system transports materials to and from the cells; and the excretory system eliminates all, or much, of the nitrogenous waste products of cellular metabolism and helps to control the water balance of the body. Much nitrogenous excretion, however, occurs through the gills of fishes. Although the digestive and respiratory systems are functionally distinct, it is convenient to study them together to some extent, for they are closely associated morphologically. The respiratory system develops embryonically as outgrowths from the digestive system.

The Coelom and Its Subdivisions

The body cavity, or **coelom**, is an epithelium-lined space containing some liquid and surrounding the viscera. Its presence permits a certain amount of movement of the organs, and their change in size and shape. Embryologically the coelom develops as a pair of mesothelial-lined spaces within the lateral plate mesoderm on either side of the digestive tract (Fig. 9–1). At first there are a right and left coelom that converge above and below the digestive tract to form a **dorsal** and a **ventral mesentery**. Much of the ventral mesentery is ephemeral. It soon disappears in the caudal part of the trunk, and the originally paired coeloms become continuous. It follows from this development that the viscera are technically outside the coelom, being separated from it by its thin mesothelial wall. However, the visceral organs are generally approached by cutting open the coelom, and it is therefore convenient to speak of them as being within the body cavity.

The coelom of fishes, amphibians and early mammalian embryos is divided into two parts. The cranial portion around the heart forms a **pericardial cavity** and the rest a **pleuroperitoneal cavity**. The separation between them is a partition known as the **transverse septum**. The ventral portion of this septum develops embryologically through the expansion of the liver in the ventral mesentery (Fig. 9–1, *B*); the dorsal portion, by a pair of folds (called **pericardiopleural membranes** in mammalian embryology) that carry the ducts of Cuvier down to the heart (Fig. 9–2). In fishes the transverse septum is a vertical partition, but the caudal migration of the heart and pericardial cavity in most tetrapods causes the septum to assume a somewhat oblique position. The paired cranial parts of

238

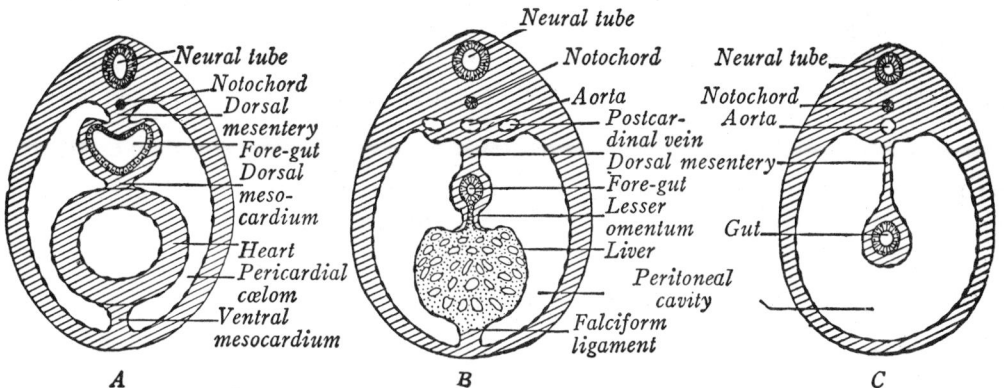

Figure 9–1. Diagrammatic cross section of a vertebrate embryo to show the relationship of the coelom and mesenteries to the visceral organs: *A*, through the level of the heart; *B*, through the liver; *C*, through the intestine. All mesenteries ventral to the digestive tract are parts of the ventral mesentery. (From Arey, Developmental Anatomy. After Prentiss.)

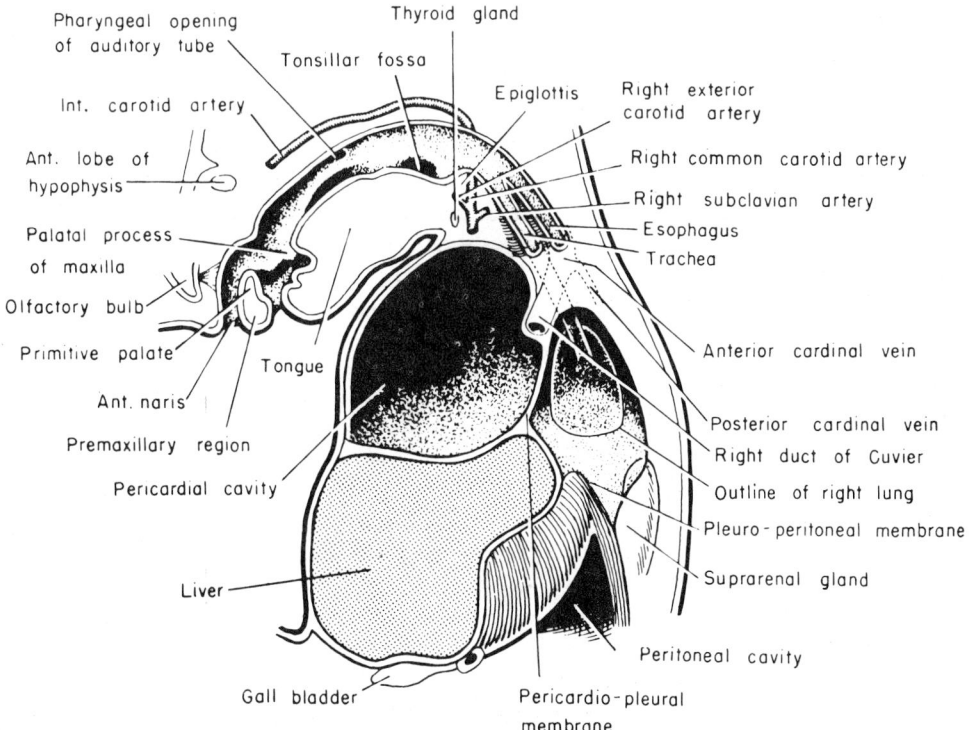

Figure 9–2. Sagittal section of a human embryo to show the subdivisions of the coelom. The pericardial cavity is already separated from the rest of the coelom, for the liver is expanding in the ventral portion of the transverse septum, and the dorsal part of the septum (pericardio-pleural membrane) has developed to carry the ducts of Cuvier. The transverse septum has an oblique position, so the future pleural cavities (only the right one is shown) lie dorsal to the pericardial cavity. A pleuroperitoneal membrane, which is shown developing, is largely responsible for the separation of each pleural cavity from the peritoneal cavity. (Redrawn from Hamilton, Boyd, and Mossman, Human Embryology, W. Heffer & Sons, Ltd.)

the pleuroperitoneal cavity thus come to lie dorsal to the pericardial cavity (Fig. 9–2). It is in these recesses that the lungs lie. The separation of this pair of recesses as **pleural cavities** (when it eventually occurs in certain reptiles) is through the development of another pair of folds, the **pleuroperitoneal membranes** (Fig. 9–2), aided by subsidiary folds from the body wall and dorsal mesentery. The **diaphragm** of mammals represents these coelomic folds, plus the ventral part of the transverse septum, plus somatic musculature of cervical origin that has invaded them (Fig. 9–3). In mammals the pleural cavities subsequently extend ventrally lateral to the pericardial cavity and thus more or less surround the pericardial cavity. The part of the original pleuroperitoneal cavity caudal to the heart and lungs now constitutes the **peritoneal cavity**.

The Development of the Digestive and Respiratory Systems

The major part of the digestive system develops from the embryonic **archenteron** and, hence, is lined with endoderm. However, variable amounts of the front and hind ends of the digestive tract are formed by ectodermal invaginations—the **stomodeum** and **proctodeum**, respectively (Fig. 9–4). The former forms the mouth or **oral cavity**; the latter contributes to the cloacal region. At first the ectodermal invaginations are separated from the archenteron by plates of tissue, but these eventually break down. It is then difficult to determine precisely where ectoderm ends and endoderm begins and the precise limits of the oral cavity.

For purposes of description, the archenteron may be divided into a **foregut** and **hindgut**. The former differentiates into the pharynx, esophagus and, generally, a stomach; the latter differentiates into the intestinal region and much of the cloaca. In most mammals the cloaca is present only in embryos. It soon becomes divided—the dorsal part contributing to the rectum; the ventral, to the urogenital passages.

In all vertebrates a series of **pharyngeal pouches** grow out from the side of the pharynx (Fig. 9–4). There are six of these in most fishes, fewer in tetrapods. The tissues between the pouches constitute the **branchial bars**, the first bar being cranial to the first pouch. The skeletal visceral arches, branchial muscles, certain nerves and the aortic arches grow into these bars. In fishes the endodermal pharyngeal pouches meet

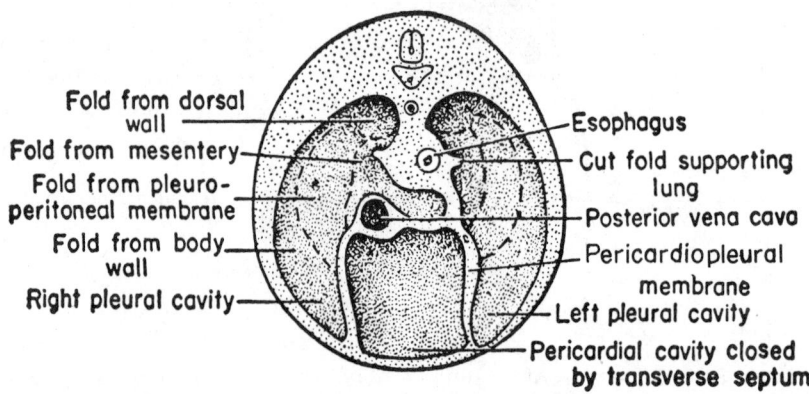

Fold from dorsal wall
Fold from mesentery
Fold from pleuro-peritoneal membrane
Fold from body wall
Right pleural cavity

Esophagus
Cut fold supporting lung
Posterior vena cava
Pericardiopleural membrane
Left pleural cavity
Pericardial cavity closed by transverse septum

Figure 9–3. Cranial view of the diaphragm of a mammalian embryo to show the various folds of which it is composed. The heart and lungs have been removed to show the caudal portions of the two pleural cavities and the pericardial cavity (darker stippling). At this stage of development, the caudal surface of the pericardial cavity lies against the diaphragm, but it later separates from it in many mammals. Also, the caval fold has not formed, but the separation of the pericardial cavity from the diaphragm would result in such a fold. The pericardiopleural membrane represents the dorsal part of the primitive transverse septum. The lighter stippling extending from dorsal to ventral is the mediastinum. (From Romer, The Vertebrate Body. After Broman and Goodrich.)

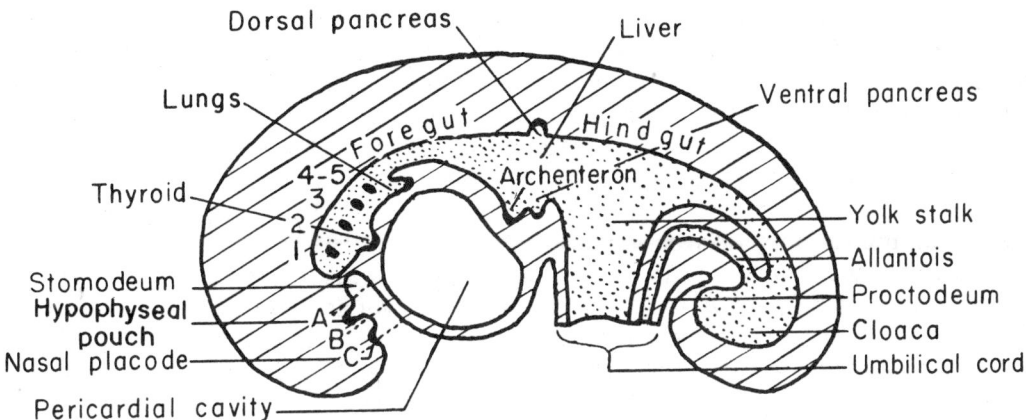

Figure 9–4. Diagrammatic sagittal section of a mammalian embryo to show the development of the digestive and respiratory systems. The points of entrance of the pharyngeal pouches are numbered. Lines *A*, *B*, and *C* indicate the comparative position of the mouth openings of *A*, an agnathous fish; *B*, a jawed fish not having internal nostrils; and *C*, fishes and tetrapods with internal nostrils.

comparable ectodermal furrows, break through to the surface, and form the **gill slits** and **pouches** on whose walls the gills develop. The first pharyngeal pouch and furrow form the **spiracle** of such a fish as *Squalus*; the others form the five definitive gill pouches. The pharyngeal pouches do not normally break through in tetrapods (except for certain ones in larval amphibians), but are present embryonically nonetheless. The first gives rise to the **tympanic cavity** and **auditory tube** in all tetrapods except urodeles; the mammalian **tonsillar fossa**, containing the palatine tonsil, develops at the site of the second pouch; the other pouches give rise to certain endocrine glands and then disappear. A **thymus** develops as epithelial thickenings of the dorsal part of most (or all) of the pharyngeal pouches in fishes and urodeles, but from the ventral part of just the third and fourth pouches in mammals. These primordia may remain as distinct glandlike structures or coalesce into a single one. The thymus has been implicated in the development of the organism's capacity to produce antibodies. **Parathyroid glands** are absent in fishes but present in all tetrapods. They develop as epithelial thickenings from the ventral part of the third and fourth pouches in urodeles, from the dorsal part of the third and fourth in mammals. Finally, **ultimobranchial bodies** develop in all vertebrates from the posterior face of the last pharyngeal pouch. Both the parathyroid glands and ultimobranchial bodies produce hormones that regulate the level of calcium in the blood.

In addition to the lateral pharyngeal pouches, certain median evaginations arise from the floor of the pharynx. An endocrine **thyroid gland** grows out from the floor between the level of the first and second pouches. In fishes, the thyroid, although losing its connection with the pharynx, tends to remain in this cranial position, but it migrates a variable distance caudad in tetrapods. The thyroid may remain a median organ, or it may bifurcate and form a pair of glands. As discussed earlier (p. 24), the development of the thyroid of the lamprey from a part of the endostylelike subpharyngeal gland of the *Ammocoetes* larva points to a possible homology between the vertebrate thyroid and the endostyle of lower chordates. The thyroid hormone increases an organism's metabolic rate. This is particularly important in amphibian metamorphosis and in endothermic vertebrates.

The **lungs** of sarcopterygian fishes and tetrapods arise as a median bilobed evagination from the floor of the pharynx just posterior to the pharyngeal pouches. The early primordia of the lungs resemble a pair of pharyngeal pouches in the embryos of some amphibians, and it is possible that they evolved from a ventrally displaced pair of caudal pouches. It is believed that lungs are very primitive. The fossils of some placoderms show evidence of the presence of a lunglike structure, and lungs are found in the living sarcopterygian and in primitive members of the actinopterygian fishes, where they serve to supplement gill respiration. In higher actinopterygians, however, the lungs have become transformed into the dorsally placed, hydrostatic **swim bladder**. In most cases, the swim bladder develops as an evagination from the roof of the posterior pharyngeal region, but in at least one fish it arises from the lateral wall. This suggests a transitional stage from the ventrally derived lungs.

No other outgrowths arise from the digestive tract of most vertebrates until the level of the front of the hindgut. At this point one finds the liver and pancreas. The **liver** arises embryologically as a prominent ventral diverticulum (Fig. 9–4), which, as mentioned earlier, grows into the ventral mesentery posterior to the heart, and by its expansion forms the ventral part of the transverse septum. During subsequent development, the liver grows caudally in the ventral mesentery and, in the adult, remains connected to the septum (or diaphragm) only by a mesentery known as the **coronary ligament**. Functionally the liver is a very diverse organ. It secretes bile (a mixture of excretory products and the fat-emulsifying bile salts), and its cells come into intimate contact with blood that is brought to it from the stomach and intestinal region by the hepatic portal system. Many metabolic conversions occur here: excess absorbed sugars are stored, largely in the form of glycogen, or deficiencies are made up so the glucose content of the blood is kept at a constant level; amino acids are deaminated and their amino groups converted to urea; absorbed toxins may be removed; etc.

The pancreas is a gross organ in all gnathostomes, but in the agnatha it is represented only by scattered cells in the walls of the intestine and liver (p. 20). The organ arises embryologically from one or more intestinal outgrowths near the liver primordium. Frequently there is both a dorsal evagination (**dorsal pancreas**) and a ventral one (**ventral pancreas**). The latter is often paired, and is associated with the base of the liver anlage and, hence, with the future bile duct (Fig. 9–4). The ventral pancreas tends to work its way around the intestine, and to fuse and grow up into the dorsal mesentery with the dorsal pancreas. All the stalks of the primordia may persist as ducts, or certain ones may be lost in the adult. A ventral pancreatic duct can be recognized by the fact that it enters the intestine in common with the bile duct; a dorsal pancreatic duct, by its independent entrance on the opposite side of the intestine. Most of the pancreas secretes digestive enzymes, which are discharged into the intestine and act on proteins, carbohydrates, fats and nucleic acids. Little islands of endocrine tissue (the **islets of Langerhans**) are scattered among the exocrine cells; these islets produce insulin and glucagon, which are vital in carbohydrate metabolism.

More caudally along the hindgut there is, in the embryos of amniotes and certain fishes, a **yolk stalk** connecting with the **yolk sac** (Fig. 9–4). The yolk stalk and sac become relatively smaller as the embryo grows and are lost by the adult stage.

A final major outgrowth is the **urinary bladder**, present in most tetrapods. It develops from the embryonic cloaca near the caudal end of the hindgut. In the embryos of amniotes, this structure expands considerably and extends beyond the limits of the embryo as the **allantois**—an extraembryonic membrane which serves for excretion and respiration in the embryos of reptiles and birds, and for vascularizing the fetal portion of the placenta in eutherian mammals.

FISHES

The coelom and the digestive and respiratory systems of *Squalus* are reasonably good examples of the condition of these structures in primitive jawed fishes. However, primitive bony fishes ancestral to tetrapods, and possibly placoderms, had lunglike outgrowths from the back of the pharynx. A stomach was probably absent in very primitive fishes. Lower chordates are, and early vertebrates may have been, filter feeders, feeding more or less continuously on minute food particles. Filter feeders need no stomach, but a change in feeding habits occurred with the evolution of jaws. Large chunks of food are taken irregularly, and the stomach is advantageous for temporary storage and preliminary physical and chemical treatment of food.

Pleuroperitoneal Cavity and Its Contents

(A) BODY WALL AND PLEUROPERITONEAL CAVITY

It is desirable to study the caudal parts of the digestive system before the mouth and pharyngeal area. The **pleuroperitoneal cavity** should be opened by a longitudinal incision slightly to one side of the midventral line, preferably the right side if the muscles were dissected on the left. Extend the incision as far forward as the pectoral girdle and as far caudad as the base of the tail. In doing the latter, cut through the pelvic girdle and continue caudally on one side of the cloacal aperture. Make a transverse incision on each side that extends from about the middle of the longitudinal cut to the lateral line. You now have four flaps of body wall that can be turned out to expose the body cavity and its contents, but do not break tissue extending between the cranial part of the liver and the ventral body wall.

Note the layers of the body wall through which you have cut. The outermost is the **skin**. This is followed by a thin and inconspicuous layer of **connective tissue** comparable to the fascia of higher vertebrates, the **hypaxial musculature**, and finally the shiny **epithelium** lining the pleuroperitoneal cavity. That portion of the coelomic epithelium adjacent to the body wall musculature is the **parietal peritoneum**; that portion covering the viscera, the **visceral peritoneum**; and, finally, that portion that extends from the body wall to the viscera contributes to the **mesenteries**. As a result of the way they develop, mesenteries consist of a double layer of epithelium between which lie the vessels and nerves to the viscera imbedded in connective tissue (Fig. 9–1).

In most vertebrates the coelom is a closed cavity having no direct communication with the outside. But in primitive vertebrates, including the dogfish, it may communicate with the exterior by a pair of **abdominal pores** (Fig. 9–5). These can be found by probing the most caudal recess of the pleuroperitoneal cavity beside the **cloaca** (the chamber receiving the intestine and genital ducts) on the side of the body that is still intact. Each opens through the lateral wall of the cloacal aperture, but sometimes the lips of the pores have grown together. Their significance is obscure. They may serve to eliminate excess coelomic fluid, or they may be vestiges of an evolutionary stage when gametes were discharged from the coelom through genital pores.

(B) VISCERAL ORGANS

A large **liver** with a pair of long, pointed lobes occupies most of the cranioventral portion of the pleuroperitoneal cavity. You may cut off the ends of

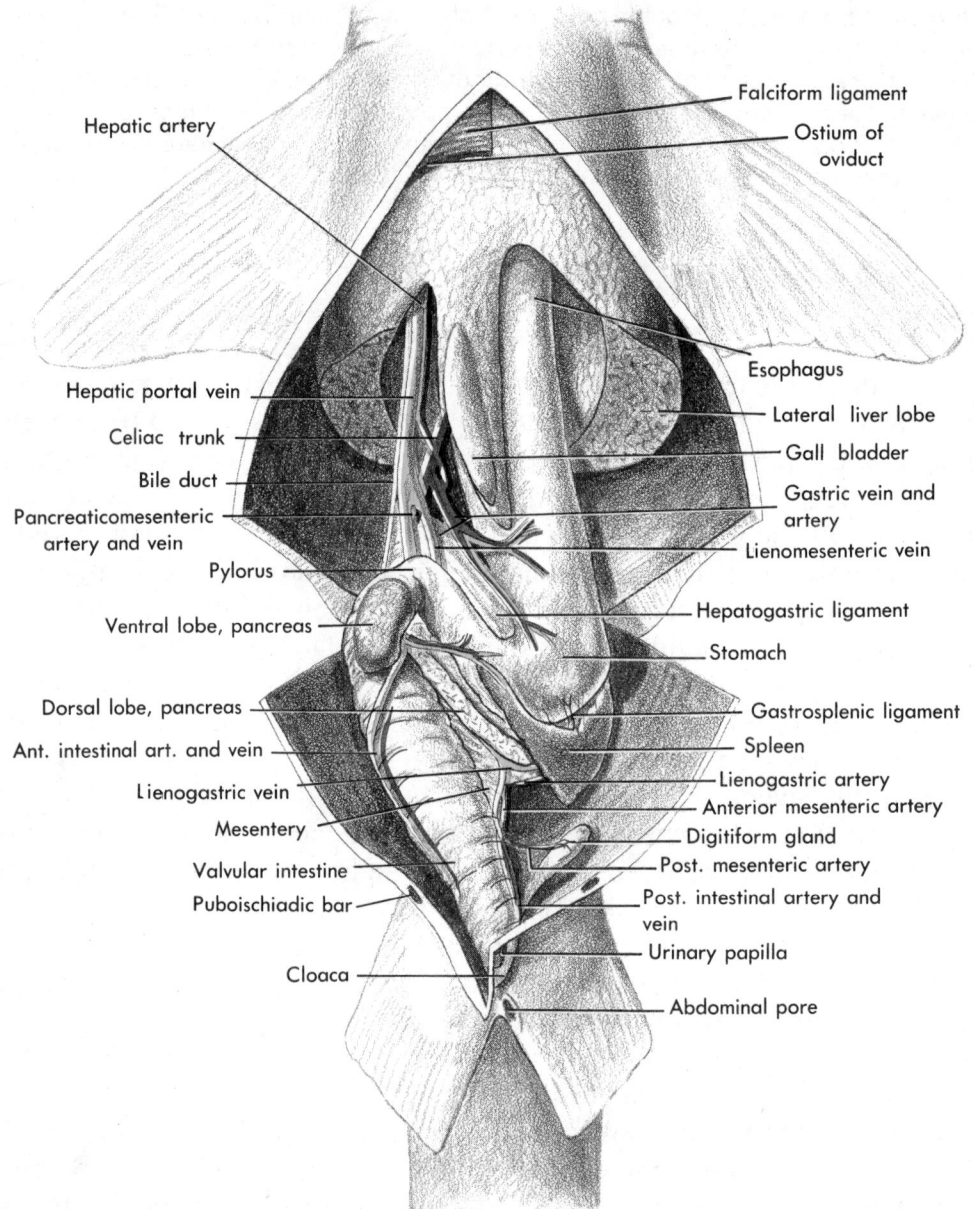

Figure 9–5. Ventral view of the abdominal viscera and blood vessels of a female *Squalus*. The distal parts of the lateral liver lobes have been cut away.

these lobes, but do not injure blodd vessels going to the liver, or the bile duct going from the liver to the intestine. The **bile duct** accompanies these vessels for most of its length but separates from the vessels near the beginning of the intestine (Fig. 9–5). Spread the lobes of the liver apart, and you will see the **esophagus** and **stomach** in a more dorsal position. Both have about the same diameter, so there is no external line of demarcation separating them. If a constriction is seen, it represents a peristaltic contraction fixed at death and during preservation of the animal, not a line of demarcation. The posterior end of the stomach curves cranially and gives the organ a **J** shape. The digestive tract then turns caudally and forms the straight, **valvular intestine** that continues to the **cloaca.** If part of the intestine has been everted through the cloaca, it must be pulled back into the body cavity.

The large, triangular-shaped organ attached to the caudal end of the stomach (at the region where the stomach turns forward) is the **spleen** — an organ related to the production and storage of blood cells. The elongate **pancreas** extends from the right side of the spleen cranially toward the beginning of the intestine. An oval-shaped lobe of the pancreas is applied to the surface of the intestine near its junction with the stomach. The fingerlike organ extending dorsally from the caudal end of the intestine is the **digitiform gland** (rectal gland). It has been shown by Burger and Hess (1959) to be a salt excreting gland, and Oguri (1964) reports that it undergoes regressive changes in sharks that enter fresh water.

Other organs within the pleuroperitoneal cavity are parts of the urogenital system. They will be considered in more detail later but should be identified at this time. A pair of large **gonads** (**testes** or **ovaries**) lie one on either side of the cranial end of the esophagus and stomach. If your specimen is a mature female, a pair of prominent oviducts will be seen dorsal to the gonads and continuing to the cloaca. Their caudal ends are enlarged, greatly so in pregnant females. The paired **kidneys** are represented by the long bands of dark material located dorsal to the parietal peritoneum on either side of the middorsal line of the pleuroperitoneal cavity. They are very long and are not uniform in diameter, for the caudal end of each is much wider than the cranial end. If the specimen is a mature male, a large, twisted excretory duct (the **archinephric duct**) will be seen on the ventral surface of each one.

(C) MESENTERIES

Pull the digestive tract ventrally and note that much of it is supported by the **dorsal mesentery.** All of it was in the embryo. That portion of the dorsal mesentery which passes from the middorsal line of the coelom to the dorsal surface of the esophagus and stomach is called the **greater omentum** or **mesogaster**; that portion which passes to the cranial part of the intestine, the **mesentery**;[13] and that portion which passes to the digitiform gland and caudal end of the intestine, the **mesocolon.** The spleen lies in the mesogaster, but the part of this mesentery between the spleen and stomach is given a special name — **gastrosplenic ligament** (Fig. 9–5). Pull the caudal end of the stomach to the animal's left and the cranial end of the intestine to the right. On looking between them, it will be seen that the mesentery (in the limited sense) does not arise from the body wall but has shifted its attachment to the mesogaster. It

---

[13]Unfortunately the term mesentery is used in two ways — in a broad sense for all membranes passing to the viscera, and in a limited sense for the mesentery supporting most of the intestine.

can also be seen that the pancreas lies in a special fold of the mesentery (Fig. 9–6). This relationship shows best on the portion of the pancreas near the spleen.

The ventral mesentery has disappeared except for that portion into which the liver has grown. The part of the ventral mesentery between the front of the liver and the midventral body wall is the **falciform ligament**; the part between the liver and the digestive tract (caudal portion of stomach and front of intestine) is the **lesser omentum** or gastrohepatoduodenal ligament. The latter mesentery, which contains the bile duct and the blood vessels going to the liver, is unusually complex in the dogfish. Near the liver the mesentery is a unit, but it divides near the digestive tract. Part of it (the **hepatogastric ligament**) passes into the angle formed by the bending of the stomach, and part of it (the **hepatoduodenal ligament**) carries the bile duct to the beginning of the intestine. A part of the dorsal mesentery supporting the pancreas extends between these two limbs of the lesser omentum and brings the larger vessels to the omentum.

Subsidiary mesenteries support certain of the genital organs. Each testis is supported by a **mesorchium**, each ovary by a **mesovarium**, and each oviduct in a mature female by a **mesotubarium**.

(D) Further Structure of the Digestive Organs

Cut open the esophagus and stomach by a longitudinal incision that extends all the way to the intestine. Remove the contents, if any, and wash out these organs. (One can often determine the feeding habits of animals by analy-

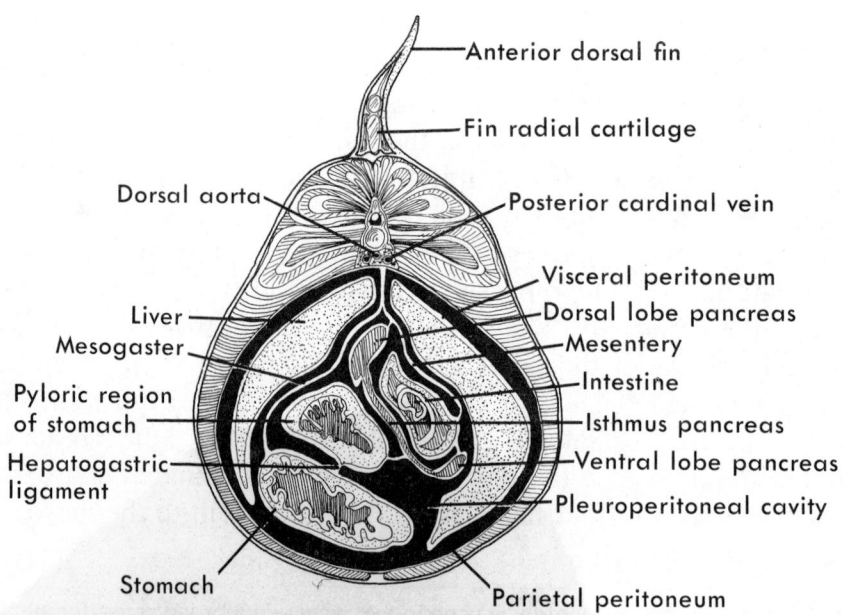

Figure 9–6. Cross section through the trunk of *Squalus* at the level of the isthmus of the pancreas; viewed from behind.

sis of the stomach contents.) The esophagus can be recognized internally by the **papillae** in its lining; the stomach, by longitudinal folds called **rugae**. If the stomach was greatly distended, the rugae will have stretched out and most of the lining will be smooth. The cranial end of the stomach, which would be adjacent to the esophagus and often near the heart, is called the **cardiac region**; the main part, the **body**; and the portion that turns forward, the **pyloric region**. Note that the last has a thicker muscular wall, especially just before the intestine, where it forms a muscular sphincter, the **pylorus**. It should be pointed out that these topographic regions in different vertebrates may or may not correspond with glandular regions given the same names. They do not in most lower vertebrates, for cardiac glands are absent. The surface of the stomach along which the mesogaster and gastrosplenic ligament attach is called its **greater curvature**; the opposite surface, the **lesser curvature**. The greater curvature represents the original dorsal side of the stomach.

The bile and pancreatic duct, which will be seen presently, enter the cranial end of the valvular intestine; the digitiform gland, the caudal end. Aside from this slight modification at either end, the intestine is undifferentiated. It contains a complex spiral fold, the **spiral valve**, which is similar to that of other primitive fishes (Fig. 9–7). The line of attachment of the valve shows as the spiral line on the surface of the intestine. If a special preparation of the valve is not available, part of it can be seen by cutting a tangential slice from the intestine wall. The spiral valve slows down the passage of food through the intestine, thus increasing digestive efficiency, and it also increases the absorptive area.

It is difficult to compare the intestine of primitive fishes with the small and large intestine of tetrapods. The cranial end of the fish intestine, which receives the bile and pancreatic ducts, would be roughly comparable to the duodenum; most of the remainder, to the rest of the small intestine; and the posterior end receiving the digitiform gland, to part, at least, of the large intestine. But in the absence of clear lines of separation between these potential regions, it is best to think of the primitive fish intestine as a unit—the valvular intestine.

Examine the liver in more detail. It consists of a long **right** and **left lobe** and a smaller **median lobe** containing the elongate, thin-walled **gall bladder** (Fig. 9–5). A part of the bladder shows on the surface, but most is imbedded within the lobe. Expose it by scraping away liver tissue and find out where the bile duct unites with it. In the dogfish the bile leaves the liver through a number of inconspicuous **hepatic ducts** that enter the gall bladder; from here it goes to the intestine through the **bile duct**. The point of entrance of the bile duct into the lumen of the intestine is some distance caudad to its point of attachment on the intestine. The duct can be traced through the wall by removing the visceral peritoneum and longitudinal muscle layer.

As in other vertebrates, the dogfish's liver is an important site for the metabolism and storage of food products. Glycogen is present, but an unusual feature is that much

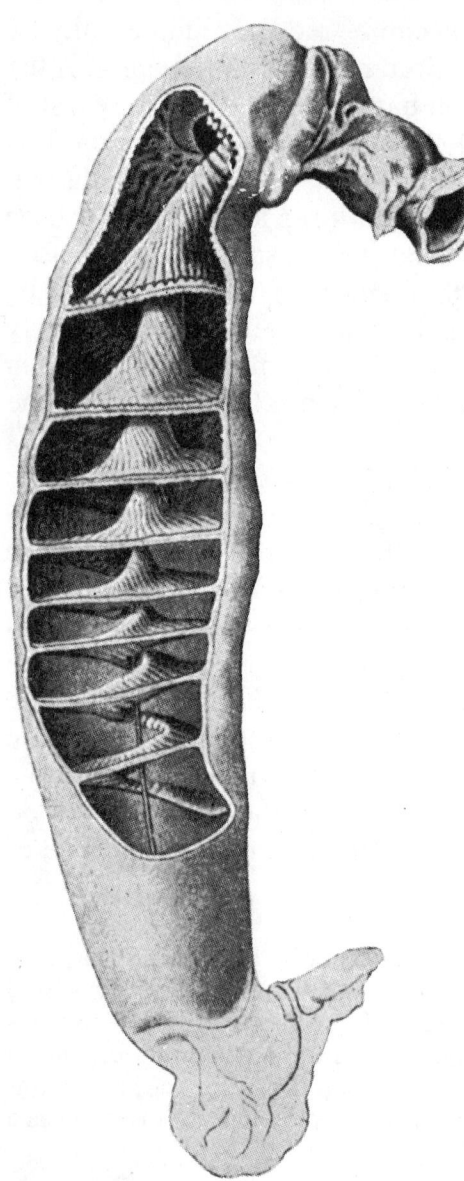

Figure 9–7. Valvular intestine of the skate, *Raja,* cut open to show the spiral valve. A bristle is shown passing through the central lumen of the intestine. (From Paul Mayer.)

of the food is stored as oil. Functionally this is somewhat similar to the swim bladder of bony fishes because the oil lowers the specific gravity of the fish and makes the animal more buoyant.

The pancreas is divided into two lobes, an elongate **dorsal lobe** that extends from the spleen cranially, passing dorsal to the pylorus, and an oval-shaped **ventral lobe** adhering to the beginning of the intestine (Fig. 9–5). The two are connected by an isthmus of pancreatic tissue. Both lobes drain by a common **pancreatic duct** that leaves the ventral lobe and travels obliquely caudad in the wall of the intestine for a short distance before entering the lumen. To see the duct, it is necessary to cut the attachment of the mesentery and carefully remove the visceral peritoneum and longitudinal muscle from the wall of the intestine adjacent to the posterior part of the ventral lobe.

Relationships at the front of the intestine are misleading, for this portion of the digestive tract has undergone a rotation of nearly 180 degrees during development. This has brought about an apparent reversal of normal relationships. In the adult the bile duct enters on what would at first seem to be the dorsal surface of the intestine, and the dorsal mesentery appears to attach along the ventral surface (Fig. 9–6). The pancreas of the dogfish is entirely a dorsal pancreas; however, its duct enters on what would seem to be the original lateral surface of the intestine.

Pericardial Cavity

The second division of the coelom, the **pericardial cavity**, is located far forward in fishes, for it lies just cranial and dorsal to the pectoral girdle and deep to the posterior hypobranchial musculature (Figs. 9–9 and 10–10, pp. 251; 294). To get at it, continue your original ventral incision forward through the pectoral girdle and the caudal hypobranchial musculature. Veer toward the midventral line as you go, but do not cut the falciform ligament. Additional transverse cuts will have to be made just cranial to the pectoral girdle.

Spread open the flaps thus formed and study the cavity. The only organ within it is the **heart**. That portion of the coelomic epithelium adjacent to the tissues surrounding the cavity is the **parietal pericardium**; that covering the heart, the **visceral pericardium**. The heart developed embryonically in the ventral mesentery beneath the pharynx (Fig. 9–1, *A*) and at one time was supported by parts of the ventral mesentery, termed mesocardia. The mesocardia disappear in the adult, and the heart remains attached only at its cranial and caudal ends.

The vertical septum separating the pericardial and pleuroperitoneal cavities is the **transverse septum**. You can now see that the liver is attached to the caudal face of the septum by the **coronary ligament**. This ligament is continuous ventrally with the cranial part of the falciform ligament. As in many other primitive fishes, the separation of the pericardial and pleuroperitoneal cavities is not complete in the dogfish, for a **pericardioperitoneal canal** connects the two ventral to the esophagus. It will be seen later after the heart is studied (p. 295).

Oral Cavity, Pharynx, and Respiratory Organs

Cut deep to esophagus

The pharynx should be opened on the same side of the body on which the pleuroperitoneal cavity was opened. With a strong pair of scissors cut through the angle of the mouth and continue caudally through the external gill slits and visceral arches. Then extend your cut medially and caudally through the pectoral girdle and body wall to intersect your previous longitudinal incisions in the body wall and esophagus. Do not cut between the liver and heart; go ventral to the pectoral fin and dorsal to the rostral end of the liver. It is not necessary to cut the liver. Swing open the floor of the oral cavity and pharynx. If the esophagus has everted into the pharynx, it must be pulled back.

The demarcation between the **oral cavity** and the **pharynx** is not clearly definable in the adult (p. 240), but the pharynx is approximately that portion of the digestive tract into which the gill slits enter. The pharynx, of course, leads to the narrower esophagus. A tonguelike structure supported by the hyoid arch lies in the floor of the mouth and front of the pharynx. It is not a true tongue, as no muscle extends up beneath its epithelium, but sometimes it is referred to as a **primary tongue**, for it is destined to contribute to the true tongue of tetrapods (Fig. 9–8).

The entrance of the **spiracle** will be seen at the front of the pharynx roof, and this is followed by the five definitive **internal gill slits**. The homologue of the spiracle may have been a complete gill slit in placoderms. A number of papillalike **gill rakers** project across the internal gill slits and act as strainers. Note that each internal gill slit leads into a large **branchial pouch**. The portion

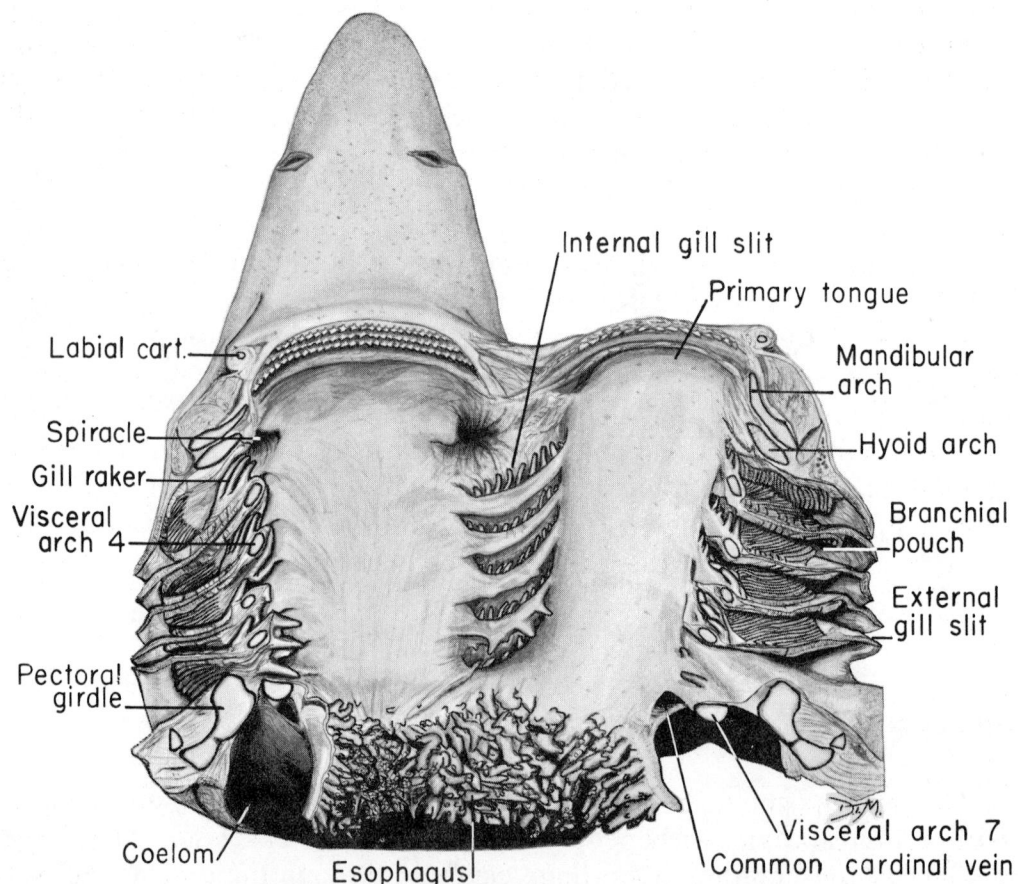

Figure 9–8. Oral cavity and pharynx of *Squalus*. The floor of the mouth and pharynx has been swung open to the right.

of the pouch lateral to the gills is the **parabranchial chamber** (Figs. 9–9 and 9–10), and it opens to the surface by the **external gill slit.** The tissue between the gill slits and pouches constitutes the **interbranchial septa.** The outermost portion of each septum is thin and flaplike and constitutes a **valve** that can close and open the external gill slits. The actual gills are composed of a number of platelike **primary lamellae,** which are attached to the surface of the septa. Examine the primary lamellae with a hand lens and note that each bears many small, closely packed, **secondary lamellae** extending perpendicularly from the surface of the primary lamella. Interbranchial septa that have lamellae on their cranial and caudal surfaces constitute a complete gill, or **holobranch.** The first gill is a **hemibranch,** for lamellae are present only on the caudal surface.

Dr. G. M. Hughes of the University of Bristol (1963) has determined the biomechanics of respiration in dogfish (Fig. 9–10). The most active phase is expiration. Contraction of the branchial muscles compresses the buccopharyngeal cavity and branchial pouches and expels water through the gill slits. During inspiration these chambers are expanded, largely by the elastic recoil of the branchial skeleton, although the lowering of the floor of the mouth and pharynx by the contraction of hypobranchial muscles does play a part. Internal pressure is reduced relative to the external pressure, the lowest pressure being found within the parabranchial chambers. The flap valves over the external gill slits close. Water enters the mouth and spiracle and follows the pressure gradient across the gills and into the parabranchial chambers. During expiration a pressure gradient is developed that closes the spiracular valves and forces water out the external gill slits. Water that entered the spiracle leaves through the more cranial gill slits, whereas water that entered the mouth leaves through the more caudal slits.

The details of the flow of water across the gills is unknown, but some of the water probably passes between the secondary lamellae in a direction counter to the flow of blood, even though such a flow would be hindered to some extent by the well developed interbranchial septum (Hughes and Hills, 1971). Teleosts have lost this septum and are able to extract a greater percentage of the oxygen from the water than cartilaginous

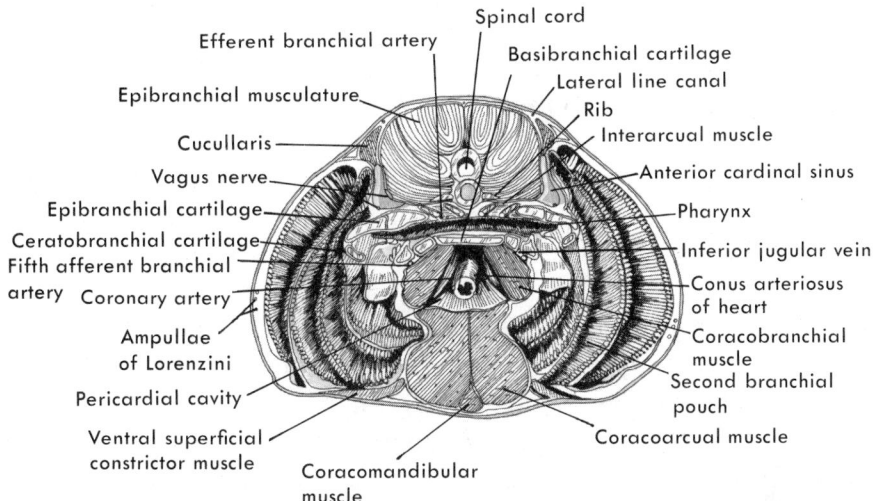

Figure 9–9. Cross section through the pharynx and front of the pericardial cavity of *Squalus;* viewed from behind.

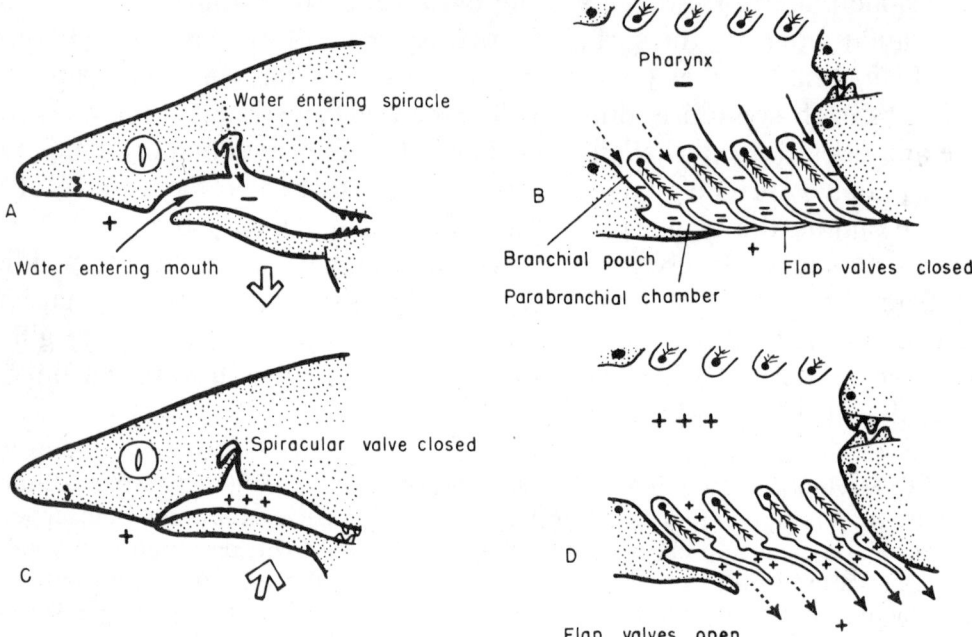

Figure 9–10. Diagrams to show the mechanics of respiration in the dogfish as seen in lateral views (*A* and *C*) and in frontal sections (*B* and *D*) of the pharynx. *A* and *B*, inspiration; *C* and *D*, expiration. Relative pressures are indicated by + and −. Open arrows indicate the direction of movement of the floor of the mouth and pharynx, solid arrows the course of the current of water entering the mouth, and dotted arrows the course of the water entering the spiracle. (Slightly modified after Hughes, Comparative Physiology of Vertebrate Respiration, Harvard University Press.)

fishes. Hughes (1972) has further shown that the heartbeat in the dogfish is coupled much of the time with the respiratory movements in such a way that there is a rapid flow of blood through the gills when water is expelled across them. This would increase the efficiency of gas exchange still more.

Cut open the spiracle on the side you have been dissecting. A minute hemibranch, known as the **pseudobranch**, can be found on the valvelike flap (**spiracular valve,** p. 33) on the cranial wall of the spiracle. Since aerated blood passes through the pseudobranch it is often regarded as vestigial, but some believe that it has an accessory respiratory function. In some marine teleosts its homologue contains secretory cells believed to be part of the salt excretory mechanism.

Examine the cut surface of a representative holobranch and note the structures of which it is composed (Fig. 9–11). A supporting **visceral arch** lies at its base. (Be sure you can identify all these arches — mandibular, hyoid and five branchial.) You may also be able to see one or more cartilaginous **gill rays** extending into the interbranchial septum at right angles to the arch. Cartilage also supports the **gill rakers.** Much of the septum is made up of branchial muscles. An **adductor** lies medial to each arch, and an **interbranchial** and a **superficial constrictor** extend out into the septum. Close examination will also reveal several blood vessels. An **afferent branchial artery** lies near the middle

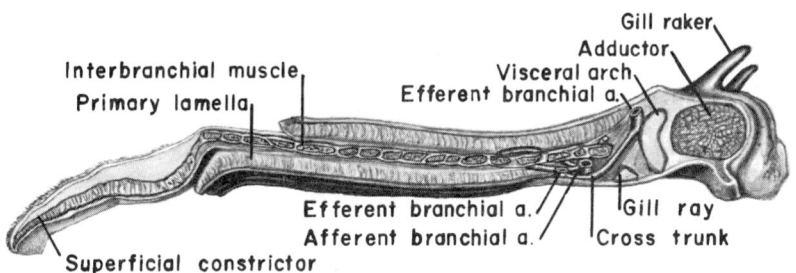

Figure 9–11. A frontal section through a holobranch of *Squalus*. The cranial surface is toward the top of the page. Twice natural size.

of the septum just lateral to the visceral arch. It brings unoxygenated blood from the heart and ventral aorta to the capillaries in the gills and probably is not injected. An **efferent branchial artery**, which will be injected if the arteries have been injected, lies at the base of the primary lamellae on each surface of the septum. These vessels, of which there are two in a typical septum, drain the gill capillaries and carry oxygenated blood to the dorsal aorta. Also recall (p. 201) that each septum would contain a **pretrematic** and **posttrematic** branch of a cranial nerve, but these are hard to find in a gross section. Which nerve, or nerves, are associated with the septum you have been studying?

A thyroid gland, a series of thymus bodies and an ultimobranchial body are derived from the pharynx but are difficult to see. The thyroid may have been seen during the dissection of the hypobranchial muscles (p. 119), and can be seen again during the dissection of the branchial blood vessels (p. 286).

PRIMITIVE TETRAPODS

The coelom of primitive tetrapods is very similar to that of fishes except that, with the beginning of neck formation, the pericardial cavity has moved caudad a short distance. This causes the transverse septum to assume a somewhat oblique position.

More conspicuous changes are seen in the respiratory and digestive systems. In respect to respiration, tetrapods retain the internal nostrils and elaborate upon the lungs of their piscine ancestors but, excepting larval amphibians, lose the gills. Larval amphibians have gill slits and gills of two types. **External gills**, which are also present in some fish larvae, develop as outgrowths from the neck surface near the gill slits in urodele and frog larvae. These are later replaced in frog larvae by **internal gills** that develop closer to the gill arch, hence may be homologous to the internal gills of adult fishes. At metamorphosis they too are lost. The lungs and methods of air exchange between the lungs and outside (a pumping action of the oral cavity and pharynx) are very primitive in amphibians, and most members of this class supplement pulmonary respiration with some other form such as cutaneous or oropharyngeal respiration.

Major changes in the digestive tract of primitive tetrapods are the evolution of small **oral glands** and a muscular **tongue**, both being correlated with the problem of food manipulation in a terrestrial environment. The amphibian tongue consists of little more than the primary tongue of fishes overlying the hyoid apparatus, plus a swelling (**gland field**) that develops between the hyoid and mandibular arches. The whole is invaded by certain prehyoid hypobranchial muscles (p. 126). The primitive spiral valve in the intestine has been lost. The resulting reduction in internal surface area is compensated for by other folds and an increase in the length of the intestine. A differentiation into **small** and **large intestines** also occurs. Finally a **urinary bladder** has evolved as an outgrowth from the ventral surface of the cloaca, but this is a part of the urinary system.

Necturus illustrates the early tetrapod stage well except for the retention of many larval features, the major ones being a vertical transverse septum, the presence of gill slits and external gills, and a poorly formed tongue. Moreover, all urodeles differ from most other tetrapods in lacking the auditory tube and tympanic cavity.

Pleuroperitoneal Cavity and Its Contents

(A) BODY WALL AND PLEUROPERITONEAL CAVITY

If your specimen of *Necturus* has been injected, a partial incision will have been made through the body wall on one side of the midventral line. Continue this incision cranially to the pectoral girdle and caudally through the pelvic girdle and along one side of the cloacal aperture. The layers of the body wall through which you have cut are similar to those of *Squalus*, i.e.; **skin, connective tissue, hypaxial muscles,** and **parietal peritoneum.** How many muscle layers were cut through? The part of the coelom opened is, as in fishes, a **pleuroperitoneal cavity.** It is lined with the **parietal peritoneum,** and the viscera are covered with **visceral peritoneum.**

(B) VISCERAL ORGANS

The **liver** is the largest of the visceral organs and lies in a cranioventral position (Fig. 9–12). Notice that it is displaced toward the right side of the cavity. It is not so obviously subdivided into lobes as it is in *Squalus*. A portion of the ventral mesentery, the **falciform ligament,** extends from the midventral line of the pleuroperitoneal cavity to attach along the entire length of the ventral surface of the liver. It will have to be cut, but do so in such a way that you can later reconstruct the major veins that pass through it. Pull the left side of the body wall and the liver apart and you will see the elongate **stomach.** Notice that it is straight as it is in larvae rather than J-shaped as it is in fully metamorphosed amphibia. The **esophagus** is a short connecting piece between the pharynx and stomach and will be seen more clearly later. The **spleen** is the elongate, oval organ attached to the left side of the stomach. The long, fingerlike left **lung** lies dorsal to the spleen; a similarly shaped right lung lies in a comparable position on the other side of the body.

The **small intestine** begins at the caudal end of the stomach and makes one cranial loop (the **duodenum**) before passing into a number of convolutions. An irregular-shaped **pancreas** lies in the vicinity of this cranial loop. Part of it is in contact with the intestine, part with the liver, and one part (the **tail**) passes dorsal to the stomach and nearly reaches the caudal tip of the spleen. Shortly before entering the **cloaca** the intestine widens and forms a short **large intestine.** The **urinary bladder,** which is probably collapsed and shriveled, lies ventral to the large intestine. It enters the cloaca independently.

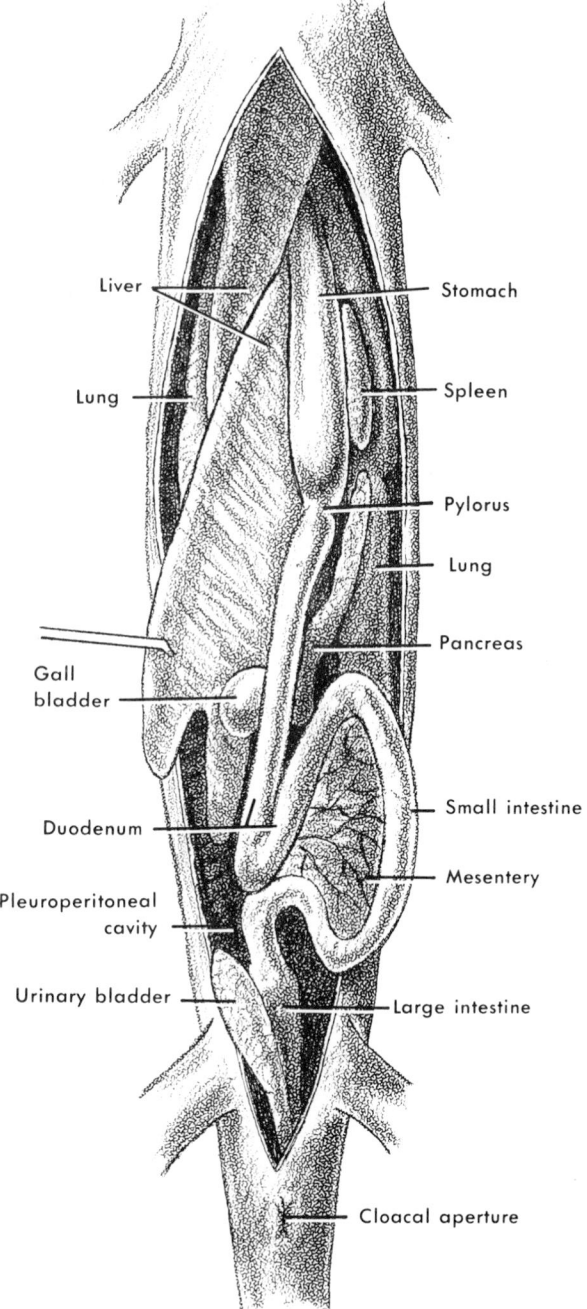

Figure 9–12. Ventral view of the digestive tract and associated organs of *Necturus.* The caudal half of the liver has been reflected, and is seen in a dorsal view.

Liver
Lung
Gall bladder
Duodenum
Pleuroperitoneal cavity
Urinary bladder

Stomach
Spleen
Pylorus
Lung
Pancreas
Small intestine
Mesentery
Large intestine
Cloacal aperture

If your specimen is a female, a pair of large, coarsely granular **ovaries** lies in a dorsal position on either side of the intestine. A pair of convoluted **oviducts** lie dorsal and lateral to the ovaries and extend nearly the length of the pleuroperitoneal cavity. Each **kidney** is an elongate, dark organ that may be seen by spreading apart an ovary and oviduct.

If your specimen is a male, a pair of large, elongate, oval-shaped **testes** lie in a dorsal position in the cranial intestinal region. The **kidneys** lie dorsal and lateral to the testes. Each has a conspicuous, convoluted **archinephric duct** along its lateral border.

(C) MESENTERIES

As in the fish, that portion of the dorsal mesentery which passes to the stomach is called the **mesogaster**; that portion passing to the small intestine the **mesentery**, in a limited sense; and that portion passing to the large intestine the **mesocolon**. The portion of the mesogaster between the spleen and the stomach is called the **gastrosplenic ligament**. Most of the left lung is connected with the mesogaster by a short **pulmonary ligament**. The right lung is also supported by a comparable but wider pulmonary ligament. However, an accessory mesentery (the **ligamentum hepatocavopulmonale**) passes from the right pulmonary ligament to the dorsal surface of the liver. The large blood vessel, the caudal vena cava, approaches the liver through the caudal margin of this mesentery.

The part of the ventral mesentery extending from the liver to the body wall (**falciform ligament**) has been seen and cut. Two other parts extend from the liver to the digestive tract—a **hepatogastric ligament** between the cranial part of the stomach and liver, and a **hepatoduodenal ligament** between the liver and the most cranial loop of the small intestine. A part of the pancreas lies in the latter ligament, but most of it extends into the dorsal mesentery. A final part of the ventral mesentery, the **median ligament of the bladder**, passes from the urinary bladder to the midventral line of the body wall.

The genital mesenteries are the same as in *Squalus*, i.e., a **mesorchium** to the testis, a **mesovarium** to the ovary, and a **mesotubarium** to the oviduct.

(D) FURTHER STRUCTURE OF THE DIGESTIVE ORGANS

Cut open the stomach and wash it out if necessary. Its lining is thrown into a number of irregular, and longitudinal folds (**rugae**). As in other vertebrates it can be divided into several gross regions (a **cardiac region** next to the esophagus, a central **corpus**, and a caudal **pyloric region**), but these are not sharply demarcated. The pyloric region ends in a sphincter, the **pylorus**.

Cut open the intestine at several points and notice that a spiral valve is lacking, but that its internal surface area is increased through many small, wavy, longitudinal folds (**plicae**).

Most of the bile is drained from the liver by several **hepatic ducts** that unite to form a **common bile duct** that enters the duodenum (Fig. 9–13). Some of the bile backs up into the **gall bladder**, where it is temporarily stored, through a **cystic duct**. The gall bladder can be found on the dorsal surface of the liver near the apex of the cranial loop of the intestine, but most of the ducts are imbedded within pancreatic tissue. Some can be found by careful dissection in this region.

The pancreas of urodeles is formed from a dorsal and a pair of ventral primordia. All fuse in the adult, but each retains its ducts (Fig. 9–13). The pancreatic ducts cannot be found by gross dissection.

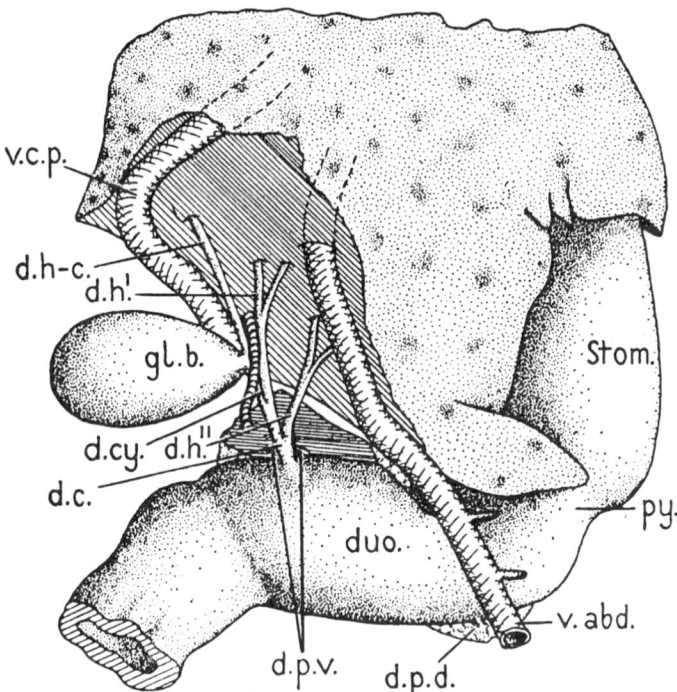

Figure 9–13.   Ventral dissection of the bile and pancreatic ducts of *Salamandra.* Part of the liver and pancreas have been cut away. Abbreviations: *d.c.,* common bile duct; *d.cy.,* cystic duct; *d.h.',* d.h.'', hepatic ducts; *d.h-c.,* hepato-cystic duct; *d.p.d.,* dorsal pancreatic duct; *d.p.v,* ventral pancreatic ducts; *duo.,* duodenum; *gl. b.,* gall bladder; *stom.,* stomach; *py.,* pylorus; *v.abd.,* ventral abdominal vein; *v.c.p.,* caudal vena cava. (From Francis, The Anatomy of the Salamander, Oxford University Press.)

Pericardial Cavity

The second division of the coelom, the **pericardial cavity**, is located farther forward in *Necturus* than in most amphibians, for it still occupies the fish position cranial to the pectoral girdle and deep to the hypobranchial musculature. Expose it by removing the musculature in this region, It is lined with **parietal pericardium** and contains only the **heart**, which is covered with **visceral pericardium.**

Cut through that portion of the ventral body wall lying between the pericardial cavity and the cranial end of your incision into the pleuroperitoneal cavity. The two divisions of the coelom are completely separated by a small, vertical **transverse septum**. A **coronary ligament** runs from the caudal surface of the transverse septum to the cranial end of the liver. A large vein (the caudal vena cava) lies within it.

Oral Cavity, Pharynx, and Respiratory Organs

Since *Necturus* is neotenic, its major respiratory organs are not the lungs but the three pairs of larval **external gills** arising from the back of the head.

Spread them apart and find the two **gill slits** between their bases. If a living specimen is available in an aquarium, notice the great vascularity of the numerous **gill filaments** and the way in which the gills are slowly moved back and forth.

Open the oral cavity and pharynx on the side on which you have been dissecting by cutting through the angle of the mouth, caudally through the gill slits, the side of the neck lateral to the pericardial cavity, the ventral portion of the pectoral girdle, and on to intersect the incision by which you opened the pleuroperitoneal cavity. The deeper part of the incision should pass dorsal to the lung and into the esophagus. Swing open the floor of the mouth and pharynx.

As explained in the introduction to this chapter, the breakdown of the plate of tissue between the stomadeum and archenteron makes it difficult to draw a sharp line between the rostral **oral cavity** and the more caudal **pharynx**. One merges with the other. Caudally, however, the pharynx leads into a somewhat constricted, short passage (the **esophagus**), which soon enters the stomach, but the longitudinal folds in the lining of the esophagus tend to be smaller than those in the stomach.

A **tonguelike fold** supported by the hyoid arch is located in the floor of the mouth (Fig. 9–14). It has developed little beyond the primary tongue of fishes, but a few hypobranchial muscle fibers enter its base (p. 126). The gland field of the tongue of adult amphibians (p. 253) is barely developed. The third to fifth visceral arches lie caudal to the hyoid arch and can be palpated. Between

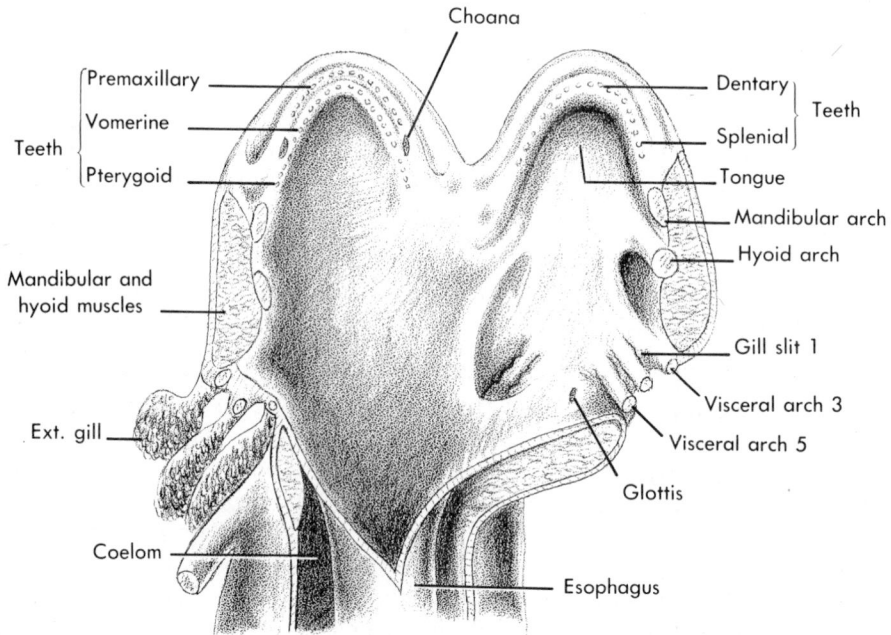

Figure 9–14. Oral cavity and pharynx of *Necturus*. The floor of the mouth and pharynx has been swung open to the right.

which do the gill slits lie? With which slits of *Squalus* are these slits homologous? Small **gill rakers** can be seen on the pharyngeal surface of the gill slits.

Necturus seldom breathes air, but the structures needed for pulmonary respiration are present. Air can enter the oral cavity and pharynx by way of the nares, nasal cavity and choanae. If a nasal cavity was not dissected in connection with the sense organs (p. 181), expose it now. A choana can be seen in the roof of the mouth lateral to the most caudal and shortest of the tooth rows (pterygoid teeth). The air leaves the pharynx through the **glottis** — a short longitudinal slit in the center of the floor of the pharynx about the level of the caudal gill slits. The glottis leads into a median **laryngotracheal chamber** from whose caudal end a pair of openings (the **bronchi**) pass to the lungs. These structures can best be seen by cutting the front of one lung, passing a blunt probe forward to the glottis, and then cutting open the floor of the esophagus and pharynx along the course of the probe. A pair of small cartilages (**lateral cartilages**), which are difficult to see but may be felt, support the wall of the laryngotracheal chamber (Fig. 4–12). They are probably derived from the sixth and seventh visceral arches (p. 64). Cut open more of the lung and observe that it is an empty sac with a smooth internal surface. In most adult amphibians the internal surface is increased by pocketlike folds. If possible, examine a demonstration of a lung of a frog.

A thyroid and a series of parathyroids and thymus glands develop from the pharynx but are difficult to find. The thyroid may be seen during the dissection of the arteries (p. 300).

MAMMALS

The evolution from amphibians through reptiles to mammals has been one of improvement upon terrestrial adaptations, including an increase in general body activity. Most mammals are active, warm-blooded animals that maintain a high and constant rate of metabolism (**endothermic**), and the organ systems concerned with metabolism are adapted accordingly.

Correlated with the development of a mobile neck and head, the pericardial cavity undergoes a caudad migration and becomes situated caudal to the pectoral girdle in the thoracic region of the body. The pleuroperitoneal cavity of lower vertebrates becomes divided into a pair of **pleural cavities** containing the lungs and a **peritoneal cavity** containing most of the remaining viscera. Muscles invade the coelomic folds that separate the pleural cavities from the peritoneal, and the whole complex forms a diaphragm whose movements, together with those of the ribs, provide an efficient means of ventilating the lungs.

The more cranial respiratory passages have also become modified. With the evolution of a distinct neck, the laryngotracheal chamber of amphibians becomes divided into a **larynx** from which a **trachea** descends to the lungs. And, as explained in connection with the skeleton, the evolution of a hard and soft palate separates a respiratory passage from the original oral cavity and rostral part of the pharynx. This permits simultaneous respiration and manipulation of food within the mouth. Food and air passages cross only in the caudal part of the pharynx, and, even here, food is normally prevented from entering the larynx through the evolution of a flaplike **epiglottis**.

The lungs themselves have increased tremendously in their internal surface area. This has been accomplished through the branching and rebranching of the respiratory passages within the lungs, so that all terminate in grapelike clusters of small, thin-walled **alveoli**. The gas exchanges with the circulatory system occur only in the alveoli.

Many changes also occur in the digestive system. Conspicuous **salivary glands** evolve from certain of the small oral glands of primitive tetrapods. In addition to providing secretions for lubricating the food, these glands secrete a digestive enzyme, **ptyalin**, that acts on carbohydrates.

The primitive tongue becomes a prominent muscular organ, for a pair of large, **lateral lingual swellings** are added rostral to the primary tongue and gland field (**tuberculum impar** of mammalian embryology), and the invasion of the organ by hypobranchial muscles has continued. The sensory innervation of the tongue correlates with the nerves of those visceral arches above which it develops. Thus the rostral part of the tongue, which develops from the lateral lingual swellings overlying the mandibular arch, is innervated by the trigeminal nerve; the caudal part (primary tongue) overlying the hyoid apparatus, by the facial and glossopharyngeal nerves; and the intermediate portion of the tongue (tuberculum impar), by both the trigeminal and facial nerves. But the muscles of the tongue, since they belong to the hypobranchial rather than to the brachiomeric group, are innervated by the hypoglossal nerve.

The most notable change in the intestinal region is a great increase in internal surface area. This is accomplished in part through an increase in the length of the intestine, and in part through the evolution of numerous, minute, fingerlike **villi**. Another major change is the division of the cloaca into dorsal and ventral portions. The ventral portion contributes to the urogenital passages; the dorsal portion forms the **rectum**. The mammalian rectum is therefore not homologous to the large intestine of lower vertebrates in which a cloaca is still present.

Superimposed upon these general features are numerous modifications of the digestive tract correlated with the diverse diets and modes of life of the various mammalian groups. The digestive tract of the cat or mink is a good example of the more primitive carnivore pattern, while that of the rabbit is illustrative of one type found in herbivores.

Digestive and Respiratory Organs of the Head and Neck

(A) Salivary Glands

The salivary glands of your specimen may be studied on the same side of the head that was used for the dissection of the muscles, provided the glands were not injured. If they were destroyed, carefully remove the skin on the opposite side of the head overlying the cheek, throat and side of the neck ventral to the auricle. The superficial cutaneous muscles (platysma and facial muscles, p. 133) must also be taken off. Be especially careful in the cheek region, for one of the salivary ducts is very superficial.

Pick away the connective tissue ventral to the auricle and you will expose the large **parotid gland**. It can be recognized by its lobulated texture. The gland is somewhat oval in the carnivores (Fig. 9–15), but shaped like a dumbbell in the rabbit. This gland is often removed with the skin in the mink. In all the mammals being considered, the **parotid duct** emerges from the front of the gland, crosses the large cheek muscle (masseter), and perforates the mucous membrane of the upper lip near the molar teeth. Frequently accessory bits of

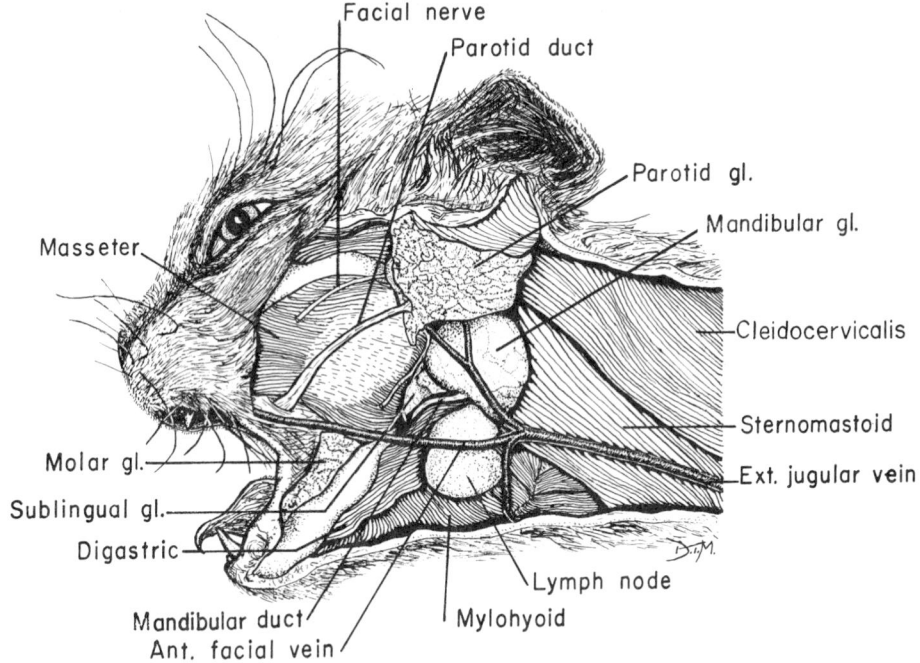

Figure 9–15. Lateral view of the salivary glands of the cat.

glandular tissues are found along the duct. Two branches of the facial nerve going to facial muscles emerge from beneath the parotid gland and cross the masseter, one dorsal and one ventral to the parotid duct. Do not confuse the duct with them.

A **mandibular gland** lies caudal to the angular process of the jaw and deep to the ventral border of the parotid. It too is often destroyed in skinning the mink. It is a large oval gland having the same lobulated texture as the parotid. Do not confuse it with smaller, smoother textured **lymph nodes** in this region. The **mandibular duct** emerges from the front of the gland and passes forward, first going lateral (cat and rabbit) or medial (mink) to the digastric muscle. The digastric is the large muscle arising from the base of the skull and inserting along the ventral border of the lower jaw (p. 165). Cut and reflect the digastric and you can see that the duct then passes medial to the caudal border of the mylohyoid — the thin transverse sheet of muscle lying between the paired digastric muscles. Cut and reflect the mylohyoid and follow the duct forward as far as you can. It is crossed rostrally by the **lingual nerve**, a branch of the trigeminal nerve returning sensory fibers from the tongue. The **hypoglossal nerve**, carrying motor fibers to the tongue musculature, lies caudal and dorsal to the mandibular duct. The mandibular ducts of opposite sides converge and enter the floor of the mouth by a pair of inconspicuous openings situated just rostral to the midventral septum of the tongue, the **lingual frenulum**. The openings are borne on flattened papillae in the cat.

A small, elongated **sublingual gland** is located beside the mandibular duct. In carnivores the sublingual lies along the caudal one third of the duct and generally abuts against the mandibular gland. It lies along the rostral one third of the duct in the rabbit. The sublingual is drained by a minute duct which parallels the mandibular duct but is hard to distinguish grossly. The ducts enter the floor of the mouth.

The parotid, mandibular and sublingual glands are the most common of the salivary glands of mammals, but others are present in certain species. The cat, mink, and rabbit have a **zygomatic gland** that has been seen if the eye was dissected (p. 178). In addition, carnivores have a small, elongate **molar gland** situated between the skin and mucous membrane of the caudal half of the lower lip, It usually is removed when the mink is skinned. Several small ducts, which cannot be seen grossly, lead from it to the inside of the lip. The rabbit has elongated **buccal glands** beneath the skin of the upper and lower lips.

Omit

(B) Oral Cavity

Open the **oral cavity** by cutting through the floor of the mouth with a scalpel. Do this from the external surface, cut on each side, and keep as close to the mandible and chin as possible. Then cut through the symphysis of the mandible with bone scissors, spread the two halves of the lower jaw apart, and pull the tongue ventrally. The rostral part of the roof of the oral cavity is formed by the bony **hard palate**; the caudal part, by the fleshy **soft palate** (Figs. 9–16 and 9–17). A pair of small openings will be seen at the very front of the hard palate just caudal to the incisor teeth. These are the openings of the **incisive ducts**. These ducts pass through the palatine fissure to the vomero-

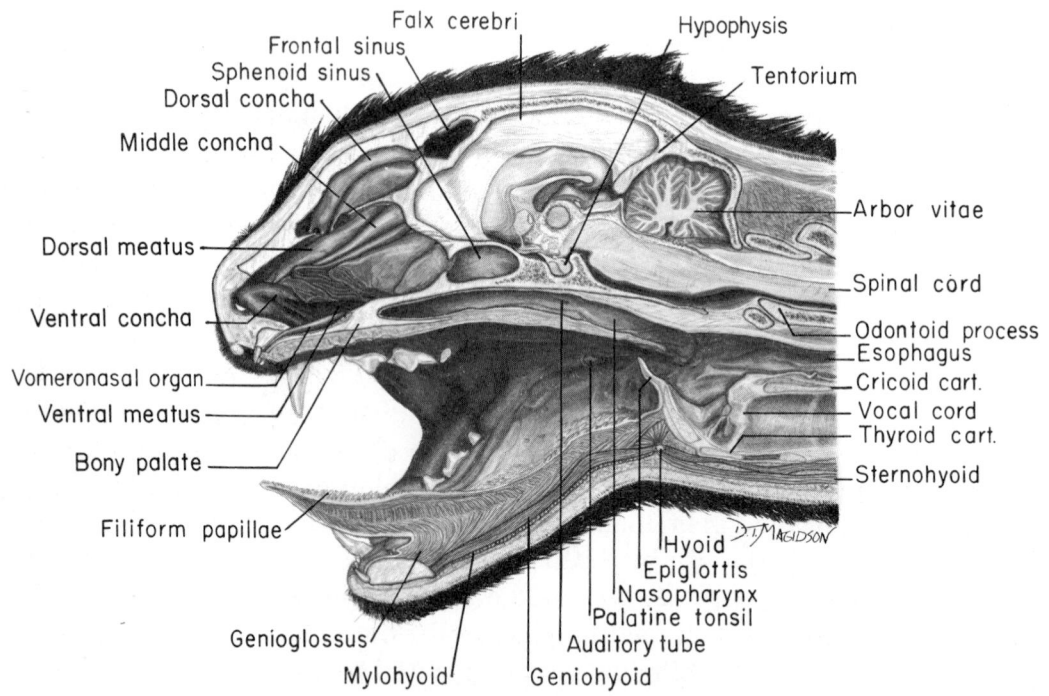

Figure 9–16. Sagittal section of the head of a cat.

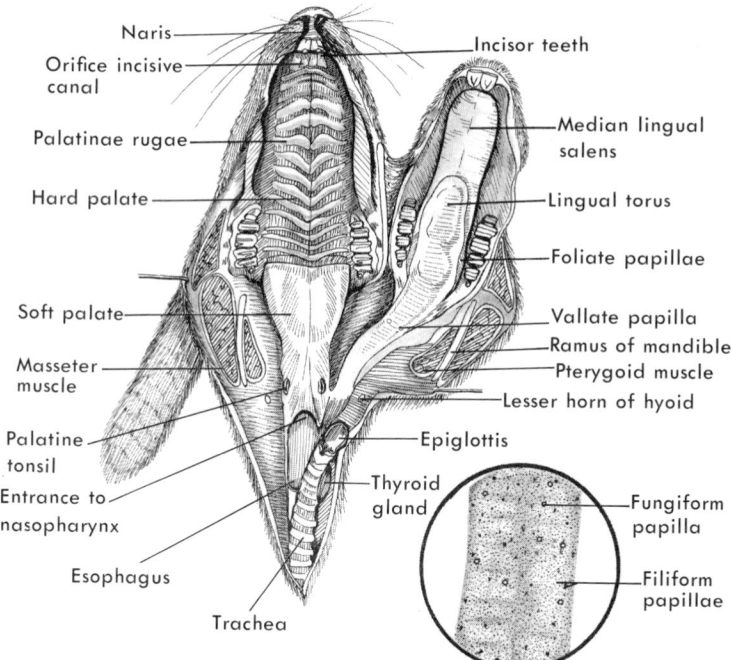

Naris
Orifice incisive canal
Palatinae rugae
Hard palate
Soft palate
Masseter muscle
Palatine tonsil
Entrance to nasopharynx
Esophagus
Trachea

Incisor teeth
Median lingual salens
Lingual torus
Foliate papillae
Vallate papilla
Ramus of mandible
Pterygoid muscle
Lesser horn of hyoid
Epiglottis
Thyroid gland
Fungiform papilla
Filiform papillae

Figure 9–17.   Oral cavity and pharynx of the rabbit. The floor of the mouth and pharynx has been swung open to the right.

nasal organs in the nasal cavities (p. 183). The lateral walls of the oral cavity are bounded by the teeth, lips and cheeks. That portion of the cavity lying between the teeth and cheeks is called the **vestibule**. A well-developed, muscular **tongue** (*lingua*) lies in the floor of the cavity and is connected to the floor by the vertical lingual frenulum previously seen.

Pull the tongue ventrally sufficiently far to tighten, and bring into prominence a pair of lateral folds that extend from the sides of the caudal portion of the tongue to the soft palate. These folds constitute the **palatoglossal arches**, and they represent the boundary between the adult oral cavity and **pharynx**. However, part of the back of the oral cavity as thus defined probably develops from the embryonic pharynx. The passage between the palatoglossal arches is called the **fauces**. Notice that the very back of the tongue lies within the pharynx.

Cut through the palatoglossal arch on one side, pull the tongue down more, and examine its dorsal surface (Fig. 9–17). In the rabbit, but not in the other mammals being considered, the back of the tongue is raised slightly, forming a **lingual torus**. This part has developed from the tuberculum impar (p. 260); the front of the tongue, which is divided by a **median lingual sulcus,** has developed from the paired lateral lingual swellings.

The dorsum of the tongue is covered with papillae, the most numerous of which are the pointed **filiform papillae**. The cranial ones of carnivores bear spiny projections with which the animal grooms its fur or rasps flesh from bones, but the caudal ones are soft. Small, rounded **fungiform papillae** are interspersed among the filiform, especially along the margins of the tongue. You may have to use a hand lens to see them. **Vallate papillae** are located near the back of the tongue. Each papilla is a relatively large, round patch set off from the rest of the tongue by a circular groove. There are two in the rabbit, four to six in the cat distributed in a caudally directed, **V**-shaped line, and half a dozen or more scattered ones in the mink. Leaf-shaped **foliate papillae** can be found along the side of the caudal part of the tongue. Those of the rabbit are collected into a pair of patches. Microscopic taste buds are found on the sides and base of all the papillae, except for most of the filiform papillae.

(C) PHARYNX

With a pair of scissors, cut caudally through the lateral wall of the pharynx on the side on which the palatoglossal arch was cut. Follow the contour of the tongue to the laryngeal region, then extend the cut dorsal to the larynx and back into the esophagus. Do not cut into the soft palate or larynx. Swing open the floor of the mouth and pharynx. The pharynx may be divided somewhat arbitrarily into oral, nasal, and laryngeal portions. The **oropharynx** lies between the palatoglossal arches and the free caudal margin of the soft palate. A pair of **palatine tonsils** lie in its laterodorsal walls. Note that each is partially imbedded in a **tonsillar fossa**. The **laryngopharynx** is the space dorsal to the enlargement, the **larynx**, in the floor of the caudal part of the pharynx. It communicates caudally with the **esophagus** and ventrally with the larynx. The slitlike opening within the larynx is termed the **glottis**. A trough-shaped fold, the **epiglottis**, lies cranial to the glottis and acts to deflect food around or over the glottis. The **nasopharynx** lies dorsal to the soft palate. Open it by making a longitudinal incision through the middle of the soft palate. Spread open the incision as wide as possible and try to shine a light down into the nasopharynx. The pair of slitlike openings in the laterodorsal walls are the entrances of the **auditory tubes** (eustachian tubes) (Fig. 9–16). What do these tubes represent phylogenetically? The choanae, or internal nostrils enter the rostral end of the nasopharynx but cannot be seen in this view.

(D) LARYNX, TRACHEA AND ESOPHAGUS

Approach the laryngeal region from the ventral surface of the neck. Several muscles will have to be removed, but do not injure any of the larger blood vessels. The **larynx** is the chamber whose walls are supported by relatively large cartilages. The **hyoid bone**, which forms a sort of sling for the support of the base of the tongue, is imbedded in the muscles cranial to the larynx; and its greater horn articulates with the front of the larynx. Its parts have been described elsewhere (p. 87). The larynx is continued caudally as the windpipe, or **trachea**, whose walls are supported by a series of cartilaginous rings, which hold the tracheal lumen open, thus permitting the free movement of air. The **esophagus** is a collapsed, muscular tube lying dorsal to the trachea. Its lumen is pushed open as food is swallowed.

The dark **thyroid gland** lies against the cranial end of the trachea. In carnivores it consists of **lobes,** one on either side of the trachea, that are connected across the ventral surface of the trachea by a very narrow band of thyroid tissue called the **isthmus.** The isthmus is frequently destroyed. The thyroid of the rabbit differs in having a wide, prominent isthmus. Two pairs of **parathyroid glands** are imbedded in the dorsomedial surface of the thyroid, but they cannot be seen grossly.

Return to the larynx and study it more thoroughly. Much of it is covered by **intrinsic laryngeal muscles** derived from the caudal branchiomeric musculature and hence innervated by the vagus nerve. Strip off these muscles from all surfaces of the larynx to expose the laryngeal cartilages. The muscles may be identified as you remove them by reference to Figure 9–18. They are named according to the cartilages between which they extend. The large cranial cartilage that forms much of the ventral and lateral walls of the larynx is called the **thyroid cartilage.** It is this cartilage that forms the projection in the neck of man known as Adam's apple. The ring caudal to the thyroid cartilage is the **cricoid cartilage.** The cricoid is shaped like a signet ring, for its dorsal portion is greatly expanded and forms most of the dorsal wall of the larynx. Careful dissection will reveal a pair of small, triangular cartilages cranial to the dorsal part of the cricoid. These are the **arytenoid cartilages.** Additional, minute cartilages are frequently associated with the arytenoids but are seldom seen. An **epiglottic cartilage** supports the epiglottis.

Cut open the larynx along its middorsal line. The pair of whitish, lateral folds that extend from the arytenoids to the thyroid cartilage are the **vocal cords.** They are set in vibration by the movement of air across them and are controlled by the movement of the arytenoids, the action of muscles within them and slight changes in the shape of the larynx. The **glottis** is the space be-

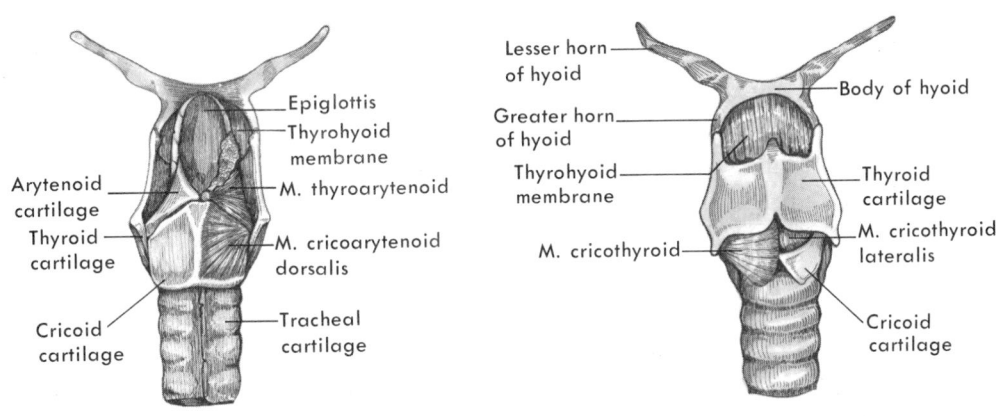

A   B

Figure 9–18. Hyoid and larynx of a mink. *A*, Dorsal view; *B*, ventral view. Intrinsic muscles of the larynx have been removed from the specimen's left side to show the cartilages.

tween them. In carnivores, an accessory pair of folds, sometimes called the **false vocal cords**, extend from the arytenoids to the base of the epiglottis. They are very small in the mink. In the rabbit a small pair of bumps, the **epiglottic hamuli**, lie at the base of the epiglottis.

Most of the laryngeal cartilages develop from certain of the visceral arches, but there is some doubt as to the precise homologies. The arytenoids and cricoid are the first to appear phylogenetically, being represented by the lateral cartilages of the laryngotracheal chamber of amphibians (p. 259). They appear to develop from the sixth, or from the sixth and seventh, visceral arches. The fourth and fifth visceral arches are incorporated in the hyoid apparatus in amphibians. In mammals the hyoid apparatus involves only the second and third visceral arches, and in most mammals the fourth and fifth arches form the newly evolved thyroid cartilage. Some believe that the tracheal rings may evolve from the splitting and multiplication of the seventh arch, but this is doubtful. The epiglottic cartilage is apparently a new structure.

Thorax and Its Contents

(A) PLEURAL CAVITIES

Open the thorax by making a longitudinal incision about two centimeters to the right of the midventral line and extending the length of the sternum. Use a pair of strong scissors. Spread open the incision and look into the **right pleural cavity**. A dome-shaped, transverse, muscular partition (the **diaphragm**) will be seen at the caudal end of the cavity. Make another cut just cranial to the diaphragm that extends laterally and dorsally to the back. Follow the line of attachment of the diaphragm but keep on the pleural side. Spread the right thoracic wall laterally, breaking the ribs near their attachment to the vertebrae.

The right pleural cavity and its **lung** (*pulmo*) are now well exposed. The coelomic epithelium lining the walls of the pleural cavity is called the **parietal pleura**; that covering the surface of the lung, the **visceral** (*pulmonary*) **pleura**. The right lung is divided into four **lobes—cranial, middle, caudal,** and **accessory** (Fig. 9–19). The accessory lobe extends dorsal to a large vein (**caudal vena cava**) and then ventrally into a pocket on the medial side of the mesentery (**caval fold**) attaching to the ventral surface of the vein (Fig. 9–20). Tear the caval fold near the vein and you can see this lobe. The lobes of the lung are attached to the medial wall of the pleural cavity by a pleural fold known as the **pulmonary ligament**. The blood vessels connected with the heart, and the **bronchi** from the trachea to the lung pass through part of this ligament, but they should not be dissected at this time. These structures constitute the **root of the lung.** Cut into a part of the lung and notice that it is not an empty organ but a very spongy one. The numerous, thin-walled, terminal pockets of the respiratory passages (**alveoli**), in which gas exchange occurs, are not visible grossly.

The medial wall of each pleural cavity (right and left) is formed only of a

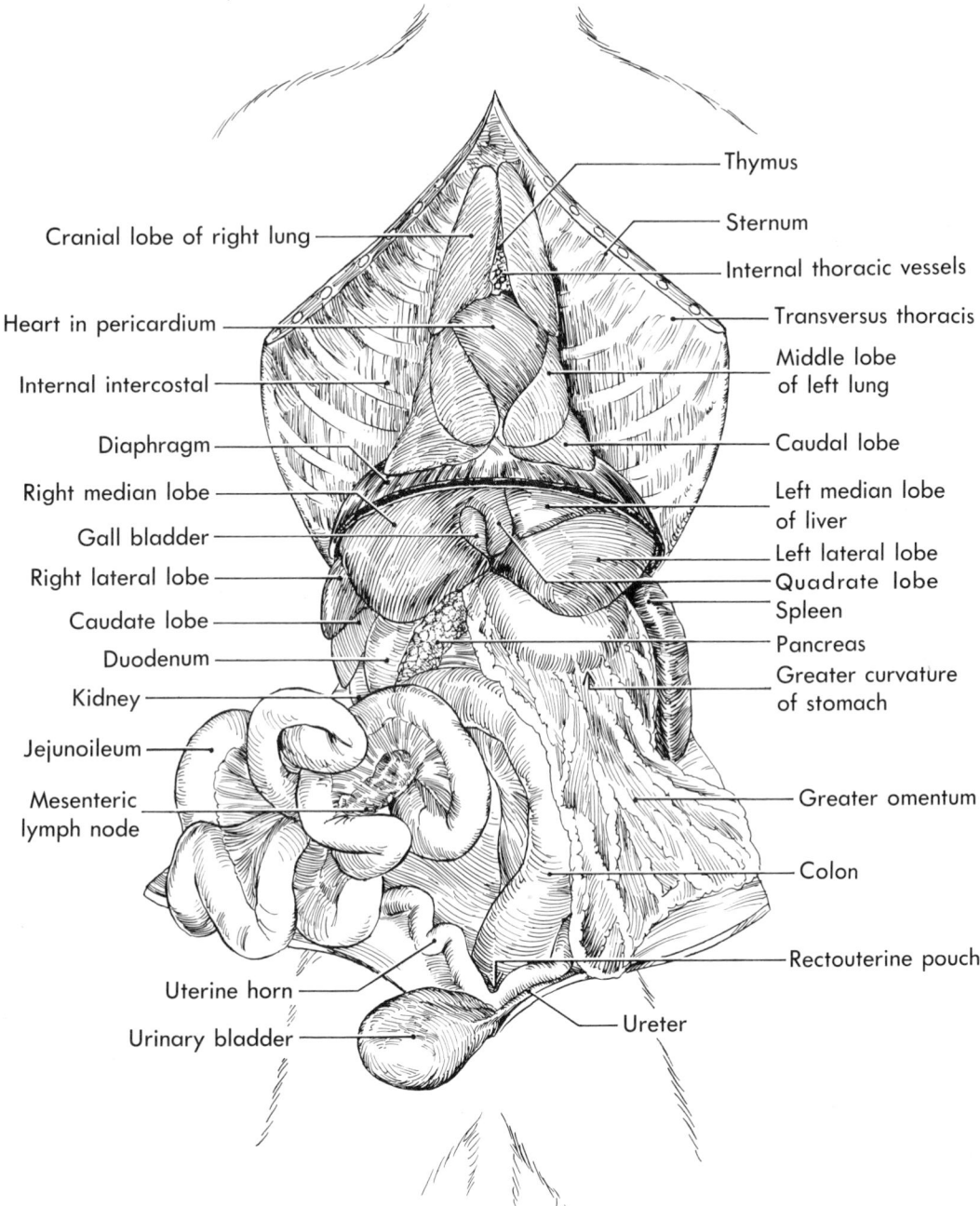

Cranial lobe of right lung

Heart in pericardium

Internal intercostal

Diaphragm

Right median lobe

Gall bladder

Right lateral lobe

Caudate lobe

Duodenum

Kidney

Jejunoileum

Mesenteric lymph node

Uterine horn

Urinary bladder

Thymus

Sternum

Internal thoracic vessels

Transversus thoracis

Middle lobe of left lung

Caudal lobe

Left median lobe of liver

Left lateral lobe

Quadrate lobe

Spleen

Pancreas

Greater curvature of stomach

Greater omentum

Colon

Rectouterine pouch

Ureter

Figure 9–19. Ventral view of the major organs of the thoracic and peritoneal cavities of a cat. (From Walker, A Study of the Cat.)

layer of parietal pleura. The space, or potential space, between the medial walls of the two pleural cavities constitutes the **mediastinum**. This space, however, is largely filled with structures that lie between the two cavities. For example, the pericardial cavity and heart, which form the large bulge medial and ventral to the lung, lie in the mediastinum. In places the medial walls of the pleural cavities meet and form a mesenterylike structure termed

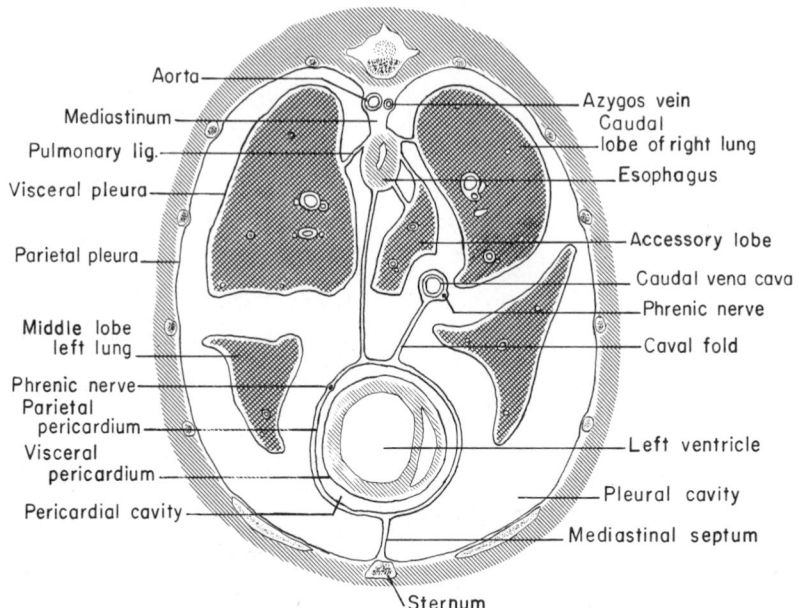

Figure 9–20. A diagrammatic transverse section through the thorax of the cat at the level of the ventricles to show the coelomic epithelium and its relation to the thoracic viscera. The section is viewed from behind so that left and right sides of the animal and drawing coincide.

the **mediastinal septum**. The mediastinal septum can be seen caudal to the heart and medial to the accessory lobe of the right lung. The caval fold is an evagination from this portion of the septum. In carnivores the mediastinal septum continues cranially ventral to the heart, but the medial walls of the pleural cavities do not meet in this region in the rabbit.

Break the line of attachment of the parietal pleura to the ventral thoracic wall. You will have to cut some blood vessels passing to the ventral thoracic wall, but do so in such a way that their ends can later be apposed. Cut laterally and dorsally through the left thoracic wall close to the diaphragm in the same way that you did on the right. Turn back the body wall and examine the **left pleural cavity**. (This will give a good view of the third muscle layer of the thorax, transversus thoracis, referred to on page 161.) The left lung does not have an accessory lobe. In the rabbit the middle lobe is not well demarcated from the cranial lobe.

Pull the left lung ventrally and examine the region dorsal to it. A large artery, the **aorta**, and the **esophagus** can be seen passing through the dorsal portion of the mediastinum. The aorta lies to the left of the vertebral column; the esophagus, more ventrally. Let the lungs fall back into place. A pair of white strands, the **phrenic nerves**, can be seen in the central portion of the mediastinum on each side of the pericardial cavity and heart. They lie ventral to the roots of the lung and pass caudad to the diaphragm. The right one follows the caudal vena cava closely; the left passes through the caudal part of the mediastinal septum. The origin of these nerves from the ventral rami of some combination of the fourth, fifth, and sixth cervical nerves is indicative of the cervical derivation of the diaphragmatic muscles. The portion of the mediastinum ventral and cranial to the heart is occupied by the dark, irregularly lobulated **thymus**. The thymus varies considerably in size, being best developed in young individuals.

(B) Pericardial Cavity

The **pericardial cavity** and **heart** are the largest structures within the mediastinum. Cut open the pericardial cavity by a midventral incision. Its wall, known as the **pericardium**, is formed of connective tissue and coelomic epithelium, the **parietal pericardium**. **Visceral pericardium** covers the surface of the heart. Parietal and visceral pericardium are continuous with each other over the vessels at the front of the heart.

Peritoneal Cavity and Its Contents

(A) Body Wall and Peritoneal Cavity

Make a longitudinal incision through the abdominal wall slightly to the right of the midventral line. Extend the cut from the diaphragm to the pelvic girdle. Then cut laterally and dorsally along the attachment of the diaphragm to the body wall. Do this on both sides, thereby freeing the diaphragm as far as the back. Reflect the flaps of the abdominal wall. What layers constitute the wall and how do they compare with those of the thoracic wall? The portion of the coelom exposed is the **peritoneal cavity**. Its walls are lined with **parietal peritoneum** and its viscera are covered with **visceral peritoneum**. Wash out the cavity if necessary.

(B) Abdominal Viscera and Mesenteries

The concave surface of the dome-shaped diaphragm forms the cranial wall of the peritoneal cavity, and the large **liver** (*hepar*) lies just caudal to it and is shaped to fit into the dome. Pull the liver and diaphragm apart. It can now be seen that the central portion of the diaphragm is formed by a tendon (the **central tendon**), into which its muscle fibers insert. A vertical **falciform ligament** extends between the diaphragm, liver and ventral abdominal wall. Sometimes a thickening, which represents a vestige of the embryonic umbilical vein, can be seen in its free edge. It is known as the **round ligament** of the liver. Diaphragm and liver are closely apposed dorsal to the falciform ligament, and the reflections of the peritoneum from the one to the other in this region constitute the **coronary ligament**.

The liver can be divided into right and left halves at the cleft into which the falciform ligament passes. Each half is divided into a lateral and a medial lobe, thus making a **left lateral**, **left medial**, **right medial**, and **right lateral lobe** (Fig. 9–19). A small **quadrate lobe** is interposed between the left medial and right medial lobes. It is partly united with the latter, being separated from it by the **gall bladder** (*vesica fellea*). A **caudate lobe** lies caudal to the right lateral lobe and abuts against the right kidney. Part of the caudate lobe extends toward the left side of the body going deep to a mesentery, the **lesser omentum**, which extends from the liver to the stomach and duodenum (Fig. 9–22). A **hepatorenal ligament** extends from the caudate lobe of the liver to the parietal peritoneum near the right **kidney**. The left kidney lies in a slightly more caudal position on the opposite side of the body.

As in the majority of vertebrates, most of the **stomach** (*ventriculus*) lies on the left side of the peritoneal cavity and is more or less J-shaped. Cut through the left side of the diaphragm to find the point at which the **esophagus** enters. There is an abrupt change in the diameter of the digestive tract at this point. The portion of the stomach adjacent to the esophagus is the **cardiac region**; the dome-shaped portion extending cranially to the left of the cardiac region, the **fundus**; the main part of the stomach, the **body**; and the narrow caudal portion, the **pyloric region**. But, as pointed out elsewhere, these gross regions do not necessarily correspond with glandular regions bearing the same names. The stomach ends in a thick muscular sphincter, the **pylorus**, which can be seen if you cut open the stomach. Do not injure any mesenteries. In carnivores you will also see longitudinal ridges in the lining that are called **rugae**. The long left and caudal margin of the stomach, which represents its original dorsal surface, constitutes its **greater curvature**; the shorter right and cranial margin, which represents its original ventral surface, the **lesser curvature**.

Notice that the lesser omentum, which represents a part of the ventral mesentery, attaches along the lesser curvature of the stomach. The mesentery that attaches along the greater curvature is the **greater omentum** or mesogaster — a part of the dorsal mesentery. The greater omentum of mammals does not extend directly to the middorsal line of the peritoneal cavity, as it does in lower vertebrates, but is modified to form a saclike structure (the **omental bursa**), which drapes down over the intestines (Fig. 9–21).

The omental bursa of the cat and mink is very large, contains considerable fat in its wall and extends over the intestine nearly to the pelvic region. It is often entwined with the intestines and must be untangled carefully. The rabbit's bursa is much smaller. In both animals the greater omentum extends caudally from its line of attachment on the stomach and then turns upon itself and extends cranially and dorsally to attach onto the dorsal wall of the peritoneal cavity. The **spleen** (*lien*), which is enormous in the cat and mink, lies in the wall of the omentum on the left side of the stomach. That portion of the greater omentum between the spleen and stomach is known as the **gastro-lienic ligament**. A small, triangular mesentery, the **gastrocolic ligament**, passes from the part of the greater omentum lying dorsal to the spleen over to the mesentery of the large intestine.

A part of the peritoneal cavity that is known as the **lesser peritoneal cavity** lies between the descending and ascending walls of the omental bursa. Tear the bursa and verify that it contains a space. At one stage of develop-

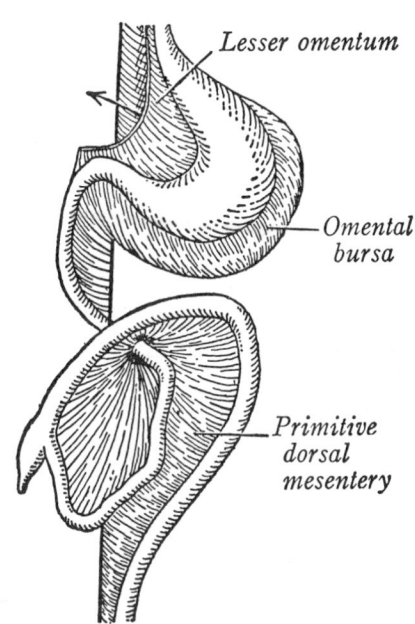

Lesser omentum

Omental bursa

Primitive dorsal mesentery

Figure 9-21. Diagrammatic ventral view of the mesenteries of a mammalian embryo to show the formation of the omental bursa from the caudad extension of the mesogaster. The liver would attach to the lesser omentum along its cut edge (double line). The arrow indicates the approximate position of the epiploic foramen. (From Arey, Developmental Anatomy.)

ment the lesser peritoneal cavity would have a wide communication with the main part of the peritoneal cavity. But subsequent adhesions of the liver to the dorsal part of the diaphragm and adjacent body wall reduce this to a relatively small **epiploic foramen** (Fig. 9–22). The epiploic foramen lies dorsal to the lesser omentum and between the candate lobe of the liver and the mesentery to the duodenum. If your specimen is large enough, you can pass a finger through the epiploic foramen and extend it dorsal to the stomach and into the omental bursa.

Carefully dissect that portion of the lesser omentum lying near the caudate lobe of the liver and the epiploic foramen in order to expose the system of bile ducts extending from the liver and gall bladder to the beginning of the duodenum. Some lymphatic vessels, which look like chains of small nodules, may have to be removed. A **cystic duct** comes down from the gall bladder and unites with several **hepatic ducts** from various parts of the liver to form a **common bile duct** (*ductus choledochus*) which passes to the duodenum. One particularly prominent hepatic duct comes in from the left lobes of the liver, and another from the right lateral lobe.

The **small intestine** extends caudad from the stomach, passes through numerous convolutions, and eventually enters a **large intestine**. Certain features of the intestinal region and the pancreas differ sufficiently in the carnivores and rabbit to warrant separate treatment, but a few common features may be observed at this time. The small intestine of mammals has differentiated into a cranial **duodenum** and a more caudal **jejunum** and **ileum**. The duodenum, as will be seen presently, is the first, approximately **U**-shaped loop of the intestine, but there is no sharp transition between the jejunum and ileum. About all that can be said in respect to their gross anatomy is that the jejunum is the cranial half of the postduodenal small intestine and the ileum

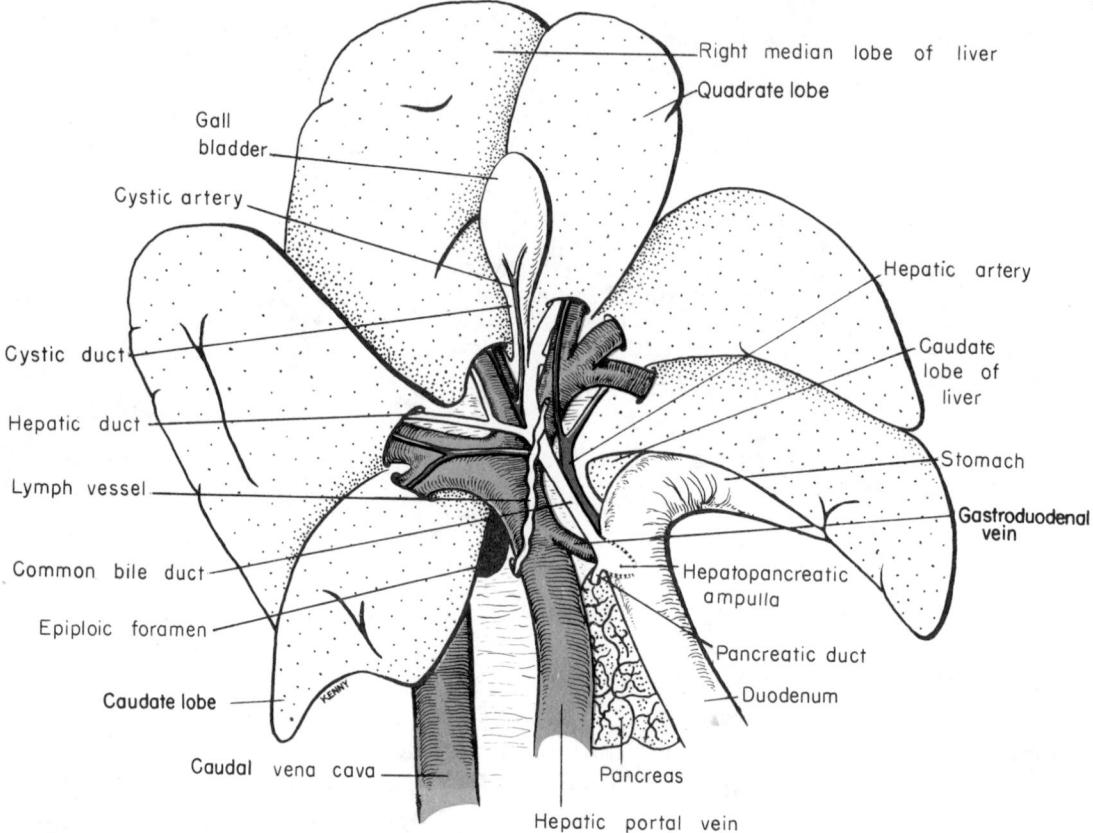

Figure 9–22. Ventral view of the vessels and ducts in the lesser omentum in the cat. The right median, quadrate, and part of the right lateral lobes of the liver have been turned forward.

the caudal half. The small intestine is supported by a part of the dorsal mesentery—that portion passing to the duodenum being the **mesoduodenum**; and that portion to the jejunum and ileum, the **mesentery**.

Cut open a part of the small intestine and examine its lumen. The lining has a velvety appearance which results from the presence of numerous, minute, fingerlike projections called **villi**. These greatly increase the internal surface area. Parasitic roundworms and tapeworms are often found in the intestine of carnivores.

The large intestine of mammals is much longer than that of lower vertebrates and generally has a greater diameter than the small intestine. Most of it constitutes the **colon**, and it is supported by a portion of the dorsal mesentery termed the **mesocolon**. Skip the pattern that the cranial part of the colon assumes for a moment and examine the caudal part. This portion extends caudad against the dorsal wall of the peritoneal cavity and enters the pelvic canal. This portion of the colon also lies dorsal to the pear-shaped **urinary bladder** and, if your specimen is a female, to the Y-shaped **uterus**. Notice that the bladder is supported by a vertical **median ligament**, which, being a part of the ventral mesentery, extends to the midventral body wall, and by a pair of lateral ligaments. The latter often contain wads of fat. Cut open the caudal part of the colon, clean it out, and notice that it lacks villi. Also notice the extension of the coelom into the pelvic canal. That portion of the coelom in the male that extends posteriorly between the large intestine and the urinary bladder is called the **rectovesical pouch**. The comparable coelomic extension in the female is divided by the uterus into a shallow **vesicogenital pouch** between the bladder and uterus, and a deep **rectogenital pouch** between the uterus and large intestine.

Deep within the pelvic canal the colon passes into the terminal segment of the large intestine, the **rectum**, which, in turn, opens on the body surface through the **anus**. The rectal region will be seen later when the pelvic canal is opened.

(C) FURTHER STRUCTURE OF THE DIGESTIVE ORGANS

Keeping these general features in mind, examine the intestine and pancreas in more detail.

Cat and Mink

The intestine and pancreas of carnivores are relatively simple. The **duodenum** curves caudad from the pylorus on the right side of the body. Then it bends toward the left and ascends toward the stomach. It is arbitrarily considered to end at its next major bend. A small, triangular-shaped peritoneal fold, the **duodenocolic fold**, extends between the caudal end of the duodenum and the mesocolon.

The **pancreas** can be recognized by its lobulated texture. A part of it lies against the descending portion of the duodenum, and a part of it extends as a tail-like process transversely across the body to the spleen. This portion lies in the dorsal wall of the omental bursa. Two pancreatic ducts are present, for the ducts of both the dorsal and the single ventral primordium are retained. The **pancreatic duct** unites with the common bile duct as the latter enters the duodenum. It can be found by carefully picking away pancreatic tissue in this region. The enlargement on the duodenum where the two unite is known as the **hepatopancreatic ampulla** (Fig. 9–22). An **accessory pancreatic duct**, which would be the duct of the dorsal pancreas, enters the duodenum about one centimeter caudal to the main duct but is small and very hard to find.

Follow the coils of the rest of the small intestine (jejunum and ileum) until it enters the colon. In the cat, a short, blind diverticulum, the **cecum**, extends caudally from the beginning of the colon. The vermiform appendix of man is located at the end of the cecum, but an appendix is absent in the cat. Cut open the wall of the cecum and colon opposite the entrance of the ileum. Notice how the ileum projects slightly into the lumen of the colon, forming an **ileal papilla**, which helps prevent the backing up of colic material into the small intestine. It is difficult to distinguish ileum and colon in the mink, for the mink lacks a cecum, and the colon is seldom wider than the ileum. Internally they can be distinguished, for the ileum is lined with villi and the colon with small irregular folds. The colon itself extends forward on the right side of the body for a short distance (**ascending colon**), crosses to the left side (**transverse colon**), and extends back into the pelvic canal (**descending colon**) to the rectum and anus.

Rabbit

The **duodenum** of the rabbit extends caudally in small convolutions nearly to the pelvic region. It then twists to the left and, continuing to convolute, ascends nearly to the stomach. It is arbitrarily considered to end where it next turns caudad. The duodenum is approximately 50 centimeters long in an adult rabbit.

Stretch out the mesentery (mesoduodenum) that lies between the descending and ascending limbs of the duodenum. The dark, lobulated tissue that you see is the **pancreas**. It is very diffuse in the rabbit (Fig. 9–23). The single **pancreatic duct**, which is the duct of the embryonic dorsal pancreas, enters the ascending limb about five centimeters cranial to the most caudal loop of the duodenum.

Follow the coils of the rest of the small intestine (jejunum and ileum) until it enters the large intestine. The caudal end of the ileum is modified to form a round, muscular enlargement known as the **sacculus rotundus**. Cut

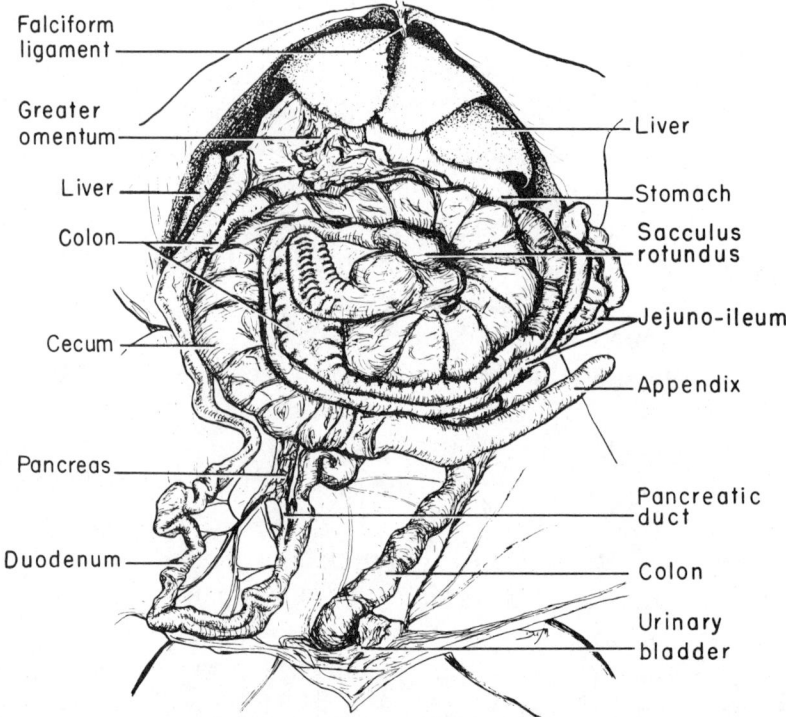

Figure 9–23. Ventral view of the abdominal viscera of the rabbit, showing in particular the large cecum and appendix characteristic of may herbivores.

open the large intestine opposite the entrance of the ileum, clean out its lumen, and find the orifice of the ileum. It is on an **ileal papilla**, which prevents material in the large intestine from backing up into the small.

A large cecum extends in one direction from the sacculus rotundus; the colon, which can be distinguished by its more wrinkled appearance, in the other. First follow the **cecum**. It is a wide, thin-walled, blind sac that extends for about 35 centimeters in a circular course and ends in a thicker-walled **vermiform appendix**. The appendix is about 12 centimeters long. Cut open the cecum and notice that the spiral line that can be seen on its surface marks the point of attachment of a **spiral valve**. The cecum contains a colony of bacteria and other microorganisms, some of which produce the enzyme cellulase which breaks down cellulose. Fermentation occurs here and many of the microorganisms, which reproduce rapidly, are themselves digested. Many of the organic acids, which result from cecal digestion, are absorbed here. Rabbits also engage in coprophagy, that is, much of the cecal contents, which is discharged as soft, mucus-covered fecal pellets, is eaten. This material is then further acted upon and absorbed in the stomach and small intestine. Cut open the appendix and note the relative thickness of its wall. Large amounts of lymphoid tissue accumulate in the wall and may help protect the body against the effects of bacterial toxins.

Return to the **colon** and follow it. Parts of the mesentery may have to be torn. The longitudinal muscle layer of the first part of the colon is limited to two or three bands, called **taeniae coli**, one of which lies along the line of attachment of the mesocolon. The wall of the colon between the taeniae protrudes to form little sacculations (**haustra coli**). More caudally the wall of the colon is smooth, bulging only where there are pellet-shaped feces within it. After a very circuitous course, the colon descends into the pelvic canal to the rectum and anus.

10

THE CIRCULATORY SYSTEM

Functions of the Circulatory System

Continuing with the organ systems that provide for the metabolic needs of the body, we will next consider the circulatory system. This system is primarily the great transport system of the body. Oxygen and food are carried from the respiratory and digestive organs to all the tissues and cells; carbon dioxide and other excretory products are carried from the tissues to sites of removal; and hormones are transported from the endocrine glands to the responding tissues. But the system has other functions. It aids in combating disease and in repairing tissues and helps to maintain the constancy of the internal environment in many ways.

Parts of the Circulatory System and the Course of the Circulation

The blood and lymph and their cells are functionally the most important part of the circulatory system, but only the vessels that propel and carry these can be studied in a course of this scope. These vessels may be subdivided in most vertebrates into (1) a **cardiovascular system**, consisting of the **heart**, **arteries**, **blood capillaries**, and **veins**; and (2) a **lymphatic system**, consisting of closed **lymphatic capillaries** and **lymphatic vessels**. In addition, the lymphatic system of higher vertebrates includes many **lymph nodes** located at strategic junctions along the course of the vessels. These nodes act as sites for the production of certain white blood cells (**lymphocytes**) and cells that transform into antibody-synthesizing cells. The **spleen** is similar to a lymph node but is interposed in the cardiovascular system. It is at different times and in different vertebrates a site for the production, storage and elimination of blood cells.

Briefly, the course of the circulation through these two systems of vessels is as follows: **Blood** leaves the heart and travels through the arteries to the capillaries in the tissues. At this point some of the blood plasma leaves the capillaries to circulate among the cells as **tissue fluid**. Much of the tissue fluid re-enters the capillary bed and returns to the heart through veins. However, in all but the most primitive vertebrates, some enters the lymphatic capillaries and is carried as **lymph** by the lymphatics to the larger veins. Primitive jawless fishes and cartilaginous fishes lack a true lymphatic system. Blood pressure in the veins is very low in these fishes, and the veins with their large sinuses adequately drain the tissues. Venous blood pressure tends to be higher in other vertebrates, and the lymphatics serve as a low pressure drainage system for the tissues. Particulate matter and plasma proteins that may be in the tissue fluid also enter the lymphatics.

276

The heart is the major pump in causing the fluids to circulate, but by the time the blood reaches the veins the pressure is relatively low. Pressures created by the heart do not directly affect the lymphatic system; thus the return of lymph, and to some extent the return of blood, are implemented by other forces. Among these, the contraction and tonus of the surrounding body muscles play a major role. The veins and lymphatics also contain **valves** that prevent a back flow of the fluids.

The lymphatic system is rather obscure and not commonly seen in gross dissections. Parts of it will be studied in the mammal, but the emphasis throughout will be on the more conspicuous cardiovascular system.

Development of the Cardiovascular System

Early in the development of a fish embryo, a system of blood vessels is established that provides for the metabolic needs of the embryo. Higher vertebrates have inherited this pattern of development, but it has been modified somewhat to fit their particular requirements. Since the early basic pattern is repeated to a large extent in all vertebrates during the ontogeny of their diverse adult patterns, it forms a necessary basis for understanding the adult cardiovascular system.

The first vessels to take definite form in the embryo are a pair of **vitelline veins** (Fig. 10–1) that lie beneath the embryonic gut and carry blood and food forward from the yolk-laden archenteron, or yolk sac, if such a sac is present. The anterior parts of these vessels fuse to form the **heart** and **ventral aorta**. Caudal to the heart they are engulfed and broken up into a capillary network by the enlarging liver. The portion of the vitelline veins lying caudal to the liver becomes the **hepatic portal system**; the portion from the liver to the heart, the **hepatic veins**. (**Portal veins** are simply veins that, after draining one capillary bed, pass to another capillary bed in a different organ. Veins going directly to the heart are referred to as **systemic veins**.)

Meanwhile, a series of six aortic arches develop in the first six branchial bars, which lie between the gill pouches, to carry blood from the heart and ventral aorta up to the **dorsal aortae**. The dorsal aortae, which are first paired but later fuse, carry the blood caudally and, by **vitelline arteries**, back to the archenteron and yolk sac. This completes one circuit.

It will be noted that the early circulation is largely a visceral circulation to and from

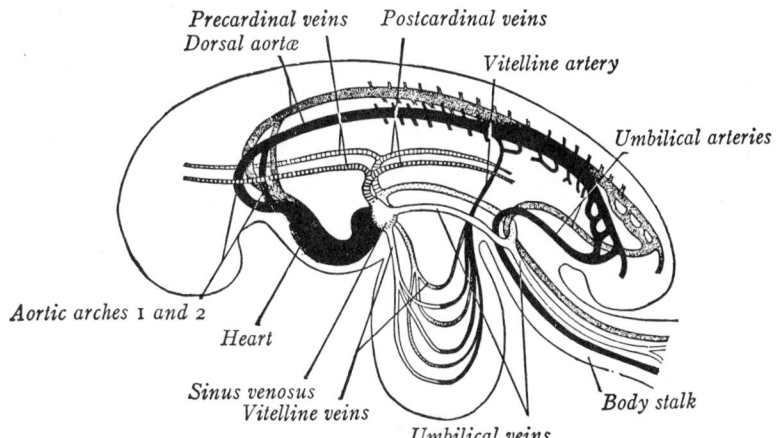

Figure 10–1. A diagrammatic lateral view of the major blood vessels of an early amniote embryo. Four more caudal aortic arches, and other vessels, would be present in later stages. The arteries have been drawn darker than the veins. Arteries are defined as vessels flowing away from the heart; veins as vessels flowing toward the heart. (From Arey, Developmental Anatomy. After Felix.)

the "inner tube" of the body, but a somatic circulation to the "outer tube" soon appears. Other branches from the dorsal aorta carry blood into the body wall and out to an allantois (by **umbilical** or **allantoic arteries**) if one is present. Blood from the dorsal portions of the body returns by way of **anterior** and **posterior cardinals**. The cardinals of each side unite anterior to the liver and turn ventrally to the heart as **common cardinals**, or ducts of Cuvier. It will be recalled that the common cardinals pass through the transverse septum which they helped to form. Blood from the more lateral and ventral portions of the body wall, and from the allantois, returns by a pair of more ventrally situated vessels. These are called **lateral abdominal veins** in lower vertebrates and **umbilical** or **allantoic veins** in amniotes. These vessels enter the base of the common cardinals in early embryos, but in the later embryos of the higher vertebrates they acquire a connection with the hepatic portal system and are drained through the liver.

As development proceeds and sites of nutrition, gas exchange, and excretion change, the pattern of vessels changes; new channels appear and some old ones atrophy. Thus in the development of an embryo, especially embryos of the higher vertebrates which have had a long and complex phylogenetic history, we see a succession of vessels. Much of the variation seen in the vessels of the adult can be attributed to the persistence of embryonic channels that normally atrophy and to the failure of certain later channels to develop. Other variation results from the enlargement of one channel over another in a primordial capillary plexus that exists in many parts of the embryo (Fig. 10–2). The relative rate of blood flow, as well as hereditary and other factors, is an important factor in determining which channels will enlarge.

The Study of Blood Vessels

While dissecting the vessels, the student should keep in mind the fact that they are subject to considerable variation and may not be just as described. Since the veins of the higher vertebrates have had a more complex ontogeny than the arteries and the

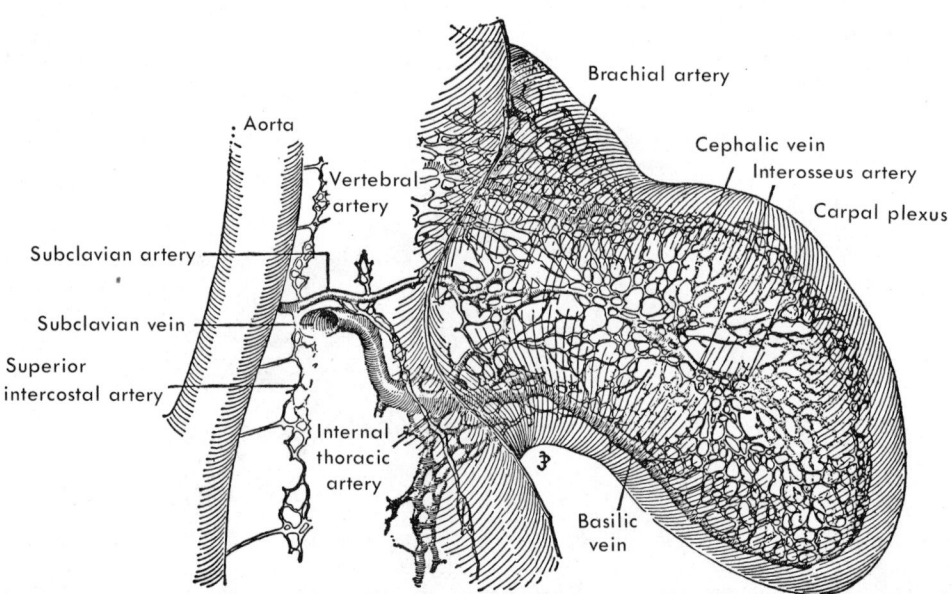

Figure 10–2. An early stage in the development of the pectoral limb bud of a pig embryo, to show the early plexus of vessels. The definitive vessels arise by the enlargement of certain channels. (From Woolard.)

venous blood has a more sluggish flow, it is to be expected that more variations will be found in the venous system of these animals than in the arterial system. Odd as it may at first seem, the peripheral parts of the vessels are subject to less variation than are certain of the more central and larger channels. For example, the left ovarian vein of a mammal always drains the ovary, but it may enter either the left renal vein or the caudal vena cava. Hence, if a vessel cannot be identified from its point of connection with a major vessel, it can be identified if its peripheral distribution can be established.

Another fact to bear in mind is that the arteries and veins tend to follow each other, especially in the higher vertebrates. For this reason, the arteries and veins of a given part of the body will often be described together. Valves frequently prevent the injection mass in veins from reaching the peripheral parts of the vessel, but the vein can usually be seen as a translucent, fluid-filled tube beside the corresponding artery.

The blood vessels of vertebrates are exceedingly numerous, and not all can be studied in a course of this scope. Emphasis has been placed on the major channels in the axis of the body and the vessels connecting with them. The blood vessels of the head and the appendages and the more distal vessels in the intestinal region are treated superficially.

FISHES

The cardiovascular system of adult fishes is in a very primitive stage, for its development does not proceed far from the early embryonic pattern described above. Fishes in general have a low pressure and sluggish circulation, for there are several capillary beds interposed along the course of the vessels and there is a pressure loss through friction in each. Thus the pressure built up by the heart is reduced almost immediately when the blood flows through the gill capillaries interposed in the aortic arches; again in the capillaries in the tissues; and, for certain circuits, still again in the liver (hepatic portal system) or kidney (renal portal system). Measurements reported by Satchell (1971) show that blood pressure in the ventral aorta of the dogfish is 25 to 39 cm. H_2O. This drops about 23 per cent to 20 to 29 cm. H_2O after blood has gone through the gills. Venous pressures are very low: 2.4 to 2.6 cm. H_2O in the caudal vein, and -0.6 to 0.2 cm. H_2O in the posterior cardinal sinuses. Venous return to the heart is facilitated by many large venous sinuses associated with the veins, for these offer very little resistance to blood flow.

The blood vessels of *Squalus* are a good example of those of most fishes, but it must be remembered that fishes ancestral to tetrapods would have had lungs as well as gills and, hence, a pulmonary circulation. The cardiovascular system should be studied on specimens in which at least the arteries and hepatic portal system have been injected. Triply injected specimens are necessary if all the veins are to be studied.

External Structure of the Heart

The pericardial cavity of the dogfish has been opened and the heart observed. Return to the cavity and examine the external features of the heart (Fig. 10–3). The heart of fishes is essentially an S-shaped tube that receives venous blood at its caudodórsal end, increases the blood pressure, and sends the blood out to the gills and body at its cranioventral end. Four

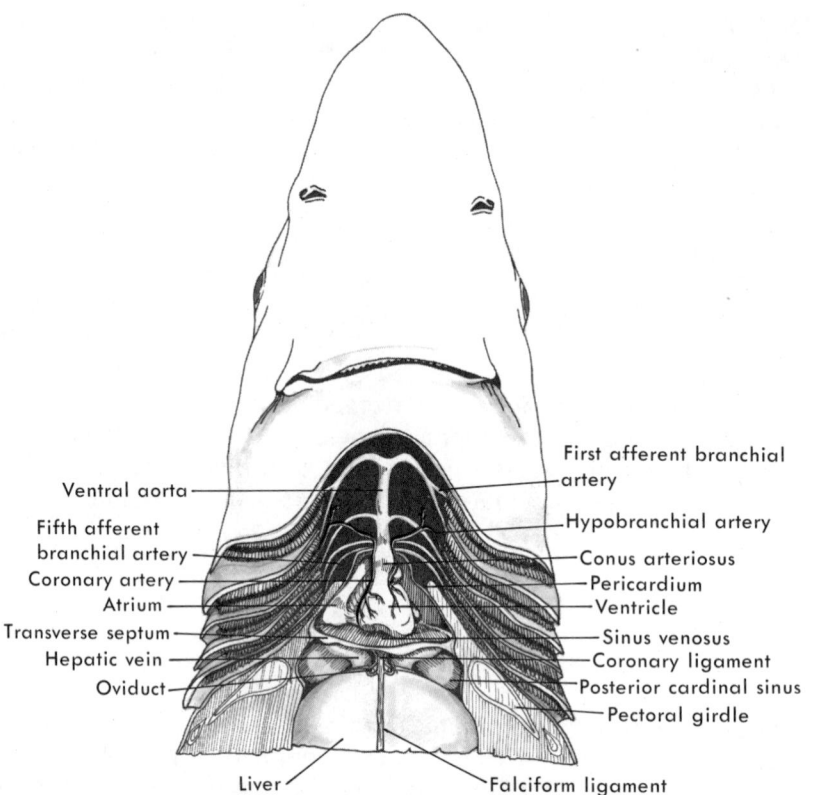

Ventral aorta

Fifth afferent branchial artery

Coronary artery

Atrium

Transverse septum

Hepatic vein

Oviduct

First afferent branchial artery

Hypobranchial artery

Conus arteriosus

Pericardium

Ventricle

Sinus venosus

Coronary ligament

Posterior cardinal sinus

Pectoral girdle

Liver

Falciform ligament

Figure 10–3. Ventral dissection of the heart and afferent branchial arteries of *Squalus*.

chambers have differentiated in linear sequence along this tube. These are, from caudal to cranial, the **sinus venosus, atrium,**[14] **ventricle,** and **conus arteriosus.** The last two lie along the ventral half of the S, and so are the first to be seen. The ventricle is the thick-walled, muscular, oval-shaped structure lying in the caudoventral portion of the pericardial cavity; the conus arteriosus, the tube-like chamber extending from the front of the ventricle to the cranial end of the cavity. Lift up the caudal end of the ventricle (the **apex** of the heart). The thin-walled, bilobed chamber dorsal to it is the atrium. The sinus venosus is the triangular-shaped chamber lying caudal to the atrium and extending between its two lobes. The caudal surface of the sinus venosus adheres to the transverse septum and receives through it the various veins draining the body.

---

[14]The terms atrium and auricle are variably used; sometimes they are used synonymously, and sometimes a subtle difference is made between them. The *Nomina Anatomica* convention will be followed: atrium is used for the entire chamber (undivided in fish, divided in mammals); auricle is an ear-shaped part of the atrium of mammals.

Venous System

(A) Hepatic Portal System and Hepatic Veins

The principal vein of the hepatic portal system is the **hepatic portal vein** going to the liver (Fig. 9–5, p. 244). It can be found lying beside the bile duct in the lesser omentum. The hepatic portal vein receives small **chole-** **dochal veins** from the bile duct but is formed by the confluence, near the craniodorsal tip of the dorsal lobe of the pancreas, of three large tributaries. A **gastric vein** comes in from the central part of the stomach; a **lienomesenteric vein,** from along the line of attachment of the mesentery to the dorsal lobe of the pancreas; and a **pancreaticomesenteric vein,** from the deep side of the ventral lobe of the pancreas. The gastric and lienomesenteric veins often come together and form a short common trunk before uniting with the pancreaticomesenteric, which is the conspicuous vein seen passing just dorsal to the pylorus. These tributaries parallel corresponding arteries and drain the abdominal viscera. Their peripheral distribution will be seen when the arteries are studied.

The hepatic portal vein can be traced into the substance of the liver where it breaks up into many branches that lead to the capillary-like sinusoids. The sinusoids, in turn, are drained by a system of hepatic veins that lead ultimately to a pair of large **hepatic veins,** or **sinuses** (Fig. 10–4). The hepatic sinuses are systemic veins. Although not injected, one of these large sinuses can be found by cutting the transverse septum down to the coronary ligament on the side of the body previously cut open, and dissecting away liver tissue just caudal to the coronary ligament. Do not injure the falciform ligament or oviduct in this region. You will soon enter the large hepatic sinus. Pass a probe forward through the vessel and it will be seen to go through the coronary ligament and transverse septum into the sinus venosus. The same is true for the hepatic sinus on the other side, but it shoud not be dissected. If desired, the hepatic sinus can also be traced caudally well into the liver.

As the portal blood passes through the liver it comes into intimate contact with the hepatic cells, for the sinusoids are not completely lined with endothelium. Excess food products in the blood coming from the digestive tract after a meal are stored in the hepatic cells largely in the form of glycogen and (in the dogfish) oil, and deficiencies in the food content of the blood between meals is made up from food stored in the cells. Numerous other metabolic conversions also occur here.

(B) Renal Portal System

Cut a cross section through the tail just caudal to the cloacal aperture. The **caudal artery** and **vein** will be seen within the hemal arch, the artery lying dorsal to the vein and being injected (Fig. 10–4). The vein will not be injected, unless it happened inadvertently during the injection of the artery. Make a series of partial cross sections through the base of the tail; begin

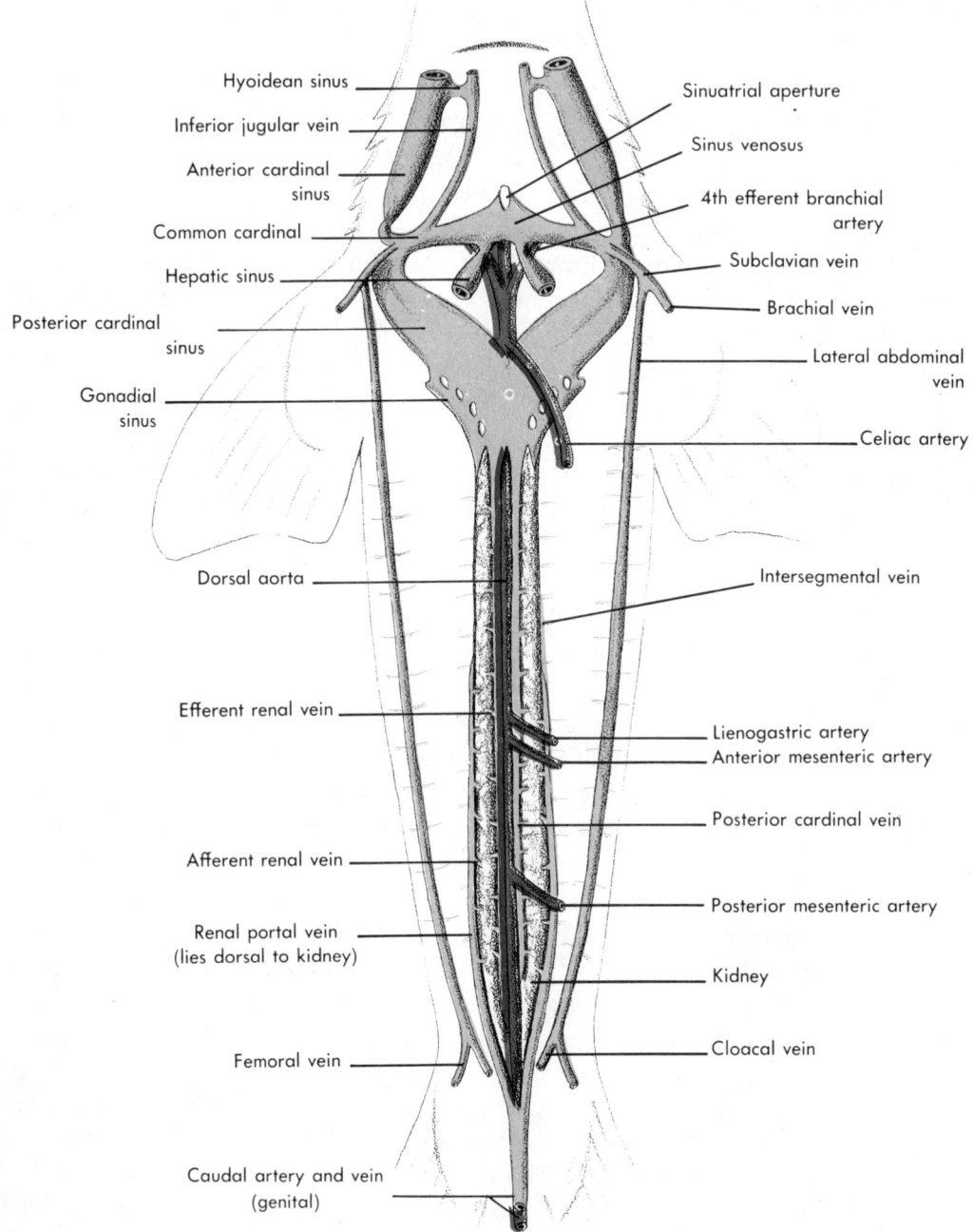

Hyoidean sinus

Inferior jugular vein

Anterior cardinal sinus

Common cardinal

Hepatic sinus

Posterior cardinal sinus

Gonadial sinus

Dorsal aorta

Efferent renal vein

Afferent renal vein

Renal portal vein (lies dorsal to kidney)

Femoral vein

Caudal artery and vein (genital)

Sinuatrial aperture

Sinus venosus

4th efferent branchial artery

Subclavian vein

Brachial vein

Lateral abdominal vein

Celiac artery

Intersegmental vein

Lienogastric artery
Anterior mesenteric artery

Posterior cardinal vein

Posterior mesenteric artery

Kidney

Cloacal vein

Figure 10–4. Semidiagrammatic ventral view of the renal portal and systemic veins of *Squalus*. Ventral visceral branches of the dorsal aorta are also shown.

your cut dorsally and extend it ventrally into, but not through, the kidneys. Space the sections about one centimeter apart and continue to make them until you find the caudal vein bifurcating into the two **renal portal veins.** The renal portals extend cranially, lying dorsal to the medial border of each kidney, but it is impractical to trace them far. They carry blood to the capillaries associated with the kidney tubules by inconspicuous **afferent renal**

veins. These capillaries are drained by **efferent renal veins** that enter the posterior cardinals (see below).

A renal portal system is not present in primitive agnathous fishes such as the lamprey, and in such forms the caudal vein leads directly into the two posterior cardinals. Such a condition is also present in the embryonic dogfish (Fig. 10–5, *A*). During subsequent development, a pair of new veins (called **subcardinals** in the embryo) appear ventral to the kidneys. These veins tap into the front of the posterior cardinals. Meanwhile a portion of the posterior cardinals caudal to this union atrophies (dotted line in Fig. 10–5, *B*). The caudal portions of the embryonic posterior cardinals are now renal portal veins, for all their blood necessarily goes to the kidneys. The adult posterior cardinals are composed of the embryonic subcardinals plus the cranial end of the embryonic posterior cardinals.

The hemal arch, caudal vein, and renal portal system are of considerable importance to the fish. The hemal arch makes possible a "tail pump," for it protects the caudal artery and vein from the waves of muscular contraction that sweep down the tail as the fish swims. Blood flows from the caudal artery through intersegmental arteries during periods of muscular relaxation, and is prevented from backing up when the tail muscles contract by valves present in the intersegmental arteries (Fig. 10–6).

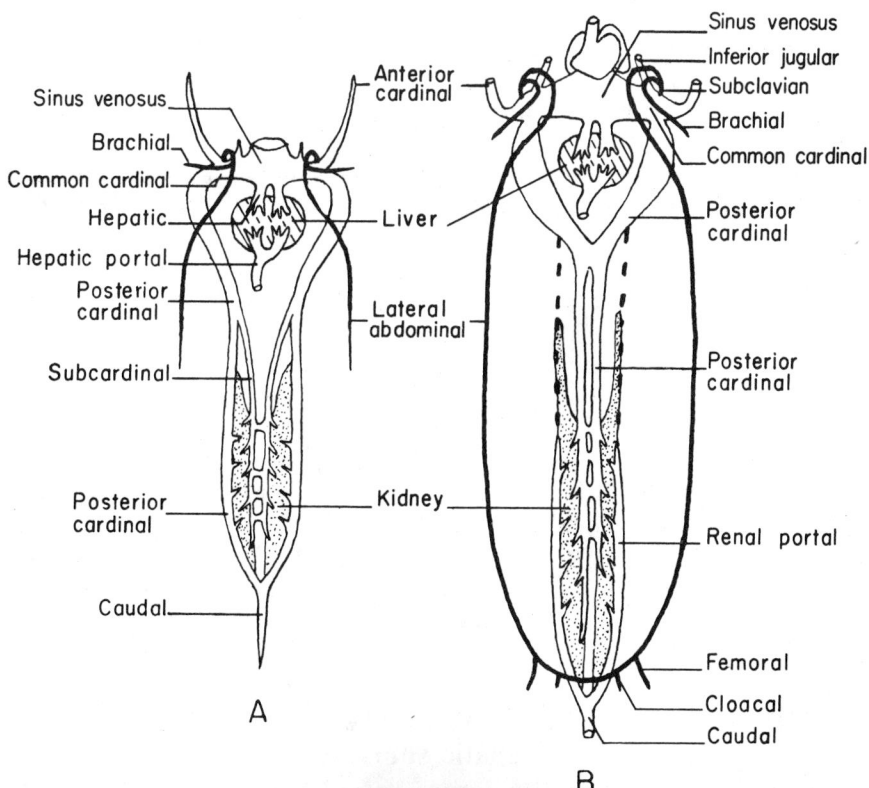

Figure 10–5. Diagrams from a ventral view of the development of the major veins of a dogfish: *A,* late embryo; *B,* adult. These diagrams show in particular the development of the adult renal portal system from the caudal parts of the embryonic posterior cardinals, and the development of the caudal parts of the adult posterior cardinals from the embryonic subcardinals. The broken lines in *B* indicate the portions of the embryonic posterior cardinals that disappear in the adult. (*A* modified after Hochstetter; *B* modified after Daniel.)

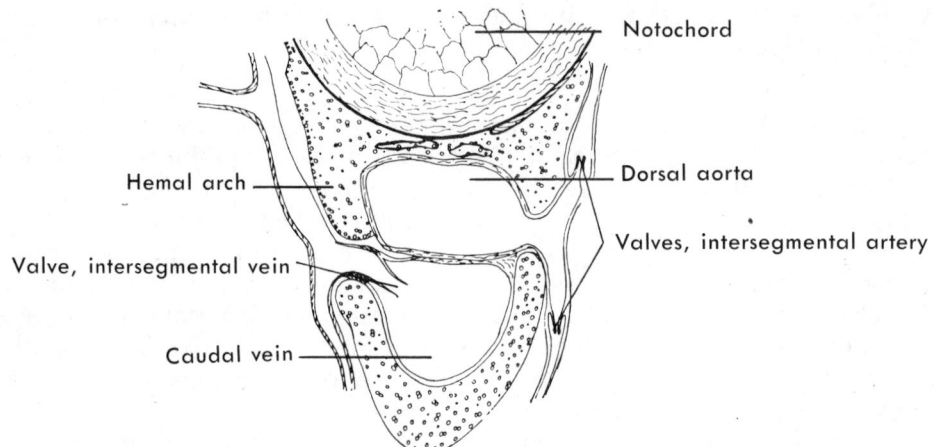

Figure 10–6. Cross section through the hemal arch and caudal blood vessels of a fish (*Clupea*). Note the valves in the intersegmental artery and vein that make possible the "tail pump." (From Satchell, Circulation in Fishes, Cambridge University Press.)

Tail muscles squeeze the peripheral veins and squirt blood from them into the caudal vein. Pressure builds up in the caudal vein because it is protected by the hemal arches. Thus, as Satchell (1971) expresses it, "Some of the power from the sigmoid waves of muscular contraction that serve to move the surrounding water backwards is deployed to move the blood in the caudal vein forwards."

This blood goes to the capillary bed surrounding the kidney tubules. This capillary bed also receives arterial blood that has already passed through the glomerular capillaries located at the beginning of the tubules. The significance of this dual blood supply to the tubules, which is characteristic of most fishes, amphibians, and reptiles, is not entirely clear. In many marine fishes, which must conserve body water, it makes possible a small blood flow through the glomeruli, where water tends to be lost from the blood, while still supplying adequate blood to the rest of the tubule for excretory augmentation and reabsorption of valuable products from the forming urine.

(C) Systemic Veins

For purposes of description, the systemic veins of fishes may be arranged in five groups: (1) the hepatic veins already seen, (2) the cardinal system draining most of the trunk and head, (3) the inferior jugulars draining the ventral portion of the head, (4) the lateral abdominal system draining the appendages and lateroventral portion of the body wall, and (5) the coronary veins draining the heart wall. Reference should be made to Figure 10–4 during the dissection of these veins, for many are hard to find unless you have a triply injected specimen. The following directions are based on specimens in which these veins have not been injected.

Open the sinus venosus by a transverse incision and wash it out. The small entrances of the pair of hepatic sinuses can now be seen in the central part of the sinus venosus. The large, round openings at the caudolateral angles of the sinus are the entrances of the paired **common cardinal veins,** or ducts of Cuvier. Pass a probe into the common cardinal on the side of the body that has been opened and cut the vessel open along the course of the probe. The common cardinal passes along the lateroventral wall of the esophagus within the transverse septum and becomes continuous with the large **posterior cardinal sinus**—the large, membrane-covered space that lies against the dorsolateral surface of the esophagus and curves toward the mid-dorsal line of the pleuroperitoneal cavity. Cut open the posterior cardinal sinus and follow it caudally. The sinuses of opposite sides are interconnected dorsal to the gonads and, in this region, receive genital vessels (**ovarian** or **testicular veins**) from **gonadial sinuses** beside the gonads, and also veins from the esophagus. It is difficult to see these vessels clearly. More caudally, each sinus narrows to form a **posterior cardinal vein** which continues caudally along the dorsomedial border of each kidney. The posterior cardinal sinus is actually just the expanded cranial part of this vein. Each posterior cardinal receives numerous **efferent renal veins** from the kidneys and **intersegmental** or **parietal veins** from between the myomeres. Renal and intersegmental veins are hard to see unless they happen to be filled with blood.

One of the paired **anterior cardinal sinuses** was seen on the large head of *Squalus* during the dissection of the nervous system (p. 202). Expose one on your present specimen by making a deep longitudinal cut extending ventrally and medially from that portion of the lateral line overlying the branchial region. Open it on the side on which you have been working. Pass a probe caudally through the vessel and it will be seen to turn ventrally and unite with the posterior cardinal sinus. The union of these two vessels marks the beginning of the common cardinal. The anterior cardinals drain the brain and all the head except the floor of the branchial region. If desired, the anterior cardinal can be traced forward by probing and cutting to a large **orbital sinus** around the eye.

The floor of the branchial region is drained by a pair of **inferior jugular veins.** The entrance of one can be seen in the cranial wall of the common cardinal just before the common cardinal enters the sinus venosus. The vein extends forward dorsal to the pericardial cavity. The anterior cardinal and inferior jugular of each side are interconnected by a **hyoidean sinus.** The hyoidean sinus can be seen lying caudal to the hyoid arch on the side of the body that was cut to open the pharynx. By probing and cutting, trace it dorsally to the anterior cardinal and ventrally to the inferior jugular. Trace the inferior jugular caudally to the common cardinal.

The appendages, and some of the lateroventral portion of the trunk, are drained by a pair of **lateral abdominal veins.** These veins are the pair of dark, longitudinal lines that one sees on the inside of the body wall beneath the parietal peritoneum. Examine the caudal end of one by probing and cutting open the vessel. After passing dorsal to the pelvic girdle, the vessel receives two tributaries. One (the **cloacal vein**) comes in from the lateral wall of the cloaca; the other (the **femoral vein**), from the pelvic fin. There is also an anastomosis between the cloacal veins of the opposite sides of the body. You may not be able to find these vessels in an uninjected specimen. Cranially, the lateral abdominal is joined by a <u>**brachial vein**</u> from the pectoral fin. The brachial can be found by cutting off the pectoral fin close to the body on the side on which you have been working. The vein lies ventral to the middle of the basal cartilages. Trace it to its union with the lateral abdominal by probing and cutting along the course of the probe. A **subscapular vein** may be seen entering the proximal end of the brachial. The vessel formed by the union of the brachial and lateral abdominal is known as the **subclavian vein.** Trace it in the same way and it will be seen to enter the front of the common cardinal beside the entrance of the inferior jugular. If it is necessary to find any of these veins on the opposite side, avoid injuring the falciform ligament.

    Coronary veins can be seen on the surface of the heart, especially the ventricle. They enter the sinus venosus by a common aperture which cannot be found at this stage of the dissection.

Arterial System

(A) BRANCHIAL ARTERIES

    The branchial arteries should be exposed on the intact side of the pharynx. If the interbranchial septa have not been mobilized, this should be done by cutting through the top and bottom of each of the five external gill slits and continuing the incision to, but not through, the internal gill slits.

    Lay the specimen on its back and note again the conus arteriosus. After this passes through the front of the pericardial cavity, the vessel is known as the **ventral aorta** (Fig. 10–3). Trace the ventral aorta cranially. It is not injected, so be careful. It passes between the hypobranchial musculature and basibranchial cartilages (Fig. 10–10, p. 294). Much of the musculature must be removed, but do not injure the major injected vessels in this region. They are derived from the efferent branchial arteries, and will be considered shortly.

    As the ventral aorta passes forward it gives off five pairs of **afferent branchial arteries** that extend up into the interbranchial septa (Figs. 9–11, 10–3, 10–7). The caudal two pairs of afferent branchial arteries (the **fourth** and **fifth**) leave the ventral aorta just anterior to the pericardial cavity, and from the dorsolateral side of the aorta. They come off very close together, and sometimes by a short, common trunk. The middle pair of afferent branchial arteries (the **third**) leaves slightly rostral to the caudal two pairs. The ventral aorta then passes rostrally for some distance without further branches. Slightly caudal to the level of the basihyal, it bifurcates. (The dark **thyroid gland** may be seen rostral to this bifurcation and ventral to the pre-hyoid muscles.) Trace one of the bifurcations and it will be seen to subdivide again. These subdivisions are the **first** and **second** afferent branchial arteries. Trace each of the five afferent branchial arteries far enough into the inter-branchial septa on the intact side of the specimen to ascertain which septa and gills they are supplying and to observe the numerous small branches that they send into the gill lamellae.

    Oxygenated blood is collected from the gill lamellae by a system of

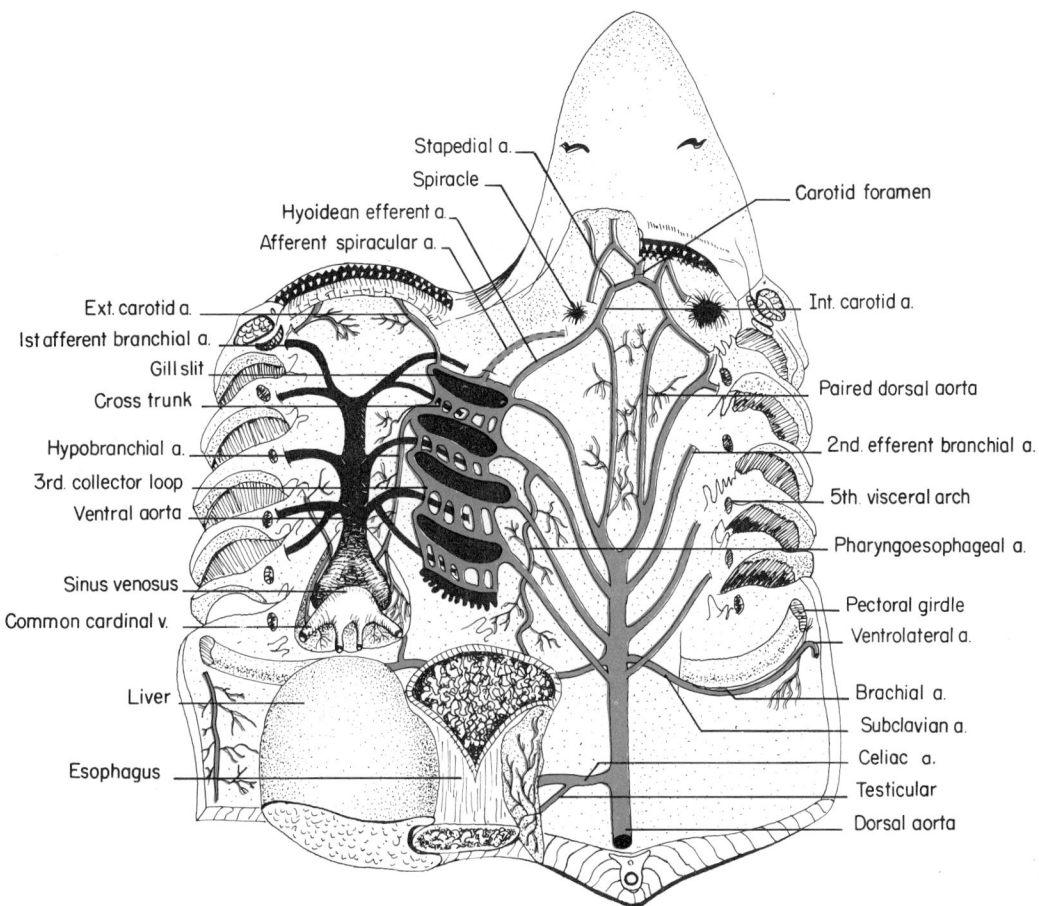

Figure 10–7. A dissection of the branchial arteries of *Squalus*. In this case the afferent branchial arteries, which have been drawn very dark, have been approached through the floor of the pharynx and are seen in a dorsal view. The efferent branchial arteries, which are shown in red, are seen in a ventral view.

efferent branchial arteries. The first portion of this system is a series of four and one-half **collector loops.** A representative collector loop can be seen by spreading open the first definitive external gill slit and looking into the first pouch. A circle of mucous membrane will be seen at the base of the gill lamellae lateral to the internal gill slit. Strip off this membrane, and a vascular circle (the first collector loop), receiving tiny vessels from the lamellae, will be seen beneath it. The vessel forming the rostral half of the loop is the **pretrematic branch** of an efferent branchial artery; it is noticeably smaller than the vessel forming the caudal half of the loop, the **posttrematic branch.** A second, third, and fourth loop can be found in the second to fourth definitive gill pouches, but only the pretrematic half of a loop is present in the last pouch.

Expose the pretrematic portion of the second collector loop. In addition to receiving tiny branches from the adjacent lamellae, it gives off many larger branches (**cross trunks**) that pass through the interbranchial septum to the posttrematic of the first loop. This is also the case for all the other pretrematics, except, of course, the first. The posttrematics are therefore functionally the most important parts of a collector loop, for each is the main vessel draining a particular holobranch. It receives not only the drainage from the lamellae adjacent to it on the cranial surface of a holobranch, but also, via the cross trunks, most of the drainage from the lamellae on the caudal surface of a holobranch.

Swing open the floor of the oral cavity and pharynx and remove the mucous membrane from the roof of these cavities. Four pairs of **efferent branchial arteries** will be seen on the caudal portion of the pharyngeal roof. They extend from the dorsal angles of the internal gill slits diagonally, medially, and caudally, and converge to form a median vessel, the **dorsal aorta.** Remove connective tissue from around the efferent branchials, and trace them on the intact side of the pharynx to the tops of the collector loops. The dorsal cartilages of some of the branchial arches must also be removed. The most rostral efferent branchial artery (the **first**) connects with the top of the first collector loop, and the last (the **fourth**) with the fourth collector loop. These vessels receive oxygenated blood from the collector loops and carry it to the more caudal parts of the body. Most goes back via the dorsal aorta, but a small **pharyngoesophageal artery** arises from the second efferent branchial and extends caudad in the roof of the pharynx and esophagus.

Arterial blood to the head travels in other vessels. A **hyoidean efferent artery** arises from the top of the first collector loop rostral to the first efferent branchial and passes forward in the roof of the pharynx. Opposite the spiracle it receives, on its medial side, a small vessel that extends forward from the medial end of the first efferent branchial. This artery and its mate of the opposite side represent some of the rostral portions of the embryonically paired dorsal aorta. They are simply called the **paired dorsal aortae.** The vessel rostral to the union of the hyoidean efferent with a paired dorsal aorta also develops embryonically from the front of the dorsal aorta. In the adult it is called the **internal carotid artery.** The internal carotid continues forward, curves toward the middorsal line, crosses or sometimes unites with its mate of the opposite side and enters the chondrocranium through the carotid foramen. Follow the internal carotids by chipping away cartilage. They soon diverge and, at the level of the hypophysis, unite with the arteries on the ventral surface of the brain. The internal carotids are the major arteries supplying the brain.

At the point at which the internal carotid curves toward the middorsal line it gives off a **stapedial** artery from its rostrolateral surface. Note the proximity of this artery to the point of union of the hyomandibular (the future stapes) with the otic region of the chondrocranium. Follow the stapedial forward. It passes dorsal to the efferent spiracular artery extending medially from the spiracle (see below) and is distributed to the orbit and snout.

Return to the pretrematic of the first collector loop. A vessel arises near the middle of this pretrematic that at first resembles a cross trunk. This is the

chip away cartilage

afferent spiracular artery. Trace it forward by cutting through the skin ventral and caudal to the spiracle (Fig. 6–4, p. 121). The afferent spiracular crosses the lateral surface of the hyomandibular cartilage and extends to the pseudobranch in the spiracle. It would of course normally carry arterial blood to this structure. An **efferent spiracular artery** continues from the pseudobranch medially to unite with the internal carotid within the cranial cavity. This portion of the vessel can be found by removing the mucous membrane lining the front of the spiracle. Approach the spiracle from its pharyngeal entrance.

An **external carotid artery** arises from the ventral end of the first collector loop and passes forward to supply the lower jaw region. It is best approached from the ventral surface of the specimen where it will be seen dorsal to the proximal end of the first afferent branchial.

Another vessel, the **hypobranchial artery,** usually arises from the ventral end of the second collector loop, but it may receive contributions from any of the other loops. The vessel supplies most of the hypobranchial musculature and then bifurcates at the front of the pericardial cavity. One branch, the **coronary artery,** is distributed to the heart. The other, the **pericardial artery,** extends caudad in the dorsal wall of the pericardial cavity.

important

Primitive fishes retain parts, at least, of all six embryonic aortic arches, but the arches are obviously modified in the adult for the interposition of the gills and the supply of blood to the head. The developmental history by which this modification comes about is complicated, but the essentials can be summarized. At an early stage in development, six complete aortic arches are present (Fig. 10–1). During subsequent development, the ventral portion of the first disappears, but the ventral portions of the remaining five form the afferent branchial arteries (Fig. 10–8, B). The dorsal part of the first arch, together with a new connection that early develops between the first and second arches (Fig. 10–8, A), forms the spiracular artery. The dorsal part of the second arch forms the hyoidean efferent; the dorsal parts of the remaining arches, the four efferent branchials. The external carotid, distal part of the internal carotid, stapedial, hypobranchial and pharyngoesophageal arteries represent new outgrowths from the dorsal aorta or various arches as shown in Figure 10–8 and Table 7.

The derivation of these vessels is reasonably straightforward, but the formation of the collector loops is more complicated. At an early stage in development the five caudal arches break near their dorsal end. The dorsal part of the second arch then grows ventrally in front of what is to be the first definitive gill slit, and the caudal four bifurcate and grow ventrally between the gill slits (Fig. 10–8, A). The caudal bifurcation of one arch unites above and below a gill slit with the cranial bifurcation of the next caudal arch to form a collector loop. This process is repeated for the other loops. Thus, each collector loop is formed from outgrowths of two successive aortic arches — the pretrematic portion of the loop from the caudal branch of the bifurcation of the cranial arch, and the posttrematic portion from the cranial branch of the bifurcation of the next caudal arch. At one stage of development each loop would be drained by the dorsal parts of both parent arches. This double drainage persists in the adult for the first collector loop (Fig. 10–8, B). But the second and succeeding loops lose their original connection (xxx in Figure 10–8, B) with their cranial parent and would be drained exclusively by the caudal parent if it were not for the development of new connections (cross trunks) with the cranial parent.

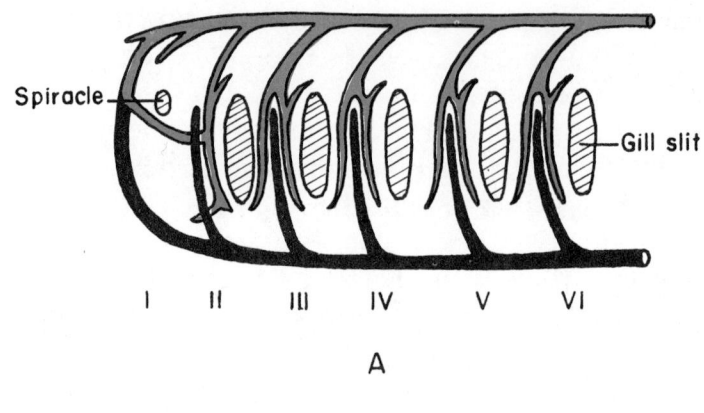

A

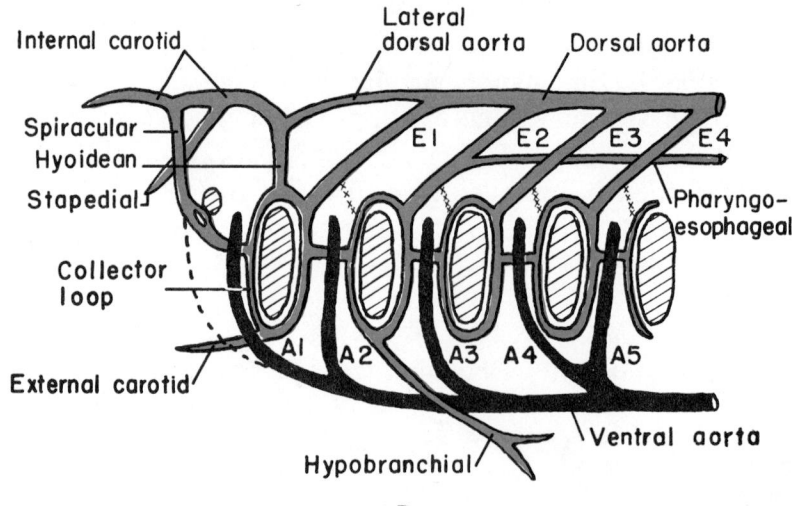

B

Figure 10–8. Diagrams of *A,* the embryonic, and *B,* the adult aortic arches and their derivatives in a dogfish as seen from a lateral view. The embryonic stage is one in which the modification of the earlier, uninterrupted aortic arches has begun. All the vessels would be paired except the caudal parts of the dorsal and ventral aorta. The afferent vessels are drawn in black; the efferent in red. Abbreviations: *A1* to *A5,* first to fifth afferent branchial arteries; *E1* to *E4,* first to fourth efferent branchial arteries; *I* to *VI,* first to sixth aortic arches.

(B) Dorsal Aorta, Its Branches and Accompanying Veins

The basic pattern of the branches of the dorsal aorta is shown in Figure 10–9. As can be seen, there are three major categories of vessels: (1) Paired **intersegmental arteries**[15] pass between each of the body segments, and each soon bifurcates into a **dorsal ramus** to the epaxial region and a **ventral ramus** to the hypaxial region. Longitudinal anastomoses may occur at various points between successive intersegmentals. Vessels to the appendages are simply enlarged ventral rami of intersegmental arteries. (2) Paired **lateral visceral arteries** pass at intervals to such dorsolateral organs as the kidneys, gonads and suprarenal glands. (3) Unpaired **ventral visceral arteries,** which develop from the embryonic vitelline arteries, pass through the dorsal mesentery to the viscera. This basic pattern of the branches of the aorta occurs in all vertebrates.

---

[15]These vessels are often called segmental arteries, but the term intersegmental is preferable, for the vessels occupy an intersegmental position.

Table 7. Derivation of the Branchial Vessels of the Dogfish

The relation of the adult branchial vessels of a dogfish to the embryonic aortic arches is shown in the following table. The derivation of each collector loop from outgrowths of two aortic arches is explained in the text, and not shown in this table.

AFFERENT VESSELS	EFFERENT VESSELS	EMBRYONIC ORIGIN
	Efferent spiracular	Aortic arch 1
	Afferent spiracular	Cross connection between aortic arches 1 and 2
Afferent branchial 1	Hyoidean efferent	Aortic arch 2
Afferent branchial 2	Efferent branchial 1	Aortic arch 3
Afferent branchial 3	Efferent branchial 2	Aortic arch 4
Afferent branchial 4	Efferent branchial 3	Aortic arch 5
Afferent branchial 5	Efferent branchial 4	Aortic arch 6
	Internal carotid	Rostral dorsal aorta and forward outgrowth
	Stapedial	Rostroventral outgrowth from dorsal aorta
	External carotid	Rostral outgrowth from the ventral extension of part of aortic arch 2
	Hypobranchial	Ventral outgrowth from ventral part of second collector loop (aortic arches 3 and 4)
	Pharyngoesophageal	Caudal outgrowth from dorsal part of aortic arch 4

Working insofar as possible from the side of the body that has been opened, trace the dorsal aorta caudad. The cranial portion of the esophagus will have to be separated from the body wall and some mesenteries will have to be torn. A pair of **subclavian arteries** arise from the aorta, generally between the third and fourth efferent branchials (Fig. 10–7). Follow one as it curves ventrally against the lateral wall of the posterior cardinal sinus. It follows the caudal margin of the scapula and gives off several branches to the adjacent musculature. Its major branches are a **brachial artery,** which enters the pectoral fin in company with the **brachial vein,** and an **anterior ventrolateral artery,** which continues cranially and ventrally along the caudal margin of the scapula and coracoid. The ventrolateral then curves caudally and extends caudad in the ventrolateral portion of the body wall. This part of the vessel can be seen beneath the parietal peritoneum between the **lateral abdominal vein** and the midventral line.

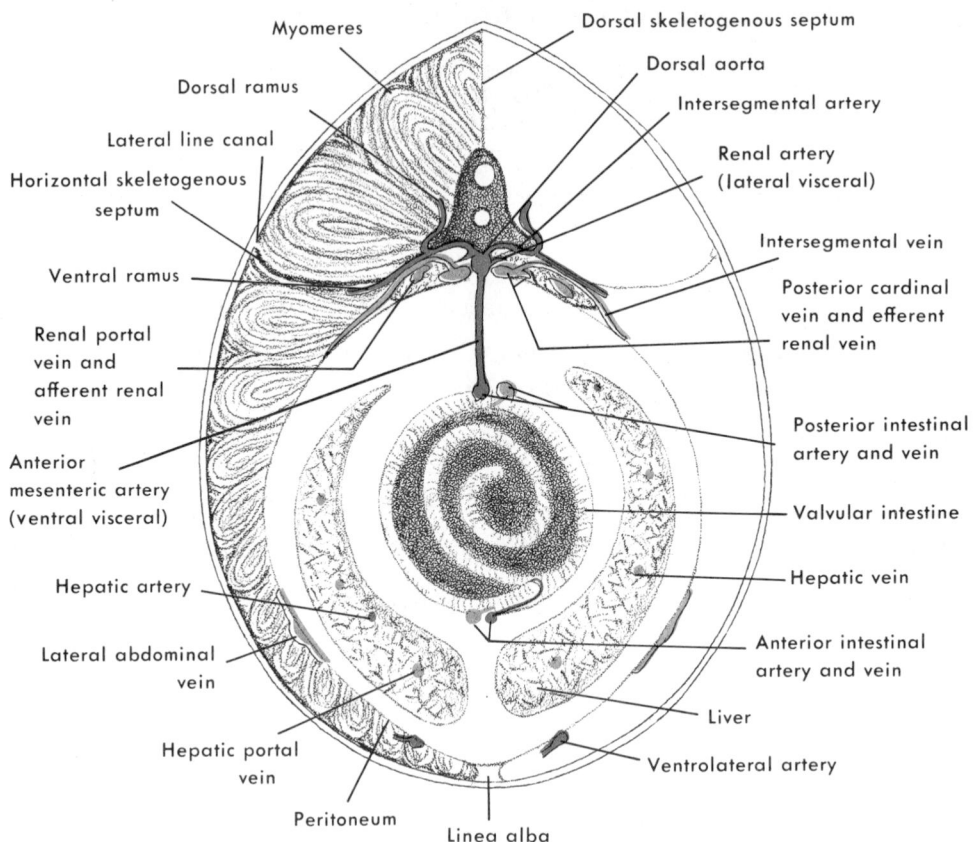

Figure 10–9. Semidiagrammatic cross section through the trunk of *Squalus* to show major arteries and veins.

The subclavians are modified intersegmental vessels. Other inter-segmentals will be seen presently, but the ventral visceral arteries should be examined next. These are accompanied by tributaries of the hepatic portal vein (p. 244). The first of these, the large **celiac artery** (Fig. 10–4), enters the front of the pleuroperitoneal cavity and extends ventrally and caudally along the right side of the stomach to the cranioventral tip of the dorsal lobe of the pancreas (Fig. 9–5). Here it bifurcates. One branch, the **pancreaticomesen-teric artery,** follows the **pancreaticomesenteric vein** dorsal to the pylorus and onto the intestine as the **anterior intestinal artery.** An **anterior intestinal vein** lies beside the artery. In addition, the pancreaticomesenteric artery sends smaller branches to the pyloric region of the stomach and the ventral lobe of the pancreas and into the spiral valve. The other branch of the celiac, the **gastrohepatic artery,** soon divides into a small **hepatic artery** that follows the **hepatic portal vein** to the liver, and a **gastric artery** that passes to the stomach following closely the **gastric vein** and its tributaries.

The next ventral visceral branches of the aorta will be found in the free caudal edge of the dorsal mesentery. Two vessels arise close together in this region and pass through the mesentery. The one going to the spleen and caudal part of the stomach is the **lienogastric artery;** the one passing to the caudal part of the intestine, the **anterior mesenteric artery** (Fig. 10–4). A final ventral branch is a small **posterior mesenteric artery** to the digitiform gland and caudal end of the intestine. These last three branches together supply the area drained by the **lienomesenteric vein** (Fig. 9–5), whose tributaries are the **lienogastric vein** and the **posterior intestinal vein.**

More caudally, a pair of **iliac arteries** arises from the aorta. Each passes ventrally in the body wall lateral to the cloaca and divides into a **femoral artery** entering the pelvic fin with the **femoral vein,** and a **posterior ventro-lateral artery.** The latter unites with the anterior ventrolateral artery which was seen extending caudad from the subclavian. The iliacs, like the sub-clavians, are modified intersegmentals. Typical **intersegmental arteries** can be seen by freeing and lifting up the lateral border of a kidney. Their most conspicuous branches pass from the aorta laterally and ventrally between the myomeres accompanied by **intersegmental veins** that drain into the posterior cardinals. Other branches extend into the epaxial region of the body (Fig. 10–9). Lateral visceral arteries include a number of small **renal arteries** and a pair of **gonadial arteries** (**ovarian** or **testicular**). The former arise from the aorta close to, or in common with, the intersegmentals and pass to the kid-neys, the latter from the very base of the celiac. Accompanying veins are tributaries of the posterior cardinals. Caudal to the iliacs, the dorsal aorta enters the tail as the **caudal artery,** which, as you have seen, lies dorsal to the **caudal vein.**

Internal Structure of the Heart

The general structure of the heart and its coronary vessels has been seen in preceding dissections. Remove the heart by cutting the attachments of the **sinus venosus** to the transverse septum and cutting across the posterior end of the ventral aorta, last afferent branchials and coronary arteries. Wash out the sinus venosus and look inside it. This thin-walled chamber receives venous blood from the body by way of the paired hepatic veins and common cardinals already observed. Venous blood then enters the **atrium** by way of the slitlike **sinuatrial aperture** at the front of the sinus. This opening is guarded by a pair of lateral folds, the **sinuatrial valve,** which prevents the backflow of blood. One or more openings of the coronary veins may be seen in the wall of the sinus near the sinuatrial aperture.

Cut forward through the dorsal end of the sinuatrial aperture and the dorsal wall of the atrium and clean out this chamber. It, too, is thin-walled, but muscular strands can be seen on the inside of its walls. Note that, despite its two lobes, its cavity is undivided. Find the **atrioventricular aperture** in the floor of the atrium. The opening is guarded by a pair of folds, the **atrioventricular valve.**

Insert a pair of scissors into the stump of the ventral aorta, cut back through the ventral surface of the **conus arteriosus,** and continue to the caudal end of the **ventricle.** The thicker, muscular walls of these chambers, especially of the ventricle, will be noted. The ventricular wall is also very spongy, for it is criss-crossed by numerous muscular strands. Notice that the lumen of the ventricle is U-shaped, for the entrance from the atrium and the exit to the conus lie nearly side by side. Backflow of blood from the front of the heart is prevented by the presence of three rows of **semilunar valves** within the conus. Each row consists of three pocket-shaped valves. One row lies at the very cranial end of the conus, and the other two lie close together near the caudal end of this chamber. The entrances to the pockets face cranially, so that they can be found and opened by probing caudally.

The dogfish heart lies in a pericardial cavity that is solidly encased by the cartilaginous pectoral girdle, the large basibranchial cartilage and the hypobranchial musculature (Figs. 9–9 and 10–10). Its location within this relatively rigid box permits it to act as an aspiratory pump. When the ventricle contracts and blood is pumped out of the heart, pressure is lowered in the pericardial cavity, for the walls of the cavity do not collapse. This, in turn, lowers the pressure within the heart, and blood is sucked, so to speak, out of the veins and into the sinus venosus. This mechanism is of considerable importance in the return of blood from veins and sinuses in which the pressure is exceedingly low.

Ventricular contraction drives blood out into the conus arteriosus under a relatively high pressure. The conus expands as it receives this surge of blood and then contracts during ventricular relaxation. In this way it buffers the delicate gill capillaries against the full pressure of ventricular contraction and tends to even out the flow of blood. However, blood flow is not perfectly even. Heart beat is synchronized with the respiratory movements of the pharynx in such a way that the largest volume of blood flows through the gill lamellae at the time when water is discharged across them. This, of course, increases the efficiency of gas exchange.

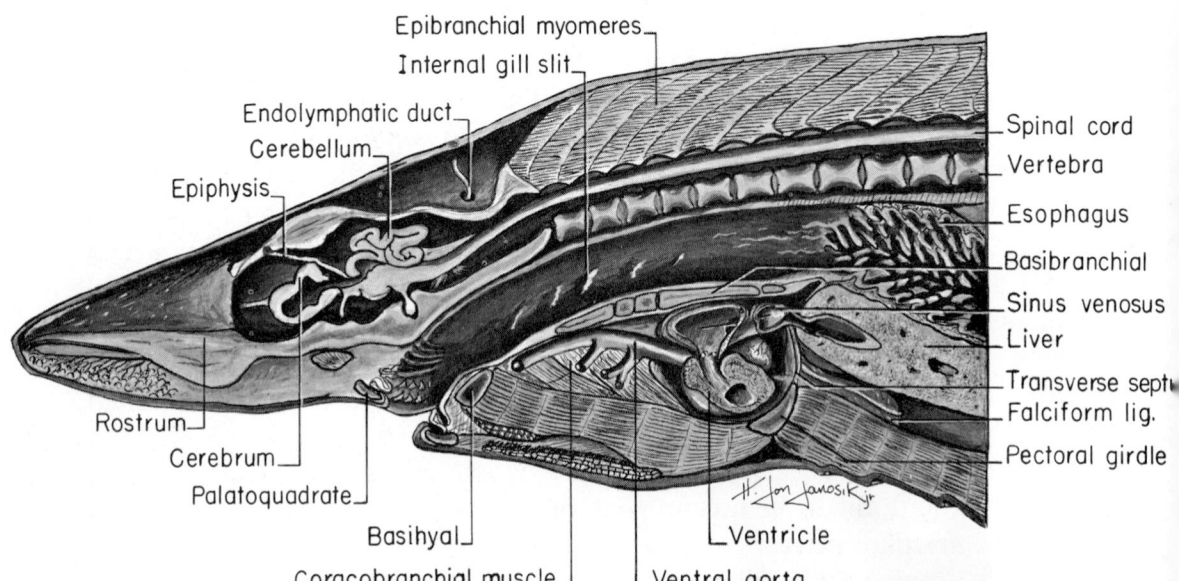

Figure 10–10. Sagittal section through the head of *Squalus*, showing the structure of the heart and its relation to surrounding structures. A bristle passes through the sinuatrial aperture.

Pericardioperitoneal Canal

Now that the heart has been removed, it is possible to find the communication between the pericardial and the pleuroperitoneal cavities that was referred to on page 249. Pass a blunt probe into the recess in the transverse septum dorsal to the line of attachment of the sinus venosus. The probe will be seen to pass into a canal beneath the visceral peritoneum on the ventral surface of the esophagus, and after passing a distance of about three centimeters it will emerge into the pleuroperitoneal cavity through a semilunar opening. This entire passage is the **pericardioperitoneal canal.** Frequently the canal bifurcates at its caudal end, in which event there would be a pair of openings into the pleuroperitoneal cavity. The canal permits liquid to escape from the pericardial cavity into the pleuroperitoneal cavity. This is important for the maintenance of the low pressure within the pericardial cavity, which is necessary for the aspiratory action of the heart. Fluid does not move in the other direction in the canal because its delicate walls collapse and act as a valve.

PRIMITIVE TETRAPODS

Important changes occur in the cardiovascular system during the transition from water to land. The most conspicuous of these are in the heart and aortic arches, for the branchial circulation is lost and a **pulmonary circulation** evolves. In this connection, the heart tends to become divided into a right and left side — the former receiving depleted blood from the body and sending it out to the lungs; the latter receiving aerated blood from the lungs and sending it out to the body. The separation is nearly complete in certain living lungfishes, but only the atrium is divided into right and left chambers in most amphibians. Reptiles also have a partial division of the ventricle. DeLong (1962) and others have shown that often there is a high degree of functional separation of the two blood streams in the heart of lower tetrapods despite an incomplete morphological separation. In frogs, the cranial aortic arches going to the head and body contain a greater percentage of left atrial blood that has returned from the lungs than does the most caudal arch. The caudal arch, going to the lungs and skin, contains a greater percentage of right atrial blood which has returned from the body, and also from the skin, where some gas exchange takes place. The significance of the incomplete morphological separation of the systemic and pulmonary circulations in amphibians and reptiles is not completely understood. It may permit the animal to shift as conditions warrant toward the single flow of blood through the heart that characterizes most fishes, or toward the double circuit through the heart that characterizes birds and mammals. A frog under water, for example, would have no need for the pulmonary circuit, because gas exchange would be through the skin. Or the incomplete separation of systemic and pulmonary circuits may help to adjust blood pressures and volumes in the two circuits. The pulmonary circuit may not be capable of handling the amount of blood that returns to the heart from the body, and some blood may by-pass the lungs by being shunted directly to the body.

As regards the aortic arches, at least the first two are lost in primitive tetrapods, but (excepting the arch to the lung) those that remain are complete and are not interrupted by gill capillaries. An important corollary of the changes in the heart and aortic arches is a relative increase in the blood pressure in the dorsal aorta and, hence, in the general efficiency of circulation. A frog, for example, has a systolic pressure in the dorsal aorta of 30 mm. Hg compared to a systolic pressure in the dorsal aorta of a dogfish of 17 mm. Hg. The frog, moreover, is a much smaller animal, so that the relative efficiency of its circulation is even greater than these figures imply.

The venous system remains somewhat fishlike in early tetrapods, but important changes are seen in the primitive lateral abdominal system, the right hepatic vein and parts of the cardinal veins. Parts of the last two have been transformed into a short **caudal vena cava.**

Necturus is a good example of the early tetrapod condition, excepting certain features of its heart and aortic arches. Being neotenic, this urodele retains larval methods of respiration. The cardiovascular system should be studied on doubly or triply injected specimens.

Heart and Associated Vessels

The pericardial cavity was noted during the study of the coelom. Open it wider, if necessary, and identify the chambers of the heart. The most conspicuous chamber in the ventral view is the large, muscular **ventricle** which occupies the caudoventral portion of the cavity. The narrow vessel that emerges from the right side of the front of the ventricle is the **conus arteriosus.** This soon expands into a wider vessel, the **bulbus arteriosus,** lying at the front of the pericardial cavity. The conus arteriosus is a chamber of the heart; the bulbus arteriosus, the modified ventral aorta, which in this case lies within the pericardial cavity. In many vertebrates that have it, the bulbus arteriosus acts to even out the pressure waves of ventricular contraction; presumably, this is its function in *Necturus*. The **atrium** lies dorsal to the conus and bulbus and appears as the lobes on either side of these structures. It is partially divided internally into **right** and **left atria.** Lift up the caudal end of the ventricle (**apex** of the heart) and you will see the small, thin-walled **sinus venosus.**

The pair of large veins that enter the sinus venosus caudally are the **hepatic sinuses. Common cardinal veins** enter the caudodorsal corners of the sinus venosus just in front of the hepatic sinuses (Fig. 10–11). A pair of **pulmonary veins** from the lungs passes dorsal to the hepatic sinuses and unites to form a single vessel which enters the left atrium. This vessel may be seen between the hepatic sinuses.

Since *Necturus* is a permanent larva respiring primarily by means of external gills, its heart is not a good example of the adult heart of a primitive tetrapod. The venous drainage of the body enters the sinus venosus, which in turn enters the right atrium. Arterial blood from the lungs enters the left atrium. Considerable mixing must occur in the heart, for the interatrial septum is poorly developed and the two atria have a common opening into the single ventricle. This mixing is unimportant in *Necturus,* for all the blood that leaves the heart passes through the gills before it is distributed to the body and lungs. Thus, the heart of *Necturus* is very similar to that of *Squalus* functionally and its structure is not sufficiently different to warrant its dissection.

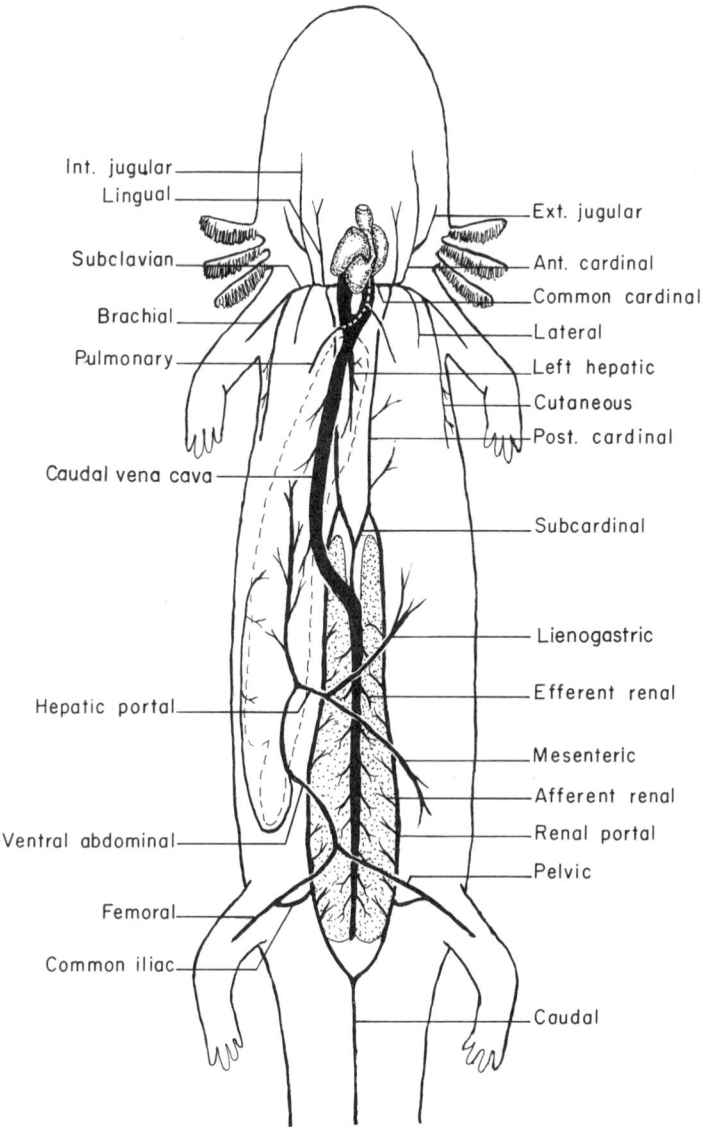

Figure 10–11. Diagrammatic ventral view of the venous system of *Necturus*. The outline of the liver is shown by broken lines.

Venous System

(A) HEPATIC PORTAL SYSTEM

Functionally, the hepatic portal system is the same in *Necturus* as in fishes, but the pattern of its tributaries differs somewhat. The major features of the pattern seen in *Necturus*, however, are very representative of those of tetrapods in general. Stretch out the mesentery of the intestine and you will see a longitudinal vessel, the **mesenteric vein,** passing forward and disappearing in the pancreas. Notice that it receives numerous **intestinal veins** from the intestine. Next look on the tail of the pancreas near the spleen. The vessel seen is the **lienogastric vein;** it is formed by the confluence of a **lienic vein** from the spleen and several **gastric veins** from the stomach. Carefully dissect away pancreatic tissue and find the point where the lienogastric and mesenteric unite. The common vessel that passes forward from here to the liver is the **hepatic portal vein** (Fig. 10–11).

(B) Ventral Abdominal Vein

The median, longitudinal vessel that lies in the falciform ligament caudal to the liver is the **ventral abdominal vein.** This vessel has evolved from the ventral migration and fusion of the paired lateral abdominal veins of fishes, and its caudal relationships are still very similar to those of the lateral abdominal. It receives several small **vesical veins** from the urinary bladder (a ventral outgrowth of the embryonic cloaca) and then bifurcates into **pelvic veins.** Each pelvic vein extends laterally and caudally and after about one-quarter inch receives, on its lateral side, a **femoral vein** from the hind limb. The vessel that continues from the pelvic and femoral caudally and dorsally to the renal portal vein (see below) is the **common iliac vein.** Blood from the leg may pass forward on one of two routes—the ventral abdominal (the primitive route) or the common iliac and renal portal (a new route).

The cranial relationships of the ventral abdominal differ from those of the lateral abdominal of fishes, for the ventral abdominal has lost its primitive connection with the common cardinals and has developed a new one with the hepatic portal system. In this respect it resembles its homologue in late amniote embryos—the umbilical vein (p. 278).

(C) Renal Portal System

The vessel that runs along the lateral margin of each kidney dorsal to the prominent archinephric duct in the male, or oviduct in the female, is the **renal portal vein.** As already described, a common iliac enters each. Trace the two renal portals caudad and try to find the point where they unite and receive the median **caudal vein** from the tail. The renal portals receive blood from the tail and some from the legs and adjacent body wall and carry it to the kidneys.

(D) Caudal Systemic Veins

At the very cranial end of the kidneys the renal portals lead into a pair of small **posterior cardinal veins** that continue forward on either side of the dorsal aorta. This continuity of renal portals and posterior cardinals is not surprising, since the renal portals develop from the caudal end of the embryonic posterior cardinals (Fig. 10–5, *A*), and *Necturus* is an incompletely metamorphosed species. Note the **intersegmental veins** that enter the posterior cardinals from the body wall. The posterior cardinals diverge at the level of the cranial end of the esophagus and unite with the anterior cardinals to form the common cardinals. This region will be studied presently.

The blood in the renal portals that passes into the kidneys, and this would be most of it, leaves through numerous, small, paired, **efferent renal veins** that are located on the ventral surface of the kidneys. These, together with **gonadial (testicular** or **ovarian) veins** from the gonads, enter the **caudal vena cava** (posterior vena cava) lying between the kidneys. After an anastomosis with the caudal ends of the posterior cardinals, the caudal vena cava extends ventrally through the ligamentum hepatocavopulmonale and enters the liver. Trace it through the liver. It receives numerous small **hepatic veins** from various parts of the liver and a particularly large **left hepatic vein** from the front of the liver. The caudal vena cava then passes through the coronary ligament and transverse septum. After this it bifurcates into the two **hepatic sinuses** that enter the sinus venosus dorsal to the ventricle.

The caudal vena cava is a new vessel compounded largely from ones previously present. The portion of it cranial to the kidneys develops from the right hepatic vein and from a caudal extension of this vein. This explains why there is only one particularly prominent hepatic vein (the left one) at the front of the liver instead of the two seen in *Squalus.* The right one is incorporated in the caudal vena cava. The caudal extension of the right hepatic taps into the embryonic subcardinals, as it does in mammal embryos (Fig. 10–28, *B*), and these (especially the right subcardinal) form the segment of the caudal vena cava between the kidneys. In fishes the embryonic subcardinals form the caudal portions of the adult posterior cardinals (p. 283). Although they now contribute to the caudal vena cava, they still retain a connection with the posterior cardinals. Blood that has passed through the kidneys may take one of two routes forward. It either can go through the posterior cardinals, its primitive route, or it can stay in the caudal vena cava. Most of it does the latter, as the posterior cardinals are reduced in size.

(E) CRANIAL SYSTEMIC VEINS

The cranial systemic veins are hard to dissect unless they are filled with blood. The common cardinal veins were seen entering the sinus venosus. Each receives a number of tributaries, most of which can best be found peripherally and then traced to the common cardinal. They should be studied on the side of the body opposite to the one on which the muscles were dissected. Do not injure arteries while dissecting the veins.

Carefully remove the skin from the lateral surface of the brachium and shoulder and you will see the **brachial vein.** It soon joins with a **cutaneous vein** from the skin to form the **subclavian vein.** Trace the subclavian forward. It turns into the musculature and enters the **common cardinal vein** near the latter's dorsal end. Now remove the skin ventral to the gill slits and separate the hypobranchial muscles from the brachial region. The longitudinal vessel lying on the hypobranchial musculature in this region is the **lingual vein.** Trace it caudally and it will be seen to enter the common cardinal beside, or in common with, the subclavian. Having located these two veins, cut through the muscles ventral to the union of these two veins with the common cardinal and thereby expose the common cardinal descending to the heart.

Next remove the skin from the side of the trunk caudal to the shoulder. The longitudinal vessel lying between the epaxial and hypaxial muscles is the **lateral vein.** Remove the scapula and its muscles and trace the vein forward. It enters the top of the common cardinal slightly dorsal to the preceding two vessels. The **posterior cardinal** can now be traced forward and it will be seen to enter the common cardinal caudal to the entrance of the lateral vein. The vessel entering the top of the common cardinal cranial to the lateral and posterior cardinals is the **anterior cardinal.** Try to trace it forward. The largest part of the vessel (**external jugular vein**) passes dorsal to the gills, but a small branch (**internal jugular vein**) goes into the roof of the mouth.

The anterior cardinal, together with the internal jugular, represents the anterior cardinal of fishes. The subclavian and brachial represent the cranial part of the primitive lateral abdominal system and still have the same essential relationships as their homologues in *Squalus*. The lingual is homologous to the inferior jugular of fishes. Although there has been some change in terminology, it is obvious that the major cranial veins of *Necturus* are very similar to those of fishes.

Arterial System

(A) AORTIC ARCHES AND THEIR BRANCHES

Return to the pericardial cavity, and carefully dissect away the muscular tissue (mostly rectus cervicis) that lies between the cavity and the external gills on the side of the pharynx that has not been cut open. Two arteries leave from each side of the front of the bulbus arteriosus. Trace them laterally on the intact side. They are probably not injected, so be careful. They cross the transversi ventrales muscles (p. 132) and then pass deep to the subarcuals (Fig. 10–12). The more cranial one, known as the **first afferent branchial artery,** follows the first branchial arch (third visceral arch) and enters the first external gill. A small, probably well injected, artery lies just in front of the distal portion of the first afferent branchial. This is the **external carotid artery.** Notice that it has numerous branches supplying the muscles in the floor of the pharynx and mouth. The external carotid is a branch of the efferent branchial system, hence its good injection, but it generally has tiny anastomoses with the first afferent branchial. One of the paired **thyroid glands** lies in the angle formed by the meeting of the first afferent branchial and external carotid. It can be recognized by its texture, for it is composed of many follicles which appear as small vesicles.

The more caudal vessel leaving the bulbus arteriosus soon bifurcates. One branch, the **second afferent branchial artery,** follows the second branchial

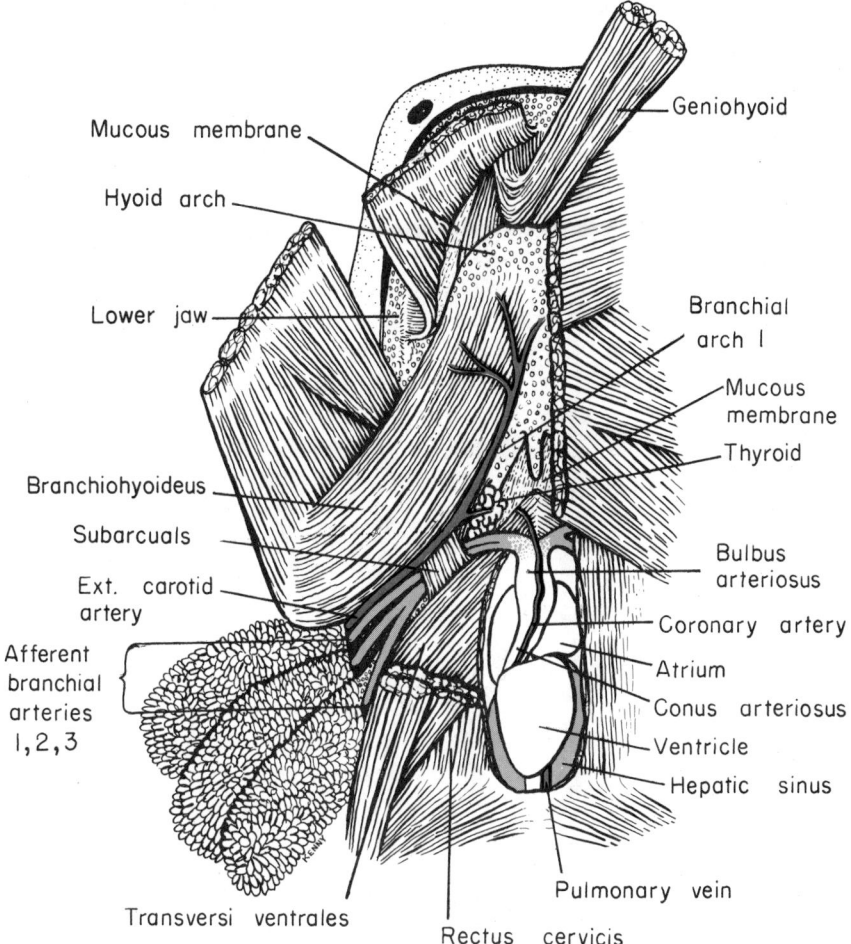

Figure 10–12. Ventral view of the afferent branchial arteries of *Necturus*. Much of the rectus cervicis has been cut away.

arch and enters the second gill. The other branch, the **third afferent branchial artery,** follows the third branchial arch and enters the third gill.

Swing open the floor of the mouth and pharynx and carefully remove the mucous membrane from the roof of the pharynx. A large pair of vessels will be seen converging toward the middorsal line where they unite to form the **dorsal aorta.** They are the **radices** of the aorta (Fig. 10–13). Carefully trace the radix on the intact side toward the gill slits. Slightly lateral to the vertebral column it gives off a small, cranial branch (the **vertebral artery**), which soon disappears in the musculature at the base of the skull. Further laterally the radix curves caudally. Another vessel lies cranial to this portion of the radix and is connected with the radix by a short, stout anastomosis called the **carotid duct.** That portion of the vessel which extends forward in the roof of the pharynx from the carotid duct is the **internal carotid artery;** that portion which extends laterally from the carotid duct represents the entrance of the **first efferent branchial artery.** The internal carotid is distributed to the facial region and enters the skull to help supply the brain. Trace the first efferent

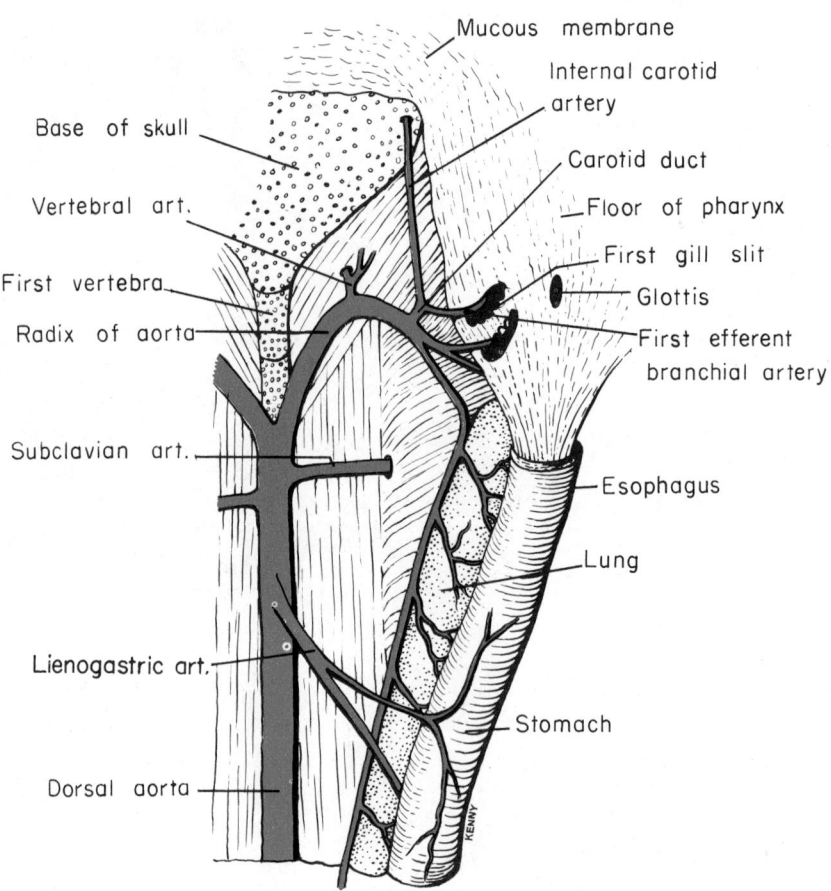

Figure 10–13.   Ventral view of the efferent branchial arteries of *Necturus.*

artery as far laterally as feasible, noting that it comes from the first external gill. Continue to trace the radix laterally from the carotid duct. You will soon see a caudal branch. This is the **pulmonary artery;** it should be traced to the lung. If difficulty is encountered, find the pulmonary artery on the lung and trace it forward. (The **pulmonary vein** lies on the opposite side of the lung. Its entrance into the heart has already been noticed.) Lateral to the pulmonary artery the radix bifurcates and receives the **second** and **third efferent branchial arteries** from the respective gills.

It is difficult to get a clear view of the efferent branchials from the roof of the pharynx. To get a better view, approach the region from the outside by carefully dissecting away the skin from the caudal surface of the base of each external gill. The three efferent branchial arteries lie near the dorsal edge of the gills. Trace them to the point reached in the previous dissection. The origin of the external carotid from the first efferent branchial should also be found during this dissection.

Necturus, like all tetrapods, has lost the first two aortic arches, but the dorsal aortae and the ventral aortae as far caudad as the third arch persist as parts of the internal and external carotids, respectively (Fig. 10–14). The third and fourth aortic arches have given rise to the first two branchial arteries, but there has been some doubt concerning the derivation of the third branchial artery. It is often assumed that the fifth aortic arch is lost and that the third branchial represents the sixth arch. The argument for this is that most tetrapods do lose the fifth arch while retaining the sixth, or all but the dorsal portion of it (ductus arteriosus), as the point of origin of the pulmonary artery. But Figge (1930) has shown that this is not the case in *Necturus.* The relationships of the third afferent branchial of *Necturus* to the third branchial arch (fifth visceral arch), its location just caudal to the second gill slit, and the fact that it leads into the third external gill (as the fifth arch does in other larval urodeles) all indicate that it represents the ventral half of the fifth aortic arch and that the ventral half of the sixth has been lost. The dorsal relationships are not so clear, but if the first part of the argument is valid, it would seem that the third efferent branchial is the dorsal part of the fifth aortic arch and that the pulmonary artery arises from the radix of the aorta by the dorsal part of the sixth aortic arch (ductus arteriosus)!

Figge has also shown that if the ventral part of the sixth aortic arch of a larval *Ambystoma tigrinum* is ligated, the larvae cannot be made to metamorphose unless they are in an unusually oxygen-rich environment. He believes that, since the lungs do not become functional, the oxygen level of the blood flowing through the gills cannot be raised to the level apparently necessary for their reabsorption. Figge further postulates that the absence of the ventral part of the sixth aortic arch in *Necturus* explains its failure to metamorphose. This may be true now but, from a phylogenetic viewpoint, its failure to metamorphose may have preceded and permitted the loss of the ventral part of the sixth arch. If the arch was lost prior to the evolution of a neoteny that rendered the arch unnecessary, *Necturus* may have become extinct.

(B) DORSAL AORTA AND ITS BRANCHES

A pair of **subclavian arteries** arises from the dorsal aorta just caudal to the union of the radices of the aorta (Fig. 10–13). Trace one laterally. At the base of the appendage it divides into a **brachial artery** that continues out into the arm, and a **cutaneous artery** to the skin and adjacent muscles. In amphibians with cutaneous respiration the cutaneous artery is very large and carries blood to the skin where some gas exchange occurs.

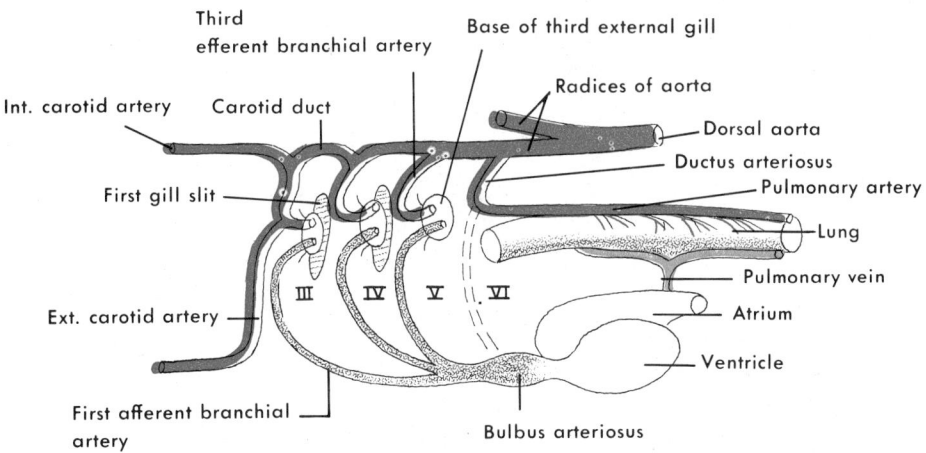

Figure 10–14. Diagrammatic lateral view of the branchial vessels of *Necturus* to show their derivation from the aortic arches. Roman numerals indicate embryonic aortic arches.

Continue to follow the aorta caudally. It next gives off ventrally a **lienogastric artery** which soon branches to go to various parts of the stomach and to the spleen. The next ventral branch, the **celiacomesenteric artery,** arises some distance caudad (Fig. 10–15). It passes ventrally to the tail of the pancreas where it divides into a **lienic artery** to the spleen, a **hepatic artery** to the liver, and a **pancreaticoduodenal artery** to the pancreas and duodenum. Much of the pancreas will have to be dissected away to see all these branches. The remaining ventral visceral arteries are a number of **mesenteric arteries** to the intestine and a pair of cloacal arteries to the cloaca (see below). You will have to separate the caudal vena cava from the aorta to see the point of origin of the mesenterics. Notice that the caudal vena cava lies toward the right of the aorta and the mesenteric arteries. This reflects the origin of this segment of the vessel from the embryonic right subcardinal.

The lateral visceral arteries consist of a number of paired **gonadial (testicular** or **ovarian) arteries** to the gonads, and very small **renal arteries** to

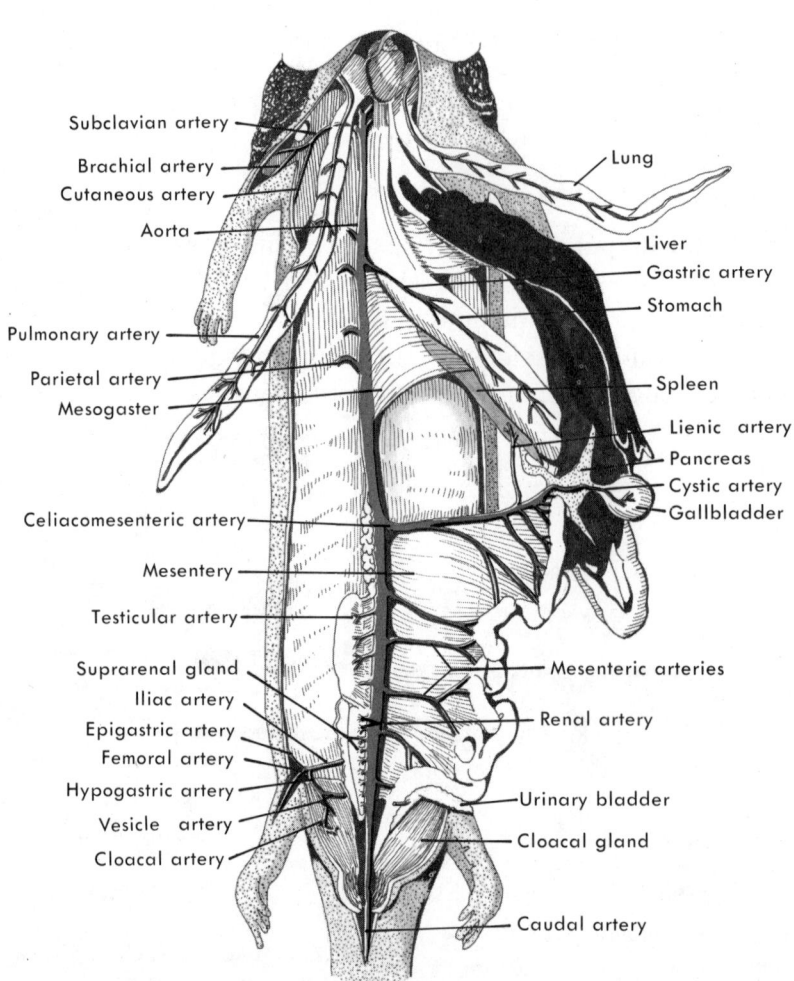

Figure 10–15.   Ventral view of the arterial system of *Necturus.*

the kidneys. The latter can be found by dissecting away the caudal vena cava between the caudal ends of the kidneys.

Paired intersegmental arteries include the subclavians already seen; a number of typical **intersegmental arteries** that arise from the dorsal surface of the aorta and pass into the body wall; and the **iliac arteries.** The iliacs can be found dorsal to the caudal ends of the kidneys. After traveling a short distance, each gives off cranially an **epigastric artery** that ascends in the body wall, caudally a **hypogastric artery** that supplies the urinary bladder and cloaca, and then the iliac continues as the **femoral artery** into the hind leg.

Cut through the body wall and muscles lateral and caudal to the cloaca and trace the aorta caudad. It gives off the paired **cloacal arteries** referred to above and then enters the hemal canal of the caudal vertebrae as the **caudal artery.**

MAMMALS

Changes that occur in the cardiovascular system during the evolution from primitive tetrapods to mammals correlate for the most part with the increase in activity and rate of metabolism. The heart becomes completely divided morphologically into a right side receiving depleted blood from the body and sending it out to the lungs, and a left side receiving aerated blood from the lungs and sending it to the body. This complete division has doubtless been one factor in greatly increasing arterial blood pressure and the general efficiency of circulation. For example, the mean blood pressure in the caudal artery of a mouse is 136 mm. Hg compared with a systolic aortic pressure in the frog of 30 mm. Hg.

The aortic arches are further reduced, for the fifth is lost on both sides as well as the first and second. The branches of the dorsal aorta, usually simply called the aorta, for mammals have no ventral aorta, continue to follow the basic pattern established in lower vertebrates.

In the cranial part of the venous system, a single cranial vena cava, or a pair of them, evolves from the anterior and common cardinals.

More caudally the renal portal system is lost and the caudal vena cava continues to the iliac and caudal veins. The loss of the renal portal system in mammals may be correlated with an increased blood pressure, for a large volume of blood now enters the kidneys directly from the aorta, and with the evolution of different mechanisms for conserving body water, it has also been suggested that the caudal migration in most mammals of a part of the embryonic kidneys with the testes (see Chapter 11) would necessitate the loss of the renal portal system. Much of the cranial portion of the posterior cardinals is also lost, but a part of them is transformed into an azygos system of veins.

The primitive lateral abdominal system of veins is represented embryonically by the umbilical veins, but these are lost in the adult. The veins from the appendages are not connected with the umbilicals but enter the venae cavae directly.

The cardiovascular system should be studied on specimens that have been at least doubly injected. Triply injected specimens are necessary to see all the hepatic portal system, and quadruply injected ones for seeing most of the lymphatic system.

Heart and Associated Vessels

Carefully cut away the pericardial sac and thymus of your specimen from around the heart and its great vessels. The **heart** (*cor*) is a large, compact organ having a pointed caudal end (its **apex**) and a somewhat flatter cranial surface (its **base**). The **right** and **left ventricles,** which are completely separated internally, form the caudal two thirds of the organ (Fig. 10–16). They are approximately conical in shape and have thick, muscular walls. The **right** and **left atria,** also completely separated internally, lie cranial to the ventricles and are set off from them by a deep, often fat-filled groove called the **coronary sulcus.** The atria are thinner-walled and darker than the ventricles. They are separated from each other on the ventral surface by the great arteries leaving the front of the ventricles. That portion of each atrium lying lateral to these arteries is called the **auricle.** The auricles are somewhat ear-shaped and tend to have scalloped margins. The separation between the ventricles appears, on the ventral surface, as a shallow groove, the **interventricular sulcus,** extending from the left auricle diagonally and toward the right.

Pick away the fat from around the large arteries leaving the cranial end of the ventricles. The more ventral vessel is the **pulmonary trunk** (Figs. 10–16, 10–17, and 10–18). It arises from the right ventricle and extends dorsally to the lungs. Trace it later. The more dorsal vessel is the **arch of the aorta.** It arises from the left ventricle deep to the pulmonary trunk, but one cannot see much of it until it emerges on the right side of the pulmonary trunk. As you continue to pick away fat and loose connective tissue from around these vessels, you will notice that they are bound together by a tough band of connective tissue, which is known as the **ligamentum arteriosum.** Try not to destroy it. Just after this connection, the pulmonary trunk bifurcates and its branches, the **pulmonary arteries,** pass to the left and right lungs. This bifurcation is most easily seen by pushing the pulmonary trunk cranially and dissecting between it and the craniodorsal portion of the heart. Two small **coronary arteries** leave the base of the arch of the aorta and pass to the heart wall. One can be found deep between the pulmonary trunk and the left auricle; the other, deep between the pulmonary trunk and the right auricle. **Coronary veins** draining the heart wall parallel the arteries.

Push the heart to the left side of the thorax and you will see the **caudal vena cava** (posterior vena cava) coming through the diaphragm and entering the right atrium. A cranial vena cava will also be seen entering this chamber from the right side of the neck. Adult carnivores normally have only the **right cranial** (anterior) **vena cava,** but the rabbit also has a **left cranial vena cava** that comes down the left side of the neck, crosses the dorsal surface of the heart and enters the right atrium. Lift up the apex of the heart and you will see this vessel. Notice that it receives the **coronary veins** from the heart wall. The coronary veins of carnivores collect into a **coronary sinus** which has a position comparable to the proximal end of the rabbit's left cranial

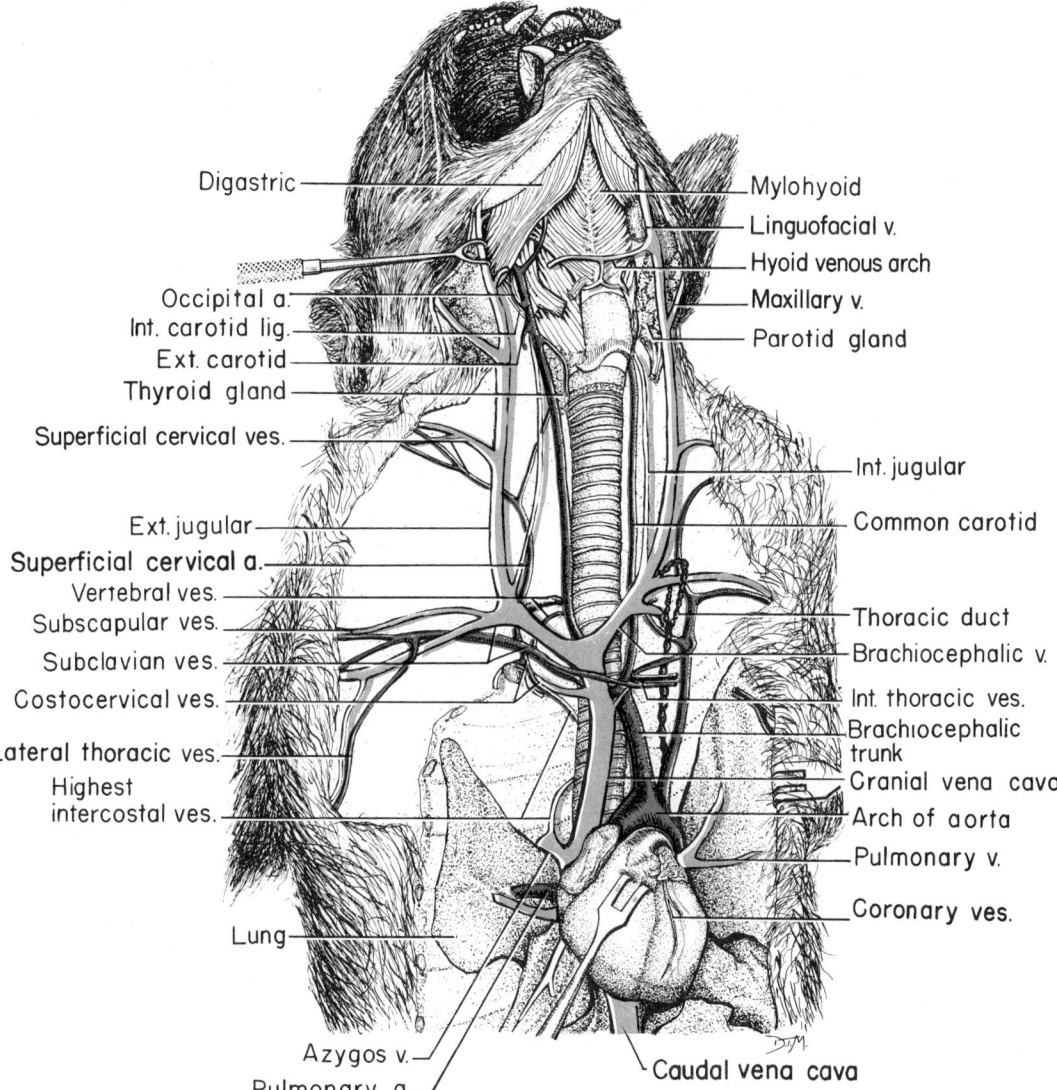

Figure 10–16. Ventrolateral view of the thoracic and caudal cervical vessels of the cat.

vena cava. Carefully pick away connective tissue dorsal to the cranial venae cavae and from the roots of the lungs. You will find the **pulmonary veins** coming from the lungs and entering the left atrium. There are several veins, but those of each side generally collect into two channels before entering the heart.

It will be noted that the mammalian heart contains only two of the primitive four chambers, but that these chambers have become completely divided. The apparently missing sinus venosus and conus arteriosus are present embryonically, but disappear as such in the adult. The sinus venosus is absorbed into the right atrium and forms that part of the atrium receiving the venae cavae. The conus arteriosus (bulbus cordis of mammalian embryology), together with the ventral aorta, splits and forms the very base of the pulmonary trunk and arch of the aorta.

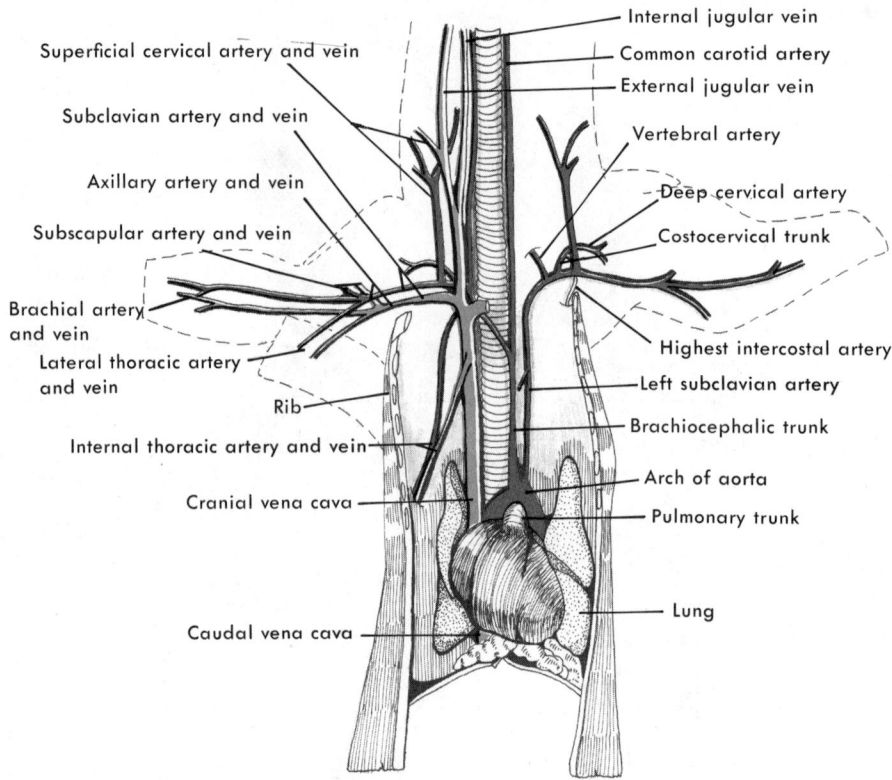

Superficial cervical artery and vein

Subclavian artery and vein

Axillary artery and vein

Subscapular artery and vein

Brachial artery and vein

Lateral thoracic artery and vein

Rib

Internal thoracic artery and vein

Cranial vena cava

Caudal vena cava

Internal jugular vein

Common carotid artery

External jugular vein

Vertebral artery

Deep cervical artery

Costocervical trunk

Highest intercostal artery

Left subclavian artery

Brachiocephalic trunk

Arch of aorta

Pulmonary trunk

Lung

Figure 10–17. Ventral view of the thoracic and cervical vessels of the mink.

In most regions of the body the arteries and veins will be described together, but the overall pattern of circulation in the cat and other mammals is evident from Figures 10–16 and 10–23. Tributaries of the caudal vena cava drain the body caudal to the diaphragm. Both venae cavae return venous blood low in oxygen to the right atrium. From here the blood goes to the right ventricle from where it is pumped to the lungs through the pulmonary trunk and arteries. After aeration in the lungs, arterial blood returns through the pulmonary veins to the left atrium. From here it goes to the left ventricle which pumps it out the aorta. Branches of the aorta carry arterial blood to all parts of the body. Blood circulating through the chambers of the heart does not supply the musculature of this organ, so a separate coronary system is necessary. As you have seen, the coronary arteries leave the base of the aortic arch and the veins return to the right atrium via the coronary sinus.

Arteries and Veins Cranial to the Heart

(A) VESSELS OF THE CHEST, SHOULDER, ARM, AND NECK

As noted in the study of the heart, carnivores have a single **cranial vena cava,** but both a left and right one are present in the rabbit. Trace it (them) forward by carefully picking away surrounding portions of the thymus, connective tissue, and fat. In all of the mammals being considered, a **subclavian**

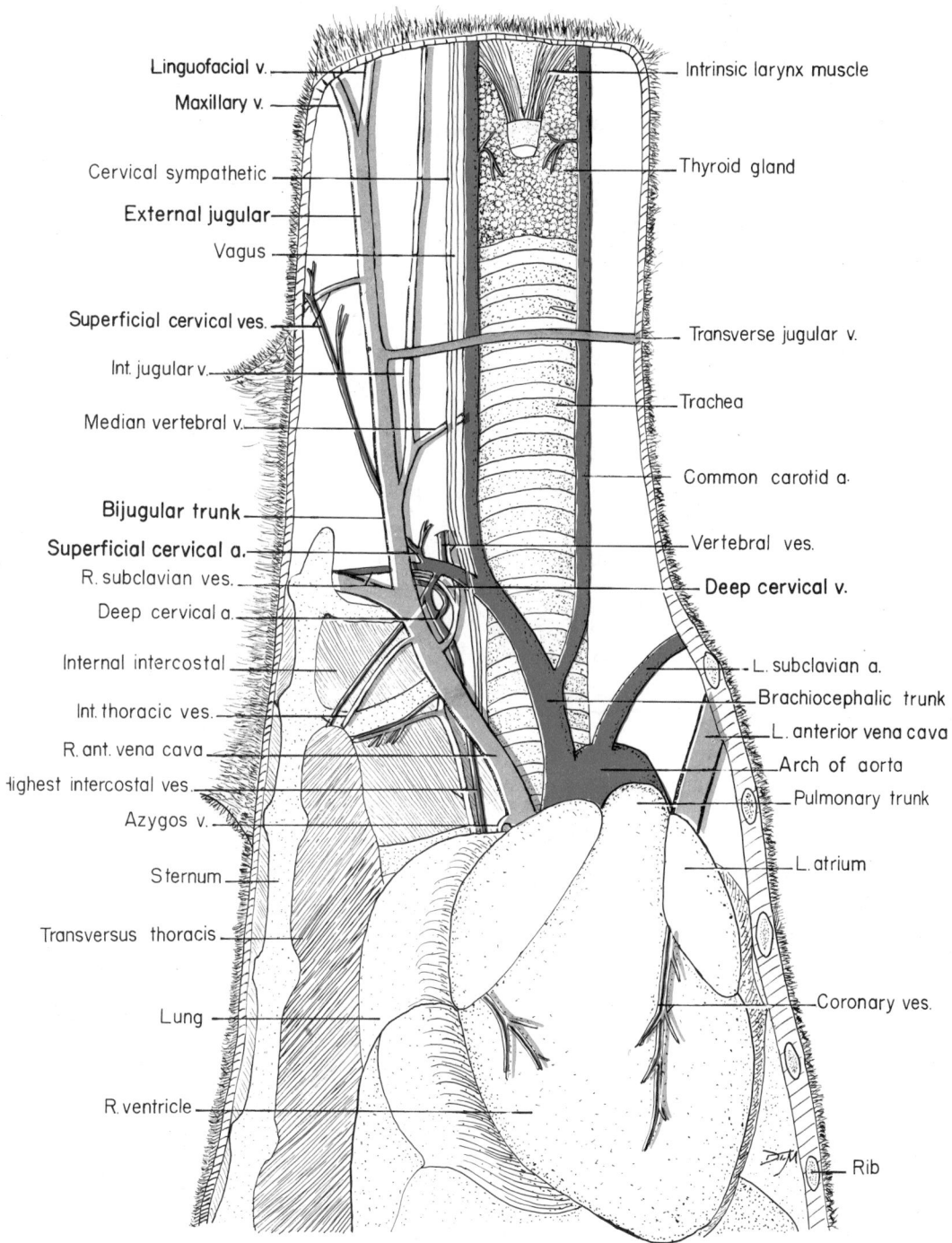

Figure 10–18. Ventral view of the thoracic and cervical vessels of the rabbit.

vein comes in from each shoulder and arm just in front of the first rib (a valve often prevents it from being injected) and joins a **bijugular trunk,** which receives the drainage from one side of the neck and head (Figs. 10–16, 10–17, and 10–18). This union forms the cranial vena cava on each side of the rabbit; in carnivores it forms the **brachiocephalic veins.** Left and right brachiocephalics, in turn, unite to form the single cranial vena cava. The carnivore vena cava is comparable to the right one of the rabbit. A left one, present in the embryo (Fig. 10–28), disappears when the left brachiocephalic develops.

Next examine certain of the tributaries of the vena cava (rabbit) or vena cava and brachiocephalic (carnivores). The most caudal tributary, entering the dorsal surface of the vena cava, is the **azygos vein,** which receives most of the **intercostal veins** from between the ribs on both sides of the body. The mammals being studied have only one azygos, on the right side of the body, but in many mammals a left one is also present. **Intercostal arteries** will be seen beside the veins; their origins will be seen later. A **highest intercostal vein,** which drains the cranial intercostal spaces, enters the azygos in the carnivores, but enters the vena cava independently and farther cranially in the rabbit.

The next cranial tributaries are several small veins from the thymus and a larger **internal thoracic vein,** which enters the ventral surface of the cranial vena cava. In the carnivores it is a single vessel at its entrance, but it bifurcates distally and drains both sides of the ventral thoracic wall. Its distal parts lie deep to the transversus thoracis muscle and are accompanied by the **internal thoracic arteries,** whose origin will be seen soon. The internal thoracic vessels continue into the cranial part of the ventral abdominal wall where they are called the **cranial epigastric arteries** and **veins.**

Return to the arch of the aorta. After giving off the coronary arteries previously described, the arch curves dorsally and to the left, disappearing dorsal to the root of the left lung. Two vessels arise from the front of the arch—a large **brachiocephalic trunk** nearest the heart and then a smaller, **left subclavian artery.** Trace the brachiocephalic forward. It sends off small branches to the thymus, and then breaks up into three vessels—two common carotid arteries that ascend the neck on either side of the trachea, and a **right subclavian artery.** These vessels continue cranially deep to the vena cava and brachiocephalic veins.

Trace one of the subclavian arteries peripherally, preferably on the side of the body in which the veins are well injected so that they can also be identified. Medial to the first rib, the subclavian artery gives rise to four branches which are most accurately identified from their peripheral distribution. The **internal thoracic artery,** previously identified, leaves the ventral surface of the subclavian and accompanies the internal thoracic vein to the ventral chest wall.

A **vertebral artery** arises from the dorsal surface of the subclavian artery nearly opposite the origin of the internal thoracic artery (cat), or somewhat cranial to the origin of the internal thoracic (mink and rabbit). Trace it and the accompanying **vertebral vein** forward. The vein normally enters the vena cava (rabbit) or brachiocephalic vein (carnivores). The vertebral vessels soon enter the transverse foramina of the cervical vertebrae through which they continue, finally to enter the cranial cavity and help supply the brain.

A short **costocervical trunk** arises from the subclavian artery just distal to the origin of the vertebral and divides almost immediately into highest intercostal and deep cervical arteries. In the rabbit, and sometimes in the carnivores, these vessels arise independently from the subclavian. The **highest intercostal artery** extends caudally across the cranial ribs supplying those intercostal spaces drained by the highest intercostal vein. This vein has been identified. The **deep cervical artery** extends dorsally to supply deep muscles of the neck. A major branch of it also passes cranial to the first rib and into the serratus ventralis muscle. The deep cervical artery is accompanied by the **deep cervical vein**, which usually drains into the vertebral vein shortly after this vein emerges from the transverse foramina. Occasionally the deep cervical vein enters the cranial vena cava independently.

The last branch of the subclavian artery is the **superficial cervical artery.** It extends deep to the subclavian vein and follows the **external jugular vein** cranially. Trace them both. The superficial cervical artery gives off one or more small branches that extend cranially, sometimes reaching the thyroid gland, but the main part of the artery continues laterally and dorsally to supply muscles on the craniolateral surface of the shoulder. A **superficial cervical vein**, a tributary of the external jugular, accompanies the distal part of the artery. One tributary of the superficial cervical vein, the **cephalic vein,** is often a conspicuous vessel, draining the lateral surface of the brachium.

After giving off these vessels, the subclavian artery and the satellite subclavian vein continue laterally into the armpit (axilla). These vessels change their names at this point to the **axillary artery** and **vein.** Major branches of the axillary artery are a **lateral thoracic artery** to the pectoral muscles and a **subscapular artery.** The latter passes between the subscapularis and supraspinatus muscles to supply deep shoulder muscles. Veins accompany the arteries but are usually not injected. When the axillary artery and vein enter the arm they are known as the **brachial artery** and **vein.**

Return to the brachiocephalic trunk, or arch of the aorta, and trace one of the **common carotid arteries** forward. It passes deep to the brachiocephalic vein, or cranial vena cava, and continues cranially, lying lateral to the trachea, supplying the trachea, thyroid gland, and other cervical structures before reaching the head. An **internal jugular vein,** which helps drain the inside of the skull, lies lateral to the common carotid through most of its course. The internal and external jugular veins unite with each other slightly cranial to the subclavian vessels to form the **bijugular trunk** previously observed. In the rabbit, a **median vertebral vein,** which courses dorsal to the esophagus, enters either jugular vein near the bijugular trunk. The rabbit also has a

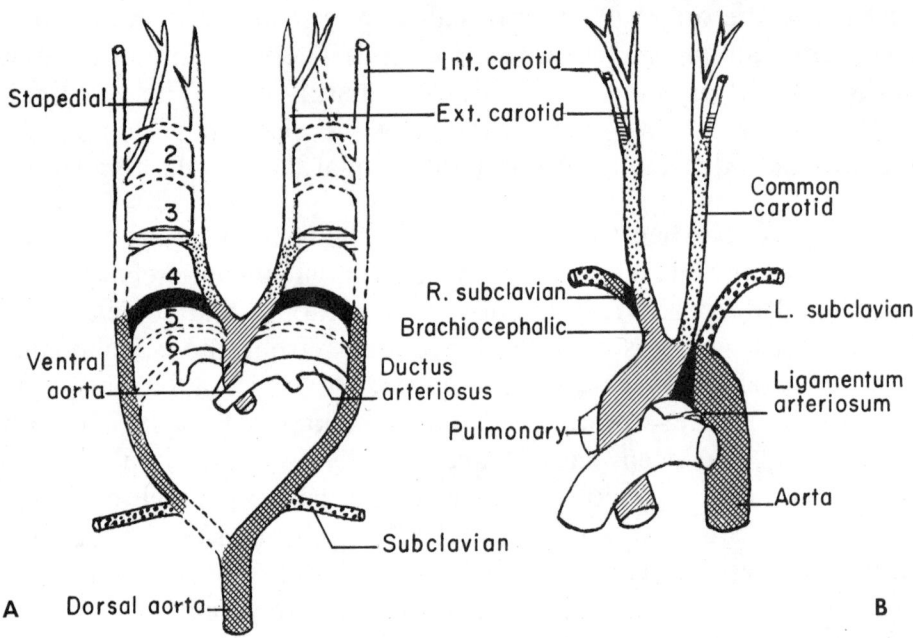

Figure 10–19. Diagrammatic ventral views of the mammalian aortic arches and their derivatives. *A,* embryonic condition; *B,* adult condition in human beings. (Slightly modified after Barry.)

transverse jugular vein that joins the external jugulars slightly cranial to the bijugular trunk.

A cervical extension of the sympathetic cord and the vagus nerve can be found between the common carotid artery and the internal jugular vein. They are bound together by connective tissue to form a **vagosympathetic trunk** in the carnivores, but can easily be dissected apart. The vagus is the larger and passes superficial to the brachiocephalic trunk.

The major arteries described are derived from the embryonic aortic arches in the manner shown in Figure 10–19. All six of the primitive aortic arches appear during embryonic development and connect the ventral aorta (paired cranial to the fourth aortic arch) with the dorsal aorta (paired to the region of the aortic arches and for a short distance caudad). The first, second, and fifth aortic arches, the dorsal part of the right sixth aortic arch, the paired dorsal aortae between arches three and four, and the right paired dorsal aortae caudad to the entrance of the right subclavian (an intersegmental artery) disappear during development. The dorsal part of the left sixth arch persists during embryonic life, as the **ductus arteriosus,** and shunts blood from the pulmonary trunk directly to the dorsal aorta. For a few hours after birth it shunts some blood in the opposite direction, thereby giving this portion of the blood a double aeration, but it soon becomes converted into the functionless **ligamentum arteriosum.**

The ventral portions of the sixth arches persist as the pulmonary arteries. The left fourth arch, together with part of the left dorsal aorta, forms the arch of the aorta. (Differential growth has the effect of shortening this arch and the adjacent dorsal aorta, so that the left subclavian of the adult leaves the arch of the adult aorta much closer to the common carotids than it does in the embryo.) The right fourth arch, plus a segment of the right dorsal aorta, forms the proximal part of the right subclavian.

A splitting of the caudal part of the ventral aorta, and of the conus arteriosus, results in the direct origin of the arch of the aorta and the pulmonary trunk from the ventricles.

The paired ventral aortae between the fourth and third arches form the common carotids. The ventral aortae rostral to the third arch become the external carotids; the third arches, plus the dorsal aortae rostral to them, the internal carotids. The internal carotid of the embryonic mammal not only supplies the intracranial part of the head but also, by its stapedial branch passing through the stapes, much of the outside of the head. However, in the adults of most mammals the external carotid taps into the stapedial and pirates most, or all, of its peripheral distribution. If the external carotid takes it all over, the stapedial disappears.

(B) Major Vessels of the Head

Skin the rest of the head on one side, if this has not been done, and also remove the auricle. Tributaries of the **external jugular vein** are superficial to other vessels in the head, so must be considered first. In the cat, mink, and rabbit, the external jugular is formed by the confluence of linguofacial and maxillary veins (Fig. 10–20). Trace the **linguofacial vein** forward. In the carnivores, it soon receives on its medial side a **hyoid venous arch,** which comes from the opposite side of the body and, in turn, receives a deep vein from the larynx (**laryngea impar**). The hyoid venous arch is not present in the rabbit. At the caudoventral border of the mandible, the linguofacial vein is formed by the joining of a lingual and a facial vein. The **lingual vein** enters and drains the tongue. It is accompanied by the hypoglossal nerve, but has probably been cut in earlier dissections. The **facial vein** continues forward along the ventral border of the masseter muscle. Its major tributaries are a **deep facial vein** from beneath the masseter, which connects with venous plexuses in the

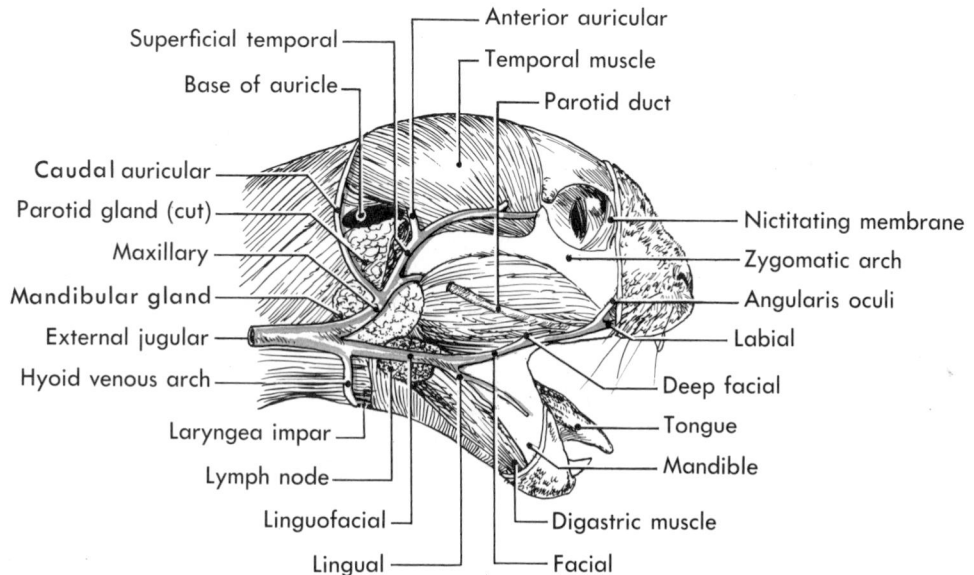

Figure 10–20. Lateral view of the tributaries of the external jugular vein of a cat. (From Walker, A Study of the Cat.)

orbit and palate; a **labial vein** from the upper lip; and a **vena angularis oculi** from the face in front of the eye.

Return to the origin of the external jugular and trace the **maxillary vein** dorsally toward the base of the auricle. It is formed by the confluence of a **caudal auricular vein** from behind the ear and a **superficial temporal vein** from in front of the ear. The superficial temporal receives tributaries from the ear, temporal muscle, and a deep branch connecting with the orbital and palatine venous plexuses. These plexuses, which will not be dissected, receive most of the drainage from inside the skull in the mammals being considered because their internal jugular veins are so small.

In order to trace the internal jugular vein and common carotid artery forward, reflect the mandibular gland and the digastric and mylohyoid muscles. At the level of the larynx, the **common carotid artery** gives off one or two **thyroid arteries** and a muscular branch. The common carotid of the mink and rabbit then divides into external and internal carotid arteries. The **internal carotid artery** goes deep toward the skull base and enters the caudal end of the carotid canal, which is located on the caudomedial side of the tympanic bulla. Together with the vertebral artery, it supplies the brain with arterial blood. An internal carotid artery is present in an embryonic cat, but as development proceeds it is reduced to a functionless **internal carotid ligament** (Figs. 10–21 and 10–22), which has the same relationships as the artery. Much arterial blood reaches the brain of the cat by way of a small ascending pharyngeal artery (see below) and a larger anastomotic branch of the external carotid.

The reduction of the internal carotid artery in the cat is related to the development

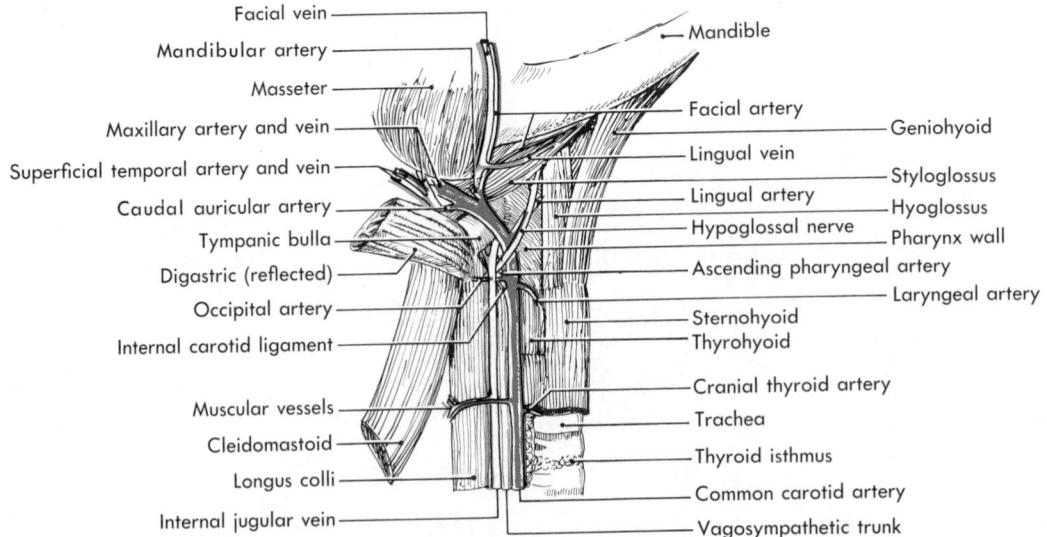

Figure 10–21. Lateroventral view of the internal jugular vein and carotid branches of a cat. Although the external jugular vein has been removed, certain of its peripheral tributaries are shown. (From Walker, A Study of the Cat.)

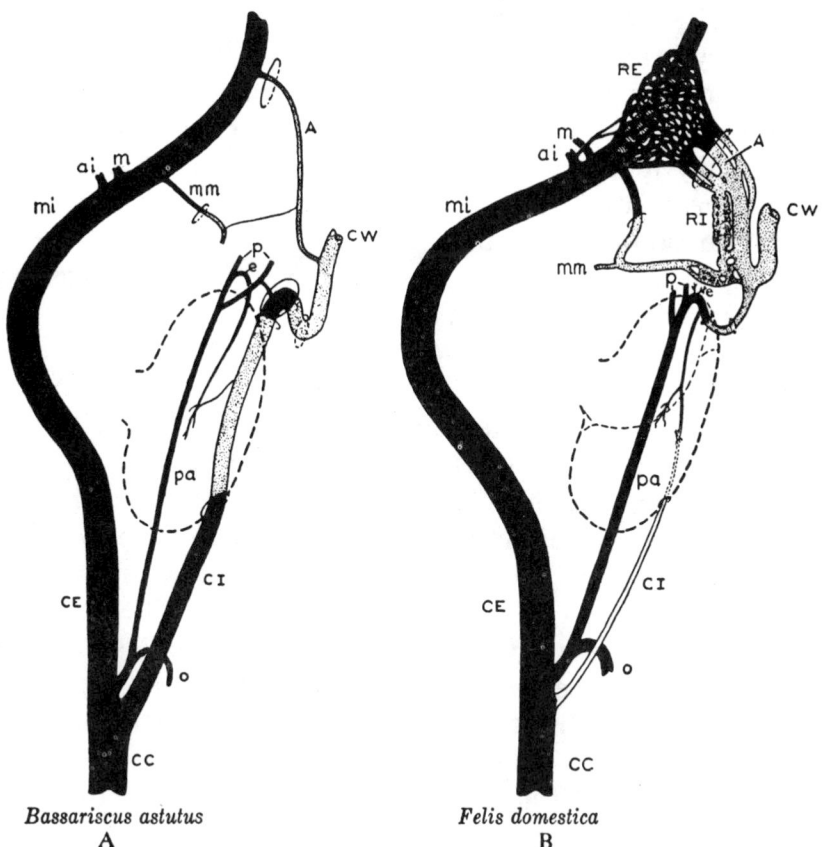

Figure 10–22. Diagrams of the carotid circulation of a primitive carnivore, a member of the raccoon family, *A,* and of the domestic cat, *B.* Stippled parts are intracranial, or run through canals. The position of the tympanic bulla is shown by broken lines. Note the vestigial nature of the proximal part of the internal carotid of the cat (shown in outline), and the way in which other vessels have enlarged to carry blood to the brain. Some other carnivores have an intermediate condition. Abbreviations: *A,* anastomotic artery; *ai,* inferior alveolar; *CC,* common carotid; *CE,* external carotid; *CI,* internal carotid; *CW,* circle of Willis located on ventral surface of brain; *e,* eustachian; *m,* masseteric; *mi,* internal maxillary; *mm,* median meningeal; *o,* occipital; *p,* pharyngeal; *pa,* ascending pharyngeal; *RE,* external rete; *RI,* internal rete. (From Davis and Story, The carotid circulation in the domestic cat, Zoological Series, Field Museum of Natural History, vol. 28.)

of a network of small arteries, known as the **carotid rete mirabile** (RE and RI, Fig. 10–22), associated with the anastomotic branch of the external carotid. These small arteries are surrounded by a network of small veins, which receive blood returning from the nasal cavities. Venous and arterial blood flow in opposite directions. The cooler venous blood absorbs heat from the arterial blood in this counter-current exchange mechanism. This permits the body temperatures of certain very active animals to rise considerably, yet prevents the delicate brain from overheating (K. Schmidt-Nielsen, 1972).

After the origin of the internal carotid artery or ligament, the common carotid is known as the **external carotid artery.** Its first branches arise so close to the origin of the internal carotid that they can be confused with it unless their peripheral distribution is established. An **occipital artery** extends dorsally to supply neck muscles in the occipital region. A small **ascending pharyngeal artery** extends deeply toward the skull base close to the internal carotid artery or ligament. This vessel is somewhat larger and more important in the cat than in the other mammals being considered because functionally it partially replaces the internal carotid. The ascending pharyngeal follows along the ventral surface of the tympanic bulla (Fig. 10–22) and finally enters the skull through the canal for the auditory tube and through the rostral portion of the carotid canal. A small **laryngeal artery** may also arise from the beginning of the external carotid.

After the origin of these vessels, the external carotid gives rise to a number of branches supplying different parts of the outside of the head. These branches accompany the corresponding veins already observed. Lingual and facial arteries arise from the ventral surface of the external carotid (Fig. 10–21). They have a common origin in the rabbit, but not in the cat or mink. The **lingual artery** enters the tongue. The **facial artery** follows the ventral border of the masseter muscle and supplies the jaws and facial structures.

Dorsal branches of the external carotid are (1) a **caudal auricular artery,** which extends dorsally behind the ear; (2) a **superficial temporal artery,** which extends dorsally in front of the ear; and (3) a **maxillary artery** that goes deep to the caudal border of the masseter to supply structures in the orbital and palatal regions. Caudal auricular and superficial temporal have a common origin in the rabbit.

Return to the **internal jugular vein** and trace it forward. It receives small tributaries from muscles at the base of the head, and then enters the skull through the jugular foramen to help drain the brain.

Arteries and Veins Caudal to the Heart

(A) VESSELS OF THE DORSAL THORACIC AND ABDOMINAL WALLS

After curving to the dorsal side of the body, the arch of the aorta is known as the **descending aorta.** Trace it caudally. As it passes through the thorax along the left side of the vertebral column, it gives off paired **intercostal arteries** to those intercostal spaces not supplied by the highest intercostals, small median branches to the esophagus, and also small branches to the bronchi, since the lungs, like the wall of the heart, need a separate arterial supply. The **thoracic portion** of the left sympathetic cord can be found at this time by carefully dissecting in the connective tissue near the heads of the ribs dorsal to the aorta. Enlargements along the cord are **sympathetic ganglia;** delicate strands passing dorsally are **communicating rami.** The left **vagus nerve** crosses the lateral surface of the arch of the aorta, passes dorsal to the root of the lung, and caudally along the esophagus. **Phrenic arteries** to the diaphragm may arise from the aorta before the aorta passes through the diaphragm, or from the last intercostals, or they may arise from vessels posterior to the diaphragm (first lumbar, cranial abdominal, celiac).

The **caudal vena cava** (posterior vena cava) was seen entering the heart. Trace it caudad. As it passes through the diaphragm, it receives several small **phrenic veins** and then disappears in the liver. Scrape away tissue from the cranial surface of the right medial lobe of the liver and find the entrance of several large **hepatic veins.** The major part of the caudal vena cava, however, passes through the right lateral and caudate lobes; it should also be exposed by scraping away liver tissue. Other hepatics, most very small, will be seen entering.

Push the abdominal viscera to the right and find the aorta emerging from the diaphragm. Just after emerging, it gives rise to two ventral vessels—first a **celiac artery** and then a **cranial mesenteric artery** (Figs. 10–23, 10–24, and

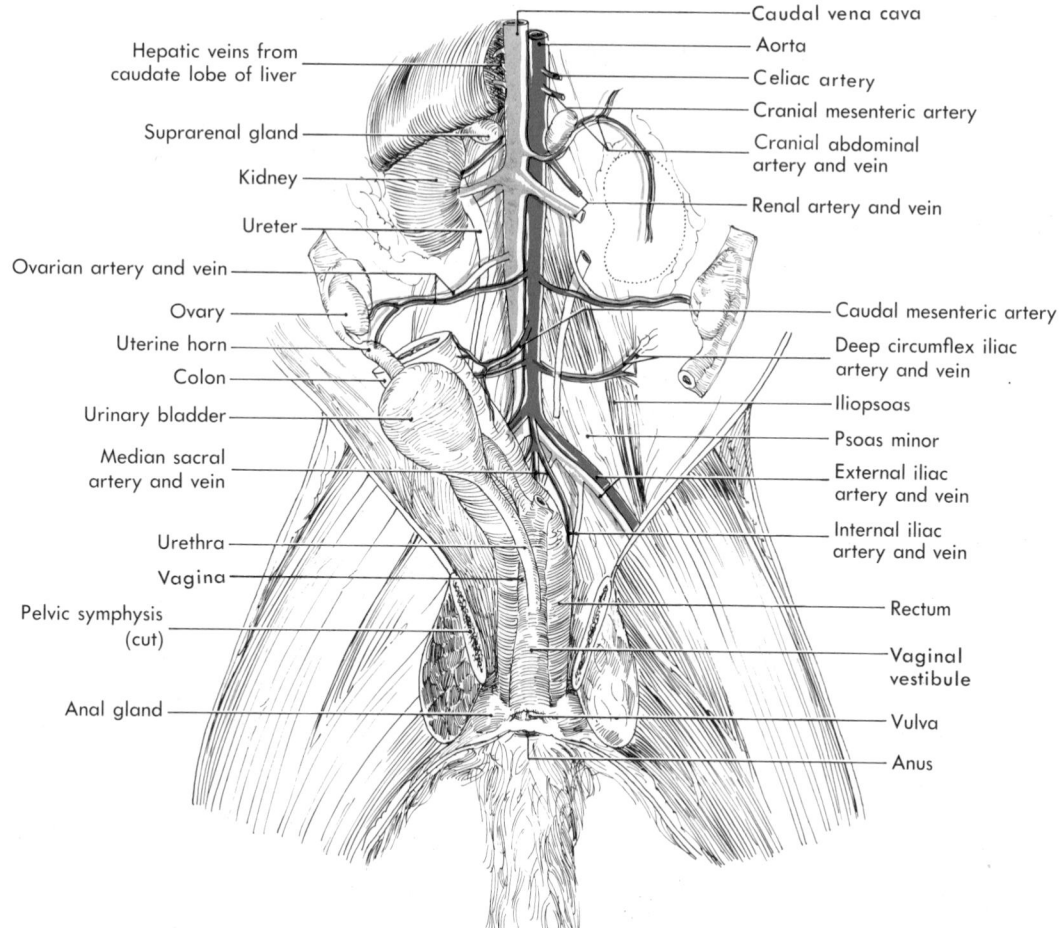

Figure 10–23. Ventral view of the abdominal portion of the aorta and caudal vena cava of a female cat. The pelvic canal has been cut open, and the left kidney and uterine horn have been omitted to show deep vessels. (From Walker, A Study of the Cat.)

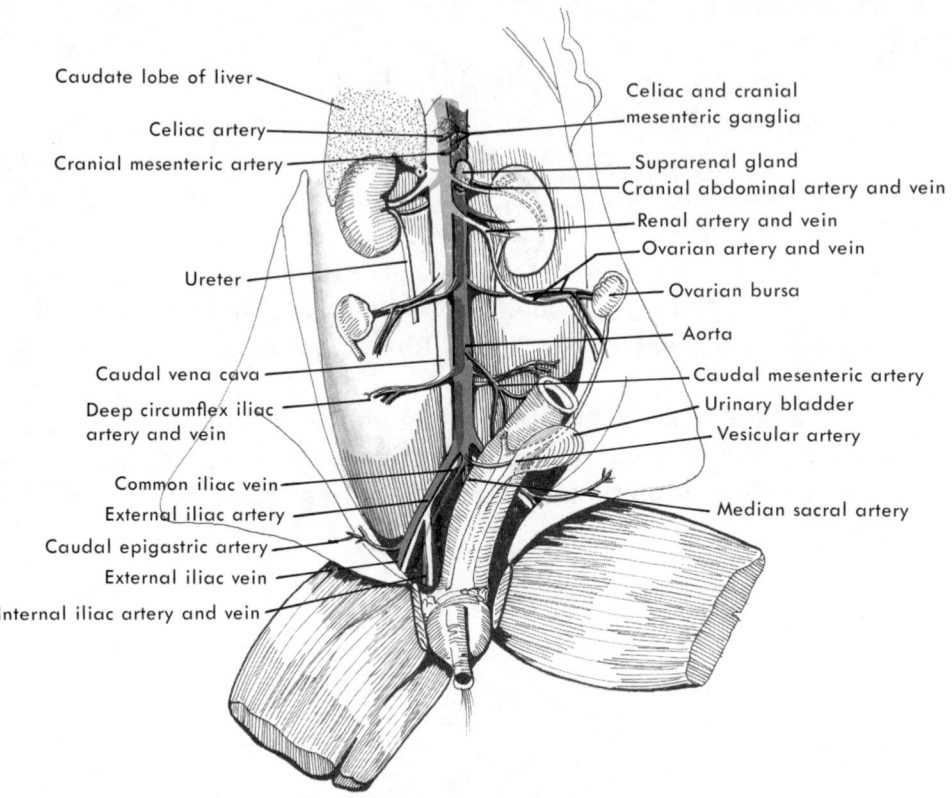

Caudate lobe of liver
Celiac artery
Cranial mesenteric artery
Ureter
Caudal vena cava
Deep circumflex iliac
artery and vein
Common iliac vein
External iliac artery
Caudal epigastric artery
External iliac vein
Internal iliac artery and vein

Celiac and cranial
mesenteric ganglia
Suprarenal gland
Cranial abdominal artery and vein
Renal artery and vein
Ovarian artery and vein
Ovarian bursa
Aorta
Caudal mesenteric artery
Urinary bladder
Vesicular artery
Median sacral artery

Figure 10–24.   Ventral view of the abdominal portion of the aorta and caudal vena cava of a female mink. The pelvic canal has been cut open.

10–25) which supply most of the abdominal viscera. Trace them later. **Celiac** and **mesenteric ganglia** lie at the base of the cranial mesenteric artery. They receive one or more **splanchnic nerves** from the sympathetic cord and send out minute branches which travel along the vessels to the viscera.

Slightly caudal to the cranial mesenteric artery, the aorta lies to the left side of the caudal vena cava. Trace these two vessels to the pelvic region. Their most cranial, paired branches are the cranial abdominal arteries and veins and the renal arteries and veins. The **renal arteries** and **veins,** which supply the kidneys, are larger and more obvious. Those of the right side of the body lie slightly cranial to those of the left side, since the right kidney is more cranially situated than the left one. Carefully dissect away fat from around each kidney so that you can lift up the lateral edge and look at the muscles dorsal to it. The vessels you see supplying the abdominal wall are the **cranial abdominal artery** and **vein.** Trace them toward the aorta and vena cava. Before joining these vessels, they pass and supply a small, hard, oval-shaped nodule embedded in the fat between the cranial end of the kidney and the aorta and vena cava. This nodule is the **suprarenal** (adrenal) **gland.** The cranial abdominal vessels usually join the vena cava and aorta just cranial to the renal vessels, but they sometimes join the renals.

The suprarenal gland is an endocrine gland of dual origin. Its medullary portion, derived from postganglionic sympathetic cells of neural crest origin, secretes hormones which assist sympathetic stimulation in adjusting the body to meet conditions of stress. Its cortical portion, of mesodermal origin, secretes numerous steroid hormones involved in many aspects of metabolism and also in sexual differentiation.

The next paired branches of the aorta are the small **testicular** or **ovarian arteries,** depending on the sex. They pass to the gonads accompanied by the **testicular** or **ovarian veins.** The ovaries are small, oval bodies lying near the cranial ends of the Y-shaped uterus. The testes have descended into the

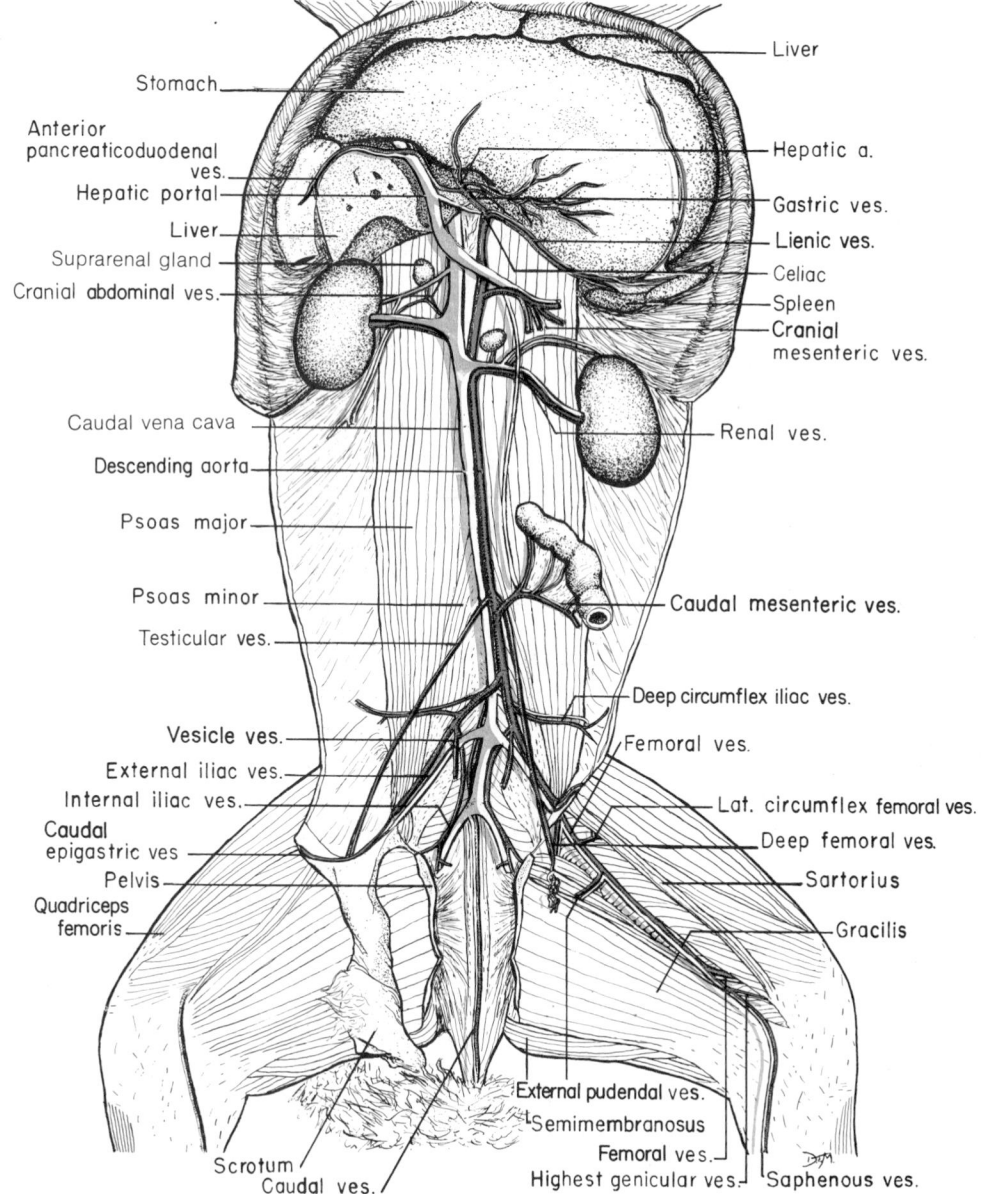

Figure 10–25. Ventral view of the posthepatic vessels of the rabbit. The pelvic canal has been cut open.

scrotum, and, in doing so, each has made an apparent hole (the **inguinal canal**) through the body wall in the region of the groin. The testicular vessels and the sperm duct (**ductus deferens**) can be seen passing through these canals (Figs. 11–7 and 11–8). The right gonadial vein enters the caudal vena cava; the left one may too, but it normally enters the left renal in the carnivores. It normally enters the vena cava in the rabbit.

A **caudal mesenteric artery** leaves the ventral surface of the aorta caudal to the gonadial arteries. Trace it later. Caudal to this vessel, the aorta of carnivores gives rise to a pair of **deep circumflex iliac arteries** which pass laterally to the musculature and body wall lying ventral to the ilia. The deep circumflex iliac arteries of the rabbit arise from a terminal branch of the aorta (common iliac). Satellite **deep circumflex iliac veins** accompany the arteries and enter the caudal vena cava. The rest of the lumbar musculature is supplied by several **lumbar arteries** and **veins** which can be found by dissecting along the dorsal surface of the aorta and vena cava between the renal and deep circumflex vessels. The lumbars are single vessels where they attach to the aorta and vena cava but they bifurcate distally. Caudal to the deep circumflex vessels, the aorta and caudal vena cava give rise to the iliac vessels supplying the pelvic region and leg. Trace them later.

(B) Vessels of the Abdominal Viscera

Return to the celiac artery and mesenteric arteries where they leave the aorta. Remove surrounding connective tissue and the sympathetic ganglia and trace the **celiac artery** a short distance until it divides into three branches (Fig. 10–26)—a **lienic artery** to the spleen, a **left gastric artery** to the lesser curvature of the stomach, and a **hepatic artery** to the liver, pancreas, duodenum, and part of the stomach. More distal parts of these vessels will be seen with the veins. The distribution of the cranial **mesenteric artery** to most of the small intestine and adjacent parts of the colon can be seen by stretching the mesentery. The **caudal mesenteric artery** supplies the descending colon and rectum (Figs. 10–23, 10–24, and 10–25).

Although not injected, the **portal vein** can be found in the lesser omentum where it lies dorsal to the bile duct and forms the ventral border of the epiploic foramen. Trace it caudad (Fig. 10–26). As it passes dorsal to the pylorus it receives a small and often inconspicuous **right gastric vein** from the pyloric region of the stomach, and a larger **gastroduodenal vein.** The latter is formed by the confluence of a **cranial pancreaticoduodenal vein** draining much of the duodenum and pancreas and a **right gastroepiploic vein** from the greater curvature of the stomach and greater omentum. **Cranial pancreaticoduodenal, right gastroepiploic, gastroduodenal,** and **gastric arteries** accompany the veins. All are derived from the hepatic artery which can be seen on the left side of the epiploic foramen. After giving rise to these arteries, the hepatic artery follows the portal vein to the liver.

Push the stomach forward and tear through the part of the greater omentum going to the spleen and dorsal body wall. Carefully dissect away the tail

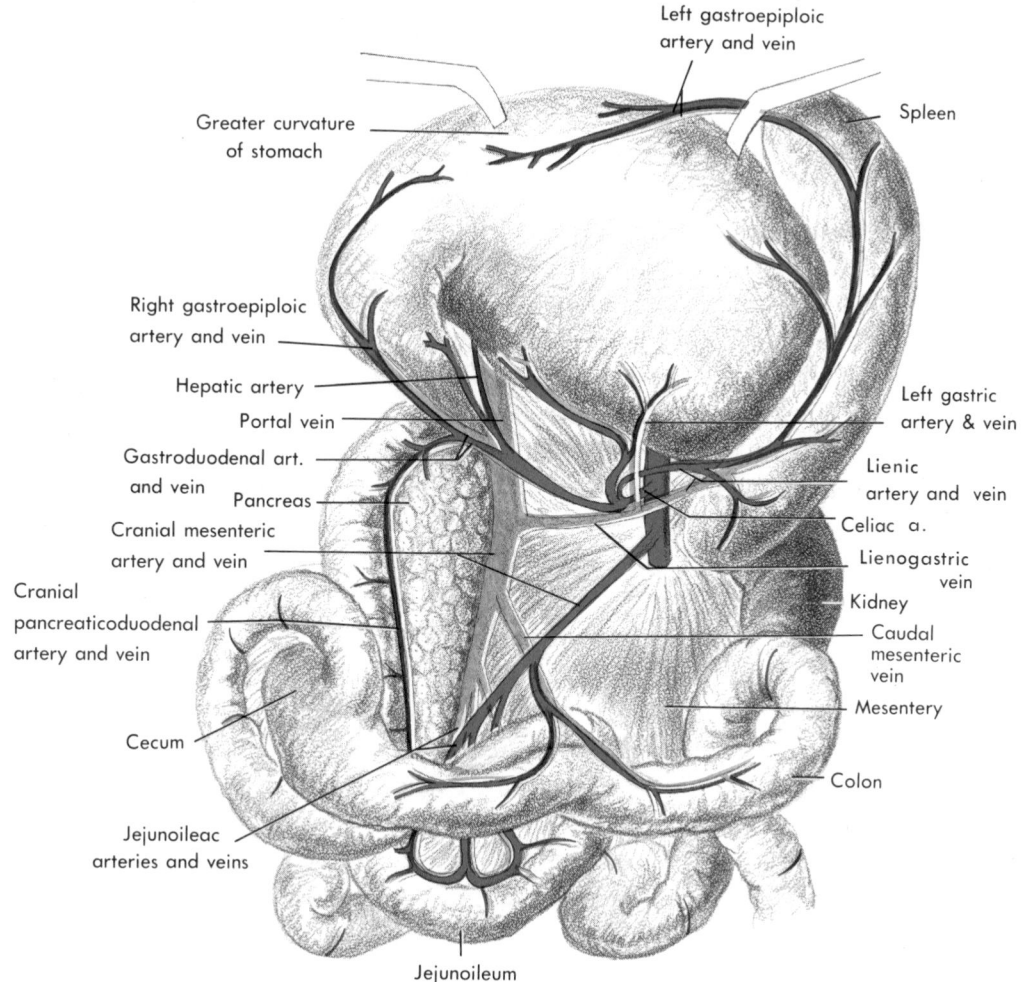

Figure 10–26. Ventral view of the hepatic portal system of veins and accompanying arteries of the cat. The stomach has been pulled forward and the tail of the pancreas dissected away.

of the pancreas, which extends toward the spleen, and notice that the portal vein is formed by the confluence of two tributaries—a lienogastric vein entering from the left side of the animal, and a much larger cranial mesenteric vein. Trace the **lienogastric vein** by continuing to dissect away pancreatic tissue. Its tributaries are a **left gastric vein,** which accompanies the left gastric artery, and drains the lesser curvature of the stomach, and a **lienic vein,** which accompanies the lienic artery to the spleen. A **left gastroepiploic artery** and **vein** can be found on the greater curvature of the stomach. They join the lienic vessels at several points.

Now trace the **cranial mesenteric vein.** One of its tributaries is the **caudal mesenteric vein** from the large intestine. Parts of this vein are accompanied by branches of the cranial mesenteric artery, and parts by branches of the caudal mesenteric artery. Other tributaries of the cranial mesenteric vein accompany branches of the cranial mesenteric artery to the caudal parts of the pancreas and duodenum (**caudal pancreaticoduodenal vein** and **artery**), and to the numerous coils of the small intestine (**jejunoileac veins** and **arteries**).

(C) Vessels of the Pelvic Region and Hind Leg

Return to the caudal ends of the caudal vena cava and aorta. The terminal branches of the aorta pass superficial to the tributaries of the vena cava as they enter the pelvis. In order to see the pelvic vessels clearly, open the pelvic canal. This is a simple procedure in the female. Cut the ventral ligament of the bladder and push it away from the cranioventral border of the pelvic girdle. Then take a scalpel, cut through the muscles on the ventral face of the girdle, and continue to cut right through the midventral symphysis. Bone scissors may be used, but this is not usually necessary if you keep in the midventral line. Now take a firm grip on the thighs and bend them as far dorsally as you can. The procedure for the male is the same, but one must use more caution to avoid reproductive ducts. First locate the cremasteric pouches that extend from the inguinal canals, across the ventral surface of the girdle and into the skin of the scrotum (Figs. 11–7 and 11–18, p. 346). They are very narrow in carnivores, but quite wide in the rabbit. They should be pushed aside before cutting. Also locate the penis emerging from the caudal end of the pelvic canal; avoid cutting it. After the canal is opened, carefully pick away fat and connective tissue from around the vessels, bladder, and rectum. Insofar as possible, confine your dissection to one side and do not injure parts of the urogenital system.

An **external iliac artery** extends from the aorta laterally and caudally toward the body wall and leg. It is accompanied distally by the **external iliac vein** (Figs. 10–23, 10–24, and 10–25). An **internal iliac artery** and **vein** enter the pelvic cavity. The iliac arteries arise independently from the aorta in the carnivores, but from a **common iliac artery** in the rabbit. The external and internal iliac veins of the carnivores unite to form a **common iliac vein** before entering the caudal vena cava; they enter the vena cava independently in the rabbit.

Trace the external iliac vessels. Usually just inside the abdominal wall, the external iliac artery and vein give off from their caudomedial surface a **deep femoral artery** and **vein,** which extend deep into the thigh (Fig. 10–27). A **caudal epigastric artery** and **vein** can be seen on the peritoneal surface of the rectus abdominis. They anastomose cranially with the cranial epigastric vessels previously seen. The caudal epigastric artery is usually a branch of the deep femoral, but it may arise directly from the external iliac near the deep femoral. An **external pudendal artery** and **vein** can be found in the mass of fat in the region of the groin. They continue through the fat and supply the external genitalia. The artery may be a branch of the caudal epigastric or of the deep femoral. The caudal epigastric and external pudendal veins normally form a short, common **pudendoepigastric trunk** before they join the deep femoral vein. After giving rise to these vessels, the external iliac vessels perforate the abdominal wall and enter the leg as the **femoral artery** and **vein.** Additional, major branches of these vessels are shown in Figure 10–27.

Now trace the internal iliac vessels. Near its origin from the aorta, the internal iliac artery gives rise to a **vesical artery** to the urinary bladder. This artery is a remnant of the large umbilical artery of the embryo which goes to the placenta. The proximal part of the embryonic umbilical artery persists as the vessel leading to the bladder, but the portion from the bladder to the umbilicus atrophies. Deeper within the pelvic cavity, the internal iliac artery gives rise to one or two **gluteal arteries** to deep pelvic muscles and to an **internal pudendal artery** to remaining pelvic viscera. **Gluteal** and **internal pudendal veins** accompany the arteries and drain into the internal iliac vein. A small **vesical vein** normally joins the internal pudendal.

After the iliac arteries have branched off, the aorta continues caudad as a very small vessel across the sacrum (**median sacral artery**) and into the tail (**caudal artery**). A **caudal vein** leads to a **median sacral vein** which normally enters a common iliac vein (carnivores), or an internal iliac vein (rabbit).

It has doubtless been noticed, during the above dissections, that parts of the mammalian venous system resemble parts of the system in other vertebrates, but some parts have changed considerably. The hepatic portal system is substantially the same as in lower tetrapods and the primitive lateral abdominal veins are represented by the

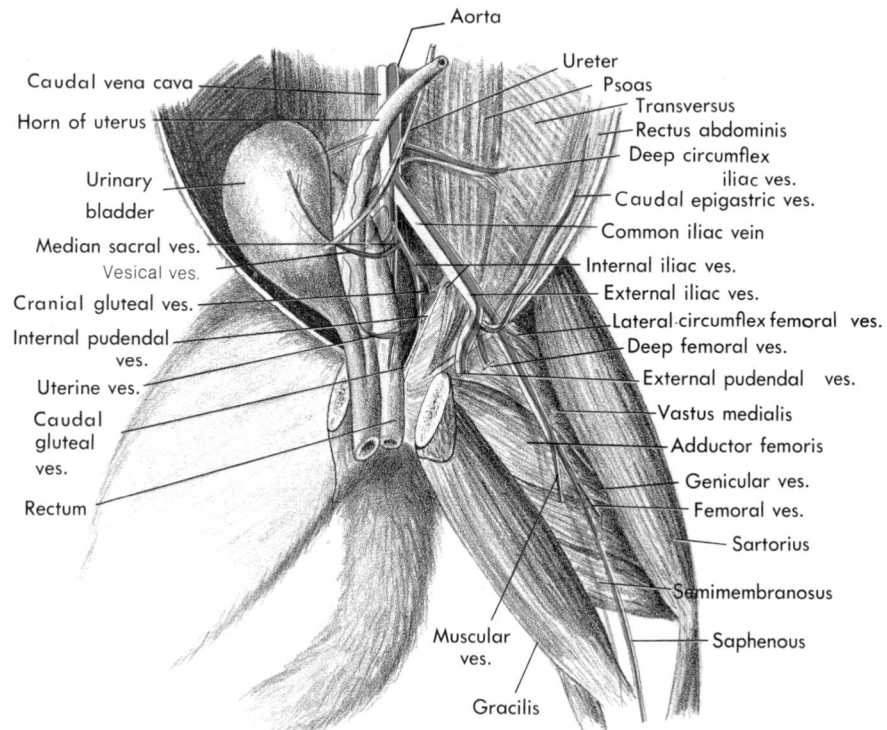

Figure 10–27. Ventral view of the distribution of the left external and internal iliac artery and vein in a female cat. The pelvic canal has been opened and the pelvic viscera pushed to the specimen's right side.

umbilical veins of the embryo (p. 278). The major change is the conversion of parts of the hepatic veins and the primitive cardinal and renal portal systems into a caval and azygos system. The way in which this comes about is best understood by recourse to the embryonic development of the veins in a mammal.

An early mammal embryo (Fig. 10–28, A) has a cardinal and an incipient renal portal system, for some of the blood in the caudal part of the posterior cardinals passes through the kidneys to a pair of subcardinals. In this stage the mammalian embryo is similar to the fish, except that in an adult fish the portion of the posterior cardinals situated just caudal to the cranial attachment of the subcardinals atrophies and the flow of renal portal blood through the kidneys and into the subcardinals is mandatory.

Later in development (Fig. 10–28, B) the right hepatic enlarges and a caudal extension of the vessel unites with the right subcardinal to form the proximal part of the caudal vena cava. The two subcardinals also unite with each other. This stage is not unlike the urodele.

Still later (Fig. 10–28, C) most of the cranial portion of the posterior cardinals atrophies, but the caudal portion on each side forms a large vessel connecting with the subcardinals. The essentially new feature, regarding the trunk veins of mammals, is the subsequent formation of a pair of **supracardinals** (Fig. 10–28, C) connecting cranially and caudally with the remnants of the posterior cardinals. The supracardinals also become connected with the subcardinals by a pair of **subsupracardinal anastomoses.** This connection makes possible the elimination of most of the caudal portion of the posterior cardinals (the renal portal system of lower vertebrates).

During subsequent development the supracardinals become divided into a cranial thoracic portion and a caudal lumbar portion (Fig. 10–28, D). The right subsupracardinal anastomosis and lumbar portion of the supracardinal enlarge, while those of the left side do not. Renal veins grow out from the subsupracardinal anastomosis to the definitive kidneys, which have migrated cranially.

By the adult stage (Fig. 10–28, E) all but the most caudal segments of the posterior cardinals are lost, the left subsupracardinal anastomosis is lost, and the posterior vena cava is extended caudad by the enlargement of the right subsupracardinal anastomosis and lumbar portions of the supracardinals. In some mammals only the right supracardinal is involved, but in the cat the right enlarges and absorbs the lumbar portion of the left supracardinal. Thus the adult caudal vena cava is formed of the right hepatic, a caudal outgrowth from the right hepatic, the middle section of the right subcardinal, the right subsupracardinal anastomosis, the lumbar portion of the supracardinals (especially the right supracardinal), and a small segment of the posterior cardinals. The renal veins are formed primarily by outgrowths from the subsupracardinal anastomosis, but the left subcardinal contributes to the left renal vein. The genital veins are formed from the subcardinals plus a small segment of the posterior cardinals; the cranial abdominals are formed from the subcardinals.

While these changes are taking place, the thoracic portion of the left supracardinal disappears. But the thoracic portion of the right supracardinal, together with the proximal end of the right posterior cardinal, forms the azygos. Carnivores do not have a left highest intercostal, but the vessel develops in such mammals as have it (rabbit) from the stump of the left posterior cardinal.

The formation of the cranial vena cava is a simpler affair. In a mammal such as a rabbit, the condition shown in Figure 10–28, C, persists. The two cranial venae cavae represent the common cardinals plus the proximal portion of the anterior cardinals. The more distal portion of the anterior cardinals is represented by the internal jugular. The external jugular is a new outgrowth. But in mammals such as the cat or mink, a cross anastomosis, which is to be the left brachiocephalic, develops between the anterior cardinals (Fig. 10–28, D). The right cranial vena cava is formed as above, but a left

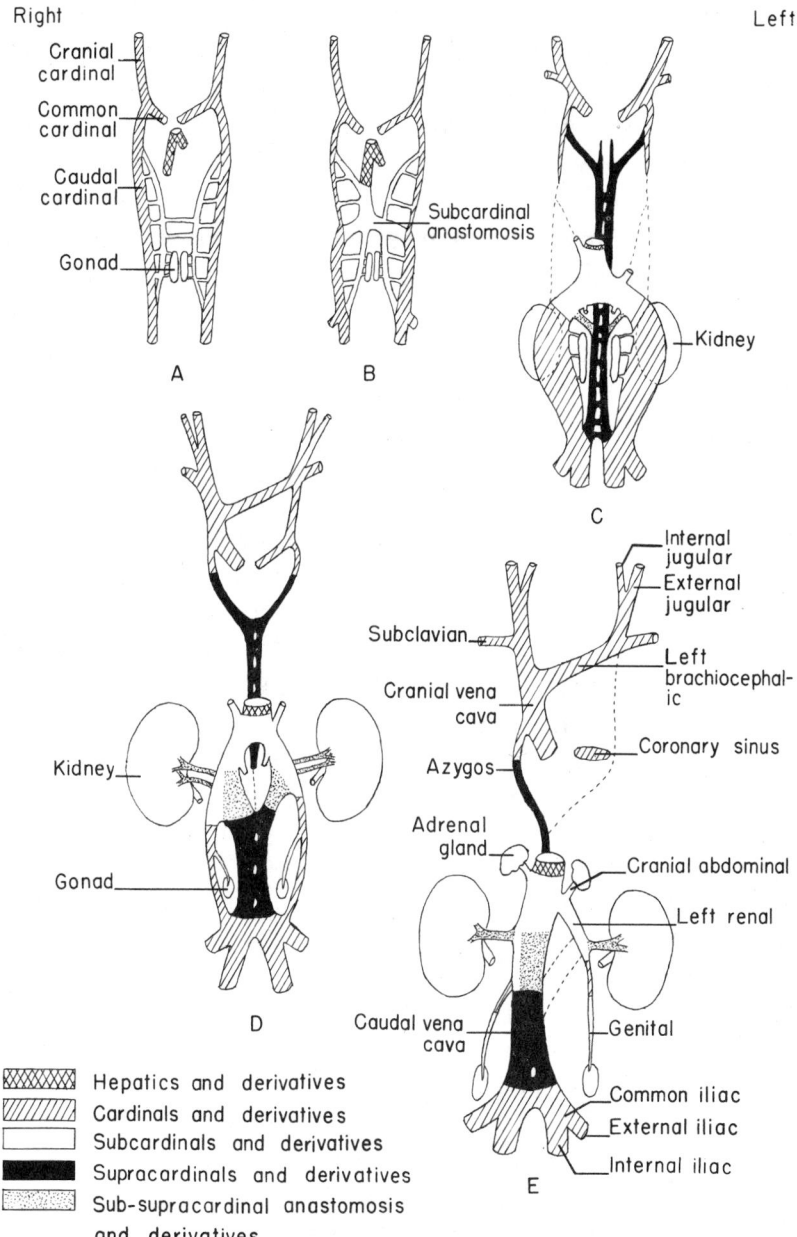

Right   Left

Cranial cardinal

Common cardinal

Caudal cardinal

Gonad

A   B

Subcardinal anastomosis

Kidney

C

Subclavian

Cranial vena cava

Azygos

Adrenal gland

Kidney

Gonad

D

Caudal vena cava

Internal jugular
External jugular

Left brachiocephalic

Coronary sinus

Cranial abdominal

Left renal

Genital

Common iliac

External iliac

Internal iliac

E

Hepatics and derivatives
Cardinals and derivatives
Subcardinals and derivatives
Supracardinals and derivatives
Sub-supracardinal anastomosis and derivatives

Figure 10–28. A series of diagrams ranging from a young embryo *A,* to an adult, *E,* to show the development of the major veins of the cat from the primitive cardinal and renal portal system. All are ventral views. For explanation, see text. (Slightly modified after Huntington and McClure, The development of the veins in the domestic cat, Anatomical Record, vol. 20.)

one does not form, for the proximal portion of the left anterior cardinal atrophies. The left common cardinal persists, however, as the coronary sinus.

Bronchi and Internal Structure of the Heart

Cut the great vessels near the heart of your specimen, remove the heart, and examine the roots of the lungs. The bifurcation of the trachea into bronchi referred to earlier (p. 266) can now be exposed. Trace a bronchus into a lung and notice that it subdivides repeatedly into smaller and smaller passages that terminate in clusters of thin-walled, microscopic sacs (the **alveoli**) where gas exchange occurs. This entire complex of passages is called the **respiratory tree.**

Again identify the chambers of the heart and the great vessels entering and leaving it as they appear in a ventral view (p. 306 and Fig. 10–16). Carefully clean the dorsal surface of the heart and identify the chambers and vessels in this view (Fig. 10–29). Internal features can be seen by dissecting either the heart of your own specimen or a separate sheep heart. The latter is preferable, if material is available, for the structures are larger and the chambers are not clogged with the injection mass. If a sheep heart is used, you will have to remove the pericardial sac and clean and identify the great vessels. They are similar to those of the mammal you have studied except that both the subclavian and common carotid arteries leave the arch of the aorta by a common brachiocephalic trunk. A small left cranial vena cava is also present in the sheep and the ligamentum arteriosum is conspicuous.

Open the right atrium by making an incision that extends from the auricle into the caudal vena cava; the left atrium, by an incision extending from its auricle through one of the pulmonary veins. To open the ventricles, first cut off the apex of the heart in the transverse plane. Cut off a sufficient amount to expose the cavities of both ventricles. Then make a cut through the ventral wall of the right ventricle and extend it from the cut surface made by removing the apex into the pulmonary artery. This will be a diagonal incision. Open the left ventricle by making an incision through its ventral wall that extends from the cut surface as far forward as the base of the arch of the

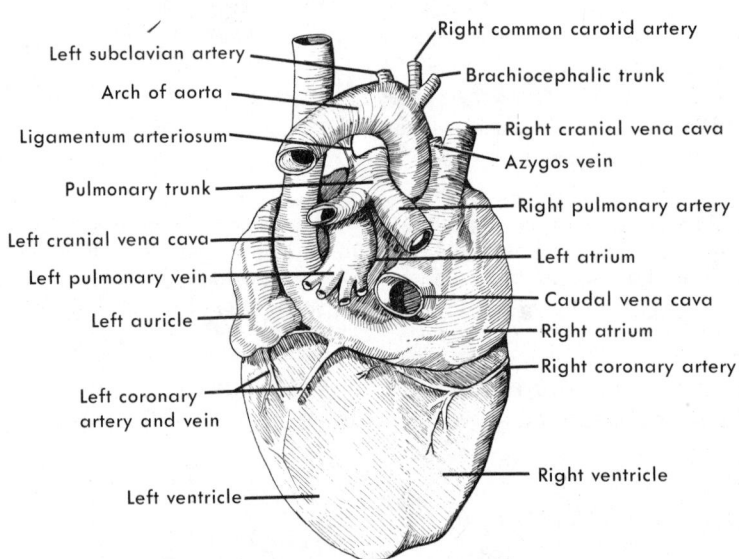

Figure 10–29. Dorsal view of the heart and great vessels of a rabbit. The base of the left cranial vena cava is represented in carnivores by the coronary sinus.

aorta. This will be a longitudinal incision. Clean out the chambers of the heart if necessary.

Find the entrance of the **right cranial vena cava** and **caudal vena cava** into the **right atrium.** The entrance of the **coronary sinus** (carnivores) or **left cranial vena cava** (rabbit) lies just caudal to the entrance of the caudal vena cava. The extent of the coronary sinus can be determined by probing. Also find the entrances of the **pulmonary veins** into the **left atrium.** The atria have relatively thin muscular walls; however, the muscles in the **auricles** form prominent bands known as **pectinate muscles** because they resemble a comb, or pecten.

The two atria are separated by an **interatrial septum.** Examine the septum from the right atrium and you will find an oval-shaped depression, the **fossa ovalis,** beside the point at which the caudal vena cava enters. Put your thumb in one atrium and forefinger in the other, and palpate this region. You will feel that the septum is unusually thin here. During embryonic life there is an opening, the **foramen ovale,** through the septum at this point and much of the blood in the right atrium (mostly blood coming in by the caudal vena cava) is sent directly to the left side of the heart and out to the body. This opening closes at birth.

The **atrioventricular openings** will be seen in the floor of the atria. The right one is guarded by the **right atrioventricular,** or **tricuspid valve,** which consists of three flaps; the left one by the **left atrioventricular,** or **bicuspid valve,** which consists of two flaps (Fig. 10–30). Since these flaps extend into the ventricles, they can be seen better from that aspect. Note that little tendinous cords **(chordae tendineae)** connect the margins of the flaps with the walls of the ventricles. Many of the chordae attach onto papillalike extensions of the ventricular muscles **(papillary muscles).** The chordae tendineae may help to open the valves, but in any case they prevent the valves from everting into the atria during ventricular contraction.

Notice that the ventricles are separated from each other by an **interventricular septum** and that the walls of the ventricles are much thicker than those of the atria. The left ventricular wall is also much thicker than the right one because the left ventricle pumps blood to all of the body except for the lungs. In addition to the papillary muscles, the inside of the ventricular walls bears irregular bands **(trabeculae carneae),** and sometimes bands that cross the lumen **(moderator bands).** There is a particularly prominent moderator band in the right ventricle of the sheep. Moderator bands are believed to prevent the overdistention of the ventricle.

Notice where the pulmonary trunk and arch of the aorta leave the ventricles. Three pocket-shaped, **semilunar valves** are located in the base of each vessel, for this part of each vessel developed from a splitting of the conus arteriosus. Those in the pulmonary artery are known as the **pulmonary valve;** those in the aorta, as the **aortic valve.** The two coronary arteries leave from behind two of the semilunar valves in the aorta. One has probably been cut through.

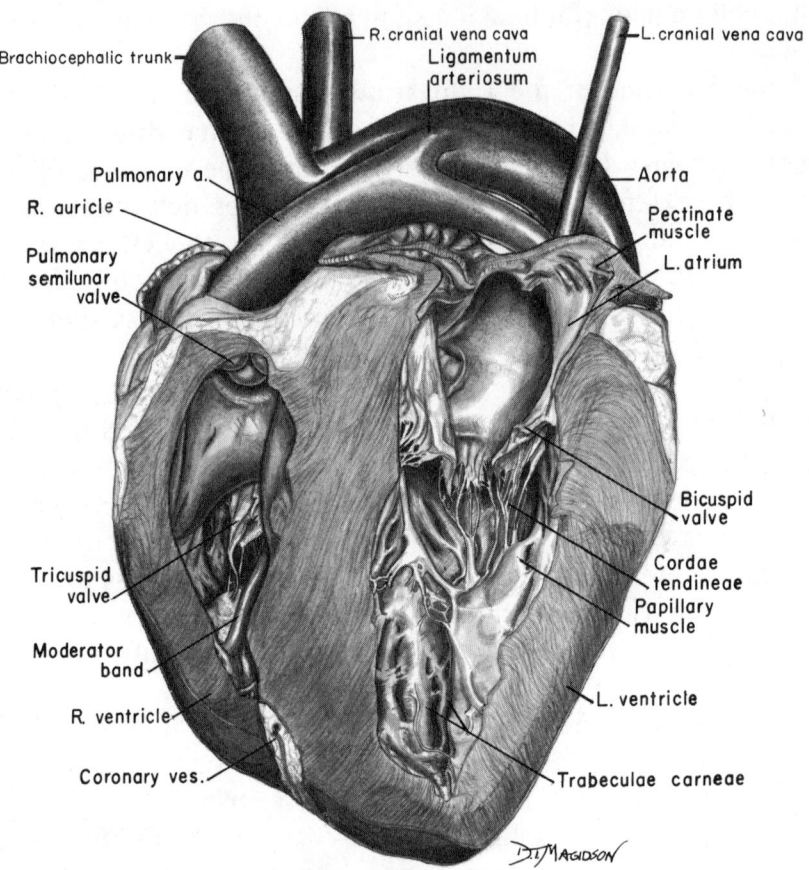

Figure 10–30. Ventral view of a dissection of the sheep heart.

Lymphatic System

The relation of the lymphatic to the cardiovascular system was considered in the introduction to this chapter (p. 276). Although the lymphatic system is not conspicuous enough in most vertebrates to be studied easily, parts, at least, of the system can be seen in mammals even though it has not been specially injected. The following directions are based on the cat, inasmuch as the lymphatics are easier to find in that animal than in the mink or rabbit, but are applicable to other mammals as well.

The major lymphatic vessel of the body is the **thoracic duct** (Fig. 10–31). This is a brownish vessel that can be found in the left pleural cavity just dorsal to the aorta. Sometimes the vessel is divided into two or more channels. Trace it forward. It passes deep to most of the arteries and veins at the front of the thorax and base of the neck, and then curves around to enter the left bijugular trunk beside the entrance of the subclavian vein (Fig. 10–16). Now trace it caudad. It passes through the diaphragm dorsal to the aorta and, dorsal to the origin of the celiac and anterior mesenteric arteries, expands into a sac called the **cisterna chyli.**

Next stretch out a section of the mesentery supporting the small intestine and hold it up to the light. Very small lymphatic vessels, in this case called **lacteals** because absorbed fat passes through them, can be seen outlined by little streaks of fat. These ultimately lead into an aggregation of **mesenteric lymph nodes** (pancreas of Aselli) located at the base of the mesentery. The mesenteric lymph nodes are drained by one or more larger lymph-

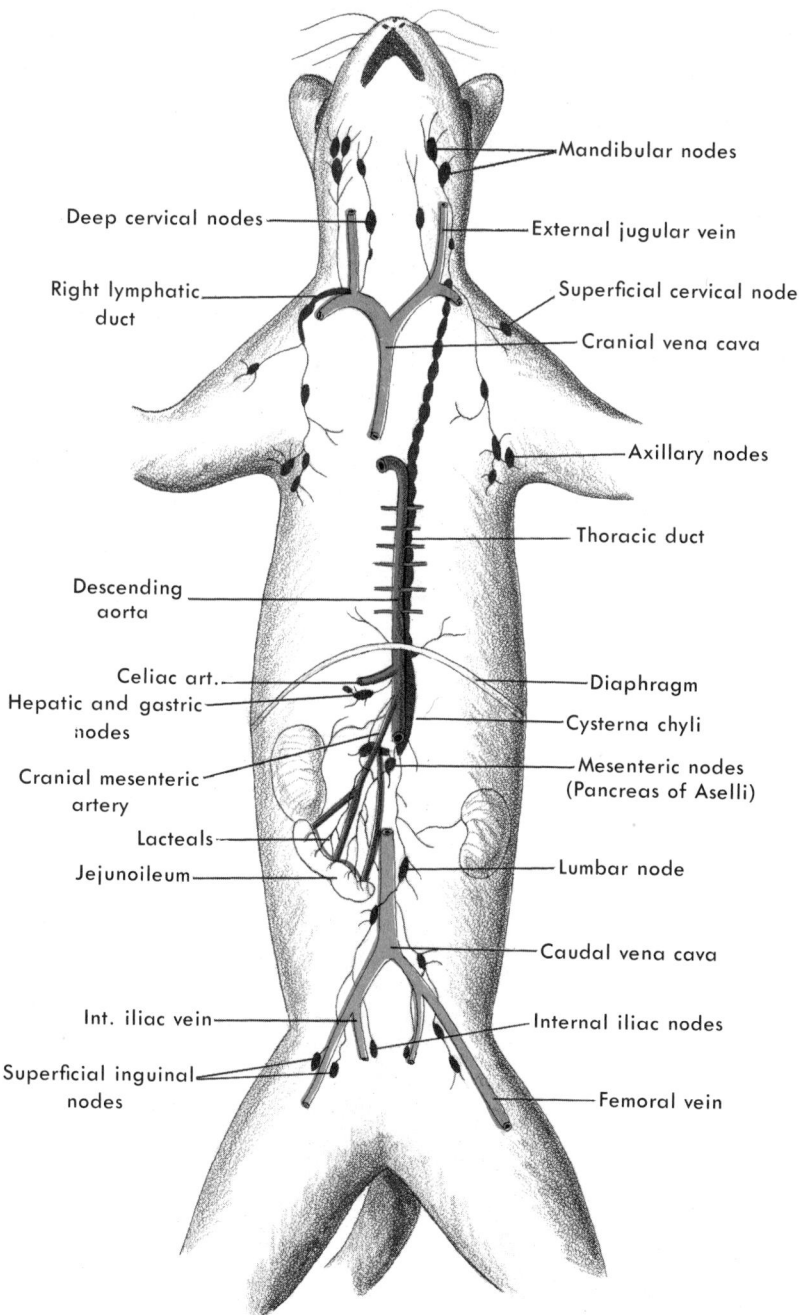

Figure 10–31. Diagrammatic ventral view of the major lymphatic vessels and groups of lymph nodes in a cat.

atics that pass along the cranial mesenteric artery to the cisterna chyli. These vessels have probably been destroyed. Lymphatic vessels from the stomach, liver, pelvic canal and hind legs also pass to the cisterna chyli, but they are hard to see. Thus, the cisterna chyli receives all the lymphatic drainage of the body caudal to the diaphragm and passes it on to the thoracic duct. The thoracic duct receives the lymphatic drainage of the thorax as it ascends through this region.

Other lymphatic vessels, which parallel the larger veins, drain the arms, neck and head. Those of the left side enter the thoracic duct, or the left bijugular trunk close to the entrance of the thoracic duct. Those of the right side enter the right bijugular trunk near its union with the subclavian, either independently or by a short common trunk (the **right lymphatic duct**).

11

THE EXCRETORY AND
REPRODUCTIVE SYSTEMS

The excretory system plays an important role in eliminating the nitrogenous waste products of cellular metabolism and helps to control the water balances of the body. The reproductive system has an entirely different function, namely, perpetuating the species. However, the two must be considered together morphologically, for in the males of most vertebrates excretory passages are utilized for the transport of the sperm, and in some cases the female genital ducts develop from excretory ducts. In view of this intimate morphological association, the two systems are sometimes referred to as the urogenital system.

History of the Kidney

A brief consideration of the history of the kidneys and their ducts is a prerequisite for an understanding of the urogenital system. The functional units of the kidneys are the **renal tubules** (nephrons). In all vertebrates they develop embryonically from a pair of **nephric ridges** located dorsal to the coelom between the somites and lateral plate mesoderm. In the ontogeny of an amniote, a pronephric kidney **(pronephros)** is succeeded by a **mesonephros** which, in turn, is succeeded by the definitive **metanephros** (Fig. 11–1). These kidneys have a linear relationship from cranial to caudal along the nephric ridge. The pronephros forms the **archinephric duct,** which extends caudad to the cloaca. The mesonephric tubules tap into this duct, but the metanephros is drained by a **ureter** which develops as a craniad outgrowth from the caudal end of the archinephric duct.

A common statement is that the ontogeny described above is a recapitulation of phylogeny, but this is an oversimplification. In anamniotes the kidney tubules develop from all the nephric ridge as in amniotes, but the adult kidney is one that occupies both the mesonephric and the potential metanephric portion of the ridge. The cranial part, at least, of such a kidney is drained by the archinephric duct, but the caudal part may be drained by one or more accessory, ureterlike ducts. Such a kidney has often been called a mesonephros but is more appropriately called an **opisthonephros,** for it is obviously somewhat different from the mesonephros of an amniote embryo. A still more primitive kidney would be one that occupies the entire nephric ridge and retains the primitive segmental arrangement of the tubules that is seen in the pronephros. (Later kidneys have a multiplication of tubules so there is more than one per segment.) Such a kidney is called an **archinephros,** or holonephros. The archinephros is largely

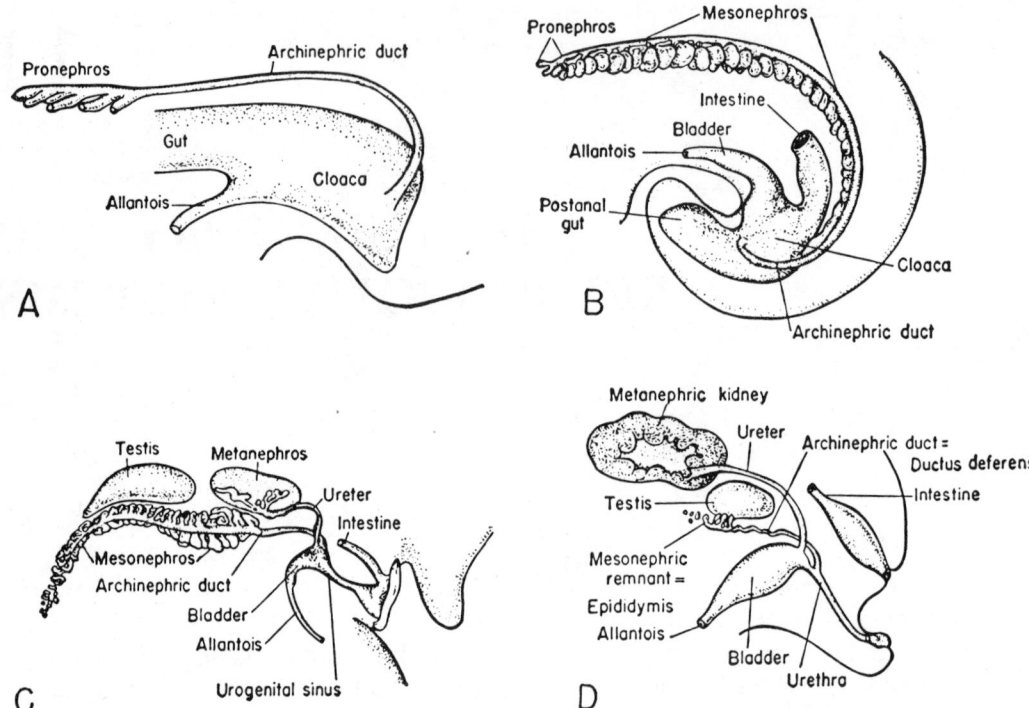

Figure 11–1. Diagrams to show the succession of kidneys in an amniote embryo. All are lateral views as seen from the left side. The genital details in these diagrams are those of a male. *A,* An early stage in which the pronephros and its duct are formed; *B,* mesonephric tubules tapping into the archinephric duct; *C,* pronephros degenerating, mesonephros functional, metanephros and ureter forming; *D,* definite stage in which the metanephros is functional and the remnants of the mesonephros and its duct are taken over by the male genital system. (From Romer, The Vertebrate Body.)

a theoretical kidney, but a close approach to it is seen in the larval hagfish in which the main kidney consists of segmentally arranged tubules and occupies all but the pronephric portion of the ridge. (The pronephric tubules form a specialized head kidney.) Thus, the evolutionary history appears to be one of a progressive shortening of the kidney and caudad concentration of its functions, with an archinephros succeeded by an opisthonephros, and the opisthonephros succeeded in amniotes by a metanephros.

Genital Ducts and Their Relation to Excretory Ducts

In very primitive vertebrates, such as the cyclostomes, the gametes of both sexes are discharged into the coelom and pass to the outside through a pair of genital pores. This may have been the ancestral condition, but in other vertebrates ducts have evolved that transmit the gametes. In the embryo of either sex, primordia for the ducts of both the male and female are laid down during a **sexually indifferent stage.** In the dogfish and some other primitive vertebrates the **oviduct** develops from a splitting of the archinephric duct; the coelomic opening of the duct (ostium) develops from the coelomic funnel **(nephrostome)** of a primitive kidney tubule. This mode of origin may be primitive, but in higher vertebrates the oviduct arises from a folding of the coelomic epithelium. In addition to the primordium of an oviduct, the same embryo acquires a series of cords, the **cords of the urogenital union** *(rete testis),* that connect the gonad with the cranial mesonephric tubules and, through them, with the archinephric duct. Thus, two potential routes are present for gamete transport.

During the subsequent differentiation of the female, the oviduct further develops, while the cords of the urogenital union and the adjacent parts of the mesonephros degenerate. In the adult female anamniote this part of the embryonic mesonephros forms the slender, more or less functionless, cranial end of the opisthonephros. In adult female amniotes, which have a metanephros and ureter, the mesonephros is represented by minute, functionless groups of tubules called the **epoophoron** (more cranial mesonephric tubules) and **paroophoron** (more caudal tubules), and the archinephric duct degenerates or forms a vestige known as the **longitudinal duct of the epoophoron** (Fig. 11–2).

During the subsequent differentiation of the male the oviduct degenerates (although some vestiges may persist), and the route through the kidney becomes the functional pathway for the sperm. In adult anamniotes the testis and the cranial end of the mesonephros (now the cranial end of the opisthonephros) are some distance apart. The sperm passes through the cords of the urogenital union (now called the **ductuli efferentes**) located in the mesorchium, through the cranial kidney tubules, and into the archinephric duct. In adult amniotes the testis and the cranial end of what was the mesonephros are close together, the pathway for the sperm is the same, but the terminology is different. The cords of the urogenital union constitute the **rete testis;** the cranial part of the mesonephros constitutes the head of the head of the **epididymis,** and its tubules are called the **ductuli efferentes** (it should be noted that different tubules, the cords of the urogenital union, are given this name in lower vertebrates); the highly coiled cranial part of the archinephric duct constitutes the body and tail of the epididymis and is called the **ductus epididymis;** and the rest of the archinephric duct is called the **ductus deferens.** The more caudal parts of the mesonephros sometimes form a functionless vestige called the **paradidymis.**

Study of the Excretory and Reproductive Systems

It is assumed that the major parts of the excretory and reproductive systems have been observed in previous dissections. In these exercises the finer aspects will be examined and related to the more conspicuous parts. In studying these systems you should not only dissect your own specimen but also examine the dissection of a specimen of the opposite sex. Since someone else, in turn, will have to examine your specimen, make a particularly careful dissection. If possible, sexually mature specimens should be studied.

FISHES

The excretory and reproductive systems of the cartilaginous fishes are a good example of a reasonably primitive vertebrate condition in most respects. The kidneys are opisthonephroi drained by archinephric ducts, which are supplemented in the male by accessory urinary ducts. The gonads are situated far forward in the body cavity. The eggs are discharged through the coelom and a pair of oviducts; the sperm through the kidneys and archinephric ducts. A cloaca is present.

In certain more subtle features, however, the urogenital system is not entirely primitive. The kidney tubules of cartilaginous fishes have large glomeruli that remove a considerable volume of water from the blood. This type of tubule is generally regarded as primitive. It is advantageous in a fresh-water environment, which may have been the ancestral vertebrate environment, but its presence in these marine fishes, in which the osmotic problem is seemingly one of conserving water, poses problems. Cartilaginous fishes compensate for water loss through the kidneys by retaining considerable urea

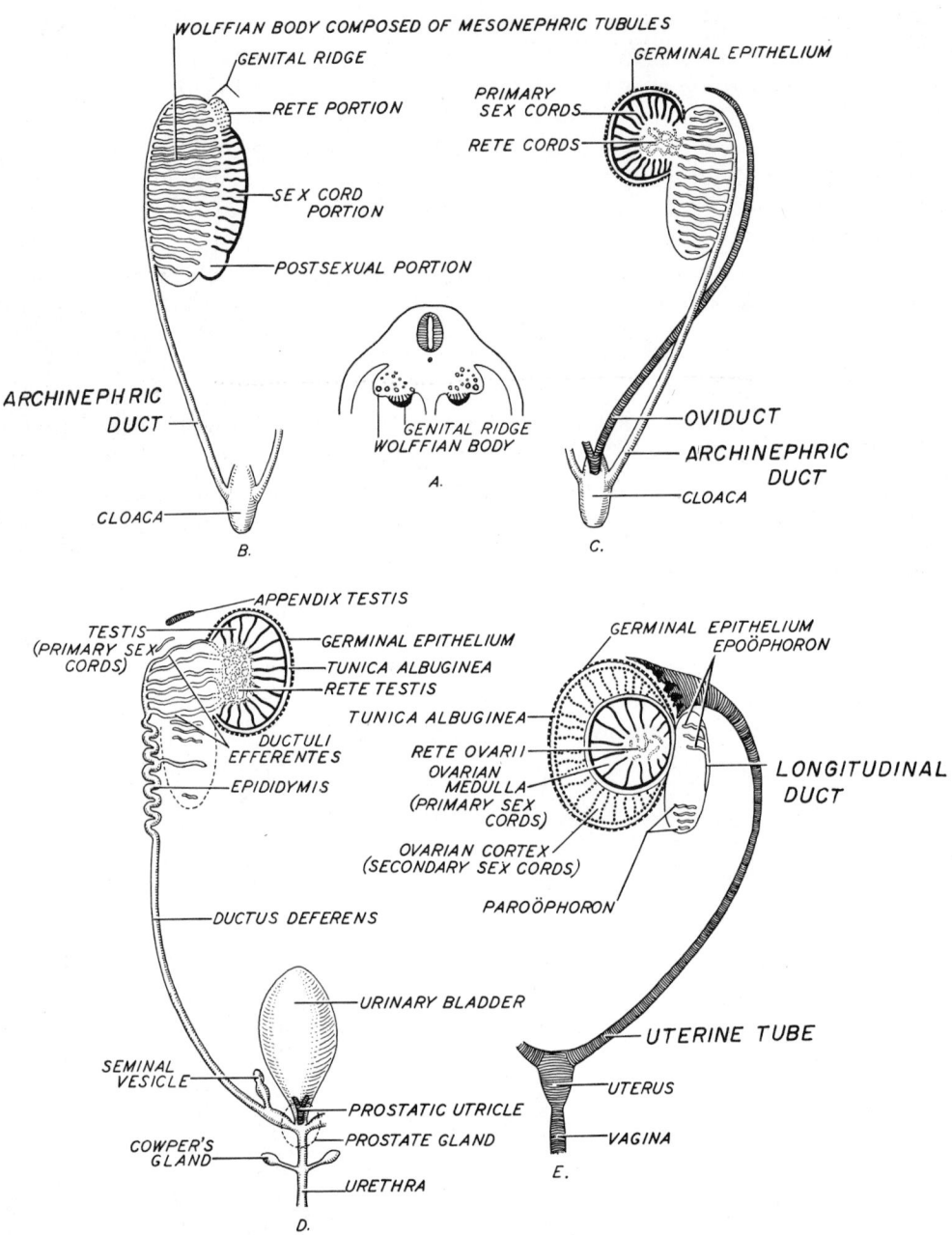

Figure 11-2. Diagrams of the development of the amniote genital system. *A,* A cross section of the early embryo showing the location of the mesonephros (wolffian body) and developing gonad (genital ridge); *B,* ventral view of an early embryo; *C,* the indifferent stage; *D,* differentiation of the male condition. The primary sex cords of the gonad of the indifferent stage become the seminiferous tubules. *E,* Differentiation of the female conditions. The primary sex cords regress, and the follicles develop from secondary sex cords. (From Turner, General Endocrinology.)

in their blood. As a consequence the blood osmotic pressure slightly exceeds that of sea water, and water enters the body by osmosis. Retention of urea results from the reabsorption of urea by the kidney tubules and a reduction in the permeability of the gill membranes to urea so that little is excreted by this route. These features are peculiar to cartilaginous fishes.

Another specialized feature of many cartilaginous fishes, including *Squalus,* is the retention of their young within a uterus until embryonic development is complete. More primitive vertebrates are egg-laying.

Define amniote & anamniote
Describe kidney devel pro mess meta
♂ ♀ differences · Opisthonephric

Kidneys and Their Ducts

The kidneys of the dogfish are a pair of bandlike organs lying dorsal to the parietal peritoneum, a position called **retroperitoneal,** on either side of the dorsal mesentery (Fig. 11–3). A conspicuous, white **caudal ligament** arises from the vertebral column between them and passes into the tail. They are **opisthonephric kidneys,** for they extend nearly the length of the pleuroperitoneal cavity; they are drained, in part at least, by **archinephric ducts** leading to the **cloaca,** and they have a relatively primitive tubule structure. (Traces of microscopic nephrostomes are associated with some of the cranial tubules.) You may have to cut the parietal peritoneum along the lateral border of a

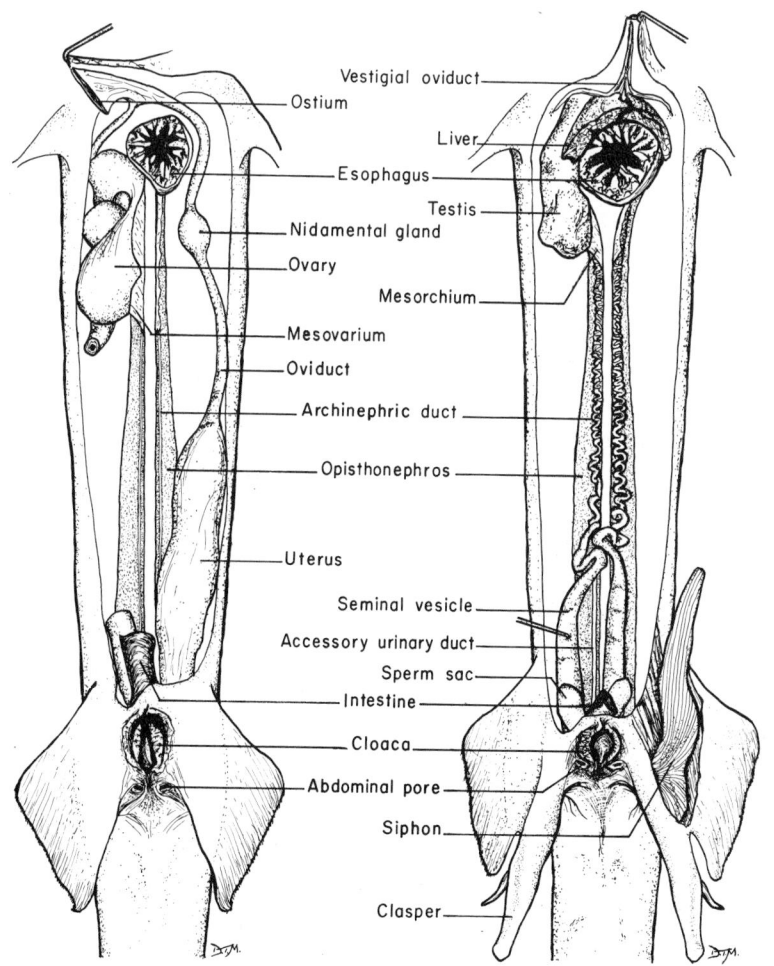

Figure 11–3. Ventral views of the urogenital system of mature specimens of *Squalus. Left,* female; *right,* male. The siphon has been dissected free on one side of the male.

kidney to trace it cranially, for the cranial two thirds of each kidney is narrower and less conspicuous than the caudal one third. It is in the caudal third that most of the urine production occurs. The cranial part is related to the reproductive system in the male; it is somewhat degenerate in the female.

The archinephric duct can easily be seen in the mature male, for it is a large, highly convoluted tube lying on the ventral surface of the opisthonephros. It is much smaller and straighter in the immature male. The duct of the female resembles that of an immature male and cannot be seen until the oviduct is studied.

Further aspects of the excretory system must be considered separately in each sex.

Male Urogenital System

Notice that the paired **testes** are located near the cranial end of the pleuroperitoneal cavity adjacent to the cranial end of the kidneys (Fig. 11–3). Each is supported by a **mesorchium** through the front of which pass several small, inconspicuous tubules called the **ductuli efferentes.** The ductuli efferentes carry the sperm from the testis to modified cranial kidney tubules. After passing through these tubules, the sperm enter the **archinephric duct** and descend through it to the cloaca. The large size of the mature male archinephric duct is attributed to its role in sperm transport.

The part of the opisthonephros receiving the ductuli efferentes is homologous to the head of the epididymis of amniotes; the highly coiled portion of the archinephric duct adjacent to this region is homologous to the ductus epididymis, which comprises the body and tail of the epididymis; and the rest of the archinephric duct is comparable to the ductus deferens.

The portion of the opisthonephros between the caudal end of the testis and the enlarged, caudal excretory region of the kidney is known as **Leydig's gland.** Most of the tubules in this region are modified to produce a secretion analogous to the seminal fluid of higher vertebrates. This secretion is discharged into the archinephric duct.

As the archinephric duct approaches the excretory portion of the kidney it straightens and enlarges to form a **seminal vesicle.** Remove the parietal peritoneum from this portion of the kidney and trace the seminal vesicle caudad. You will also have to open the cloaca on the side on which you are working by cutting through the side of the cloacal aperture and into the lateral wall of the intestine. The caudal end of the seminal vesicle passes dorsal to a **sperm sac,** whose blind cranial end should be freed from the seminal vesicle. Secretions of the seminal vesicle and sperm sac contribute to the seminal fluid. The caudal end of the sperm sac and that of the seminal vesicle unite to form a **urogenital sinus.** The sinus develops from the archinephric duct. Cut open the ventral surface of the sperm sac and sinus, clean them out, and notice the papilla that bears the opening of the seminal vesicle.

The urogenital sinuses of opposite sides unite caudal to the entrances of the seminal vesicles and extend into the **urogenital papilla** located dorsally in the cloaca. Probe caudally through the urogenital sinus and notice the emergence of the probe through the tip of the urogenital papilla.

The sperm sacs appear to be diverticula from the urogenital sinus, but they develop from the caudal ends of the oviducts, which are present in the sexually indifferent stage of the embryo. Other remnants of the paired **oviducts** are tubular folds which can be found cranially on either side of the falciform and coronary ligaments. These portions of the oviducts unite in the falciform ligament and have a common entrance, the **ostium,** into the coelom. The ostium can be found along the caudodorsal edge of the ligament. The intervening portions of the oviducts are lost in adult males.

The tubules of the excretory part of the kidney do not enter the seminal vesicle, but rather enter an **accessory urinary duct** which lies against the dorsomedial edge of the seminal vesicle. This duct can be found by freeing the lateral edge of the seminal vesicle and dissecting dorsal to the vesicle. The accessory urinary duct, despite its name, carries virtually all the urine, for little if any urine is excreted by the cranial parts of the kidney in a mature male. Trace the accessory duct caudad; it enters the urogenital sinus caudal to the entrance of the seminal vesicle.

It can now be appreciated that the cloaca is a sort of sewer (Latin, cloaca = a sewer), for it receives the feces from the digestive system, the urine from the excretory system, and the gametes from the reproductive system. The cloaca is not divided in the male, but the excretory and genital products enter more dorsally and caudally than the feces.

Unlike most fishes, fertilization in the dogfish and other cartilaginous fishes is internal. During copulation one of the **claspers** on the pelvic fins of the male is turned forward and inserted into the cloaca and oviduct orifice of the female. The sperm proceed from the cloaca of the male into the groove on the dorsal surface of the clasper, and thence into the female. A sac-shaped muscular walled **siphon** is associated with each clasper. One can be found by skinning the ventral surface of the pelvic fin, if this has not already been done (p. 121). Cut open the siphon, and you can pass a probe through it into the groove on the clasper. It had long been thought that sea water enters the siphon and then is forcibly ejected during copulation, thus propelling the sperm along the clasper groove. But Heath (1956) finds no evidence that water is taken up. Rather, the siphons secrete copious amounts of a mucopolysaccharide which may lubricate the claspers and contribute to the seminal fluid.

Female Urogenital System

The **ovaries** are a pair of large organs located near the front of the pleuroperitoneal cavity adjacent to the cranial ends of the kidneys (Fig.

11–3). Each is supported by a **mesovarium** and contains eggs in various stages of maturity. When the eggs are mature, they attain a diameter of nearly three centimeters and contain an enormous amount of yolk. Cut into one of the larger eggs to see the yolk. Each egg is surrounded by a sheath of follicular cells, but this cannot be seen grossly.

When the eggs are mature, they break through the wall of the follicle and ovary (a process called **ovulation**), pass into the coelom, and enter the front of the paired **oviducts.** In mature females each oviduct is a prominent tube suspended by a **mesotubarium** from the ventral surface of the kidney. In immature specimens the oviducts are small tubes lying against the kidneys, and mesotubaria are lacking. Trace an oviduct cranially. It passes dorsal to the ovary and then curves ventrally and caudally in front of the liver to enter the falciform ligament. The oviducts of opposite sides unite within the falciform ligament and have a common opening, the **ostium,** into the coelom. The ostium is located on the caudodorsal edge of the ligament, and can be opened by spreading its lips apart. Its location appears to be an adaptation for the reception of the unusually large, heavy eggs. In most lower vertebrates the ostia are separate and situated dorsally near the front of the pleuro-peritoneal cavity.

The oviduct is narrow in diameter throughout much of its length but enlarges in two regions. One enlargement lies dorsal to the ovary. This is the **nidamental gland**—a gland that secretes a thin, horny shell known as the candle around groups of two or three eggs as they come down the oviduct. There is also evidence that sperm is stored in this region for some time before fertilization. The other enlargement, the **uterus,** occupies approximately the caudal one third to one half of the oviduct. It is very large in pregnant females, for the embryos develop here.

Open the cloaca by cutting through the side of the cloacal aperture and into the lateral wall of the intestine. The two oviducts enter the caudodorsal part of the cloaca just ventral to a **urinary papilla.** The urogenital portion of the cloaca, known as the **urodeum,** is partially separated by a horizontal fold from the anteroventral, fecal portion of the cloaca (**coprodeum**).

The kidneys of the female *Squalus* are drained by the **archinephric ducts,** for there are no accessory urinary ducts as there are in the male. Some female elasmobranchs, however, have accessory ducts. An archinephric duct can be found by making an incision through the parietal peritoneum along the lateral border of the caudal part of the kidney and very carefully reflecting the parietal peritoneum from the surface of the kidney. The archinephric duct lies on the ventral surface of the kidney directly dorsal to the attachment of the mesotubarium. If you do not see it on the kidney, it probably adhered to the dorsal surface of the parietal peritoneum and can be picked off the peritoneum. The archinephric duct is much smaller than in the male and is not convoluted. Trace it caudad. The caudal ends of the ducts of opposite sides enlarge slightly and unite to form a small **urinary sinus** which opens through the tip of the urinary papilla. The sinus is too small to be easily dissected.

Reproduction and Embryos

As the eggs develop within the follicles in the ovary, certain ones begin to increase in size through the accumulation of yolk. At the time of ovulation each ovary contains two or three ova averaging three centimeters in diameter. After ovulation the follicular cells are converted into a corpus luteum. This is a large body two centimeters in diameter in an early pregnancy, but it gradually regresses. The eggs enter the oviduct, which can stretch greatly in a living specimen. As they pass through the nidamental gland, they are fertilized and a thin, horny shell (the candle) is deposited around several of them. This mass then passes to the uterus, in which it may be seen in specimens in an early stage of pregnancy.

After several months the shell breaks down and the embryos develop within the uterus. Hisaw and Albert (1947) find that the gestation period lasts nearly two years. Pups that are slightly over a year old, which is a stage often seen in pregnant specimens obtained from biological supply houses, range in length from 12 to 20 centimeters; much of the yolk is contained within an **external yolk sac** suspended from the underside of the embryo (Fig. 11–4). This is a **trilaminar yolk sac** (Fig. 11–6), for it contains all three germ layers. The rest of the yolk is carried in an **internal yolk sac,** which can be found within the pleuroperitoneal cavity. Pups just before parturition range in length from 23 to 29 centimeters. The yolk in the external sac has been consumed, but a small reserve remains in the internal sac.

Numerous vascular **uterine villi** line the uterus and are applied to the surface of the embryo, especially its thin-walled, vascular, external yolk sac. The embryo doubtless obtains water from the mother by this pseudo-placental relationship, but most, if not all, of the other nutritional needs are supplied by the yolk, which is moved by ciliary action into the embryo's intestine where it is digested and absorbed. During the course of development there is a steady decrease in the weight of organic matter in the embryo and yolk but a 78 per cent gain in weight because of a great increase in water and minerals. This mode of development, in which the young are born as miniature adults but do not receive much nutrition from the mother, is referred to as **ovoviviparous** development. In a few sharks, such as *Mustelus laevis,* there is an intimate union between the yolk sac and maternal tissues, and the embryos derive most of their nutritional requirements from the mother through this yolk sac placenta. Such animals are said to be **viviparous.** Sharks regarded as reproductively more primitive than *Squalus* are **oviparous,** for they lay eggs and the embryos develop in the surrounding water. The eggs of many oviparous vertebrates are provided with only a moderate amount of

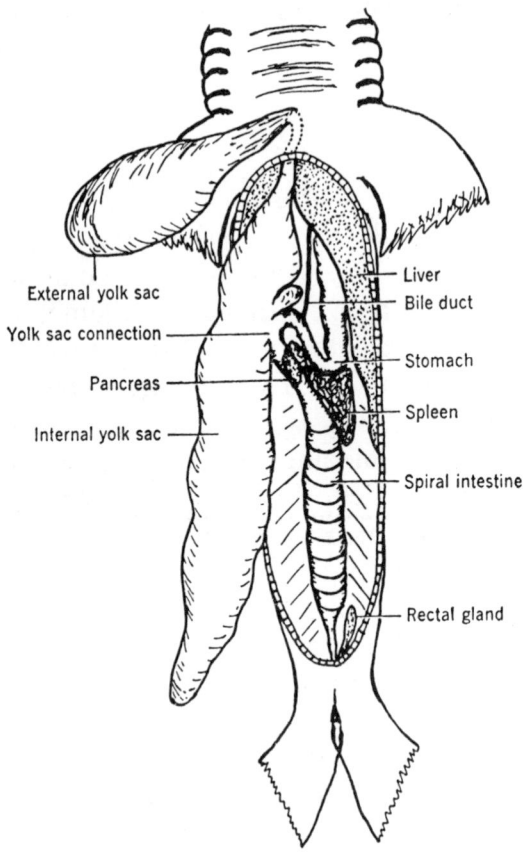

External yolk sac

Yolk sac connection

Pancreas

Internal yolk sac

Liver

Bile duct

Stomach

Spleen

Spiral intestine

Rectal gland

Figure 11–4. External and internal yolk sacs in a 220 mm. embryo of *Squalus suckleyi*. (From Brown, Physiology of Fishes, Academic Press. After Hoar.)

yolk, and the embryos soon hatch as free swimming larvae that feed for themselves. In the case of oviparous elasmobranchs, the eggs are provided with a large amount of yolk and the embryos develop within a horny egg case. The candle of the dogfish is regarded as a vestige of such a case.

PRIMITIVE TETRAPODS

The most primitive tetrapods, the amphibians, are still anamniotes and their excretory and reproductive systems have not changed significantly from the condition of these systems in primitive fishes. The only new feature of any consequence is a relatively large urinary bladder formed as a ventral outgrowth of the cloaca. In many other respects these systems in amphibians are even closer to what is believed to be the primitive vertebrate condition than they are in *Squalus*. Amphibians are living in the primitive environment, fresh water, and the tubules in their opisthonephros are of the primitive type that eliminate excess water as well as nitrogenous waste products. Amphibians also retain the primitive mode of reproduction. They are oviparous, laying their eggs in the water or in very moist areas. In most cases, the eggs hatch into free-swimming larvae that later metamorphose to adults. Thus, while amphibians have made a start in adapting to terrestrial conditions in most of their organ systems, they are limited, as a group, to moist habitats because of their inability to conserve water and reproduce under terrestrial conditions. *Necturus,* although a permanent larva, is a good example of this level of tetrapod evolution.

Kidneys and Their Ducts

The kidneys of *Necturus* are **opisthonephric.** They lie in the caudal half of the pleuroperitoneal cavity on either side of the dorsal mesentery (Fig. 11–5). They are easily seen in the male, but you will have to push the ovary and oviduct apart to see one in the female. Notice that the kidneys have bulged into the body cavity, so that both their dorsal and ventral surfaces are covered by visceral peritoneum. Also notice that the caudal part of a kidney is much larger than the cranial part. The cranial part of the male kidney is related to the reproductive system, while that of the female is somewhat degenerate. The cranial part of the kidney is best seen by lifting up the lateral border of the organ and looking on its dorsolateral surface.

The kidneys are drained in both sexes exclusively by the **archinephric ducts,** for accessory urinary ducts are absent in *Necturus*. The archinephric duct of the male is a large, convoluted tube extending down the lateral border of the kidney to the cloaca. The archinephric duct of the female is similarly

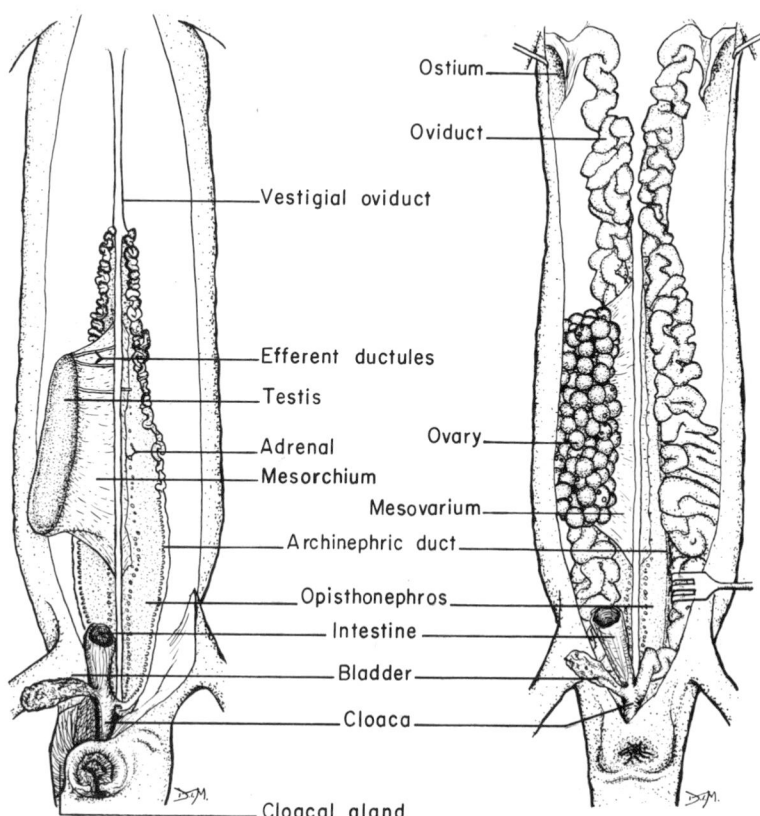

Ostium
Oviduct
Vestigial oviduct
Efferent ductules
Testis
Adrenal
Ovary
Mesorchium
Mesovarium
Archinephric duct
Opisthonephros
Intestine
Bladder
Cloaca
Cloacal gland

Figure 11–5. Ventral views of the urogenital system of *Necturus*. *Left,* male; *right,* female.

located but is much smaller and is not convoluted. Small **collecting tubules** may be seen entering the archinephric duct from the caudal, excretory portion of the kidney.

A **urinary bladder** lies ventral to the large intestine and is connected to the midventral body wall by a mesentery known as the **median ligament of the bladder.** The bladder enters the cranioventral wall of the cloaca. Urine reaches the bladder from the cloaca, for there is no direct connection between the archinephric ducts and bladder.

The **suprarenal glands** of primitive tetrapods consist of cortical and medullary cells clustered together in irregular patches. You may see a number of small, well-vascularized patches of suprarenal tissue along the ventral surface of the kidney of *Necturus*.

Male Urogenital System

The **testes** are a pair of oval organs located near the cranial ends of the kidneys. Each is supported by a **mesorchium.** Several inconspicuous **ductuli efferentes** pass through the front of the mesorchium carrying sperm from the testes to the modified cranial kidney tubules. This portion of the kidney thus functions as the head of the epididymis. From here, the sperm pass to the cloaca through the **archinephric duct,** which is therefore serving as a ductus deferens as well as an excretory duct (Fig. 11–5). Notice that the cranial part of the archinephric duct, a part comparable to the ductus epididymis of amniotes, is much more convoluted than the rest of the duct. The small black line along the edge of the cranial part of the archinephric duct, and extending forward along the side of the posterior cardinal vein, is a remnant of the oviduct that has persisted from the sexually indifferent stage of the embryo. A shorter but similar remnant of the oviduct can sometimes be seen along the very caudal end of the archinephric duct.

Cut through the body musculature lateral to the cloaca and find the point at which the archinephric duct joins the dorsal wall of the cloaca just caudal to the large intestine. Open the cloaca by making an incision that extends from the cranial end of the cloacal aperture into the lateral wall of the intestine. The archinephric ducts enter the craniodorsal wall of the cloaca just caudal to a transverse ridge formed by the entrance of the intestine, but their openings probably will not be seen.

Skin one side and the ventral surface of the cloaca, if this has not been done, and observe the large **cloacal gland** consisting of many small tubules. The secretions of this gland, together with the secretions of a less conspicuous **pelvic gland** in the dorsal wall of the cloaca, agglutinate the sperm into clumps called **spermatophores.** The spermatophores of salamanders generally are deposited in the water to be picked up by the female, but in a few species they are transmitted directly to the cloaca of the female. Finally, notice the numerous **papillae** on the cloacal lips, which are a characteristic feature of the male.

Female Urogenital System

The **ovaries** are a pair of large, granular-appearing organs located on either side of the dorsal mesentery adjacent to the cranial part of the kidneys (Fig. 11–5). Each is supported by a **mesovarium** and contains eggs within follicles in various stages of maturity. The oviducts are a pair of large, convoluted tubes lying along the lateral side of each kidney and extending forward nearly to the front of the pleuroperitoneal cavity. Each terminates cranially in a funnel-shaped opening called the **ostium,** and caudally in the cloaca. Cut through the musculature lateral to the cloaca to see that an oviduct attaches to the craniodorsal wall of the cloaca just caudal to the entrance of the large intestine. At ovulation the eggs pass into the coelom, are carried by the action of cilia on the coelomic epithelium to the ostia, and descend the oviducts to the cloaca. The oviducts are not differentiated grossly, but part of their lining is glandular. The oviduct glands secrete the jelly layers around the eggs, but the jelly does not swell until it takes up water when the eggs are deposited.

Open the cloaca by cutting from the cranial end of the cloacal aperture into the lateral wall of the intestine. The oviducts enter the craniodorsal wall of the cloaca through a pair of **genital papillae.** If these are not clear, slit an oviduct near its caudal end and pass a probe through it into the cloaca. The caudal end of each **archinephric duct** can be seen to leave the kidney and pass onto the wall of the oviduct, with which it becomes intimately united. However, the archinephric duct enters the cloaca independently beside the opening of the oviduct.

The **cloacal gland,** which helped to produce the spermatophores in the male, is present in a much reduced state in the female. It can be found by skinning one side and the ventral surface of the cloaca. The homologue of the pelvic gland of the male is transformed into a **spermatheca** within whose tubules the sperm are stored. In *Necturus* this is from the breeding season in the fall until egg laying in the spring. The spermatheca is located in the dorsal wall of the cloaca but cannot be seen grossly. Finally, notice that the lips of the female cloaca bear smooth folds rather than the papillae characteristic of the male.

MAMMALS

In the course of their evolution through reptiles to mammals, vertebrates have become well adapted to an active, terrestrial mode of life. There has been a large increase in the number of kidney tubules and this makes possible the elimination of the large volume of nitrogenous wastes produced by the high level of metabolism of mammals. Whereas a kidney of *Necturus* contains approximately 50 tubules in its caudal excretory portion, the number in a small mammal such as a mouse is on the order of 20,000, and human beings have an estimated 1,000,000 to 4,000,000 per kidney. Nitrogen is eliminated without an excess loss of body water. In mammals a special segment of the kidney tubule, the **loop of Henle,** concentrates sodium ions in the inter-cellular fluid deep within the kidney and thereby makes an osmotic gradient that causes water to be reabsorbed from the terminal portion of the tubule. The kidney is now a **metanephros** drained by a **ureter,** and what is left of the mesonephros and archinephric duct is taken over by the male genital system. Thus, there is a more complete separation of excretory and genital functions in amniotes than in anamniotes. There is also a division of the cloaca in most mammals that separates the urogenital tract from the digestive tract.

Reptiles and mammals can also reproduce upon the land and do not need to return to water. Mating on land is facilitated by the evolution in the male of a **copulatory organ** and **accessory genital glands** that secrete the seminal fluid in which the sperm are carried. These organs are found in only a few anamniotes in which internal fertilization occurs.

The high body temperature of most mammals poses one reproductive problem because the final stages of sperm formation cannot occur at intraabdominal temperatures. The testes of most mammals undergo a marked posterior migration (descent) and come to lodge outside the body cavity in a **scrotum** where the temperature is several degrees lower than the abdominal temperature.

Terrestrial reproduction also necessitates a method of suppressing a free-swimming aquatic larva. Primitive reptiles remain oviparous but suppress the larva by the evolution of a **cleidoic egg** (Fig. 11–6). This is an egg in which provision is made for all the requirements of the embryo, so that it can develop directly into a miniature adult capable of living on the land. The egg is supplied with a large store of yolk, which eventually becomes suspended in a **yolk sac.** As it descends the oviduct, **albumin,** or similar secretions which supply other metabolic needs, and a protective **shell** are added to the egg. The embryo itself early develops **extraembryonic membranes** that fulfill other needs. A protective **chorion** and a fluid-filled **amnion** (which provides a local aquatic environment) evolve from ectodermal and mesodermal layers that covered the yolk sac in such an animal as the dogfish. The yolk sac of amniotes is therefore **bilaminar** with a wall of just mesoderm and endoderm. Finally, a respiratory and excretory allantois evolves from the urinary bladder of the Amphibia.

Primitive prototherian mammals lay this type of egg, but in therian mammals this egg, minus its shell and albumin, is retained in the female reproductive tract and a

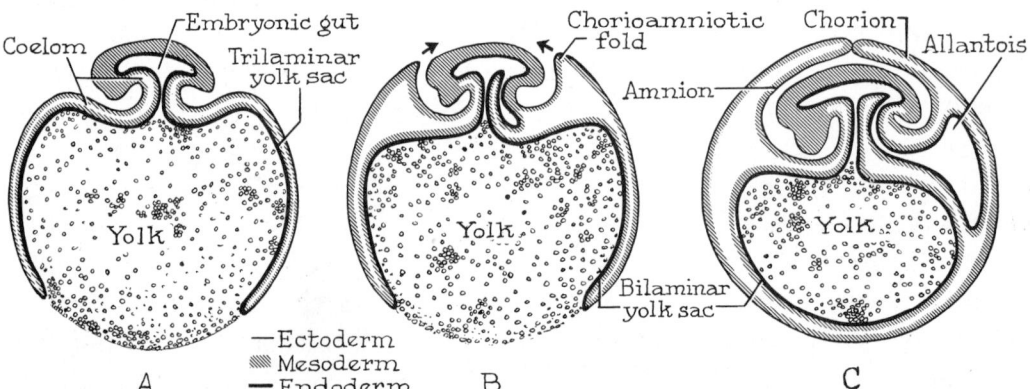

Figure 11–6. Diagrammatic sections of vertebrate embryos to show the extraembryonic membranes. *A,* Trilaminar yolk sac of a large yolked fish embryo; *B,* hypothetical derivation of the chorioamniotic folds of the amniote embryo from the superficial layers of the trilaminar yolk sac; *C,* extraembryonic membranes of an amniote embryo. (From Villee, Walker, and Barnes, General Zoology.)

placenta evolves. In eutherian mammals the placenta is simply a union of the chorion and allantois (**chorioallantoic membrane**) on the one hand with tissues of the female on the other. As will be seen, this mode of reproduction necessitates changes in the female reproductive tract, notably the evolution of a **uterus.**

Excretory System

The **kidneys** (**renes**) of mammals, which are **metanephroi,** are located against the dorsal wall of the peritoneal cavity in a retroperitoneal position. Each is surrounded by a mass of fat (**adipose capsule**), which should be removed, and each is closely invested by a **fibrous capsule.** Notice that a kidney is bean-shaped. The indentation on the medial border is called the **hilus.** Carefully remove connective tissue from the hilus and you will see the **renal artery** and **vein** entering the kidney, and, caudal to them, the **ureter** that drains the kidney.

Remove one of the kidneys and cut it longitudinally through the hilus. Study the half that includes the largest portion of the ureter. The hilus expands within the kidney into a chamber called the **renal sinus.** Pick away fat in the sinus in order to expose the blood vessels and the proximal end of the ureter. The sinus is largely filled with these structures, so its size is often not appreciated. If the vessels and ureter were removed, the space left would be the sinus. The portion of the ureter within the sinus is expanded and is termed the **renal pelvis** (Figs. 11–7 and 11–8). The substance of the kidney converges in carnivores and in the rabbit to form a single, nipple-shaped **renal papilla** which projects into the renal pelvis. (In some mammals, human beings among them, there are many renal papillae, and the proximal portion of the renal pelvis is subdivided into chambers for them, called **calyces.**) Notice that the substance of the kidney can be subdivided into a peripheral, light **cortex** and a deeper, darker **medulla.** The medullary substance of the carnivores and rabbit constitutes one large **renal pyramid** whose apex is the renal papilla; in species with many papillae, there is a pyramid for each one.

Trace one of the ureters. It extends caudally dorsal to the parietal peritoneum and then turns ventrally in the lateral ligament of the bladder to enter the caudal part of the **urinary bladder.** As the ureter enters the lateral ligament of the bladder, it passes dorsal to the ductus deferens (male), or horn of the uterus (female). The urinary bladder itself is a pear-shaped organ with a broad, rounded, cranial end (its **vertex**) and a narrow, caudal part (its **body**). Cut open the bladder and you may be able to see the points of entrance of the ureters in the dorsal wall of the body. Clean away connective tissue from around the bladder. It gradually narrows caudal to the entrance of the ureters and passes into the pelvic canal. This narrow passage, which carries urine to the outside, is called the **urethra.** The urethra begins just caudal to the entrances of the ureters. Its more caudal parts will be considered with the reproductive organs.

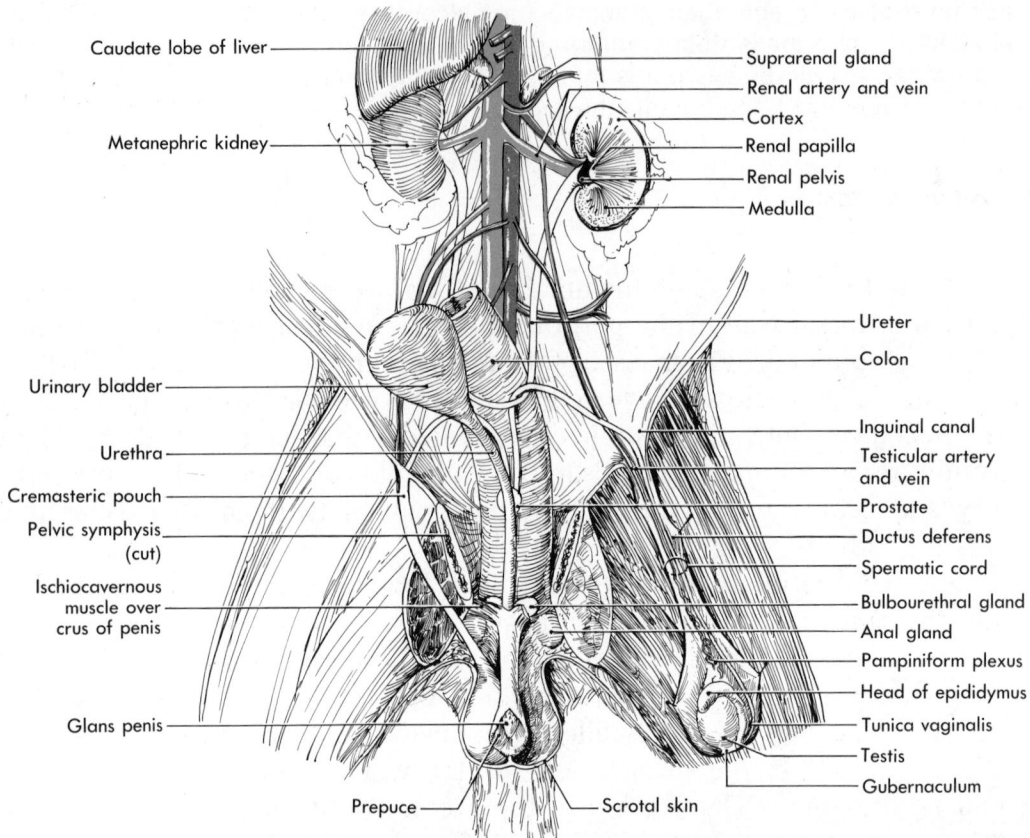

Figure 11–7. Ventral view of the urogenital system of a male cat. That of a male mink is very similar. One kidney has been sectioned to show its internal structure. The pelvic canal has been cut open and the cremasteric pouch has been dissected on the right side of the drawing. (From Walker, A Study of the Cat.)

Male Reproductive System

For reasons stated in the introduction to the mammalian urogenital system, the testes of most mammals have undergone a descent so that they lie in a sac, the **scrotum,** outside the body cavity (Figs. 11–7 and 11–8). The scrotum of the cat and mink is situated caudal to the ventral surface of the pelvic girdle and, as in man, caudal to the penis. But in the rabbit the scrotum lies on the ventral surface of the pelvic girdle cranial to the penis. This position, incidentally, occurs only in lagomorphs among placental mammals, although a comparable location is seen in marsupials.

Carefully cut through the scrotal skin on each side and separate the **scrotal skin** from the deeper layers of the scrotum. A dense layer of connective tissue containing some smooth muscle fibers, the **dartos tunic,** is closely associated with the skin and will come off with it. The dartos forms the septum between the left and right sides of the scrotum. Deeper layers of the scrotum take the form of a pair of cordlike sacs that extend caudad from the abdominal wall, cross the ventral surface of the pelvic girdle, and enter the skin sacs. These cordlike sacs are composed of a layer of muscle (**cremasteric muscle**) fascia and coelomic epithelium drawn down from the abdominal wall during the descent of the testes. Although considered to be a part of the scrotum, each sac may be called a **cremasteric pouch**.[16] The

[16]The term scrotum properly includes all the layers separating the coelomic space in which each testis lies from the outside. It is useful, however, to have a separate term to designate the complex of tissues, apart from the skin, enfolding each testis and its duct and vessels, because this complex is clearly defined and must be separated from the skin in order to see fully the anatomical relationships that result from the descent of the testes. I am following Prof. Gerard (in P.-P. Grassé, Traité de Zoology, Masson et Cie., Paris, 1954, Vol. 12) in calling this part of the scrotum the cremasteric pouch.

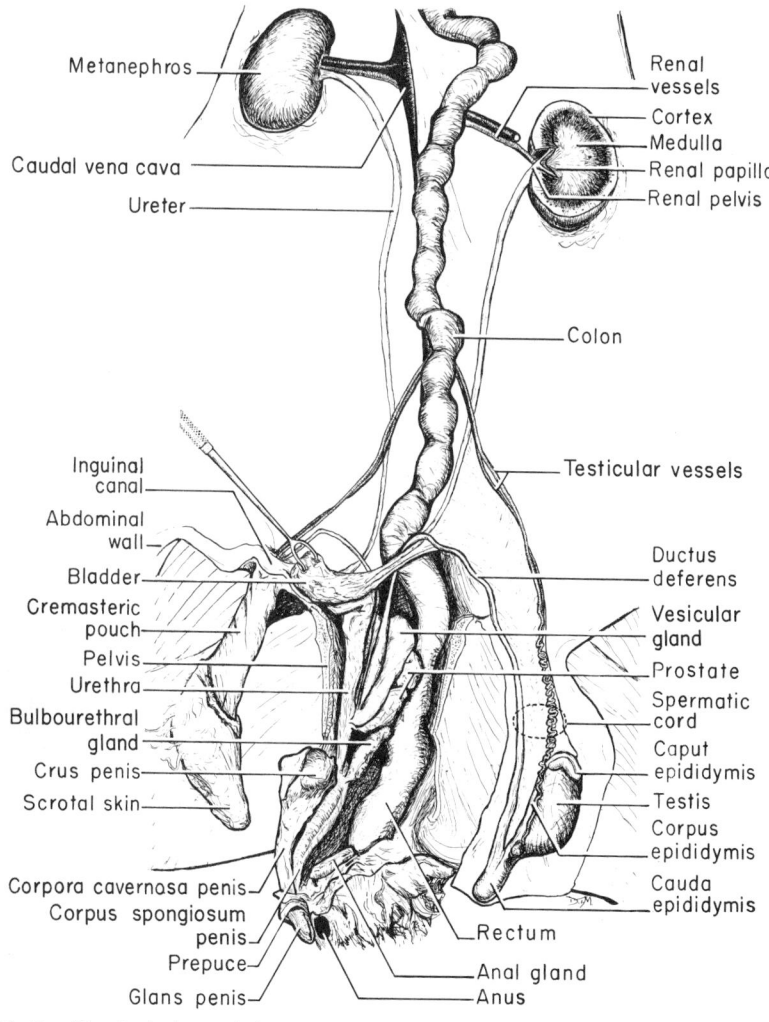

Figure 11-8. Ventral view of the urogenital system of a male rabbit. One kidney has been sectioned to show its internal structure. The pelvic canal has been opened and the urethra and penis twisted to one side to show the accessory genital glands. The scrotum is shown intact on the left side of the drawing and opened on the right.

proximal parts of the cremasteric pouches lie just beneath the skin and should have been seen and saved when the pelvic canal was opened (p. 322). The testes lie within the caudal ends of the cremasteric pouches. This portion of each pouch is quite large in carnivores, while the rest is a constricted tube, for the testes remain permanently in the scrotum. But all the pouch is wide and of nearly uniform diameter in the rabbit, for the testes move back and forth—into the scrotum during the breeding season, back into the abdomen the rest of the time. The atrophy of the cremasteric muscle in the cat and its hypertrophy in the rabbit is correlated with whether the testes remain permanently in the scrotum or are migratory.

Leave the cremasteric pouch intact on one side, but cut open the other one along its ventral surface. Extend the cut from the caudal end of the pouch to the body wall but do not cut through the body wall. Notice that the cremasteric pouch contains a cavity, the cavity of the **processus vaginalis,** or *tunica vaginalis* (Fig. 11-9, *B*). Pass a probe forward through this cavity near the body wall and the probe will enter the peritoneal cavity. The processus vaginalis is a coelomic sac that descends with the testis. Thus, the wall of the cremasteric pouch is lined with coelomic epithelium (the **parietal layer** of the processus vaginalis), and the structures within it (testis, epididymis, ductus deferens, testicular vessels and nerves) are also covered with coelomic epithelium (the **visceral layer** of the processus vaginalis). The complex of ductus deferens and associated vessels and nerves, together with their covering of coelomic epithelium, is known as the **spermatic cord.** Notice that the cord is supported by a mesentery, a part of the **mesorchium,** passing from the dorsal wall of the pouch and that the most caudal structures in the pouch (testis and epididymis) and this portion of the mesorchium are united to the caudal end of the pouch by a band of tissue called the **gubernaculum.** In many mammals, man included, only the distal portion of the processus vaginalis around the testis persists in the adult; the rest of the processus vaginalis atrophies. But this is not the case in the mammals being considered.

The contents of the cremasteric pouch can now be examined in more detail. The testis is the relatively large, round (cat and mink), or elongate (rabbit) body lying in the caudal part of the pouch. Testicular blood vessels and nerves attach to its cranial end. Notice that the testicular artery coils upon itself before it reaches the testis and is closely invested by a venous network, the **pampiniform plexus,** formed by the testicular vein. It is possible that this is a mechanism that permits heat exchange from the arterial to the venous blood, which would help to keep testicular temperature low. The **epididymis** is a band-shaped structure closely applied to the surface of the testis. It can be divided into three regions—a **head** at the cranial end of the testis, a **body** on the lateral surface of the testis, and a **tail** at the caudal end of the testis. The head of the epididymis is functionally connected with the testis and is made up of modified kidney tubules homologous to those of the cranial end of the opisthonephros of anamniotes. The rest of the epididymis

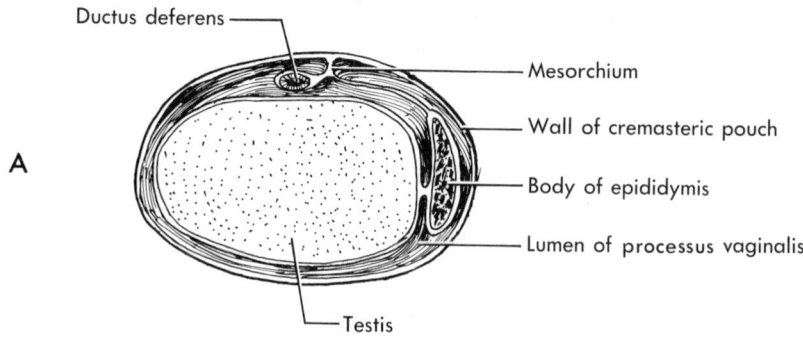

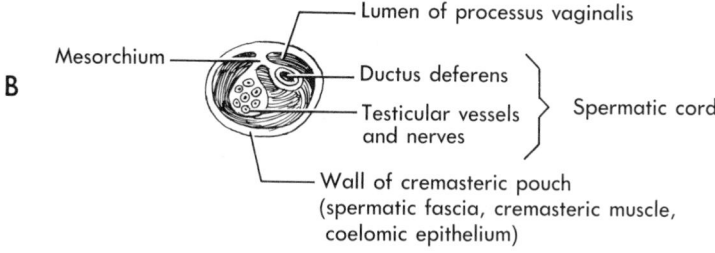

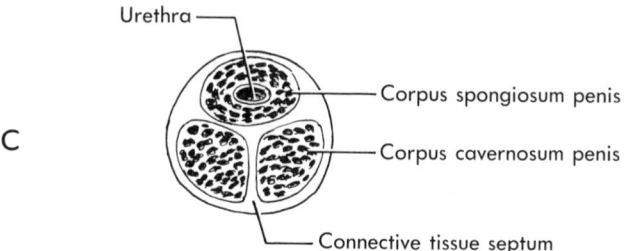

Figure 11–9. Semidiagrammatic, enlarged transverse sections through parts of the reproductive system of a male cat: *A*, through the scrotum at the level of the testis; scrotal skin removed; *B*, through the middle of the cremasteric pouch; *C*, through the middle of the penis, skin removed. (From Walker, *A Study of the Cat*.)

consists of the highly convoluted **ductus epididymis** (former cranial end of the archinephric duct) imbedded in connective tissue.

The **ductus deferens,** the first part of which is somewhat convoluted, leaves the tail of the epididymis and, in company with the testicular vessels and nerves, ascends the cremasteric pouch and passes through the abdominal wall. The passage through the body wall is known as the **inguinal canal.** The canal is not long in the mammals being considered, its cranial end (**internal inguinal ring**) being the entrance into the peritoneal cavity; its caudal end (**external inguinal ring**), the attachment of the outermost layer of the wall of the cremasteric pouch to the external surface of the abdominal muscles.

In man the inguinal canal passes diagonally through the abdominal wall, hence is much longer.

Continue to follow the ductus deferens. It passes forward in the peritoneal cavity for a short distance, then loops over the ureter and extends caudad into the pelvic canal between the urethra and large intestine. The ductus deferentes of opposite sides then converge and soon enter the urethra. The portion of the urethra distal to this union carries both sperm and urine. Various accessory genital glands, which secrete the seminal fluid, are associated with the ends of the ductus deferentes and adjacent parts of the urethra; however, these glands, and the details of the union of the ductus deferentes with the urethra, are different in cat and mink (Fig. 11-7) and rabbit (Fig. 11-8).

In the cat and mink the two ductus deferentes enter the urethra independently, and a small **prostate** surrounds their point of entrance and the adjacent urethra. At the caudal end of the pelvic canal, a pair of **bulbourethral glands** (Cowper's glands) enter the urethral canal. These glands lie dorsal to a pair of processes (crura of the penis) that extend from the base of the penis to the ischia.

In the rabbit the two ductus deferentes pass between the urethra and a heart-shaped **vesicular gland** (Fig. 11-8). Carefully separate these structures from each other. The ductus deferentes enter the narrow caudal end of the vesicular gland which, in turn, enters the urethra. The dorsal wall of the vesicular gland is rather thick and includes the **prostate.** It is possible, by very careful dissection, to free the cranial end of the prostate and turn it caudad. Further dissection will reveal overlapping cranial and caudal lobes of the prostate. Both enter the urethral canal just caudal to the entrance of the vesicular gland. A bilobed **bulbourethral gland** (Cowper's gland) enters the dorsal surface of the urethral canal caudal to the prostate. It lies craniodorsal to a pair of processes (crura of the penis) that extend from the base of the penis to the ischia.

In the cat, mink, and rabbit, the **crura of the penis** extend from the base of the penis to the ischia and each is covered by muscular tissue (**ischiocavernosus muscle**). If the crura were not torn when the pelvic canal was opened, you may have to cut one now to see the paired (carnivore) or bilobed (rabbit) bulbourethral gland clearly.

The **penis** encloses the part of the urethra lying outside the pelvic canal. The free end of the penis, **glans penis,** lies in a pocket of skin called the **prepuce** (*preputium*). Cut open the prepuce to better see the glans and the opening of the urethral canal. A number of small spines are borne on the glans of the cat. The glans penis of the mink is quite long. The rest of the penis is a firm, cylindrical structure which should be exposed by removing the skin and surrounding loose connective tissue. Make a cross section of this portion of the penis and examine it with a hand lens. The urethral canal lies along the dorsal surface of the penis (if the organ is flaccid) imbedded in a column of spongy tissue called the **corpus spongiosum penis** (Fig. 11-9, C).

A pair of columns of spongy tissue separated by a septum, which is often indistinct, lie along the opposite surface and are surrounded by a ring of dense connective tissue. These columns are the **corpora cavernosa penis.** The glans penis is simply a caplike fold of the corpus spongiosum penis that covers the distal ends of the corpora cavernosa penis. The crura of the penis are the diverging proximal ends of the corpora cavernosa penis. The spongy tissue of which all these corpora consist is known as **erectile tissue,** and the spaces within it become filled with blood during erection. Make a cross section through the glans penis and look for a small bone, the **os penis** (baculum), that lies on one surface of the urethra and helps to stiffen this part of the penis. It is a relatively long bone, hooked on the distal end, in the mink. (The baculum of the whale was prized by whalers as a walking stick!)

Before leaving the urogenital system, dissect beneath the skin on either side of the rectum near the anus and find a pair of **anal glands.** These are elongate in the rabbit, round in the carnivores. They produce an odoriferous secretion that enters the anus and presumably is for sexual attraction, or stimulation.

Female Reproductive System

The **ovaries** are a pair of small, oval bodies (Fig. 11–10). In the adult they lie slightly caudal to the kidneys, because they have undergone a partial descent and the metanephroi have shifted cranially during development. The small size of the mammalian ovaries is correlated with the uterine development of the embryos. Fewer eggs are produced, and they do not contain much yolk. The eggs are microscopic, but you may see small vesicles, the **graafian follicles** (*folliculi vesiculosi*), each of which contains an egg, protruding on the surface of the ovary. The ovary of the cat and rabbit protrudes into the body cavity. When eggs are discharged from the follicles (**ovulation**) they pass through a part of the coelom to the entrance of the reproductive duct, which is not far away. The ovary of the mink and some other mammals is completely enfolded in a mesenteric sac, the **ovarian bursa** (Fig. 11–11), from which the reproductive duct emerges, so eggs never can enter the main part of the coelom. It is necessary to cut open the ovarian bursa to see the ovary of the mink. After ovulation the follicles are transformed into **corpora lutea,** which may also be seen protruding from the surface of the ovary, especially in pregnant specimens.

Typical oviducts are present in early mammalian embryos, but they differentiate into several regions during development and their caudal ends fuse in varying degrees (Fig. 11–12). Thus, the adult reproductive tract is more or less Y-shaped. The cranial part of each wing of the Y forms a narrow, convoluted **uterine tube** (fallopian tube) lying lateral to the ovary. Notice that a uterine tube curves over the front of the ovary and forms a hoodlike expansion (**infundibulum**) with fringed (fimbriated) lips. The infundibulum is

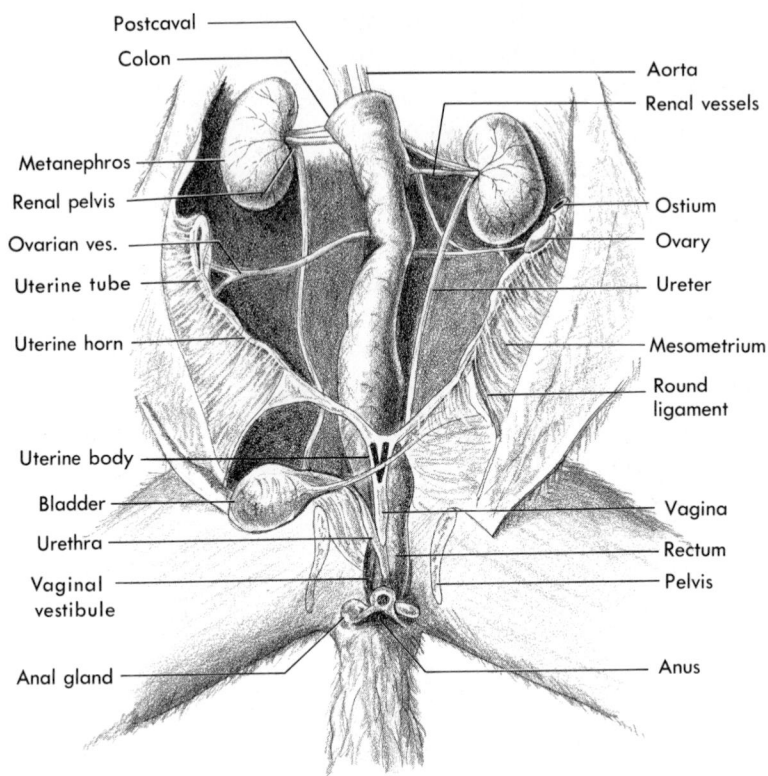

Figure 11–10. Ventral view of the urogenital system of a female cat. The pelvic canal has been cut open. The uterine body has been sectioned to show its bipartite nature.

smaller in the mink and lies within the ovarian bursa. Spread open the lips and you will see the coelomic opening of the tube, the **ostium.**

The rest of each wing of the Y lies caudal to the ovary and forms a much wider tube—the **horn of the uterus** (cat and mink), or **uterus** (rabbit). This

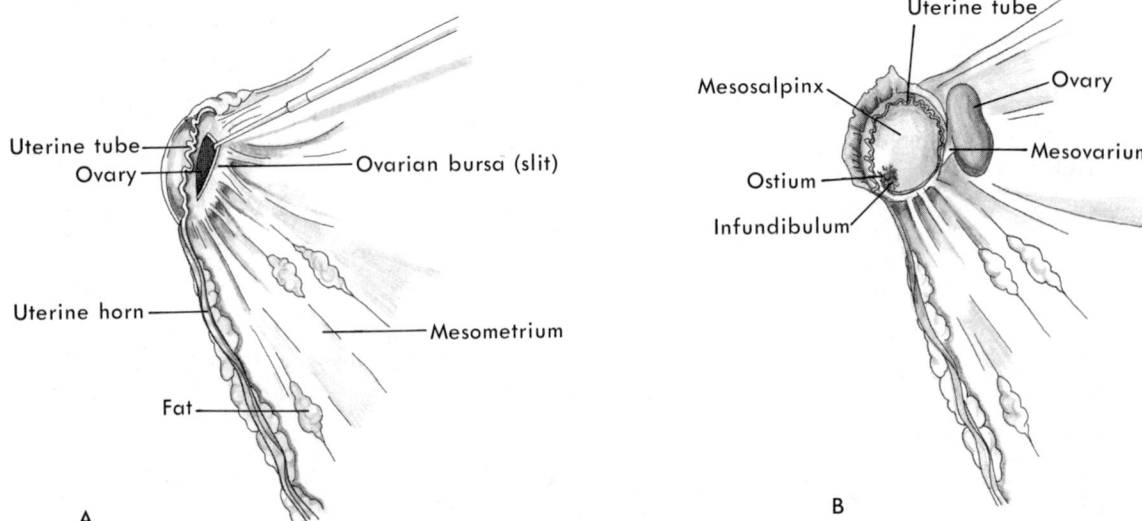

Figure 11–11. Ventral views of the right ovary of the mink. *A,* Ovary enclosed within the ovarian bursa; *B,* ovary rolled out of opened bursa.

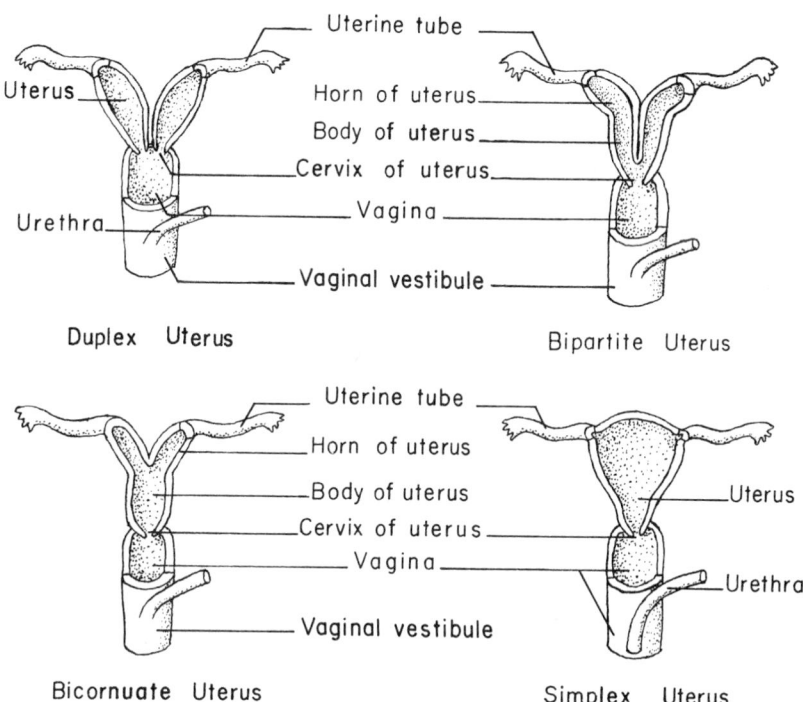

Figure 11–12. Diagrams to show the progressive fusion of the posterior ends of the oviducts in placental mammals. The uterus and part of the vagina have been cut open. The duplex type of uterus, in which the lower ends of the oviducts have united to form a vagina but the uteri remain distinct, is found in rodents and lagomorphs. In the bipartite uterus of carnivores, the lower ends of the uteri also have fused to form a median body from which uterine horns extend, but a partition is present in the body of the uterus. This partition is lost in the bicornuate uterus of ungulates. In the simplex uterus of primates, the uteri have completely united to form a large median body from which the uterine tubes arise. In primates, the vaginal vestibule also divides, so that the vagina and urethra open independently on the body surface. (Modified after Wiedersheim, Comparative Anatomy of Vertebrates, The Macmillan Company.)

section is straight in carnivores but somewhat convoluted in the rabbit. It is very large in pregnant specimens, for the embryos develop within it.

The ovary and reproductive tract of the female are supported by a mesentery known as the **broad ligament.** Often a great deal of fat lies within it. The portion of the broad ligament attaching to the uterus is the **mesometrium;** that attaching to the uterine tube, the **mesosalpinx;** and that attaching to the ovary, the **mesovarium** (Fig. 11–11). Pull the uterine horn (carnivores) or uterus (rabbit) toward the midline, thereby stretching the mesometrium. The mesenteric fold extending diagonally across the mesometrium from a point near the cranial end of the uterine horn (or uterus) to the body wall, and lying perpendicular to the broad ligament, is the **round ligament.** Notice that the round ligament attaches to the body wall at a point comparable to the location of the inguinal canal in the male. The round ligament is the female counterpart of the male gubernaculum—a strand that plays an important role in the descent of the testis.

The two uterine horns (carnivores) or uteri (rabbit) converge cranial to the pelvic canal and enter a common median passage. This would be the stem of the Y, and it is formed in part by the fusion of the lower ends of the oviducts and in part by the division of the cloaca (see page 355). In the cat and mink the cranial part of this median passage is the **body of the uterus.** The body is not long and soon leads into the **vagina.** The demarcation between these two will be seen presently when the tract is cut open. In the rabbit the two uteri enter the vagina directly, for there is no median body to the uterus. The vagina proceeds through the pelvic canal, lying between the urethra and large intestine. Carefully separate these structures from each other and find the point where the vagina and urethra unite. The common passage from here to the body surface is the **vaginal vestibule.** It is a relatively long passage in quadrupeds. The comparable area in women is known as the **vulva,** but the vulva is very shallow, for vagina and urethra are independent nearly to the body surface. The opening of the vaginal vestibule, or vulva, is flanked by skin folds, the **labia,** but these are not conspicuous in quadrupeds.

Cut through the skin around the opening of the vaginal vestibule and completely free the vestibule and vagina from the rectum. A pair of **anal glands,** which are elongate in the rabbit and round in carnivores, can be found by dissecting beneath the skin on the lateral surface of the vestibule (rabbit), or of the rectum near the anus (carnivores). These glands produce an odoriferous secretion that enters the anus and presumably is used for sexual attraction, or stimulation. Now open the median portion of the genital tract by making a longitudinal incision through its dorsal wall that extends from the vestibule to the horns of the uterus (carnivores), or uteri (rabbit). Veer away from the middorsal line toward one of the uterine horns as you open the body of the uterus in carnivores. A small bump may be seen in a pocket of tissue in the midventral line of the vestibule near its orifice. This is the **glans clitoridis.** The clitoris consists of only the small glans in carnivores, but is large and continues cranially in the rabbit. In the rabbit, the rest of it can be exposed by removing the skin from the ventral surface of the vestibule. The clitoris and its glans develop from a phalluslike structure which is present in the sexually indifferent stage of the embryo. Make a cross section through the organ, and it will be seen to consist of a pair of columns of spongy tissue, the **corpora cavernosa clitoridis,** homologous to the corpora cavernosa penis.

More cranially in the vaginal vestibule, you will see the entrance of the urethra. The genital passage cranial to this union is the vagina, and it continues forward to the neck or **cervix of the uterus** (Fig. 11-13). In the cat and mink the single cervix lies about halfway between the urethral orifice and the horns of the uterus and appears as a pair of folds constricting the lumen of the reproductive tract. The body of the uterus lies between the cervix and the horns. Notice that the cranial part of the body is subdivided into right and left sides by a vertical partition. The uterus of carnivores is therefore bipartite (Fig. 11-12). In the rabbit the vagina extends forward to the two uteri, and each uterus has a papillalike **cervix** extending into the vagina. This type of uterus is called **duplex** (Figs. 11-12 and 11-13).

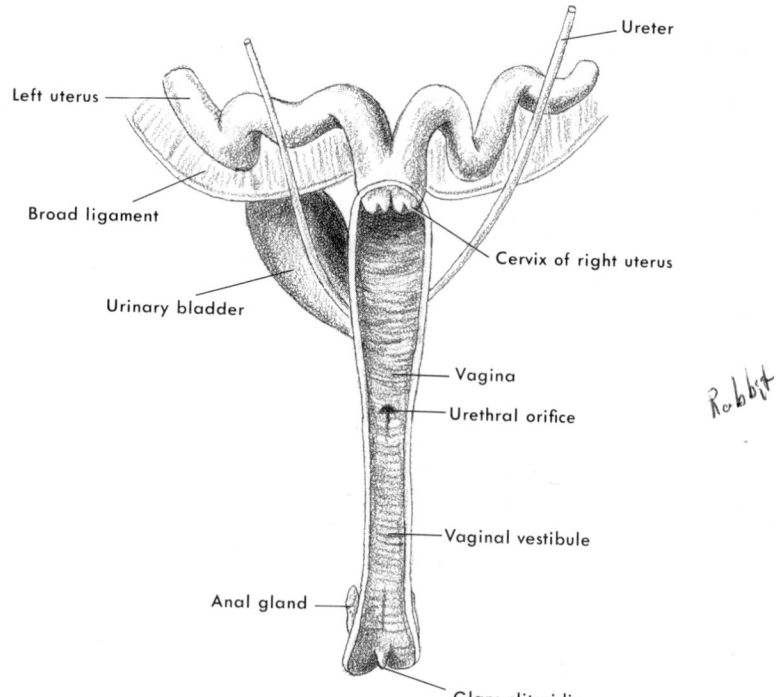

Figure 11–13. Reproductive tract of a female rabbit, cut open and viewed from the dorsal side.

The origin of most of the mammalian male and female urogenital tracts from those of the lower vertebrates has been indicated. However, a consideration of the cloacal region and its fate in mammals has been deferred until the terminal portions of the urogenital passages could be studied in the mammal.

A cloaca was seen in the lower vertebrates, and one is still present in monotremes and in the embryos of the higher mammals. But it becomes divided and contributes to the intestinal and urogenital passages in the adults of the higher mammals. In an early, sexually indifferent, eutherian embryo the cloaca consists of a chamber derived from the enlargement of the caudal end of the hindgut (Fig. 11–14, A). At first this endodermal cloaca is separated from the ectodermal proctodeum by a plate of tissue, but this plate soon breaks down and the proctodeum contributes to the cloaca.

The cloaca receives the intestine dorsally and the allantois ventrally. Even at an early stage (Fig. 11–14, A) the cranial portion of the cloaca is partly divided into a dorsal **coprodeum** receiving the intestine, and a ventral **urodeum** receiving the allantois, the ureters, and the archinephric ducts. Finally, a small **phallus** is present on the ventral surface of the body cranial to the cloaca. This stage is similar to the cloaca of lower vertebrates except for the more ventral entrance of the urogenital ducts.

Later in the sexually indifferent period (Fig. 11–14, B), the urodeum and coprodeum become completely separated from each other by a fold of tissue (**urorectal fold**) and form the urogenital sinus and rectum, respectively. Oviducts now enter the front of the urogenital sinus beside the archinephric ducts, but the attachments of the ureters shift onto the allantois and developing urinary bladder.

In the subsequent differentiation of a male (Fig. 11–14, C and D) the constricted neck of the bladder (allantois) and the urogenital sinus form that portion of the urethra that is not included in the penis (segments 1 and 2). The archinephric ducts form the ductus deferens and enter the urethra. Their point of entrance is a landmark that

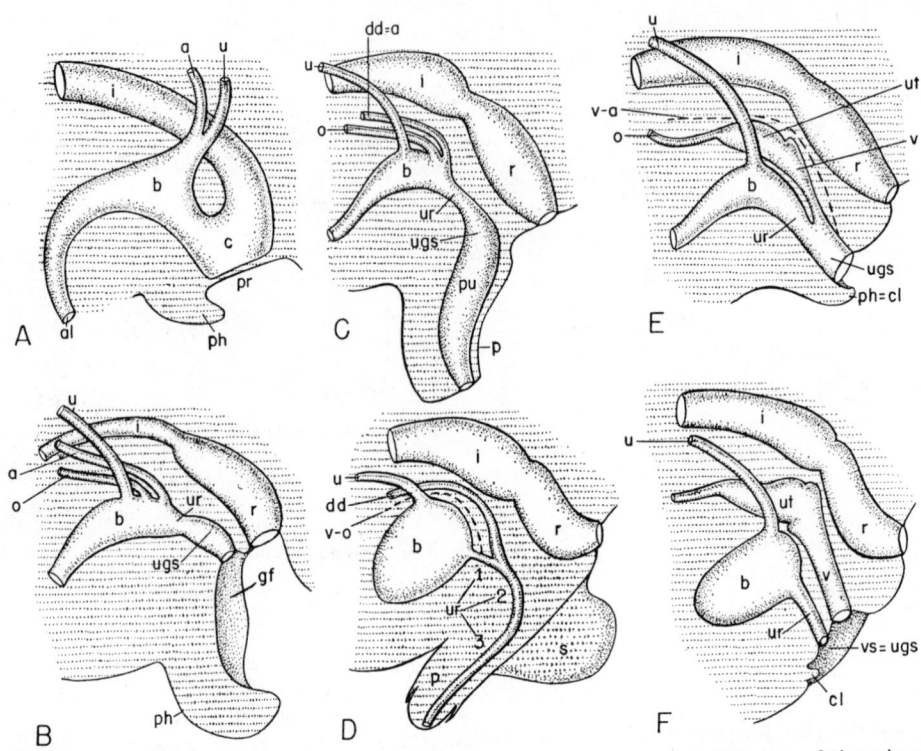

Figure 11–14. A series of diagrams in lateral view showing the division of the cloaca that occurs during the embryonic development of a placental mammal. Cranial is toward the left, and only one member of the paired ducts is shown. A, An early sexually indifferent stage in which the cloaca is undivided; B, a later sexually indifferent stage in which the cloaca has become divided into a dorsal rectum and a ventral urogenital sinus; C, early male condition; D, adult male; E, early female condition; F, adult female. (See text for further explanation.) Abbreviations: a, archinephric duct; al, stalk of the allantois; b, urinary bladder or proximal part of the allantois; c, cloaca; cl, clitoris; dd, ductus deferens; gf, genital fold; i, intestine; o, oviduct; p, penis; ph, phallus; pr, proctodeum; pu, penile urethra; r, rectum; s, scrotum; u, ureter; ugs, urogenital sinus; ur, urethra; ut, uterus; v, vagina; v-a, vestige of archinephric duct; v-o, vestige of oviduct; vs, vestibule. (From Romer, The Vertebrate Body.)

separates the portion of the urethra derived from the allantois from the portion derived from the urogenital sinus. The oviducts disappear in the male, although their point of entrance into the urethra may form a small sac (**prostatic utricle**) within the prostate. The phallus enlarges to form the penis and a groove on its ventral surface closes over, by the coming together of the **genital folds,** to form the penile portion of the urethra (segment 3).

In the subsequent differentiation of most female mammals (Fig. 11–14, E) the constricted neck of the bladder forms the entire urethra. The female urethra is thus comparable to only a small portion (the allantoic segment) of the male urethra. The urethra and the two oviducts, whose lower ends have fused to form the vagina and uterus (Fig. 11–12), enter the urogenital sinus, which becomes the vaginal vestibule. In most female mammals the vaginal vestibule remains undivided. But in primates it, too, becomes divided and continues the urethra and vagina nearly to the surface as separate passages. This distal part of the vaginal vestibule forms the shallow vulva. The archinephric ducts disappear, and the phallus forms the small remnant known as the clitoris. The genital folds of the indifferent stage form the labia minora in such mammals as have these labia. The labia majora are skin folds comparable to the scrotum of the male.

Reproduction and Embryos

As stated earlier, eutherian mammals are viviparous; the embryos develop within the uterus and are born as miniature adults. If any of the specimens are pregnant, cut open the uterus and examine the embryos. Mammalian embryos produce the various extraembryonic membranes characteristic of amniotes (Fig. 11–6). Since the outermost membrane is the chorion, the whole complex of embryo and extraembryonic membranes is often called the **chorionic sac. Chorionic villi** arise from the surface of the chorioallantoic membrane in eutherian mammals and penetrate or unite in various ways with the uterine lining. This combination of uterine lining and villi constitutes the **placenta.** In many mammals, including the ones considered, the union of villi and uterine lining is intimate and some of the uterine lining is discharged at birth. Such a placenta is said to be **deciduous,** in contrast to a **nondeciduous** placenta in which the union is not intimate and maternal tissue is not discharged at birth.

Deciduous placentas have different shapes according to the distribution of the villi that make an intimate union with the uterine lining. In carnivores a belt-shaped band of villi unites with the uterine lining to form a definitive placenta of the **zonary** type. This band is easily seen on the surface of the chorionic sac. In the rabbit a disc-shaped patch of villi unites, and the placenta is said to be **discoidal.**

Still other terms describe the microscopic details of the union. In carnivores the surface epithelium lining the uterine portion of the placenta disappears. The villi penetrate the substance of the uterine lining and come in contact with the endothelial walls of the maternal capillaries (**endothelio-chorial placenta**). This happens in the rabbit but, in addition, the maternal blood vessels break open and the epithelium on the surface of the villi disappears. The circulation of fetus and mother is then separated only by the endothelial walls of the fetal capillaries (**hemoendothelial placenta**). To summarize, carnivores have a chorioallantoic, deciduous, zonary, endothelio-chorial placenta; the rabbit has a chorioallantoic, deciduous, discoidal, hemoendothelial placenta.

Cut open the chorionic sac and placenta and you will see the embryo enclosed by the **amnion.** Open the amnion. The cord of tissue extending from the underside of the embryo to the chorionic sac and placenta is the **umbilical cord.** It contains the allantoic stalk, umbilical or allantoic blood vessels, and a vestige of the yolk sac stalk. Its surface is covered by the amnion.

APPENDIX 1

THE PREPARATION OF SPECIMENS

It is common practice to purchase specimens from biological supply houses, but students are frequently interested in the way they are prepared, and it is sometimes necessary to prepare a specimen. The following directions are presented with these needs in mind. Only the simpler and more common procedures are described. Further details can be found in such a book as Wilder and Gage, *Anatomical Technology,* or can be obtained by writing to a biological supply house.[17] Directions for preparing many interesting and valuable teaching demonstrations can be found in Dr. Milton Hildebrand's *Anatomical Preparations.*

1. Killing the Specimen

After a specimen is caught it must, of course, be killed. And, incidentally, some specimens, cats for example, should be obtained from the proper sources. Long years have passed since Wilder and Gage could write, "There is usually no difficulty in taking a cat when it is wanted. Such as will not come when called may be secured by means of a strong net...."!

(A) Aquatic Vertebrates. The best way to kill aquatic specimens is with chloroform or ether, or by immersing them in a weak solution of chloretone. One ounce of a saturated chloretone solution added to one quart of water will kill most specimens and leaves them relaxed.

(B) Terrestrial Vertebrates. Amphibians and reptiles should be killed by being placed in a container with chloroform or ether. This method may be used for mammals, but injection of a barbiturate, such as sodium pentobarbitol, is more humane if barbiturates can be obtained from the institutional physician or a local veterinarian. One may also gas mammals, by putting the specimen in a container and inserting a hose from a gas line. Turn the gas off when the specimen becomes quiet, but leave the specimen in the container for 5 to 10 minutes.

---

[17]I am indebted for much of the following information to various "Turtox Service Leaflets" published by the General Biological Supply House, Chicago, Illinois.

2. **Preserving Specimens**

(A) **Simple Preserving.** Many animals can be preserved simply by cutting open the body cavity and immersing the specimen in a solution of alcohol or formalin. The preserving solution can also be injected into the body cavity and into the larger muscle masses. For most vertebrates use a 70 per cent solution of alcohol or an 8 to 10 per cent solution of formalin. Commercial formalin is a saturated, aqueous solution (40 per cent) of formaldehyde gas. An 8 per cent formalin solution is a mixture of 8 parts of commercial formalin and 92 parts water. It is best to fix the specimen in 10 per cent formalin for a day or two and then to transfer it to an 8 per cent solution.

Formalin-preserved specimens are generally unpleasant to work with for any length of time. Several things may be done to alleviate the situation. Lanolin may be applied to one's hands before and after working on such specimens. Pure lanolin is better than the "toilet" variety, as it makes a more lasting coating on one's hands. The specimens may be soaked for several hours in water before using them, or the specimens may be soaked for a while (the longer the better) in a deformalinizing solution. Such solutions remove some but not all of the formalin. One formula suggested by Fort, Wilson, and Goldberg (*Science,* vol. 94, pp. 169–170, 1941) is an aqueous solution containing 5.7 per cent by weight of sodium bisulfate and 3.8 per cent by weight of sodium sulfite.

(B) **Embalming.** Better results can be obtained on large specimens, however, by injecting the preserving fluid into the circulatory system—a process called embalming. After a cat, for example, has been killed, it should be tied out on a spreading board in the position in which one desires it to harden. Then make an incision through the skin on the medial surface of the thigh to expose the femoral artery. The artery is smaller and lighter in color than the accompanying vein. Mobilize the artery and slip a thread beneath it. Then make a V-shaped cut halfway through the vessel with a pair of scissors. Insert a 16- or 18-gauge injection needle into the vessel, and tie it in. Force a wad of cotton into the animal's mouth to prevent the embalming solution from leaking.

An embalming solution that is often used is made according to the following formula:

Carbolic acid (melted crystals)	5 parts
Formalin (40%)	5 parts
Glycerin	5 parts
Water	85 parts

The formalin is the chief preservative: the carbolic acid is a disinfectant, and also helps to preserve the color of the tissues, and gives a sanitary odor to the specimen; the glycerin prevents the specimen from drying out too quickly.

The solution can be injected with a gravity bottle or with a large hand syringe. Inject sufficient fluid so that the specimen appears bloated and the fleshy parts are firm. It should not be possible to move the appendages or head easily if enough fluid has been used. An average-sized cat takes about 1 liter. If a syringe is used, it must obviously be refilled. Be sure to cap the base of the needle with your finger when the syringe is taken off, or the fluid will escape. After injecting, remove the needle, and tie the artery.

If embalmed specimens become unusually dry while being dissected they may be moistened with a wetting fluid made as follows:

Carbolic acid crystals	30 g.
Glycerin	250 cc.
Water	1000 cc.

3. Injection of the Cardiovascular System

The blood vessels of an animal can be rendered more conspicuous by injecting a colored solution into them that will later harden. Colored latex is usually used. If specimens are simply to be preserved, the injection should be made before preservation. But embalmed specimens should be left for several days before injecting the vessels in order to let the embalming solution penetrate the tissues. Injection is made with an injecting syringe and with the needle inserted and tied into a cut in one of the vessels. Avoid injecting air. After injecting, the vessle should be tied off with a piece of string on both sides of the cut. The vessel through which the injection is made varies with the animal, as described below. Ordinarily the arteries are injected before the veins. The success of the injection should be checked by seeing whether or not the smaller peripheral vessels are filled. It must be remembered, however, that veins contain valves, and these generally prevent some of the veins from filling—notably veins in the appendages.

(A) **Dogfish.** To inject the arteries of the dogfish, cut across the tail and inject forward through the caudal artery. Inject the hepatic portal system through the hepatic portal vein or one of the intestinal veins.

(B) *Necturus.* In *Necturus* the systemic veins should be injected first. Insert the needle into the caudal part of the caudal vena cava and inject forward until the vessels in the liver are filled (8 to 10 cc.). Just before finishing the injection, press against the caudal vena cava near the liver and thus force some injection mass into the posterior cardinals. The arteries are injected by inserting the needle into the dorsal aorta about two centimeters cranial to the celiacomesenteric artery. Inject forward until the gills assume the color of the mass (2 to 3 cc.), and caudally until the smaller intestinal arteries are filled (2 to 3 cc.). The hepatic portal system is injected through the mesenteric vein (8 to 10 cc.).

(C) **Mammal.** Inject the arteries of a mammal caudally through one of the common carotids until you fill the vessels in the intestine or those beneath the skin (30 cc. for a cat). The systemic veins are injected caudally through the external jugular (30 to 50 cc.), and the hepatic portal system through one of the intestinal veins in the mesentery (12 to 18 cc.). If only the arteries are to be injected, the blood may be allowed to accumulate in the veins, but if the veins are to be injected as well, the blood should be drained off during the arterial injection through a cut in the external jugular vein.

4. Injection of the Lymphatic Vessels

If the lymphatics of the mammal are to be studied, they can be injected in a freshly killed specimen. A saturated solution of Berlin blue should be injected subcutaneously in various parts of the body, for this substance will enter the lymphatic capillaries but not those of the cardiovascular system. Injections should be made in all the foot pads, in the pad at the end of the nose, and in the lips. It is necessary to inject slowly and to maintain a pressure on the syringe for 15 to 30 minutes at each site. Also massage the limbs and neck, working toward the center of the body, to aid the flow of the injection fluid. To inject the lymphatics at the base of the mesentery, open the body and inject into the peripheral parts of the mesenteric lymph nodes. The efferent lymphatic from any lymph node may be injected in the same way.

If a more permanent preparation is desired, a weak solution of warm gelatin may be mixed with the Berlin blue. In this case, the injection must be made while the body and the gelatin are still warm. Injecting with gelatin solution is more difficult.

5. Preparation of Skeletons

Skeletons should be prepared from mature specimens that have been freshly killed, or from specimens preserved in brine. Fresh specimens are better. First skin and dismember the specimen, removing the head and legs. Then cut off the larger masses of flesh from the bones. The rest of the flesh can easily be scraped off after it has been loosened in one of several ways. (1) Let it macerate (decay) in a closed container of water at room temperature for several months. (2) Allow insects to remove most of the soft parts. Dermestid beetles are particularly good, and some institutions maintain a colony for this purpose. (3) Simmer the specimen in water or in a soap solution made by diluting one part of the following stock solution with three or four parts of water:

Ammonia (strong)	150 cc.
Hard soap	75 g.
Potassium nitrate (saltpeter)	12 g.
Water	2000 cc.

The brain can be removed through the foramen magnum with a wire having a flattened loop at one end.

If disarticulated bones are desired, one can let the bones simmer for a considerable time (an hour or two). But if a skeleton in which the bones are held together by ligaments is the objective, one must look at the specimen frequently to be sure that the muscles are soft enough to scrape off, but that the ligaments are not so soft that they detach easily. It is also necessary to cut some of the larger ligaments and tendons, especially those on the underside of the paws of mammals, or they will shorten and distort the specimen.

After the flesh has been removed, it is generally desirable to degrease the bones by leaving them for a day or so in turpentine, benzine or carbon tetrachloride. Carbon tetrachloride is the most effective, but its fumes are poisonous, so the work should be done under a fume hood. The bones should then be bleached for a day or two in a solution of hydrogen peroxide. When the specimen is cleaned and bleached satisfactorily, the parts of the skeleton must be nailed out in the desired position before the preparation dries.

To get a disarticulated skull, simply continue the boiling process until the sutures are soft enough to permit pulling the bones apart. Another method is to fill the cranial cavity with dried peas, tightly cork the foramen magnum, and put the skull in water. The peas will swell and loosen the bones. Young adults should be used for disarticulated skulls, as some of the skull bones grow together in old specimens.

APPENDIX 2

REFERENCES

The references given below include those cited in the text, those of particular value for laboratory studies in comparative anatomy and certain key references on the functional significance and interrelationships of the various organs. More inclusive bibliographies can be found in standard textbooks and in many of the works cited below.

GENERAL

Alexander, R. McN.: Animal Mechanics. London, Sidgwick and Jackson, 1968.

Bolk, L., and others: Handbuch der vergleichenden Anatomie der Wirbeltiere. 6 vols. Berlin and Vienna, Urban und Schwarzenberg, 1931–1938. Reprinted in 1967 by A. Asher and Co., Amsterdam.

Bronn, H. G., and others: Klassen und Ordnungen des Thier-Reichs. Abteilung VI (Vertebrates). Leipzig and Heidelberg, C. F. Winter'sche Verlagshandlung, 1874–1938.

DeBeer, G. R.: The Vertebrate Skull. Oxford, Clarendon Press, 1937.

DeBeer, G. R.: Vertebrate Zoology. Revised ed. New York, The Macmillan Company, 1953.

Edgeworth, F. H.: The Cranial Muscles of Vertebrates. London, Cambridge University Press, 1935.

Fort, W. B., Wilson, H. C., and Goldberg, H. G.: The daily removal of formalin from preserved biological specimens used in class work. Science, vol. 94, pp. 169–70, 1941.

Foxon, G. E. H.: Problems of the double circulation in vertebrates. Biological Reviews of the Cambridge Philosophical Society, vol. 30, pp. 196–228, 1955.

Gans, C.: Biomechanics, An Approach to Vertebrate Biology. Philadelphia, J. B. Lippincott Company, 1974.

Goodrich, E. S.: Studies on the Structure and Development of Vertebrates. New York, Dover Publications, 1958.

Grassé, P.-P. (ed.): Traité de Zoologie. Vols. 11–27 deal with protochordates and vertebrates. Paris, Masson et Cie., 1948–1973.

Gray, J.: Animal Locomotion. London, Weidenfeld and Nicolson, 1968.

Hildebrand, M.: Anatomical Preparations. Berkeley, University of California Press, 1969.

Hildebrand, M.: Analysis of Vertebrate Structure. New York, John Wiley and Sons, 1974.

Hughes, G. M.: Comparative Physiology of Vertebrate Respiration. Cambridge, Harvard University Press, 1963.

Hyman, L. H.: Comparative Vertebrate Anatomy. 2nd ed. Chicago, University of Chicago Press, 1942.

International Committee on Veterinary Anatomical Nomenclature: Nomina Anatomica Veterinaria, 2nd ed. Vienna, Adolf Holzhausen's Successors. Distributed in U.S.A. by Department of Anatomy, New York State Veterinary College, Ithaca, New York.

Kappers, C. U. A., Huber, G. C., and Crosby, E. C.: The Comparative Anatomy of the Nervous System of Vertebrates, Including Man. 2 vols. New York, The Macmillan Company, 1936.

Nelsen, O. E.: Comparative Embryology of the Vertebrates. New York, The Blakiston Company, 1953.

Parker, T. J., and Haswell, W. A.: A Text-Book of Zoology. Vol. 2. 6th ed., revised by C. Forster-Cooper. London, The Macmillan Company, 1940.

Patt, D. I., and Patt, G. R.: Comparative Vertebrate Histology. New York, Harper and Row, 1969.

Prosser, C. L. (ed.): Comparative Animal Physiology. 3rd ed. Philadelphia, W. B. Saunders Company, 1973.

Romer, A. S.: Vertebrate Paleontology. 3rd ed. Chicago, University of Chicago Press, 1966.

Romer, A. S.: The Vertebrate Body. 4th ed. Philadelphia, W. B. Saunders Company, 1970.

Schmidt-Nielsen, K.: How Animals Work. Cambridge, Cambridge University Press, 1972.

Walls, G. L.: The Vertebrate Eye and Its Adaptive Radiation. Bloomfield Hills, Cranbrook Institute of Science, Bull. No. 19, 1942.

Webster, D., and Webster, M.: Comparative Vertebrate Morphology. New York, Academic Press, 1974.

Wessells, N. K. (ed.): Vertebrate Adaptations. San Francisco, W. H. Freeman and Company, 1969.

Wiedersheim, R.: Comparative Anatomy of Vertebrates. 3rd English ed., adapted by W. N. Parker. London, The Macmillan Company, 1907.

Wilder, B. G., and Gage, S. H.: Anatomical Technology as Applied to the Domestic Cat. New York, A. S. Barnes and Company, 1882.

Williams, E. E.: Gadow's arcualia and the development of tetrapod vertebrae. Quarterly Review of Biology, vol. 34, pp. 1–32, 1959.

Young, J. Z.: The Life of Vertebrates. 2nd ed. London, Oxford University Press, 1962.

LOWER CHORDATES AND FISHES

Barrington, E. J. W.: The supposed pancreatic organs of *Petromyzon fluviatilis* and *Myxine glutinosa*. Quarterly Journal of Microscopical Science, vol. 85, pp. 391–417, 1945.

Barrington, E. J. W.: The Biology of Hemichordata and Protochordata. San Francisco, W. H. Freeman and Company, 1965.

Bigelow, H. B., Schroeder, W. C., and Farfante, I. P.: Fishes of the Western North Atlantic. Part I. Lancelets, Cyclostomes and Sharks. New Haven, Sears Foundation for Marine Research, Yale University, 1948.

Burger, J. W., and Hess, W. N.: Function of the rectal gland in the spiny dogfish. Science, vol. 131, pp. 670–671, 1960.

Cahn, P. H. (ed.): Lateral Line Detectors. Bloomington, Indiana University Press, 1967.

Cohen, D. H., and Duff, T. A.: Electrophysiological identification of a visual area in shark telencephalon. Science, vol. 182, pp. 492–494, 1973.

Daniel, J. F.: The Elasmobranch Fishes. 3rd ed. Berkeley, University of California Press, 1934.

Dean, B.: Fishes, Living and Fossil. New York, The Macmillan Company, 1895.

Gans, C., and Parsons, T. S.: A Photographic Atlas of Shark Anatomy. New York, Academic Press, 1964.

Gibbs, S. P.: The anatomy and development of the buccal glands of the lake lamprey (*Petromyzon marinus* L.) and the histochemistry of their secretion. Journal of Morphology, vol. 98, pp. 429–470, 1956.

Gilbert, S. G.: Pictorial Anatomy of the Dogfish. Seattle, Washington University Press, 1973.

Goodrich, E. S.: A Treatise on Zoology, edited by E. Ray Lankester. Part IX. Vertebrata Craniata, Fasc. I. "Cyclostomes and Fishes." London, Adam and Charles Black, 1909.

Goodrich, E. S.: On the development of the segments of the head of Scyllium. Quarterly Journal of Microscopical Science, vol. 63, pp. 1–30, 1918.

Heath, G. W.: The siphon sacs of the smooth dogfish and spiny dogfish. Anatomical Record, vol. 125, p. 562, 1956.

Hisaw, F. L., and Albert, A.: Observations on the reproduction of the spiny dogfish, *Squalus acanthias*. Biological Bulletin, vol. 92, pp. 187–199, 1947.

Hoar, W. S., and Randall, D. J.: Fish Physiology. Vols. 1–5. New York, Academic Press, 1969–1971.

Hughes, G. M.: The relationship between cardiac and respiratory rhythms in the dogfish, *Scyliorhinus canicula* L. Journal of Experimental Biology, vol. 57, pp. 415–434, 1972.

Hughes, G. M., and Hills, B. A.: Oxygen tension distribution in water and blood at the secondary lamella of the dogfish gill. Journal of Experimental Biology, vol. 55, pp. 399–408, 1971.

Liem, K. L.: A probable homologue of the clavicle in the holostean fish *Amia calva*. Journal of Zoology (London), vol. 170, pp. 521–532, 1973.

Marinelli, W., and Strenger, A.: Vergleichende Anatomie und Morphologie der Wirbeltiere. I. Lieferung. Lamperta fluviatilis L., III. Lieferung. Squalus acanthias L. Vienna, Franz Deuticke Verlag, 1954 and 1959.

Norris, H. W., and Hughes, S. P.: The cranial, occipital, and anterior spinal nerves of the dogfish, *Squalus acanthias*. Journal of Comparative Neurology, vol. 31, pp. 293–402, 1920.

Nursall, J. R.: Swimming and the origin of paired appendages. American Zoologist, vol. 2, pp. 127–141, 1962.

O'Donoghue, C. H., and Abbot, E. B.: The blood vascular system of the spiny dogfish, *Squalus acanthias* Linne, and *Squalus sucklii* Gill. Transactions of the Royal Society of Edinburgh, vol. 55, pp. 823–894, 1928.

Oguri, M.: Rectal glands of marine and fresh water sharks, comparative histology. Science, vol. 144, pp. 1151–1152, 1964.

Satchell, G. H.: Circulation in Fishes. Cambridge, Cambridge University Press, 1971.

Willemse, J. J.: The way by which flexures of the body are caused by muscular contractions. Koninklijk Nederland Akademie Wentschap. Proceedings Series C, vol. 62, pp. 589–593, 1959.

Young, J. Z.: The autonomic nervous system of selachians. Quarterly Journal of Microscopical Science, vol. 75, pp. 571–624, 1933.

AMPHIBIANS AND REPTILES

Barclay, O. C.: The mechanics of amphibian locomotion. Journal of Experimental Biology, vol. 23, pp. 177–203, 1946.

Chase, S. W.: The mesonephros and urogenital ducts of *Necturus maculosus* Rafinesque. Journal of Morphology, vol. 37, pp. 457–532, 1923.

DeLong, K. T.: Quantitative analysis of blood circulation through the frog heart. Science, vol. 138, pp. 693–694, 1962.

Figge, F. H.: A morphological explanation for the failure of Necturus to metamorphose. Journal of Experimental Zoology, vol. 56, pp. 241–265, 1930.

Francis, E. B.: The Anatomy of the Salamander. London, Oxford University Press, 1943.

Gans, C., and Parsons, T. S. (eds.; Volume 1 also edited by A. d'A. Bellairs.): Biology of the Reptilia. Vols. 1–4. London, Academic Press, 1969–1973.

Gilbert, S. G.: Pictorial Anatomy of the Necturus. Seattle, University of Washington Press, 1973.

Harris, J. P., Jr.: Necturus papers: The skeleton of the arm. The pelvic musculature. The muscles of the forearm. The levator anguli scapulae. The musculus depressor mandibulae. Natural history. Field and Laboratory, vols. 20, 21, 22, 25, 27, 1952–1959.

Herrick, C. J.: The Brain of the Tiger Salamander, *Ambystoma tigrinum.* Chicago, University of Chicago Press, 1948.

Kingsbury, B. F.: On the brain of *Necturus maculatus.* Journal of Comparative Neurology, vol. 5, pp. 138–205, 1895.

Lombard, R. E., and Straughan, I. R.: Functional aspects of anuran middle ear structures. Journal of Experimental Biology, vol. 61, pp. 1–23, 1974.

Miller, W. S.: The vascular system of *Necturus maculatus.* University of Wisconsin Science Series, vol. 2, pp. 211–226, 1900.

Moore, J. A. (ed.): Physiology of the Amphibia. New York, Academic Press, 1964.

Noble, G. K.: The Biology of the Amphibia, New York, Dover Publications, 1954.

Romer, A. S.: Osteology of Reptiles. Chicago, University of Chicago Press, 1956.

Wilder, H. H.: The skeletal system of Necturus maculatus Rafinesque. Memoirs of the Boston Society of Natural History, vol. 5, pp. 387–439, 1903.

Wilder, H. H.: The appendicular muscles of *Necturus maculosus.* Zoologische Jahrbücher, sup. 15, part 2, pp. 383–424, 1912.

MAMMALS

Abell, N. B.: A comparative study of the variations of the postrenal vena cava of the cat and rat and a description of two new variations. Denison University Journal of the Science Laboratory, vol. 40, pp. 87–117, 1947.

Arey, L. B.: Developmental Anatomy. 7th ed., revised. Philadelphia, W. B. Saunders Company, 1974.

Barone, R., and others: Atlas d'Anatomie du Lapin. Paris, Masson et Cie., 1973.

Barry, A.: The aortic arch derivatives in the human adult. Anatomical Record, vol. 111, pp. 221–238, 1951.

Bloom, W., and Fawcett, D. W.: A Textbook of Histology. 10th ed. Philadelphia, W. B. Saunders Company, 1975.

Crouch, J. E.: Text-Atlas of Cat Anatomy. Philadelphia, Lea & Febiger, 1969.

Davis, D. D.: The giant panda, a morphological study of evolutionary mechanisms. Fieldiana, Zoological Memoirs, vol. 3, pp. 1–340, 1964.

Davis, D. D., and Story, H. E.: The carotid circulation in the domestic cat. Zoological Series Field Museum of Natural History, vol. 28, pp. 1–47, 1943.

Field, H. E., and Taylor, M. E.: An Atlas of Cat Anatomy. Chicago, Chicago University Press, 1954.

Gilbert, S. G.: Pictorial Anatomy of the Cat. Seattle, University of Washington Press, 1968.

Huntington, G. S., and McClure, C. F. W.: The development of the veins in the domestic cat. Anatomical Record, vol. 20, pp. 1–31, 1920.

Jayne, J.: Mammalian Anatomy. Part I. The Skeleton of the Cat. Philadelphia, J. B. Lippincott Company, 1898.

Kerr, N. S.: The homologies and nomenclature of the thigh muscles of the opossum, cat, rabbit, and rhesus monkey. Anatomical Record, vol. 121, pp. 481–493, 1955.

Klingener, D.: Laboratory Anatomy of the Mink. Dubuque, Wm. C. Brown Company, 1972.

Miller, M. E., Christensen, G. C., and Evans, H. E.: Anatomy of the Dog. Philadelphia, W. B. Saunders Company, 1964.

Northcutt, R. G., Kenneth, L. W., and Barber, R. P.: Atlas of the Sheep Brain. 2nd ed. Champaign, Illinois, Stiles Publishing Company, 1966.

Ranson, S. W.: The Anatomy of the Nervous System. 10th ed., revised by S. L. Clark. Philadelphia, W. B. Saunders Company, 1959.

Rasmussen, A. T.: The Principal Nervous Pathways. 4th ed. New York, The Macmillan Company, 1952.

Reighard, J. E., and Jennings, H. S.: Anatomy of the Cat. 3rd ed., Revised by R. Elliott. New York, Henry Holt and Company, 1935.

Sisson, S.: The Anatomy of Domestic Animals. 4th ed., revised by Grossman, J. D. Philadelphia, W. B. Saunders Company, 1953.

Weber, M., Burlet, H. M. de, and Abel, O.: Die Saugetiere. 2 vols., 2nd ed. Jena, G. Fischer, 1927–1928.

Yoshikawa, T.: Atlas of the Brains of Domestic Animals. University Park, Pennsylvania State University Press, 1968.

Young, J. Z.: The Life of Mammals. New York and Oxford, Oxford University Press, 1957.

INDEX

Supracondylar foramen, 104
Supracoracoideus, 128
Supraoccipital bone, 67
Supraorbital canal, of *Squalus,* 171
Supraorbital crest, 49
Suprarenal glands, 318, 342
Suprascapular cartilage, of *Necturus,* 94
 of *Squalus,* 90
Suprascapular nerve, 233
Supraspinatus, 142
Supraspinous fossa, 104
Supratemporal canal, of *Squalus,* 171
Surangular, 69
Sweat glands, 40
Swim bladder, 242
Sylvius, aqueduct of. See *Cerebral aqueduct.*
Sympathetic cord, 192, 237
Sympathetic ganglia, 237, 316
Sympathetic nervous system, 192
Symphyses, of cat, 106
 of *Chelydra,* 99
Synapsid reptiles, 30
Synapsid temporal fenestra, 76
Synotic tectum, 47
 of *Necturus,* 62
Systemic veins, 277
 caudal, of mammals, 316–325
  of *Necturus,* 298
 cranial, of mammals, 308–316
  of *Necturus,* 299
 of *Squalus,* 284

Tachyglossus, 30
Taeniae coli, 275
Tail, of mammals, 39
 of *Necturus,* 35
 of *Squalus,* 32
  myomeres of, 117
Talus, of cat, 106
 of *Chelydra,* 101
Tapetum lucidum, of mammals, 179
 of *Squalus,* 175
Tarsal bones, of cat, 106
 of *Chelydra,* 100
 of *Necturus,* 96
 of primitive tetrapods, 93
Tarsus, of cat, 106
 of *Necturus,* 36
Tectum, of sheep, 219
 of *Squalus,* 196
Teeth, horny, of lamprey, 15
 of cat, 86
 of dermal bone, 53
 of *Necturus,* 63
 of primitive tetrapods, 58
Tela choroidea, of sheep, 221
 of *Squalus,* 195
Telencephalon, 194
 of sheep, 216–217
Teleostei, 29
Temporal artery, 316
Temporal fenestra, 76
Temporal fossa, 76, 80

Temporal line, 80
Temporal neocortex, 231
Temporal vein, 314
Temporalis, 166
Tendon, definition of, 111
Tensor fasciae antebrachii, 144
Tensor fasciae latae, of cat and mink, 150, 151
 of rabbit, 150, 151, 155
Tensor tympani, 189
Tentacles, of *Amphioxus,* 10
Tentorium, 83, 216
Teres major, 143, 234
Teres minor, 143
Terminal filament, 232
Testes, of lamprey, 23
 of *Necturus,* 255, 342
 of *Squalus,* 245, 336
Testicular arteries, of mammals, 319
 of *Necturus,* 304
 of *Squalus,* 293
Testicular veins, of mammals, 319
 of *Necturus,* 299
 of *Squalus,* 285
Tetrapods, primitive. See *Primitive tetrapods.*
Thalamocortical fibers, 225
Thalamus, of sheep, 218
 of *Squalus,* 195
Thecodont, 87
Therapsida, 30
Theria, 30
Thigh, of mammals, 39
  muscles of, 149–153
Third ventricle of brain, of sheep, 218, 223
 of *Squalus,* 195, 208
Thoracic artery, 311
Thoracic duct, 328
Thoracic trapezius muscle, 139
Thoracic vertebrae, 69, 70
Thoraciscapularis, 128
Thoracolumbar fascia, 134
Thorax, 39, 266–269
Thymus, 241, 269
Thyroarytenoid muscle, 168
Thyrohyal bone, 88
Thyroid artery, 314
Thyroid cartilage, 265
Thyroid gland, 241
 lobes of, 265
 of mammals, 265
 of *Necturus,* 300
 of *Squalus,* 119, 286
Tibia, of cat, 106
 of *Chelydra,* 100
 of *Necturus,* 96
 of primitive tetrapods, 93
Tibial nerves, 237
Tibialis caudalis, 160
Tibialis cranialis, 160
Tissue, notochordal, 46
Tissue fluid, 276
Tongue, of lamprey, 15
 of mammals, 263
 of primitive tetrapods, 253
 primary, of *Squalus,* 250
Tonguelike fold, of mouth of *Necturus,* 258